W0042829

Nonlinear Phenomena in Physics and Biology

NATO ADVANCED STUDY INSTITUTES SERIES

A series of edited volumes comprising multifaceted studies of contemporary scientific issues by some of the best scientific minds in the world, assembled in cooperation with NATO Scientific Affairs Division.

Series B: Physics

Recent Volumes in this Series

Volume 69 – Photovoltaic and Photoelectrochemical Solar Energy Conversion
edited by F. Cardon, W. P. Gomes, and W. Dekeyser

Volume 70 – Current Topics in Elementary Particle Physics
edited by K. H. Mütter and K. Schilling

Volume 71 – Atomic and Molecular Collision Theory
edited by Franco A. Gianturco

Volume 72 – Phase Transitions: *Cargèse 1980*
edited by Maurice Lévy, Jean-Claude Le Guillou, and Jean Zinn-Justin

Volume 73 – Scattering Techniques Applied to Supramolecular and Nonequilibrium Systems
edited by Sow-Hsin Chen, Benjamin Chu, and Ralph Nossal

Volume 74 – Rigorous Atomic and Molecular Physics
edited by G. Velo and A. S. Wightman

Volume 75 – Nonlinear Phenomena in Physics and Biology
edited by Richard H. Enns, Billy L. Jones, Robert M. Miura, and Sadanand S. Rangnekar

Volume 76 – Metal Hydrides
edited by Gust Bambakidis

Volume 77 – Nonlinear Phenomena at Phase Transitions and Instabilities
edited by T. Riste

This series is published by an international board of publishers in conjunction with NATO Scientific Affairs Division

A Life Sciences B Physics	Plenum Publishing Corporation London and New York
C Mathematical and Physical Sciences	D. Reidel Publishing Company Dordrecht, Boston, and London
D Behavioral and Social Sciences E Applied Sciences	Sijthoff & Noordhoff International Publishers Alphen aan den Rijn, The Netherlands, and Germantown, U.S.A.

Nonlinear Phenomena in Physics and Biology

Edited by

Richard H. Enns
Simon Fraser University
Burnaby, B.C., Canada

Billy L. Jones
Simon Fraser University
Burnaby, B.C., Canada

Robert M. Miura
University of British Columbia
Vancouver, B.C., Canada

and

Sadanand S. Rangnekar
Simon Fraser University
Burnaby, B.C., Canada

PLENUM PRESS • NEW YORK AND LONDON
Published in cooperation with NATO Scientific Affairs Division

Library of Congress Cataloging in Publication Data

Main entry under title:

Nonlinear phenomena in physics and biology.

(NATO advanced study institutes series. Series B, Physics ; v. 75)
"Proceedings of a NATO Advanced Study Institute on Nonlinear Phenomena in Physics and Biology, held August 17-29, 1980, at the Banff Center, Banff, Alberta, Canada" —T.p. verso.
Bibliography: p.
Includes index.
1. Nonlinear theories—Congresses. 2. Physics—Congresses. 3. Biological physics—Congresses. I. Enns, Richard H. II. NATO Advanced Study Institute on Nonlinear Phenomena in Physics and Biology (1980 : Banff, Alta.) III. Series.
QC20.7.N6N66 530.1'5 81-15882
ISBN 978-1-4684-4108-6 ISBN 978-1-4684-4106-2 (eBook) AACR2
DOI 10.1007/978-1-4684-4106-2

Proceedings of a NATO Advanced Study Institute on Nonlinear Phenomena in Physics and Biology, held August 17-29, 1980, at the Banff Center, Banff, Alberta, Canada

© 1981 Plenum Press, New York
Softcover reprint of the hardcover 1st edition 1981
A Division of Plenum Publishing Corporation
233 Spring Street, New York, N.Y. 10013

All rights reserved

No part of this book may be reproduced, stored in a retrieval system, or transmitted in any form or by any means, electronic, mechanical, photocopying, microfilming. recording, or otherwise, without written permission from the Publisher

PREFACE

The Advanced Study Institute (ASI) on Nonlinear Phenomena in Physics and Biology was held at the Banff Centre, Banff, Alberta, Canada, from 17 - 29 August, 1980. The Institute was made possible through funding by the North Atlantic Treaty Organization (who supplied the major portion of the financial aid), the National Research and Engineering Council of Canada, and Simon Fraser University. The availability of the Banff Centre was made possible through the co-sponsorship (with NATO) of the ASI by the Canadian Association of Physicists.

12 invited lecturers and 82 other participants attended the Institute. Except for two lectures on nonlinear waves by Norman Zabusky, which were omitted because it was felt that they already had been exhaustively treated in the available literature, this volume contains the entire text of the invited lectures. In addition, short reports on some of the contributed talks have also been included.

The rationale for the ASI and this resulting volume was that many of the hardest problems and most interesting phenomena being studied by scientists today are nonlinear in nature. The nonlinear models involved often span several different disciplines, a simple example being the Volterra-type model in population dynamics which has its analogue in nonlinear optics and plasma physics (the 3-wave problem), in the discussion of the social behavior of animals, and in biological competition and selection at the molecular level. With the rapid growth of interest in nonlinear phenomena, as judged by the current literature, it was felt to be most timely to have an interdisciplinary Institute in this subject area.

As can be seen from the list of lecturers and participants, the ASI attracted scientists from many different disciplines. Some of the physical scientists and mathematicians were delving into biological problems while the biologically oriented researchers were interested in learning about the theoretical techniques and ideas of the physical scientists. The selection of topics to be covered in the Institute was made on the basis of trying to maximize the cross-fertilization of ideas. The lecturers and participants of one group would be the students of the other and vice-versa. To maintain a coherent and in-depth programme, certain topics, eg., nonlinear quantum field theory, were deliberately omitted. The programme of the ASI was split into four interconnected streams of lectures, two streams (lectures given by Norman Zabusky, Alwyn Scott, Mark Ablowitz, David Kaup and Bengt Fornberg) concentrating on nonlinear wave phenomena, one on the nonlinear behavior of open systems far from equilibrium (lectures given by G. Nicolis, Lou Howard, John Rinzel, Robert Miura and Stuart Kauffman), and the fourth on competition and evolution in biophysical systems (Peter Schuster and Don Ludwig). The contents of this volume have been grouped accordingly.

Special thanks are due to many individuals whose assistance, cooperation and collaboration were essential at various stages of the organization of the ASI and the preparation of this volume:

- to Dr. Mario di Lullo for his encouragement and support from the very beginning;

- to Doreen Young for the typing of this entire volume, a job which was cheerfully and meticulously done;

- to my Co-Directors, Robert Miura and Billy Jones, for helping to turn the idea of such an ASI into reality;

- to Sada Rangnekar who was my right-hand man in editing this volume, drawing figures, and numerous other chores;

- to Ralph Kerr who eased the administrative burden of running the ASI;

- to the personnel of the Banff Centre who did so much to make our stay a pleasant, as well as professionally rewarding, one;

and to all of the lecturers and participants who ultimately were the real reason for the success of the ASI.

Richard H. Enns
Director of the Institute

ORGANIZING COMMITTEE

Enns, Professor R.H.	Director Department of Physics Simon Fraser University Burnaby, B.C., Canada
Jones, Professor B.L.*	Co-director Department of Physics Simon Fraser University Burnaby, B.C., Canada
Miura, Professor R.M.	Co-director Department of Mathematics University of British Columbia Vancouver, B.C., Canada
Young, Miss D.	Secretary Department of Physics Simon Fraser University Burnaby, B.C., Canada
Rangnekar, Dr. S.S.	Department of Physics Simon Fraser University Burnaby, B.C., Canada

****

***During the preparation of this volume, our colleague and friend, Billy Jones, sadly passed away. We dedicate this work to him.*

****

CONTENTS

Seminars

COMPUTATION AND INNOVATION IN THE NONLINEAR SCIENCES

Norman J. Zabusky

Department of Mathematics and Statistics
University of Pittsburgh
Pittsburgh, Pennsylvania 15260

COMPUTATION AND INNOVATION IN THE NONLINEAR SCIENCES

Norman J. Zabusky

Department of Mathematics and Statistics
University of Pittsburgh
Pittsburgh, Pa. 15260, U.S.A.

As the lead-off speaker I would like to reflect on the previous School of Nonlinear Mathematics and Physics [Zabusky, 1968] that Martin Kruskal and I organized 14 years ago. Several members of the present faculty or their teachers were there. W. Heisenberg, well-known for his pioneering contributions to quantum mechanics, opened the School with a talk "Nonlinear Problems in Physics" [Heisenberg, 1967] that I take the liberty of distilling into four aphorisms:

1. Simplification by symmetrizing.
2. Initial progress by linearizing.
3. General features by statistical methods.
4. Apparently coherent phenomena with long-time unpredictability.

The first three represent techniques and methods employed to understand nonlinear phenomena in "point and continuum mechanics". The last is one of the surprises of nonlinear systems. He concluded his remarks with:

> "I wish to emphasize again that the progress of physics certainly will depend to a large extent on the progress of nonlinear mathematics. ... It may be that every such problem is individual and requires individual methods. Yet as I have said, there are definitely some common features and therefore one can learn by comparing different nonlinear problems."

S.M. Ulam, trained as a pure mathematician and also known for his creative contributions to applied problems, also addressed the School. He noted amusingly that Fermi once said, "The Bible does not state that the ultimate fundamental theories in physics should be linear", a fact that elementary particle physicists have been confronting more in recent years.

He emphasized what he saw of the increasing role that electronic digital computers would play in mathematics and the natural sciences by synergizing understanding of experiment and theory. That is, posing and studying the graphical output of judiciously chosen numerical experiments enhance the insight and intuition of working scientists.

He discussed several problems, two of which have begun to flourish in the last decade: the Fermi-Pasta-Ulam problem and near-recurrent phenomena which led to the discovery of the soliton; and the problem of iterated nonlinear algebraic transformations which M. Feigenbaum found with the aid of the computer to have certain universal properties. Ulam also expressed an overwhelming feeling for the impact that mathematics would have in the biological sciences. Taken together, his remarks pointed the direction for future work and we are here to review progress in some of these areas.

I would like to emphasize particularly the role of computers and graphical output as I have come to appreciate it over the last twenty years.

Von Neumann in 1946 [see Taub, 1963] said it first with penetrating clarity, viz.,

> "To what extent can human reasoning in the sciences be more efficiently replaced by mechanisms?"
>
> "Our present analytical methods seem unsuitable for the important problems arising in connection with nonlinear partial differential equations and in fact with virtually all types of nonlinear problems in pure mathematics. The truth of this statement is particularly striking in fluid dynamics ..."
>
> "The advance of analysis is, at this moment stagnant along the entire front of nonlinear problems. That this phenomenon is not of a transient nature but we are up against an important conceptual difficulty ... [which tends] to obscure the great physical and mathematical regularities that do exist."

> [The problems of turbulence and shock-waves] "give us the first indication regarding the field of nonlinear partial differential equations when a mathematical penetration into this area, that is so difficult to access will at last succeed. Without understanding them to one's thinking even from the strictly mathematical point of view, it seems futile to attempt that penetration."
>
> "... really efficient high-speed computing devices may, in the field of nonlinear partial differential equations as well as in many other fields which are now difficult or entirely denied of access, provides us with those heuristic hints which are needed in all parts of mathematics for genuine progress."
>
> "[This is] a type of physical experimentation which is really computing ... [and] should ultimately lead to important analytical advances."

The computing machine is being used in mathematics and the natural sciences in three related and non-routine ways:

1. Aiding in proofs: eg., the results of Appel and Haken in searching for "reducible configurations" helped to prove that a planar map can be colored with four colors so that no two adjacent regions have the same color.

2. Validating: checking analyses and "joining" regions where asymptotical analyses have been performed.

3. Heuristic: discovering the essential properties of systems of equations through judiciously chosen runs. When counter-intuitive phenomena are encountered and elucidated we may have a mathematical breakthrough.

It is the latter theme that I will weave through my lectures and try to communicate the meaning of the proper choice of graphical representations, [Zabusky, 1981].

W. Pauli, the theoretical physicist who made contributions to many areas was known to have a mystical side. When he visited a laboratory, smooth running apparatus almost always developed abberations, a phenomenon that his friends called the "Pauli effect". One day some experimentalists at Göttingen were mystified by the lack of repeatability in their experiments; later they were relieved to find out that Pauli had passed through the vicinity on an express train.

Pauli [Jung and Pauli, 1955] in reviewing Kepler's contributions noted,

> "...What is the nature of the bridge between the sense perceptions and concepts? All logical thinkers have arrived at the conclusion that pure logic is fundamentally incapable of constructing such a link. ... The process of understanding nature as well as the happiness that man feels ... in the conscious realization of new knowledge, seems thus to be based on a correspondence, a "matching" of inner images ... with external objects and their behaviour ... images [called by Kepler archetypal ("archetypalis")] with strong emotional content, not thought out, but beheld, as it were, while being painted. ... As ordering operators and image-formers in the world of symbolical images, the archetypes [or "primordial images" of C.J. Jung] thus function as the sought for bridge ..." (I have underlined words.)

I found his remarks exciting but somewhat distant and have taken the liberty of transforming them to a more contemporary form, namely,

> "The discovery of new knowledge in the natural sciences is a manifestation of a "congruence", that is a linkage or a resonance between data and an image of that knowledge in deeper levels of our consciousness. To date, and for many, deductive mathematics has often bridged the two domains. A proper picture or graph in the external domain can synergize a connection to the images in our "nonconscious" mind and provide an alternative circuit for discovery. Many present abstract fields of mathematics had their origin in making this linkage."

Discovery favors the "prepared" mind. Pauli noted, "One should never declare that theses laid down by rational formulation are the only possible presuppositions of human reason." With computers, this intuitive - geometric approach should be developed, taught to our students and become part of the scientists modus operandi.

REFERENCES

Heisenberg, W., 1967, Nonlinear problems in physics, Physics Today, May issue, p. 27. Also published [Zabusky, 1968].

Jung, C.J., and Pauli, W., 1955, "The Interpretation of Nature and the Psyche," Bollingen Series LI, Pantheon Books.

Taub, A., (ed.), 1963, "Collected Work of John von Neumann," Vol. 5, McMillan, New York.

Zabusky, N.J., ed., 1968, "Topics in Nonlinear Physics," Sections by W. Heisenberg, C. Truesdell, I. Prigogine, M. Baus, N. Bloembergen, P.G. Saffman and J.A. Wheeler, Springer-Verlag, New York.

Zabusky, N.J., 1981, Computational synergetics and mathematical innovation, J. Comp. Phys., to be published.

INTRODUCTION TO NONLINEAR WAVES

Alwyn C. Scott

Center for Nonlinear Studies
Los Alamos National Laboratory
Los Alamos, New Mexico 87545

"There is no better, there is no more open door by which you can enter into the study of natural philosophy than by considering the physical phenomena of a candle."

- Michael Faraday

"In the foundations of any consistent field theory, the particle concept must not appear in addition to the field concept. The whole theory must be based solely on partial differential equations and their singularity-free solutions."

- Albert Einstein

TABLE OF CONTENTS

INTRODUCTION TO NONLINEAR WAVES

Alwyn C. Scott

Center for Nonlinear Studies
Los Alamos National Laboratory
Los Alamos, New Mexico 87545

I. A HISTORY OF THE SOLITARY WAVE

A funny thing happened to solitary wave research over the past decade: it became respectable. No longer is it possible for all soliton buffs to meet in a small room; nor can one now read the important papers in a few weeks. The early, innocent days are gone, and (as Fig. 1 shows) soliton research output has entered a period of exponential growth with a doubling time of about 18 months. The solitary wave concept has emerged as a widely accepted paradigm for exploring and modeling the dynamics of the real world.

More than simple curiosity leads us to ask about the gestation process preceeding this explosive event. We do not wish it to be said of us: "If we seem to see farther than others, it is because we are standing on the faces of giants."

A. From the Beginning to the Great War

Hydrodynamic solitary waves (like Lorentz contraction, universal gravitation and the survival of the fittest) must have been around since the dawn of time; but the noticing of them began only with that accidental observation so fluently recorded by John Scott Russell (see Fig. 2) in his "Report on Waves" to the British Association in 1844 [Russell 1844; see also Robinson and Russell 1838].

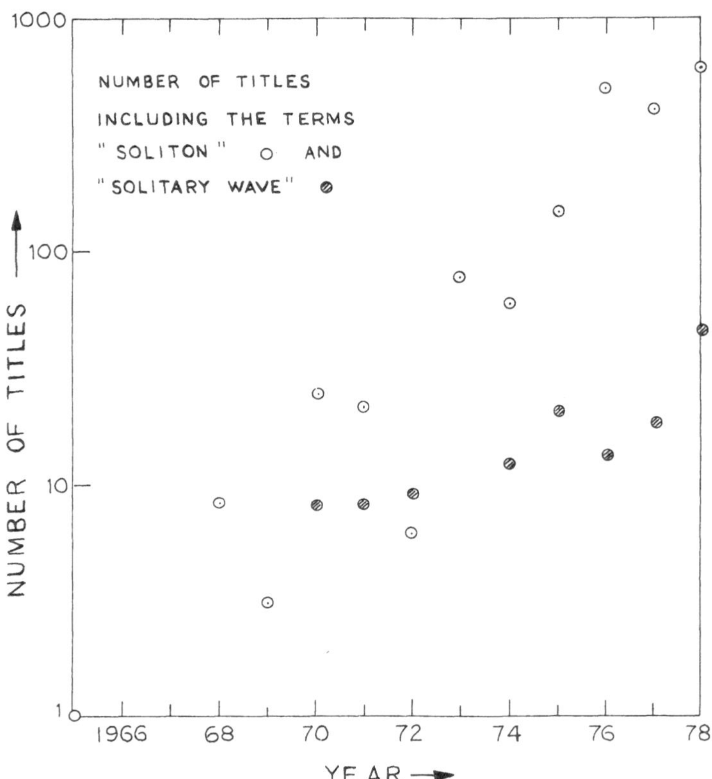

Fig. 1. Number of scientific journal articles including the term soliton and solitary wave for each year (data from the "Permuterm" Subject Index).

Fig. 2. Photograph of John Scott Russell (ca. 1860).

> I was observing the motion of a boat which was rapidly drawn along a narrow channel by a pair of horses, when the boat suddenly stopped - not so the mass of water in the channel which it had put in motion; it accumulated round the prow of the vessel in a state of violent agitation, then suddenly leaving it behind, rolled forward with great velocity, assuming the form of a large solitary elevation, a rounded, smooth and well-defined heap of water, which continued its course along the channel apparently without change of form or diminution of speed. I followed it on horseback, and overtook it still rolling on at a rate of some eight or nine miles an hour, preserving its original figure some thirty feet long and a foot to a foot and a half in height. Its height gradually diminished, and after a chase of one or two miles I lost it in the windings of the channel. Such, in the month of August 1834, was my first chance interview with that singular and beautiful phenomenon.....

The "narrow channel" was (and still is, thanks to the efforts of local canal enthusiasts)[1] the Union Canal linking Edinburgh with Glasgow, and the "rapidly drawn boat" was not there by accident. It was part of a careful series of experiments through which Russell was studying the force-velocity characteristics of variously shaped boat hulls in order to determine design parameters for conversion from horse- to steam-power. Extensive wave-tank measurements in 1834 and 1835 established the following important properties of hydrodynamic solitary waves. [Russell, 1844, Robinson and Russel, 1838.]

1. These waves are independent and localized dynamic entities that move with fixed shape and velocity.

2. A wave of height h travels in water of depth d with velocity

$$u = \sqrt{g(d + h)} \tag{1.1}$$

   (where g is the acceleration of gravity).

[1] Readers wishing to learn more of efforts to protect this historic canal can contact the Linlithgow Union Canal Society c/o Bruce Jamieson, 121 Baronshill Ave., Linlithgow EH 49 7JQ, Scotland.

3. A sufficiently large initial mass of water will produce two or more solitary waves as indicated in Fig. 3.

4. Solitary waves of depression are not observed.

5. Solitary waves cross each other "without change of any kind".

His understanding of hydrodynamic solitary waves provided a scientific basis for Russell's "wave line hull" (or "solid of least resistance") which profoundly altered the contemporary tendency to build up a ship's bow "with the shape of a duck's breast" and led him to become one of the great naval architects of the nineteenth century[2]. Although subsequent scientific discussions of hydrodynamic solitary waves involved Airy, Boussinesq, Rayleigh and Stokes (to name a few of the most prominent [Miles, 1980]) attention was focused upon the correctness rather than the significance of Russell's observations. The collective judgement of nineteenth century science is quantitatively recorded in Lamb's opus [Lamb, 1932] on hydrodynamics which allots only 3 of 730 pages to the solitary wave. Evidence that Russell himself retained a much broader appreciation of the ultimate importance of his discovery is provided by an almost unnoticed posthumous work in which (among many provocative ideas) he correctly estimated the height of the earth's atmosphere from Eq. (1.1) and the fact that the sound of cannon fire travels faster than the command to fire it [Russell, 1885].

---

[2] See John Scott Russell - A Great Victorian Engineer and Naval Architect by George S. Emmerson (John Murray, London, 1977) for a timely and sympathetic biography of this "zealous educator, idealistic social reformer, and would-be peacemaker between nations, as well as the undisputed and respected leader of his profession." Emmerson's account is particularly valuable since L.T.C. Rolt's Isambard Kingdom Brunel (London, 1957) ascribes the difficulties between these two brilliant engineers to base elements in Russell's character. Rolt's view of Russell is not expressed by Brunel's son [see I. Brunel, The Life of Isambard Kingdom Brunel, London (1870)] and it has been rejected by both A.M. Robb [see "John Scott Russell, the 'Great Eastern' and some other matters", College Courant, Glasgow, Martinmas 1958, Whitsun 1959] and by G.P. Mabon [J. Roy. Soc. Arts, pp. 204-8 and 299-302 (1967)]. The reader is left to judge whether Rolt's taste in human conduct is of service even to the memory of Brunel.

WAVES - Order 1

"The genesis by a large low column of fluid of a compound or double wave of the first order, which immediately breaks down by spontaneous analysis into two, the greater moving faster and altogether leaving the smaller."

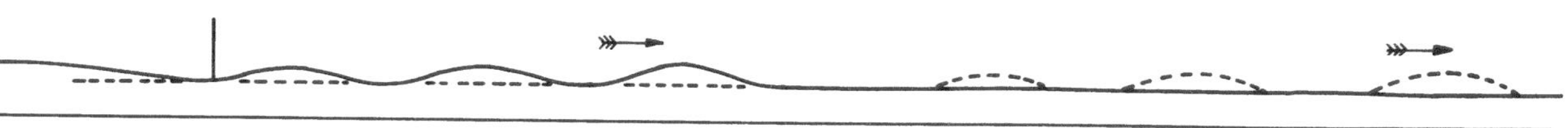

"Instead of genesis of a compound wave, the added mass sends off a series of single waves, the first being the greatest: these however do not remain together, but speedily separate as shown in the dotted lines, and become the further apart the longer they travel."

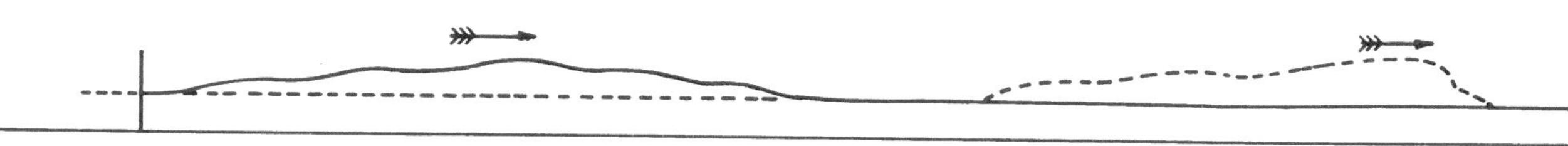

"The genesis of a compound wave by impulsion of the plate with a variable force and velocity, which variations have produced corresponding variations on the wave form. After propagation the wave breaks down by spontaneous analysis, so that after the lapse of a considerable period the compound wave is resolved into single separate waves."

Fig. 3. Multiple solitary waves generated in the tank experiments of John Scott Russell.

In 1895 Korteweg and deVries presented a study of hydrodynamic wave motion that, because it reduced the problem to its essential elements, was to play a key role in subsequent theoretical developments [Koretweg and deVries, 1895]. Their results can be summarized by the equation (subscripts denote partial derivatives)

$$\phi_t + u_o\phi_x + A\phi\phi_x + B\phi_{xxx} = 0 \qquad (1.2)$$

where ϕ is vertical displacement, u_o is the velocity given in (1.1), $A \equiv 3u_o/2d$ and $B \equiv u_o[d^2/b - T/2\rho g]$ (also T and ρ are the surface tension and the density of water). They found traveling wave solutions

$$\phi(x,t) = \tilde{\phi}(x - ut) \qquad (1.3)$$

to be a family of periodic elliptic functions, called <u>cnoidal waves</u>, that could be interpreted as Russell's solitary wave in the infinite wavelength limit. From a physical perspective this solitary wave maintains a dynamic balance between the <u>ying</u> of nonlinearity (expressed by the term $A\phi\phi_x$) and the <u>yang</u> of dispersion (expressed by the term $B\phi_{xxx}$).

Solitary waves can also be found that maintain a dynamic balance between nonlinearity and diffusion; a simple example is the ordinary candle. The heat from the flame diffuses into the wax, vaporizing it at the rate required to provide fuel for the flame. If the energy stored in the wax is E(joules/cm) and the power input required by the flame is P(joules/sec), the flame will propagate at the velocity u for which

$$P = uE \quad . \qquad (1.4)$$

The flame digests energy at the same rate it is eaten. The <u>ying</u> of nonlinear energy release is balanced by the <u>yang</u> of energetic diffusion. The reader is cautioned not to dismiss this as a trivial example; Michael Faraday [1910] didn't.

Perhaps the most important class of systems to display nonlinear diffusion is found in the nerve fibers of animal organisms, but during the first half of the nineteenth century physiologists assumed nervous activity to be propagated with the velocity of light. Contrary to the arguments of Luigi Galvani, there was nothing special about "animal electricity". Young Hermann Helmholtz arrived on the scene with a calling to become a physicist but, through financial necessity, with training as a physician; and he chose the direct measurement of nerve signal propagation speed as one of his first experimental adventures. Using an apparatus (of which that shown in Fig. 4 is a refined version) he reported

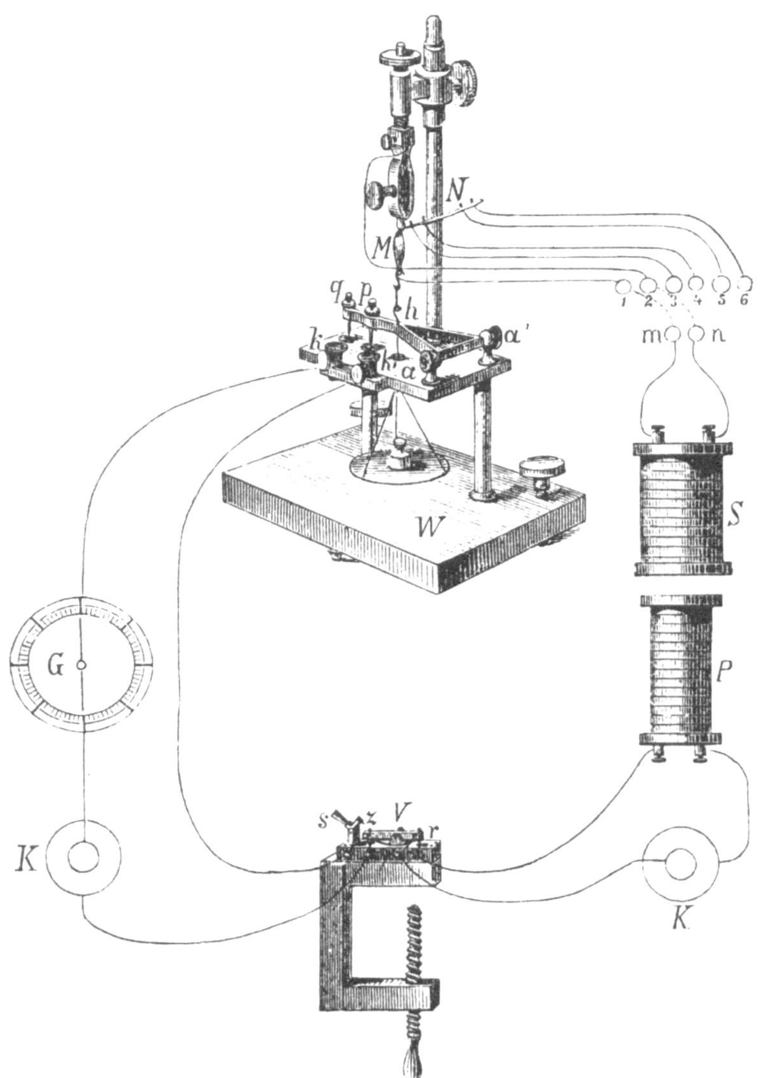

Fig. 4. Apparatus used by Helmholtz to measure the speed of propagation on a frog's nerve-muscle (NM) preparation. Closure of switch (V) induces a pulse on the nerve and starts a time measurement on a ballistic galvanometer (G). When the muscle twitches, a mercury contact (h') is broken and the time measurement stops (from L. Hermann, Handbuch der Physiologie [1879]).

on January 15, 1850 a velocity of some 32 meters per second along the frog's sciatic nerve [Helmholtz, 1850]. He missed, however, the essential explanation of nonlinear diffusion by supposing that the relatively small velocity implied that signals were carried by the motion of material particles; a notion that was to confuse neurophysiologists until well into the twentieth century. In retrospect it seems difficult to understand the failure of nineteenth century mathematical science to appreciate the descriptive power of nonlinear diffusion. It is certainly not a question of skill when one considers the names of Boussinesq and Rayleigh and Helmholtz. Indeed Helmholtz's analytical study of hydrodynamic vortex motion [Helmholtz, 1858, 1868] was an early and prophetic contribution to ultimate developments in solitary wave theory.

It must have been that linear diffusion is so "unwavelike". Any pulse-like solution of $\phi_t = \phi_{xx}$ will fall where it has downward curvature and rise where it curves upward, and therefore spread itself out. How could this qualitative effect be altered by the presence of nonlinearity? Consider augmenting the linear diffusion equation to

$$\phi_{xx} - \phi_t = F(\phi) \tag{1.5}$$

where $F(\phi)$ is chosen to be the simple cubic $\phi(\phi - a)(\phi - 1)$. A traveling wave solution, as is indicated in (1.3), was found by Huxley to be the solitary wave[3]

$$\tilde{\phi} = \left[1 + \exp\left(\frac{x - ut}{\sqrt{2}}\right)\right]^{-1} \tag{1.6}$$

which propagates at the fixed velocity

$$u = (1 - 2a)/\sqrt{2} \tag{1.7}$$

Simple and important as this solution is, it was not published until 1965!

---

[3] By J. Nagumo et al [1965] and attributed to A.F. Huxley as a private communication. Analysis of a nerve fiber model equivalent to (1.5) with $F(\phi)$ a piecewise linear function was first published by F.F. Offner et al [1940]; unfortunately an algebraic error renders their final results incorrect. The first correct solitary wave solution to (1.5) was published by the present author [1962] without knowledge of Offner's work and only a vague notion of the relation between (1.5) and nerve conduction.

Nonetheless if we had a time machine, it would be interesting to return to Paris (say) at the turn of the century for a few years of study. The literature of this golden age of applied mathematics is difficult to come by nowadays[4], but some important aspects have been reviewed recently by George Lamb [1976]. Of particular interest is the "Bäcklund transform" which can be viewed as a technique for constructing solutions to a partial differential equation (pde) via the Pfaffian form: $d\phi = Pdx + Qdt$. Clearly one must require

$$\phi_x = P; \quad \phi_t = Q \tag{1.8}$$

and the condition of integrability

$$P_t = Q_x \quad . \tag{1.9}$$

If P and Q can be found as functions of ϕ and a known solution, ϕ_0, then a new known solution (ϕ_1) can be generated by integrating the first order pair

$$\phi_{1,x} = P(\phi_1, \phi_0) \tag{1.10}$$

$$\phi_{1,t} = Q(\phi_1, \phi_0) \tag{1.11}$$

The generation process is therefore

$$\text{"known" solution } (\phi_0) \xrightarrow{\text{Bäcklund T.}} \text{"new" solution } (\phi)$$

after which, of course, the "new" solution is "known" and can be used to generate another new solution in an hierarchial order. It's easy to find a Bäcklund transform (BT) for <u>any</u> linear pde for which each "turn of the crank" introduces a new eigenfunction into the total solution. Only certain nonlinear pde's are found to have BT's, but Bäcklund showed that these include

$$\phi_{\xi\tau} = \sin\phi \tag{1.12}$$

which arose in connection with research on surfaces of constant negative curvature. The BT for (1.12) is

---

[4] [A substantial portion of the French literature can be found in the library of Clark University, Worcester, Mass.]

$$\phi_{1,\xi} = 2a \sin\left(\frac{\phi_1 + \phi_0}{2}\right) + \phi_{0,\xi}$$

$$\phi_{1,\tau} = \frac{2}{a} \sin\left(\frac{\phi_1 - \phi_0}{2}\right) - \phi_{0,\tau} \tag{1.13}$$

where a is an arbitrary constant. (To check this the reader needs merely to demonstrate that $\phi_{0,\xi\tau} = \sin\phi_0$ implies that $\phi_{1,\xi\tau} = \sin\phi_1$.) A known solution of (1.12) is clearly the "vacuum" $\phi_0 = 0$. Integration of (1.13) then gives

$$\phi_1 = 4 \tan [\exp(a\xi + \tau/a)] \tag{1.14}$$

a function that was familiar enough in the nineteenth century to carry a special name: the gudermannian. Generation of (1.14) from the vacuum can be represented by the diagram

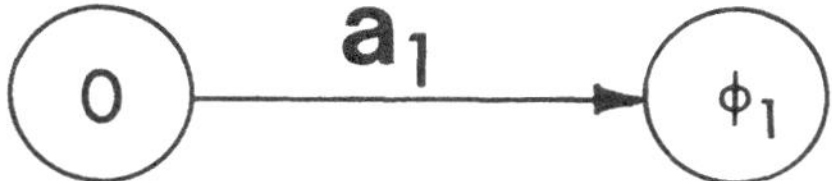

and subsequent development of a hierarchy of solutions by

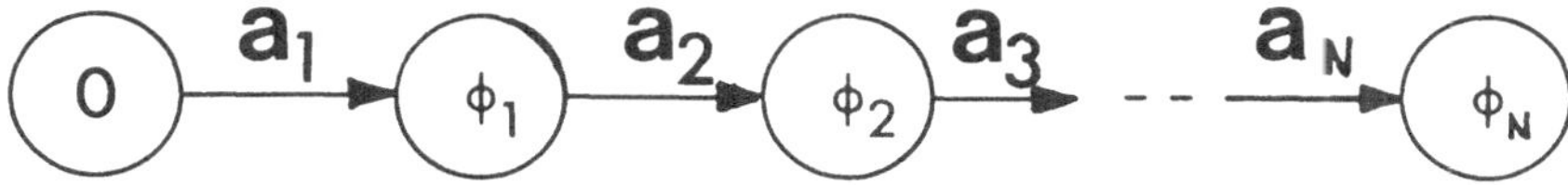

Each hierarchial level in such a diagram includes an additional component (or nonlinear eigenfunction) with the basic form indicated in (1.14). Direct integration to obtain higher level solutions becomes increasingly tedious, but Bianchi showed how these could be found by simple algebraic techniques obviating the need for integration beyond the first level [Lamb, 1976]. The use of such analytical power to find solitary wave solutions for the nonlinear diffusion equation (1.5) would have been like cracking an almond with John Scott Russell's cannon.

Another thread of our story begins in 1912 when Gustav Mie began to publish his "theory of matter" [Mie, 1912a, 1912b, 1912c]. In this brilliant series of papers, Mie suggested a nonlinear augmentation of the Maxwell equations from which the elementary particles (e.g. the electron) would arise in a natural way. To this end he defined a "world function" (Φ) as an energy functional depending upon electric field intensity ($\overline{E}$), magnetic flux density ($\overline{B}$), and the four components of electromagnetic potential ($\overline{A}$, $i\phi$). Requiring Φ to be a function of the parameters $\eta \equiv \sqrt{(\overline{E}^2-\overline{B}^2)}$ and $\chi \equiv \sqrt{(\phi^2-A^2)}$ insured Lorentz invariance; and the specific choice

$$\Phi = -\frac{1}{2}\eta^2 + \frac{1}{6}a\chi^6 \tag{1.15}$$

led to a static, spherically symmetry electric potential (ϕ) satisfying $r\phi'' + 2\phi' + ar\phi^5 = 0$ with a solution

$$\phi \doteq \frac{[3r_o^2/a]^{1/4}}{\sqrt{r^2 + r_o^2}} \tag{1.16}$$

Setting $4\pi[3r_o^2/a]^{1/4} = e$ (the electronic charge) yields a spherically symmetric model for the electron with "radius" r_o and electric potential

$$\phi \rightarrow \frac{e}{4\pi r} \tag{1.17}$$

for $r >> r_o$. The Lorentz invariance that is built into (1.15) permits this solution to travel with any speed up to the limiting velocity of light exhibiting appropriate Lorentz contraction.

This brief sketch can hardly do justice to the depth and scope of Mie's ideas. Especially for discussions of the relationship of nonlinear electrodynamics with quantum theory and gravity the reader is urged to consult the original work.

B. Between the Wars

Bäcklund transformations ceased to be of research interest after World War I. George Lamb believes that this was due, at least in part, to the untimely deaths of many young scientists active in the field. Be that as it may, all solitary wave research seemed to sleep until 1934 when Max Born began to reconsider Mie's nonlinear electromagnetics. Born was particularly concerned with establishing a "gauge invariant theory" (for which solutions would be independent of the electromagnetic potential) that would be compatible with the requirements of quantum theory. To this end he eliminated the χ dependence in Mie's functional ansatz [see (1.15)] and took instead a Lagrangian density of the form

$$\mathcal{L} = E_o^2\left[\sqrt{1 + (\overline{H}^2 - \overline{E}^2)/E_o^2} - 1\right] \tag{1.18}$$

where E_o is a nonlinear limit to the magnitude of field intensities, and $\overline{H}$, of course, is magnetic field intensity [Born, 1934; Born and Infeld, 1934a, 1934b, 1935; Frenkel, 1934; Feenberg, 1935; Barbashov and Chernikov, 1967]. In the limit of low field amplitudes ($|\overline{E}|^2$ and $|\overline{H}|^2 << E_o^2$) this clearly reduces to the classical $\mathcal{L} \doteq \frac{1}{2}(\overline{H}^2 - \overline{E}^2)$. Born and Infeld found a spherically symmetric model electron for which the electric field ($\overline{E}$) was everywhere finite although electric displacement ($\overline{D}$) exhibited a singularity at the origin. For plane wave solutions the Lagrangian density (1.18) is of the general class

$$\mathcal{L} = \mathcal{L}\left(\phi_t^{\,2} - \phi_x^{\,2}\right) \tag{1.19}$$

which implies a solitary wave solution of arbitrary shape but with a speed equal to the velocity of light.

An entirely unrelated event of the mid-thirties was the first mathematical study of the nonlinear diffusion equation (1.5) by Kolmogoroff, Petrovsky and Piscounoff in the Soviet Union [Kolmogoroff et al, 1937; Fisher, 1937]. Since this study was motivated by the problem of genetic diffusion, rather than nerve propagation, the nonlinear function $F(\phi)$ was chosen to be of the form $\phi(1 - \phi)$ which does not lead to a solitary wave solution. How much more rapidly might neurophysiology have advanced if these researchers had been aware of the nerve problem? It was in 1936 that discovery of the giant axon of the squid was announced by J.Z. Young and in 1938 that Cole and Curtis published the classic cathode ray recording of an impulse on the squid nerve which is shown in Fig. 5.

Also from the Soviet Union there emerged during this period an equally unrelated line of research motivated by a fundamental problem in solid state physics: the relation between dislocation dynamics and plastic deformation of crystalline material. In this study Frenkel and Kontorova [1939] showed that a basic equation to describe dislocation motion takes the form

$$\phi_{xx} - \phi_{tt} = \sin\phi \tag{1.20}$$

This can be viewed as a nonlinear augmentation of the linear wave equation ($\phi_{xx} - \phi_{tt} = 0$) in the same sense that (1.5) is a nonlinear augmentation of the linear diffusion equation ($\phi_{xx} - \phi_t = 0$). Furthermore (1.20) is identical to (1.12) under the independent variable transformation

$$\begin{aligned} \xi &= \tfrac{1}{2}(x - t) \\ \tau &= \tfrac{1}{2}(x + t) \end{aligned} \tag{1.21}$$

The gudermannian solution (1.14) corresponds, under this transformation, to the propagation of a single dislocation described by

$$\phi = 4\,\tan^{-1}\left[\exp\left(\frac{x - ut}{\sqrt{1 - u^2}}\right)\right] \tag{1.22}$$

where the velocity

$$u = \sqrt{(1 - a^2)/(1 + a^2)}$$

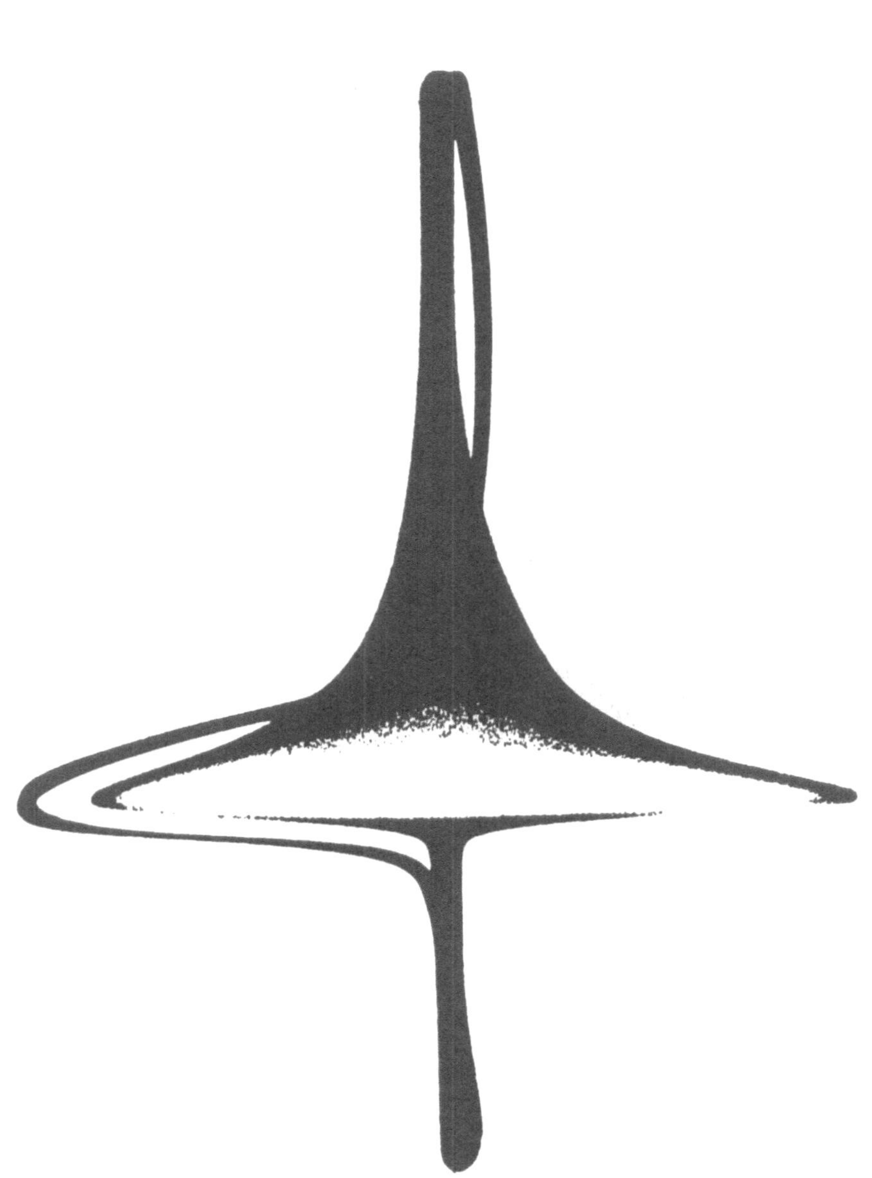

Fig. 5. Direct measurement of the increase in membrane conductance (band) during the action potential (line) on the squid giant axon. Time marks are 1 millisecond [K.S. Cole and H.J. Curtis, *Nature*, 142:209 (1938)].

To the extent that it existed at all between the wars, nonlinear wave research was fragmented beyond belief. Old knowledge was unavailable for application to new problems. The fundamental relationships between studies in nonlinear electrodynamics, solid state physics, hydrodynamics, neurodynamics, genetic diffusion and applied mathematics were entirely overlooked. The prophetic insights of John Scott Russell were completely forgotten.

C. Soliton Research from 1945 to 1974

If scientific research had suffered from the spilling of its young blood in the first world war, this mistake was not to be repeated during the second. Poets, perhaps, could still be driven into the guns but not engineers and scientists; they were drafted for more important tasks. U.S. science, in particular, was greatly strengthened by the war. Not only did we receive several dozen of the very best European scientists as permanent citizens, researchers and teachers; but the forced development of our technological power (in such diverse areas as electronics, microwaves, communications and control, and the manipulation of elementary particles) left us in a state of high morale. We could do anything.

One thing we did was to develop the digital computer, and Enrico Fermi suggested one of the very first scientific problems to which it was addressed: the dynamics of energy equipartition in a slightly nonlinear multimode mechanical system. This system consisted of 64 equal mass particles connected by slightly nonlinear springs. It was expected that if all the initial energy were put into a single mode, the slight nonlinearity would cause a gradual progress toward equipartition of this energy among all 64 modes in what could be considered a thermalized (or ergodic) state. The computer results obtained by Pasta and Ulam were surprising; no tendency toward thermalization was observed. If the energy was originally put into the lowest frequency mode, it returned almost entirely into that mode after a period of interaction with a few other low frequency modes [Fermi, Pasta and Ulam, 1955; Fermi, 1965; Newell, 1974]. John Pasta has commented that, shortly before his untimely death, Fermi felt this to be one of the most important problems he (Fermi) had studied.[5] It was certainly one of the most mysterious. Pursuit of this mystery led Zabusky and Kruskal to approximate the nonlinear spring-mass system by a rediscovered Korteweg-deVries (KdV) equation [see (1.2)] (which they found to apply also to wave motion in a collisionless plasma) and to observe numerically that KdV solitary waves pass through each other with no change in shape or speed [Zabusky and Kruskal, 1965]. It should be noted that Zabusky and Kruskal were not

[5] During anecdotal reminiscences presented at the Como Conference on Nonlinear Stochastic Problems organized by J. Ford and G. Casati in June, 1977. It is probably correct to consider Fermi a casualty of the war on the scientific front.

the first to observe nondestructive interactions of solitary waves. Apart from the above mentioned tank measurements by Scott Russell, Perring and Skyrme [1962] had published results of a numerical study of (1.20) in which two solitary waves of the form (1.22) underwent a collision. They observed perfect recovery of wave shapes and speeds after the collision and were led to the analytic description

$$\phi = 4 \tan^{-1} \left[\frac{u \sinh (x/\sqrt{1-u^2})}{\cosh (ut/\sqrt{1-u^2})} \right] . \tag{1.23}$$

This result would have been no surprise to Bäcklund or to Bianchi; it is merely the second (i.e. ϕ_2) in the hierarchy of solutions generated from the vacuum by the BT given above in (1.13). Nor would it have been a surprise to Seeger, Donth and Kochendörfer [1951, 1953] who had noted in 1953 the connection between this early work and the paper by Frenkel and Kontorova [1939]. But Perring and Skyrme were interested in (1.20) as a one-dimensional model for an elementary particle, and, in this context, one supposes that the complete absence of scattering might have been a bit disappointing. Throughout the 1960's, Eq. (1.20) arose in a wide variety of problems in applied science (including the propagation of ferromagnetic domain walls, the "self-induced transparency" effect in nonlinear optics, and propagation of magnetic flux quanta on Josephson transmission lines) and eventually became known as the "sine-Gordon" equation.[6]

The important contribution by Zabusky and Kruskal [1965] in their 1965 publication was to recognize the relation between nondestructive solitary wave collisions and the mystery of Fermi-Pasta-Ulam (FPU) recurrence effect. The solitary wave solutions of the KdV equation were viewed as independent and localized dynamic entities (called solitons) out of which more complex behavior could be constructed. By 1967 this insight had led Gardner, Greene, Kruskal and Miura to a truly brilliant scheme for developing a general solution to the KdV equation through a series of linear calculations [Gardner et al, 1967]. This method was soon expressed in the following elegant and general form by Lax [1968].

---

[6] This catchy name first appeared in the paper: J. Rubinstein, "Sine-Gordon equation", [1970] although rumor has it that the coinage was actually by Kruskal. Some physicists have rather immoderately objected to the term, which seems strange because in general physicists lead (the academic world at least) in the invention of whimsical jargon. For additional reviews of sine-Gordon applications during the 1960's see: G.L. Lamb, Jr., [1971] and A. Barone, et al [1971].

We are interested in a general nonlinear wave equation, $\phi_t = N(\phi)$, where $N(\cdot)$ denotes a nonlinear operator on some suitable space of functions. Suppose we can find two linear operators, L and B, which depend upon $\phi(x,t)$ (a solution of the nonlinear pde) and which satisfy the operator equation

$$iL_t = BL - LB \quad . \tag{1.24}$$

If L is viewed as a scattering operator with potential $\phi(x,t)$, its eigenvalues (λ) are found from study of

$$L\psi = \lambda\psi \quad . \tag{1.25}$$

Now if the time dependence of the scattered waves is taken to be

$$i\psi_t = B\psi \quad , \tag{1.26}$$

Eq. (1.24) implies that the eigenvalues in (1.25) are independent of time. Thus the computation

$$\phi(x,0) \longrightarrow \phi(x,t) \tag{1.27}$$

can be effected through the following three steps:

1. Direct problem. Calculate scattering parameters (such as the reflection and transmission coefficients of L) for ψ at $|x| = \infty$ and $t = 0$ from a knowledge of $\phi(x,0)$.

2. Time evolution of the scattering data. Use (1.26), together with the asymptotic form of B at $|x| = \infty$, to calculate the time evolution of the scattering data.

3. Inverse problem. From a knowledge of the scattering data of L as a function of time, construct $\phi(x,t)$.

Each bound state eigenfunction of (1.25) corresponds to a particular soliton component in the general solution. In the small amplitude (linear) limit there are no solitons present in the solution and the above described procedure degenerates into the usual Fourier transform method for linear pde's. This it has become known as the "inverse scattering transform method" (ISTM) wherein we see soliton components of the solution acting as generalized Fourier components.

Another extremely important development of the 1960's was the discovery by Morikazu Toda of exact two soliton interactions on a nonlinear-spring mass system in which the spring potential took the form [Toda, 1967a, 1967b, 1969, 1973, 1975].

$$\text{potential} = \frac{a}{b}\left[\exp(-br_n) - 1\right] + ar_n \quad . \tag{1.28}$$

This model is of great flexibility because it can be varied between the harmonic limit ($a \to \infty$ and $b \to 0$ with ab finite) and the hard sphere limit ($a \to 0$ and $b \to \infty$ with ab finite).

Up to this point we have been concentrating our attention on the dynamics of solitary waves. In 1965, however, Whitham began a series of papers that investigated the dynamics of periodic travelling waves [Whitham, 1965a,b, 1967a,b, 1970, 1974]. These take the general form

$$\phi(x,t) = \tilde{\phi}(kx + \omega t) \equiv \tilde{\phi}\,(\theta) \tag{1.29}$$

and include the cnoidal waves obtained by Korteweg and deVries for the equation that carries their names. Whitham supposed the solution to be locally periodic (though not sinusoidal) and assumed that the underlying pde could be derived from a Lagrangian density functional, $\mathcal{L}$. From this he obtained an averaged Lagrangian over a period of the wave by

$$<\mathcal{L}> \equiv \frac{1}{2\pi}\int_0^{2\pi} \mathcal{L}\; d\,\theta \tag{1.30}$$

as a function of the local values of k, ω, and an amplitude parameter, A. Euler variation of the averaged Lagrangian with respect to A and θ then gives two equations

$$\frac{\delta <\mathcal{L}>}{\delta A} = 0 \Rightarrow \frac{\partial <\mathcal{L}>}{\partial A} = 0 \tag{1.31}$$

$$\frac{\delta <\mathcal{L}>}{\delta \theta} = 0 \Rightarrow \frac{\partial}{\partial x}\,\frac{\partial <\mathcal{L}>}{\partial k} + \frac{\partial}{\partial t}\,\frac{\partial <\mathcal{L}>}{\partial \omega} = 0 \tag{1.32}$$

These, together with the pulse conservation equation,

$$\frac{\partial \omega}{\partial x} = \frac{\partial k}{\partial t} \tag{1.33}$$

suffice to fix the slow x and t variation of k, ω and A. Eq. (1.31) can be viewed in this context as a nonlinear dispersion equation

$$D(k, \omega, A) = 0 \tag{1.31'}$$

which depends upon the local value of the wave amplitude, A.

The above listed items are only the most important features in a growing panorama of nonlinear wave activities which became increasingly less parochial throughout the nineteen-sixties. Solid state physicists began to see some relationship between their solitary waves (domain walls, self-shaping light pulses, magnetic flux quanta) and those from classical hydrodynamics, and applied mathematicians began to suspect that the ISTM might apply to a broader set of nonlinear wave equations. It was in the context of this growing excitement and self-awareness that Alan Newell organized a research conference for three and a half weeks during the summer of seventy-two in which the participants "ranged over a wide spectrum of ages (from graduate students to senior scientists), background interests (biology, electrical engineering, geology, geophysics, mathematics, physics) and countries of origin (United States, Canada, Great Britain, Australia)" [Newell, 1974]. It is difficult to overemphasize the importance of this conference to solitary wave research in the english-speaking world. Countless cross-disciplinary bonds of collaborative interaction and friendship were formed, and a sense of the existence of nonlinear wave study as a broad and vigorous activity was established. Solitary wave research came out of the closet.

One of the most significant inputs to Newell's conference was from the Soviet Union. In a paper first published in 1971, Zakharov and Shabat [1972] showed that the "Lax-operators" L and B [see (1.24)-(1.26)] could be found for the "nonlinear Schrödinger" (NLS) equation

$$i\phi_t + \phi_{xx} = k\,|\phi|^2\phi \tag{1.34}$$

and proceeded in an elegant and systematic way to develop an ISTM and derive an infinite set of conservation laws corresponding to those previously found for the KdV equation by Robert Miura [1968a,b]. It should perhaps be emphasized that (1.34) is not a precious object invented by Zakharov and Shabat to display their analytical wizardry; it had arisen in practice to describe envelope waves in hydrodynamics, nonlinear optics, nonlinear acoustics and plasma waves [Benney and Newell, 1967; Bespalov and Talanov, 1966; Kelley, 1965; Tappert and Varma, 1970, Ichikawa et al, 1972]. The Zakharov and Shabat paper was widely circulated and read and discussed (formally and in the evenings over Newell's endless supplies of beer), and everyone left the conference realizing that four of the most fundamental nonlinear wave systems [KdV (1.2), the "sine-Gordon" equation (SGE) (1.20), NLS (1.34), and the nonlinear spring mass system described by Toda] displayed solitary waves with the special behavior that had led Zabusky and Kruskal to coin the term soliton.

Within the next two years the basic ingredients (i.e. Lax operators) for the ISTM had been constructed for the SGE [Ablowitz et al, 1973; Takhtadzhyan and Faddeev, 1974] and the Toda lattice [Henon, 1974; Flashka, 1974]. Following the discovery, through a very general approach based on differential geometry, of a Bäcklund transform for the KdV equation [Wahlquist and Estabrook, 1973], Newell demonstrated the equivalence of the BT and ISTM for a rather wide class of nonlinear wave systems [Ablowitz et al, 1974; Chu and Scott, 1974].

D. Particle Physics to the Present

Einstein's conviction that a consistent theory for particle physics must be based on singularity free solutions of partial differential equations [Einstein, 1954] was shared by some of the most distinguished of his colleagues. In addition to the above mentioned interest of Born and Infeld in this question, both Heisenberg [1966] and de Broglie [1960, 1963] have described non-linear field theories which, in their simplest representations, can be viewed as the augmentation of classical e.m. field equations by a nonlinear term of the form $|\phi|^2\phi$ as in (1.34). The ideas of de Broglie bear an interesting relationship to the ISTM. In his "theory of the double solution" the real particle is a singularity free solution of a nonlinear wave equation $\phi = \Phi \exp(i\,\theta')$, but associated with it is a corresponding solution of a linear wave equation $\psi = \Psi \exp(i\,\theta)$ for which

$$\theta = \theta'$$

except inside a small sphere surrounding the particle. The function ψ is taken to be a solution of Schrödinger's equation and the above phase condition allows the particle to be "guided" by ψ. In the same sense the nonlinear solution of (1.24) can be viewed as being guided through space-time by the linear asymptotic solution of (1.26).

Several investigators during the 1960's gave consideration to the sine-Gordon equation (1.20) as a (1 + 1) dimensional field theory for elementary particles [Perring and Skyrme, 1962; Enz, 1963; Hobart, 1963; Derrick, 1964; Rosen 1965; Skyrme, 1971; see also references in Footnote 6] but this work was of little general interest until the early 1970's when it became clear that the special properties of the SGE allowed it to be completely quantized [Dashen et al, 1974, 1975; Goldstone and Jakiw, 1975; Faddeev, 1975].

This showed that classical solutions [e.g. (1.22)] "survive quantization". The classical field energy was found to be a useful first approximation for the soliton mass with quantum effects introducing a second order correction. Following this line there has been a dramatic rise in research related to the solitary wave picture of elementary particles as shown in Fig. 6. Several useful

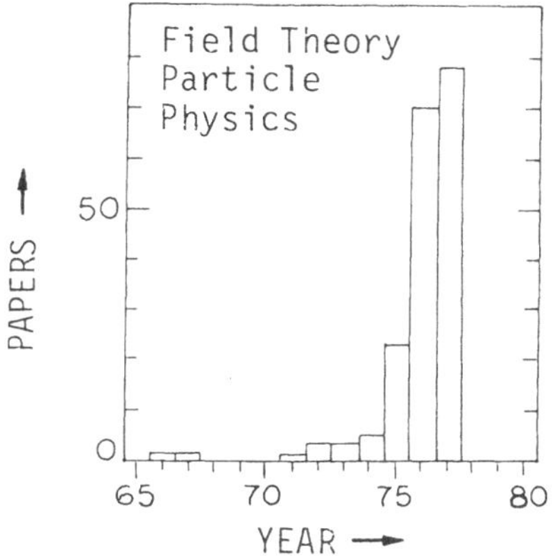

Fig. 6. Annual number of publications on solitary waves in elementary particle physics (Y.H. Ichikawa, private communication).

reviews are available for readers interested in following recent developments [Rebbi, 1979; Parsa, 1979; Jakiw, 1977; Makhankov, 1978]. It is interesting to observe that the "topological solitons" currently being studied are three dimensional analogs of the vortex which was originally described by Helmholtz [1858, 1868].

E. Neurodynamics to the Present

Post World War II research in electrophysiology was galvanized by the substantial increase in electronic measurement technology developed in support of military communications and RADAR. Peacetime dividends were not long in coming; a theoretical basis for pulse propagation on the giant axon of the squid was soon developed. The general outlines for a theory of nerve fiber dynamics had been known since the mid-thirties and can be sketched as follows [see Fig. 7]. If v(x,t) is transmembrane voltage and i(x,t) is axial current, then

$$v_x = -ri \tag{1.35a}$$

$$i_x = -cv_t - j_i. \tag{1.35b}$$

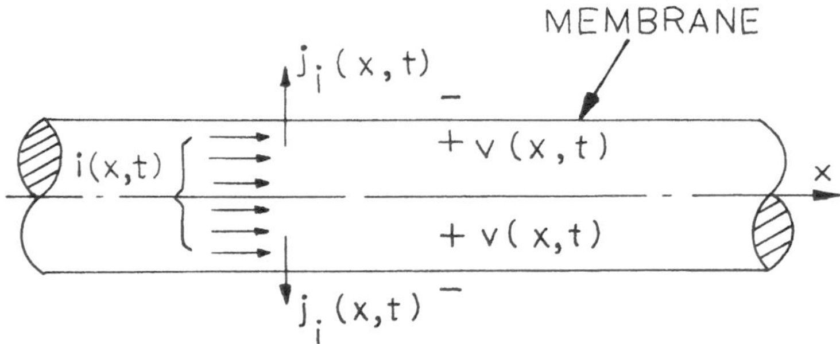

Fig. 7. Currents and voltage in a nerve fiber.

In (1.35a) the parameter r is the series resistance per unit length of the ionic core (axoplasm); thus (1.35a) is merely a statement of Ohm's law. Eq. (1.35b), on the other hand, is a conservation law for electrical charge inside the fiber for which c is the capacitance per unit length of the membrane and j_i is the ionic current per unit length crossing the membrane. If r is independent of x, (1.35) takes the form

$$v_{xx} - rc\, v_t = rj_i \tag{1.36}$$

which is a linear diffusion operator augmented by the nonlinear effects of transmembrane ionic current (j_i). The main problem facing post-war researchers was to represent the dynamics of j_i, and a rather definitive answer to this question was provided in 1952 by Hodgkin and Huxley [1952]. Their theoretical picture introduced three additional phenomenological variables to represent the opening and closing of various ionic channels across the membrane, but they managed to obtain an excellent prediction of both shape and speed of the solitary wave shown in Fig. 5.

A simpler dynamical description for j_i was introduced in 1962 by Nagumo, Arimoto and Yoshizawa [1962] and was based upon previous work by FitzHugh. In this picture [called the FitzHugh-Nagumo equation (FNE)],

$$j_i = F(v) + R \tag{1.37a}$$

$$R_t = \varepsilon(V + a - bR) \tag{1.37b}$$

where R is a "recovery variable" and ε, a and b are adjustable constants. In the limiting case $\varepsilon = 0$ it is clear from (1.37b) that R remains constant; thus substitution of (1.37a) into (1.36) gives a result that is identical to (1.5). In fact, R does remain relatively constant during the leading edge of a typical pulse so analysis of (1.5) yields useful first approximations to both the shape of the leading edge and the speed of the entire pulse.

It is interesting to emphasize at this point how our basic picture of the nerve impulse emerges as a singularity-free solution of a nonlinear pde (1.36) just as Einstein had prescribed for the elementary particles of matter. In this context we might refer to the nerve impulse as an "elementary particle of thought".

The dynamic behavior of nerve impulses as they collide with others, propagate past irregularities in the fiber structure, interact with others at points where a fiber branches, etc. is currently a subject of considerable analytical and experimental interest. For a survey of this work in the Soviet Union the reader should consult the fine book by Khodorov [1974]. Somewhat more recent surveys have been published by the present author [1975,1977] in which it has been suggested that physical scientists (physicists and electrical engineers in particular) should become seriously interested in neurodynamics. An excellent recent discussion of mathematical results in the study of "reaction-diffusion systems" (of which the nerve fiber system is a special case) has been provided by Paul Fife [1979].

F. Current Solitary Wave Research

After a gestation period of almost a century and a half, the solitary wave concept has finally been established. It was a difficult birth, but a new paradigm is now a part of our collective scientific thought. The "thingness" of solitary waves in general (and solitons in particular) is widely accepted as a structural basis for viewing and understanding the dynamic behavior of complex nonlinear systems.

But the sheer bulk of current solitary wave research prohibits a survey with any degree of bibliographic detail. Furtunately some books are becoming available that should help the interested reader find a way to activities and results [Calogero, 1978; Miura, 1976; Lonngren and Scott, 1978; Caudrey and Bullough, to appear; Hermann, 1977a,b,c,; Barut, 1978; Bishop and Schneider, 1979; Lamb, to appear]. Also a new review and research journal [Physica D - Nonlinear Phenomena, edited by H. Flaschka, J. Ford, A.C. Newell and A.C. Scott and published by the North-Holland Publishing Co. The first issue appeared in March of 1980.] has recently been established that will draw together nonlinear wave research from the widely scattered scientific workshops.

From Fig. 1 we see that the solitary wave paradigm is being exercised in an impressive number of studies throughout science. One of the most fundamental areas of application is in statistical mechanics where soliton (or solitary wave) modes are found, from both experimental and theoretical investigations, to participate in energy equipartiation. Such studies are related to soliton perturbation theory wherein the dynamic behavior of solitary waves that are "almost solitons" is calculated as a slow variation in soliton parameters (speeds and location) induced by the difference between the actual pde and some approximation that is integrable via the ISTM. The discovery of solitary wave solutions to pde's of physical interest is now being vigorously pursued using the tools of "modern" differential geometry which in turn is directly in the nineteenth century tradition that died (or was badly wounded) in the first world war. Finally there is a vast array of specific solitary wave applications: from hydrodynamics to meteorology, from computer technology to shock wave dynamics, from nonlinear filter theory to the propagation of crystal dislocations and domain walls, from the elementary particles of matter to the elementary particles of thought.

With our present perspective, John Scott Russell's last book [Russell, 1885], far from being a catalog of "extraordinary and groundless speculations", seems rather conservative.

II. SOME GENERAL COMMENTS

In this section are recorded some introductory comments (or points of view) on nonlinear wave dynamics that I have found helpful.

A. Conservation Laws

Consider a dynamical system in one space dimension for which some quantity (let's call it Q) is conserved. It is convenient to define F as the flow of Q past a point, and D as the density of Q. Then for two points separated by the differential distance dx, we expect that $F(x) - F(x + dx) = d(D \cdot dx)/dt$ which leads to the first order p.d.e.

$$F_x + D_t = 0 \quad . \tag{2.1}$$

This is a conservation law for the quantity Q. It is particularly interesting when $F = F(D)$ for then it becomes

$$\frac{dF}{dD} D_x + D_t = 0 \tag{2.2}$$

with the solution

$$D = D[x - u(D)t] \quad , \tag{2.3}$$

where

$$u(D) \equiv dF/dD \quad . \tag{2.4}$$

Thus the local value of velocity for a fixed density is given by the slope indicated. If F(D) is concave upward (as in hydrodynamic waves) higher levels of D move faster than lower levels and a shock appears on the leading edge. If F(D) is concave downward on the other hand (as in automobile traffic), shocks appear on the trailing edge. See Chapter 2 of Whithams book [1974] for additional details.

A particularly simple conservation law is

$$\phi_t + \phi\phi_x = 0 \tag{2.5}$$

for which $F = \frac{1}{2}\phi^2$ and $D = \phi$ so

$$F = \tfrac{1}{2} D^2 \tag{2.6}$$

and shocks should be expected on the leading edge of a density pulse. When shocks begin to form, we expect the physical approximations upon which (2.5) is based to break down. Higher-order effects such as dispersion or dissipation should then be included in the description. Adding dispersion to (2.5) leads to the Korteweg-deVries equation

$$\phi_t + \phi\phi_x = \phi_{xxx} \quad . \tag{2.7}$$

Adding dissipation leads to the Burgers equation

$$\phi_t + \phi\phi_x = \phi_{xx} \quad . \tag{2.8}$$

A conservation law always implies an associated linear problem since (2.1) allows us to write the differential form

$$d(\ell n\,\psi) = F dt - D dx \quad . \tag{2.9}$$

Indeed (2.1) is just the condition for integrability of (2.9). But from (2.9)

$$\psi_x = -D\psi \ , \tag{2.10a}$$

$$\psi_t = F\psi \ , \tag{2.10b}$$

which is an associated linear problem.

If (for convenience) we write Burgers equation in the special form

$$\phi_t = 2\phi\phi_x + \phi_{xx} \ , \tag{2.11}$$

then $D = -\phi$ and $F = \phi_x + \phi^2$ so the associated linear problem becomes

$$\psi_x = \phi\psi \ , \tag{2.12a}$$

$$\psi_t = (\phi_x + \phi^2)\psi = \phi_x\psi + \phi\psi_x = (\phi\psi)_x = \psi_{xx} \quad . \tag{2.12b}$$

In this case (known as the Hopf-Cole transformation)(2.12a) can be used to transform initial conditions $\phi(x,o)$ to $\psi(x,o)$ whereupon (2.12b) gives the linear evolution of $\psi(x,o)$ to $\psi(x,t)$. Then $\phi(x,t)$ can be obtained again through (2.12a).

B. Power Balanced Solitary Waves

Consider augmenting the linear diffusion equation $\phi_{xx} - \phi_t = 0$ with a nonlinear function to yield

$$\phi_{xx} - \phi_t = F(\phi) \quad . \tag{2.13}$$

If

$$F(\phi) = \phi(\phi - a)(\phi - 1) \tag{2.14}$$

with $0 < a < 1$, the system has two stable stationary points: $\phi = 0$ and $\phi = 1$. Assuming a traveling wave solution

$$\phi(x,t) = \tilde{\phi}(x - ut) \tag{2.15}$$

where u is an (as yet) undetermined wave velocity, (2.13) reduces to a pair of first-order o.d.e.'s (prime denotes differential)

$$\tilde{\phi}' = w \tag{2.16a}$$

$$w' = F(\tilde{\phi}) - uw \tag{2.16b}$$

For this system the singular points at $(\phi,w) = (0,0)$ and at $(\phi,w) = (1,0)$ are both saddles. Thus it is convenient in general to use a "shooting method" to find the precise value of velocity at which a trajectory leaving (0,0) as $x - ut$ increases from $-\infty$ becomes identical to a trajectory approaching (1,0) as $x - ut \to \infty$.

With the simple nonlinearity of (2.14), this trajectory can be determined by choosing

$$w = K\tilde{\phi}(\tilde{\phi} - 1) \tag{2.17}$$

where K is to be determined. Then from (2.16)

$$\frac{dw}{d\tilde{\phi}} = \frac{\tilde{\phi} - a}{K} - u$$

while from (2.17)

$$\frac{dw}{d\tilde{\phi}} = 2K\tilde{\phi} - K$$

whereupon $K = 1/\sqrt{2}$ and

$$u = (1 - 2a)/\sqrt{2} \tag{2.18}$$

as was previously noted.

C. Conservative Solitary Waves

Consider augmenting the linear wave equation $\phi_{xx} - \phi_{tt} = 0$ with a nonlinear function to yield

$$\phi_{xx} - \phi_{tt} = F(\phi) \quad . \tag{2.19}$$

The traveling wave assumption (2.15) then leads to the nonlinear o.d.e.

$$\tilde{\phi}''(1-u^2) = F(\tilde{\phi}) \tag{2.20}$$

which is readily integrated for a particular choice of $F(\cdot)$. The special case

$$F(\phi) = \sin\phi$$

leads to a first integration

$$\tfrac{1}{2}(\tilde{\phi}')^2(1-u^2) = E - \cos\tilde{\phi} \tag{2.21}$$

where E is a constant of integration. A second integration is obtained by arranging (2.21) into the elliptic integral form

$$\int_0^{\tilde{\phi}} \frac{d\phi}{\sqrt{2(E-\cos\phi)}} = \frac{x - ut - x_o}{\sqrt{1-u^2}} \tag{2.22}$$

where the second constant of integration (x_o) is taken as the place where $\tilde{\phi} = 0$ at $t = 0$.

For the special case $E = 1$, (2.22) reduces to the single kink solution

$$\tilde{\phi} = 4\tan^{-1}\left[\exp\left(\frac{x - ut - x_o}{\sqrt{1-u^2}}\right)\right] \quad . \tag{2.23}$$

III. INVERSE SCATTERING TRANSFORM THEORY

Consider the sine-Gordon equation

$$\phi_{xx} - \phi_{tt} = \sin\phi \quad . \tag{3.1}$$

The independent variable transformation

$$\begin{Bmatrix} x \\ t \end{Bmatrix} \to \begin{Bmatrix} \xi & = & \tfrac{1}{2}(x - t) \\ \tau & = & \tfrac{1}{2}(x + t) \end{Bmatrix} , \tag{3.2}$$

brings (3.1) to the form

$$\phi_{\xi\tau} = \sin\phi \quad . \tag{3.3}$$

A. Lax Operators

In 1973, Ablowitz, Kaup, Newell, and Segur [1973] showed that Lax operators [see eqs. (1.24)-(1.26) of Section I] for (3.3) are

$$L \equiv i \begin{bmatrix} \partial_\xi & \tfrac{1}{2}\phi_\xi \\ \tfrac{1}{2}\phi_\xi & -\partial_\xi \end{bmatrix} \tag{3.4}$$

$$B = \frac{-1}{4\gamma} \begin{bmatrix} \cos\phi & \sin\phi \\ \sin\phi & -\cos\phi \end{bmatrix} \tag{3.5}$$

with

$$\psi \equiv \begin{bmatrix} \psi_1 \\ \psi_2 \end{bmatrix} \quad . \tag{3.6}$$

Thus the scattering equation $L\psi = \gamma\psi$ becomes

$$\psi_{1,\xi} + i\gamma\psi_1 = -\tfrac{1}{2}\phi_\xi\psi_2 \quad , \tag{3.7a}$$

$$\psi_{2,\xi} - i\gamma\psi_2 = \tfrac{1}{2}\phi_\xi\psi_1 \quad , \tag{3.7b}$$

and the time evolution equation $i\psi_t = B\psi$ becomes

$$\psi_{1,\tau} = \left(\frac{i}{4\gamma}\right)(\psi_1\cos\phi + \psi_2\sin\phi) \quad , \tag{3.8a}$$

$$\psi_{2,\tau} = \left(\frac{i}{4\gamma}\right)(\psi_1\sin\phi - \psi_2\cos\phi) \quad . \tag{3.8b}$$

The "Lax condition" for time independent eigenvalues

$$iL_t = BL - LB \iff \phi_{\xi\tau} = \sin\phi$$

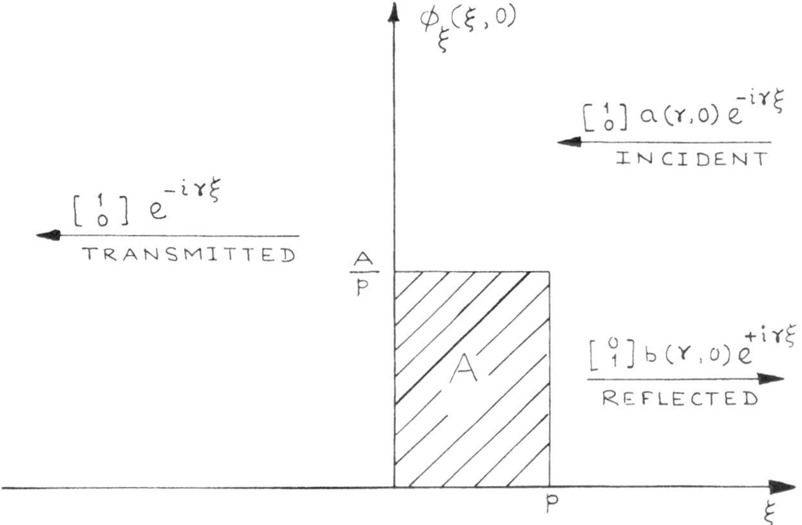

Fig. 8. Scattering problem for the square well potential.

is identical to the cross derivative condition

$$(\psi_\tau)_\xi = (\psi_\xi)_\tau \iff \phi_{\xi\tau} = \sin\phi \quad .$$

B. Forward Scattering

A somewhat different inverse scattering transform (IST) formalism was published at about the same time by Takhtadzhyan and Faddeev [1974].

To become familiar with the "go" of this IST method it is helpful to study (3.7) for a particular initial condition. To this end we choose

$$\phi_\xi(\xi,0) = \begin{cases} A/p & \text{for} \quad 0 < \xi < p \\ 0 & \text{for} \quad \xi < 0 \; , \; \xi > p \end{cases}$$

as is shown in Figure 8. The reader should easily convince his- or herself that incident, transmitted and reflected waves with ξ variation as indicated are obtained. For the square well potential of Figure 8 it is straightforward to show that the incident and reflected wave amplitudes are, respectively,

$$a(\gamma,0) = \exp(i\gamma p)[\cos(mp) - (i\gamma/m)\sin(mp)] \quad , \tag{3.9}$$

$$b(\gamma,0) = \exp(-\gamma p)(A/2mp)\sin(mp) \quad , \tag{3.10}$$

where for typographical convenience we have defined [Scott et al, 1976]

$$m^2 \equiv \gamma^2 + A^2/4p^2 \quad . \tag{3.11}$$

Consider next the upper half of the complex γ-plane indicated in Figure 9. At zeros of $a(\gamma)$, the incident wave amplitude is (by definition) zero and in the upper-half plane (UHP) the "transmitted wave" decays in the $-\xi$ direction while the "reflected wave" decays in the $+\xi$ direction. Thus we have the

> IMPORTANT FACT: Upper half plane zeros $\{\gamma_j\}$ of $a(\gamma)$ correspond to bound state eigenvalues of the scattering equations (3.7).

The Lax condition insures that these bound state eigenvalues $\{\gamma_j\}$ are independent of time. From (3.9) and (3.11) it is readily seen that zeros of $a(\gamma)$ are roots of

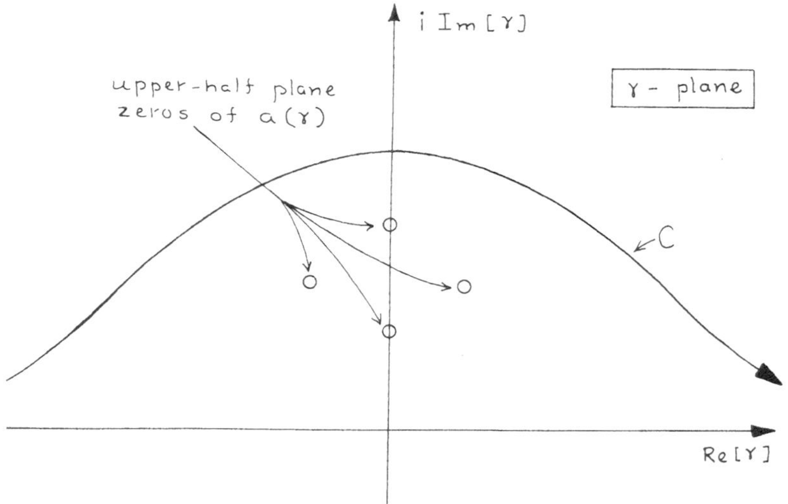

Fig. 9. Zeros of a(γ) in the upper-half complex γ-plane.

$$\cot(mp) = -\left[\frac{A^2}{(2mp)^2} - 1\right]^{\frac{1}{2}} , \tag{3.12}$$

and if

$$A = 2\pi N \quad , \quad N = 1, 2, \ldots \quad , \tag{3.13}$$

then these bound-state eigenvalues lie only on the imaginary axis and their number is equal to N.

The remainder of the scattering data at $\tau = 0$ consists of the reflection coefficient

$$\frac{b(\gamma,0)}{a(\gamma,0)}$$

evaluated for real γ, and the residues of the poles of the reflection coefficient evaluated at the $\{\gamma_j\}$. For simple poles these are

$$C_j(0) \equiv \frac{b(\gamma_j,0)}{a'(\gamma_j,0)}$$

C. Time Evolution of the Scattering Data

The time evolution of the bound-state eigenvalues $\{\gamma_j\}$ is trivial: they remain constant. The time evolution of the reflection coefficient and its poles can be obtained by observing, from condition (3.13), that as $\xi \to \pm\infty$, $\sin\phi \to 0$, $\cos\phi \to 1$ and (3.8) takes the asymptotic form

$$\begin{aligned} \psi_{1,\tau} &= +\frac{i}{4\gamma}\psi_1 , \\ \psi_{2,\tau} &= -\frac{i}{4\gamma}\psi_2 . \end{aligned} \tag{3.8'}$$

Thus ψ varies with τ (as $\xi \to +\infty$) in the following way.

$$\psi(\xi,\tau) = \begin{bmatrix} f_1(\xi)\exp(i\tau/4\gamma) \\ \\ f_2(\xi)\exp(-i\tau/4\gamma) \end{bmatrix} , \tag{3.14}$$

whereupon the reflection coefficient varies with τ as

$$\frac{b(\gamma,\tau)}{a(\gamma,\tau)} = \frac{b(\gamma,0)}{a(\gamma,0)} \exp\left(\frac{-i\tau}{2\gamma}\right) . \tag{3.15}$$

The residues of the UHP (simple) poles of the reflection coefficient have the time variation

$$C_j(\tau) = \frac{b(\gamma_j\tau)}{a'(\gamma_j,\tau)} = \frac{b(\gamma_j,0)}{a'(\gamma_j,0)} \exp\left(\frac{-i\tau}{2\gamma_j}\right) \tag{3.16}$$

D. <u>The Inverse Scattering Calculation</u>

Given the scattering data at time τ

$$\{\gamma_j\} \quad , \quad \frac{b(\gamma,\tau)}{a(\gamma,\tau)} \quad , \quad \text{and } \frac{b(\gamma_j,\tau)}{a'(\gamma_j,\tau)} \quad ,$$

the "potential" in (3.7), $\phi_\xi(\xi,\tau)$, can be obtained from the Marchenko type integral equation (where the notation of τ variation has been suppressed for convenience)

$$K(\xi,y) = F(\xi+y) - \int_\xi^\infty \int_\xi^\infty K(\xi,\alpha)F(\alpha+\alpha')F(\alpha'+y)\,d\alpha\, d\alpha' \quad ,$$

$$\xi - y < 0 \quad , \tag{3.17}$$

where

$$F(\cdot) \equiv \frac{1}{2\pi}\int_{-\infty}^{\infty} \frac{b(\gamma)}{a(\gamma)} \exp[i\gamma(\cdot)]d\gamma - i \sum_{j=1}^{N} C_j \exp[i\gamma_j(\cdot)]. \tag{3.18}$$

From (3.18) and the scattering data, $F(\xi + y)$ is readily obtained. Then (3.17) is solved for $K(\xi, y, \tau)$. Finally the potential is calculated as

$$\boxed{\phi_\xi(\xi,\tau) = 4K(\xi,\xi,\tau) \ .} \tag{3.19}$$

At this point the reader is probably asking at least two questions:

QUESTION # 1: Is it any easier to effect a solution of (3.17) than to study directly the original p.d.e. (3.3)?

QUESTION # 2: Where (in the world) does such an integral equation as (3.17) come from?

The answer to the first question is: "Yes". Suppose that the reflection coefficient, b/a, is zero along the real axis of the γ-plane. Then in (3.18) F consists only of a finite sum associated with the N bound-state eigenvalues $\{\gamma_j\}$. Each of these corresponds to a soliton with velocity (in x-t space)

$$u_j = \frac{4\gamma_j^2 + 1}{4\gamma_j^2 - 1} , \tag{3.20}$$

and (3.17) can be solved directly to obtain an explicit N-soliton formula.

The second question can be answered by introducing a "pseudotime" y through the Laplace transform

$$\Psi_i(\xi, y) = \frac{1}{2\pi} \int_C \psi_i(\xi, \gamma) \exp(-i\gamma y) d\gamma . \tag{3.21}$$

Then the scattering equations (3.7) become

$$\Psi_{1,\xi} - \Psi_{1,y} = -\tfrac{1}{2} \cdot \phi_\xi \Psi_2 , \tag{3.22a}$$

$$\Psi_{2,\xi} + \Psi_{2,y} = +\tfrac{1}{2} \phi_\xi \Psi_1 . \tag{3.22b}$$

If the contour C for the definition of this Laplace transform is chosen (as indicated in Fig. 9) to lie above all poles of the reflection coefficient, then eqs. (3.22) are <u>causal in pseudotime</u>. Choosing the asymptotic conditions as $\xi \to +\infty$

$$\Psi_1 \to \delta(x + y) , \tag{3.23a}$$

$$\Psi_2 \to F(\xi - y) , \tag{3.23b}$$

implies that the incident wave is a delta function propagating in pseudotime and the reflected wave is the Laplace transform of the reflection coefficient. This leads directly to (3.18) and (3.17) is simply the Laplace transform of the scattering equation for (3.23a) assuming causality in pseudotime. Additional details are presented in Scott et al [1976].

IV. PERTURBATION THEORY

Pure solitons are rather dull objects. They pass through each other with no change in final shape and speed; only a phase shift indicates that an interaction has taken place. But there is considerable experimental and numerical evidence that solitons are structurally stable or "robust" to small changes in the dynamical system. It is interesting to ask how such small changes could affect the motion of solitons, to accelerate them (say) or to slow them down. For such questions a perturbation theory often provides convenient answers.

A. General Situation

Given the nonlinear wave equation

$$NL(\phi_o) = 0 \tag{4.1}$$

where ϕ_o is an N-soliton solution with 2N parameters fixing the soliton speeds and positions, it is often of interest to study

$$NL(\phi) = \varepsilon f \tag{4.2}$$

where εf is a small (one hopes!) structural perturbation of the original "pure soliton" system given by (4.1). Expanding in a series

$$\phi = \phi_o + \varepsilon\phi_1 + \ldots \tag{4.3}$$

we find that ϕ_1 must satisfy the linear equation

$$\mathcal{L}\,\phi_1 = f \tag{4.4}$$

where $\mathcal{L}$ is the linearization of the nonlinear operator NL(•) about the zero-order solution ϕ_o.

Formally (4.4) can be solved as

$$\phi_1(x,t) = \int_o^t dt' \int_{-\infty}^{\infty} dx' G(x,t|x',t') f(x',t') \tag{4.5}$$

where G is a Green's function for the linear operator

As a function of x and t, G is composed of solutions of the homogeneous equation

$$\mathcal{L}\tilde{\Psi} = 0 \ . \tag{4.6}$$

Said another way, the x and t dependence of G lies in

$$\mathcal{N}(\mathcal{L}) \quad ,$$

the "null space" of the operator $\mathcal{L}$. Similarly the x' and t' dependence of G lies in

$$\mathcal{N}(\mathcal{L}^+) \quad ,$$

the null space of the "adjoint" of $\mathcal{L}$.

In general, G has two components: a "discrete" part, G_d, composed of the discrete spectrum of $\mathcal{L}$ and $\mathcal{L}^+$ and a "continuous" part, G_c, composed from the continuous spectrum. Thus

$$G = G_d + G_c \quad . \tag{4.7}$$

Now it is certainly convenient to have

$$\int_{-\infty}^{\infty} G_d(x,t|x',t')f(x',t')dx' = 0 \tag{4.8}$$

otherwise, from (4.5), ϕ_1 will grow linearly with time and the first two terms in (4.3) will soon cease to be a good approximation to ϕ.

How is one to achieve condition (4.8)? The x' dependence of both G_d and f appear to be determined. The answer is to allow the 2N parameters in ϕ_o to change slowly (order ε) with time. This gives an additional contribution to the source from the time derivative of each parameter. Thus

$$f \rightarrow F \equiv f - \sum_{j=1}^{2N} \phi_{o,p_i} \dot{p}_i \tag{4.9}$$

and condition (4.8) which can be written

$$F \perp \mathcal{N}(\mathcal{L}^+) \tag{4.10}$$

yields just 2N first-order equations for the $\{p_i\}$.

This set of 2N first-order o.d.e.'s for the time dependence of the $\{p_i\}$ can be integrated to obtain an appropriate set $\{p_i(t)\}$ which gives the time dependence in ϕ_o to avoid unbounded temporal growth (secularity). Then ϕ_1 is calculated from

$$\phi_1 = \int_0^t dt' \int_{-\infty}^{\infty} dx' G_c(x,t|x',t') F(x',t') \qquad (4.11)$$

which is an estimate of the radiation associated with the speed modulations computed from condition (4.10). From our present perspective, (4.3) takes the form

$\phi = \phi_0$ ← N-soliton formula with speeds and phases modulated to avoid secularity.

$+\varepsilon\ \phi_1$ ← Radiation.

$+ \ldots$ ← Higher-order corrections.

It is important to note that the null spaces of $\mathcal{L}$ and $\mathcal{L}^+$ can be readily generated from the parameter dependence of ϕ_0. Given that

$$NL[\phi_0(x,t,p)] = 0 \qquad (4.12)$$

differentiation with respect to p gives

$$\mathcal{L}\,\phi_{0,p} = 0 \qquad (4.13)$$

an element in $\mathcal{N}(\mathcal{L})$. The radiation spectra of these operators can be obtained easily by generating ϕ_0 from vacuum radiation, differentiating with respect to radiation amplitude and setting that amplitude to zero. Details are discussed in McLaughlin and Scott [1978] and Scott [1979].

B. Specific Examples

This approach has been used to study the motion of solitons of the sine-Gordon equation ($\phi_{xx} - \phi_{tt} - \sin\phi = 0$) which has been perturbed to [McLaughlin and Scott, 1978]

$$\phi_{xx} - \phi_{tt} - \sin\phi = \gamma + \alpha\phi_t + \sum_n \mu_n \delta(x - na_n)\sin\phi\ . \qquad (4.14)$$

The motivation for this study was the problem of propagating flux quanta (or fluxons) on a Josephson junction transmission line. The various perturbative terms in (4.14) can be identified as follows:

γ - represents the effect of transverse current bias which applies a force in one direction to a kink and in the other to an antikink.

α - represents the effect of dissipation.

μ_n - represents the effect of local regions of increased Josephson tunnel current current.

With the $\mu_n = 0$, for example, condition (4.10) leads directly to the o.d.e.

$$\frac{du}{dt} = \frac{\pi}{4}\gamma\left(1-u^2\right)^{3/2} - \alpha u(1-u^2) \tag{4.15}$$

for the speed of a single kink.

A similar approach has been used to study transverse interactions between nerve pulses on parallel fiber bundles [Scott and Luzader, 1979]. These interactions can lead to "locked states" in which a bundle or "assembly" of pulses may propagate with the same speed.

V. ENERGY TRANSPORT ON THE ALPHA HELIX PROTEIN BY SOLITONS

A. Introduction

"How can energy be transmitted in biological systems?" is a basic question that was discussed in depth at a 1973 meeting of the New York Academy of Sciences amid talk of a "crisis in bioenergetics" [Green, 1973; Ann. N.Y. Acad. Scie. 1974]. A central issue in the "crisis" is that the attractive mechanism of energy transduction via excited molecular vibrations is presumed (on the basis of a linear dynamic analysis) to have an unacceptably short lifetime; but a promising answer to this objection has been proposed by Davydov [to appear a,b; 1976, 1977]. He suggests that the nonlinear character of interatomic forces (e.g. the hydrogen bond) can lead to the formation of robust solitary waves (often called "solitons"[7]) which exhibit greatly increased radiative lifetimes and a correspondingly increased ability to transport energy over large distances.

---

[7] Since the term "soliton" was coined by Zabusky and Kruskal, thousands of papers have appeared in a wide variety of research areas. Several recent books include those by Hermann, [1977]; K. Lonngren and A. Scott [1978]; Bishop and Schneider [1979]; Caudry and Bullough [to appear]; Lamb [to appear].

As a specific context for the development of his idea, Davydov has concentrated upon the α-helix protein and has chosen the relatively isolated amide-I (or C = O "stretch"; note that this notation is chemical and not mathematical. C corresponds to a carbon atom, O to an oxygen atom, and = to a double bond between them.) vibration of the peptide group as the main "basket" in which energy is carried. According to a linear analysis, energy transported by this means should spread out from the effects of dispersion and rapidly become disorganized and lost as a source for biological mechanisms. But in the nonlinear analysis of Davydov, propagation of amide-I vibrations is retroactively coupled to longitudinal sound waves of the α-helix, and the coupled excitation propagates as a localized and dynamically self-sufficient entity called a soliton (a solitary wave). The amide-I vibrations generate longitudinal sound waves which, in turn, provide a potential well that prevents vibrational dispersion. Thus the soliton holds itself together.

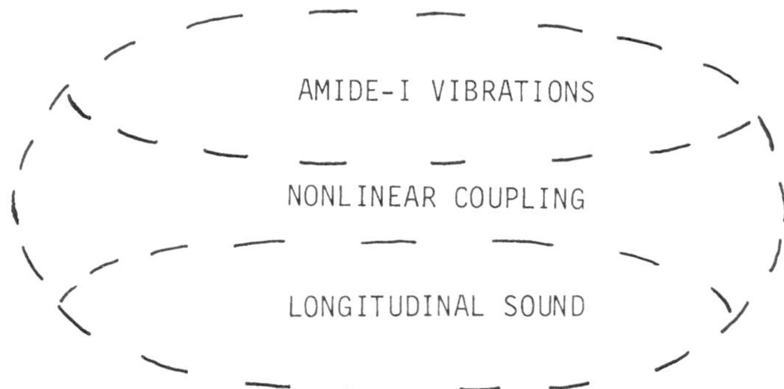

For such a coupled excitation to be viable, certain "threshold" conditions must be satisfied. The nonlinear coupling between amide-I vibrations and nonlinear sound waves must be sufficiently strong and the amide-I vibrations must be energetic enough for the retroactive interaction to "take hold". Below this threshold, a soliton cannot form and the dynamic behavior will be essentially linear. Above threshold the soliton is a possible mechanism for lossless energy transduction.

A recent numerical study of Davydov's fundamental equations confirms his analytical results. A sharp threshold between linear (dispersive) behavior and nonlinear (soliton) formation is clearly seen and this threshold is related to fundamental physical parameters describing the α-helix protein. In the following section we attempt to describe, as carefully as possible for the general reader, the basis for these numerical computations. To this end each term in Davydov's model is physically described with reference to the basic atomic structure. The next section displays our main numerical observations with emphasis upon their physical significance. Finally, we summarize our results and discuss some important open questions in this new area of nonlinear biomolecular dynamics. All mathematical discussions are presented in appendices not because we feel these are unimportant but to make the scientific logic of Davydov's theory as clear and as widely understandable as we can. This theory may, after all, help to resolve the "crisis in bioenergetics."

B. Davydov's Model for the Alpha-Helix Protein

The atomic structure of α-helix protein is sketched (as a stereogram) in Figure 10. The basic helix follows the sequence:

etc. - N - C - C - N - C - C - N - etc.

with a pitch of 5.4 Å. Superimposed on this basic structure are three "spines" which are almost longitudinal with the sequence:

etc. --- N - C = O --- N - C = O --- N - C = O --- etc.

where "O = C" represents the locus of the amide-I vibration and "O --- N" is the longitudinal hydrogen bond that holds the structure in its helical form. Davydov's equations describe propagation along these three spines of amide-I bond energy and longitudinal sound waves. Nonlinearity of the hydrogen bond leads to coupling of these two propagating systems and, if certain threshold conditions are satisfied, the formation of a soliton.

Let us begin by considering the equations that Davydov has derived to describe propagation along the three spines. From Davydov et al [1978] these are:

$$\begin{aligned} i\hbar \frac{da_{n\alpha}}{dt} &= [E_0 + W + \chi_1(\beta_{n+1,\alpha} - \beta_{n-1,\alpha})]a_{n\alpha} \\ &- J(a_{n-1,\alpha} + a_{n+1,\alpha}) + L(a_{n,\alpha+1} + a_{n,\alpha-1}) \\ &+ \chi_2[\beta_{n+1,\alpha}\, a_{n+1,\alpha} - \beta_{n-1,\alpha}\, a_{n-1,\alpha} \\ &\qquad - \beta_{n\alpha}(a_{n+1,\alpha} - a_{n-1,\alpha})] \end{aligned} \tag{5.1}$$

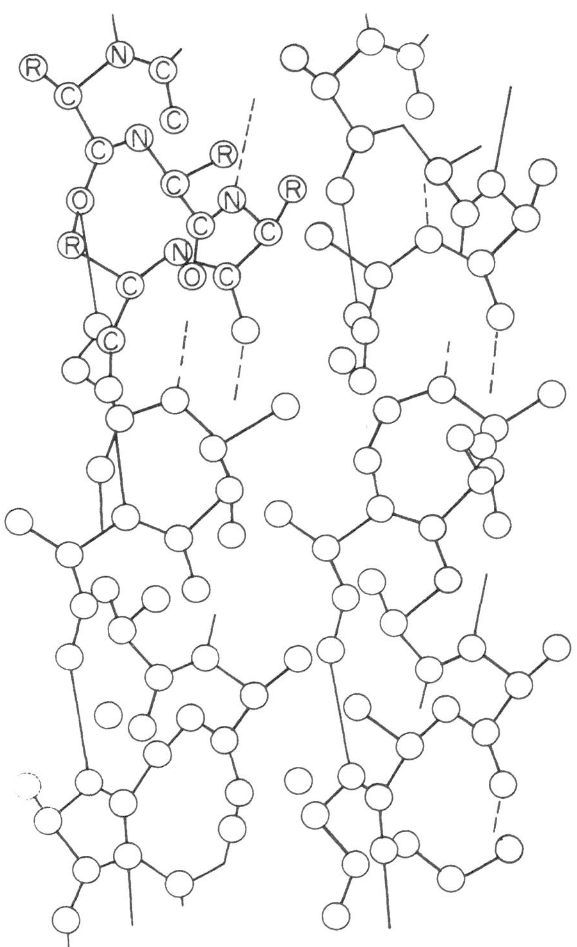

Fig. 10. Stereogram of α-helix protein.

$$M \frac{d^2\beta_{n\alpha}}{dt^2} - w(\beta_{n+1,\alpha} - 2\beta_{n\alpha} + \beta_{n-1,\alpha})$$

$$= \chi_1(|a_{n+1,\alpha}|^2 - |a_{n-1,\alpha}|^2)$$

$$+ \chi_2[a^*_{n\alpha}(a_{n+1,\alpha} - a_{n-1,\alpha})$$

$$+ (a^*_{n+1,\alpha} - a^*_{n-1,\alpha})a_{n\alpha}] \quad , \tag{5.2}$$

$$W \equiv \frac{1}{2} \sum_{n,\alpha} \left[M\left(\frac{d\beta_{n\alpha}}{dt}\right)^2 + w(\beta_{n\alpha} - \beta_{n-1,\alpha})^2\right]. \tag{5.3}$$

Broadly speaking, (5.1) describes the propagation of amide-I vibrations via dipole-dipole interactions and (5.2) represents the propagation of longitudinal sound. The total longitudinal sound energy is defined in (5.3). Please don't be frightened; each term, when individually considered, is quite plausible. Let's do this.

1) Subscripts. There are two subscripts to the dynamical variables, n and α. These run over the ranges:

$$n = -1, 0, 1, 2, \ldots n_{max} \quad , \quad \alpha = 1, 2, 3 \quad .$$

Thus n specifies a particular unit cell along a spine and α chooses a particular spine.

2) Bond occupation amplitude, $A_{n\alpha}$. Consider (5.1) with the nonlinear coefficients, χ_1 and χ_2, the dipole-dipole coupling coefficients, J and L, and the sound energy, W, set equal to zero. Then (5.1) becomes

$$i\hbar \frac{da_{n\alpha}}{dt} \doteq E_0 a_{n\alpha} \tag{5.1'}$$

In this equation

$$|a_{n\alpha}|^2 = a_{n\alpha} a^*_{n\alpha}$$

represents the probability of finding a quantum of bond energy E_0 at unit cell n on spine α. If the sum of such probabilities

$$\sum_{n,\alpha} |a_{n\alpha}|^2 = 1 \quad ,$$

a single quantum of amide-I bond energy is present on the helix.

Equation (5.1') is the quantum dynamical description of a simple harmonic oscillator. It says that the magnitude of $a_{n\alpha}$ remains constant and its phase progresses linearly with time as

$$a_{n\alpha}(t) = a_{n\alpha}(0) \exp(-i E_0 t/\hbar) .$$

From (5.5), $E_0 \doteq 1650\ \mathrm{cm}^{-1}$ in spectrographic units so[8]

$$E_0 = 0.205 \text{ ev.}$$

$$= 0.328 \times 10^{-19} \text{ joule.}$$

3) Longitudinal displacement, $\beta_{n\alpha}$. Consider (5.2) with the nonlinear coefficients, χ_1 and χ_2, set equal to zero. Then

$$M \frac{d^2\beta_{n\alpha}}{dt^2} - w(\beta_{n+1,\alpha} - 2\beta_{n\alpha} + \beta_{n-1,\alpha}) = 0 \tag{5.2'}$$

This is a linear equation for longitudinal sound propagation on the helix, where $\beta_{n\alpha}$ is the displacement of unit cell n on spine α from its equilibrium position and M is the mass of (see Fig. 10)

$$2C + O + N + H + R \quad .$$

For the computations to be reported here we (quite arbitrarily) take R to be CH_3. Thus

$$M = 70 \times \text{mass of proton.}$$

$$= 1.17 \times 10^{-25} \text{ kilograms.}$$

The parameter w in (5.2') gives the linear restoring force per unit of hydrogen bond stretching. From (5.6) a somewhat similar bond is said to have a force constant of 0.76 millidynes per angstrom. Thus we take

$$w = 76 \text{ newtons per meter.}$$

---

[8] One electron volt = $8065.5\ \mathrm{cm}^{-1}$ = 1.602×10^{-19} joule.

From (5.2') the longitudinal sound speed is $[w/M]^{\frac{1}{2}}$ times the longitudinal distance between unit cells. Since the pitch of the helix is 5.4 Å corresponding to 3.6 spines, the length of a single unit cell along one spine is 4.5 Å. Thus, the

sound speed = 1.15×10^4 meters per second.

4) Dipole-dipole coupling. If (5.1) is considered in the approximation that the sound energy, W, and the nonlinear coefficients, χ_1 and χ_2 are zero, it can be written in the form

$$i\hbar \frac{da_{n\alpha}}{dt} + J(a_{n-1,\alpha} - 2a_{n\alpha} + a_{n-1,\alpha}) - E_0 a_{n\alpha} =$$

$$- 2Ja_{n\alpha} + L(a_{n,\alpha+1} + a_{n,\alpha-1}) . \qquad (5.1'')$$

These terms with coefficients J and L represent the effects of dipole-dipole couplings between the amide-I vibrations. The particular form presented in (5.1") is chosen to emphasize that the effect of the "J-term" is to provide a mechanism for longitudinal propagation of bond energy. Indeed if the right hand side of (5.1") were zero, it would be satisfied by a plane wave of probability amplitude propagating in a dispersive manner.

The "J-term" represents dipole-dipole coupling between a particular amide-I bond and its next neighbors in the longitudinal direction. The "L-term" represents a corresponding coupling to lateral neighbors. Fortunately for our numerical studies, the values for these coupling coefficients have been calculated (and checked for their effects on infrared spectra) as [Nevskaya and Chirgadze, 1976]:

$$J = 7.8 \text{ cm}^{-1} = 1.55 \times 10^{-22} \text{ joule}$$

and

$$L = 12.4 \text{ cm}^{-1} = 2.46 \times 10^{-22} \text{ joule.}$$

5) Nonlinear coefficients, χ_1 and χ_2. These terms represent anharmonicity in the longitudinal hydrogen bonds. Their effect is to provide nonlinear coupling between the longitudinal sound waves (5.2') and dispersive propagation of amide-I bond energy (5.1") as was described in the Introduction. This coupling permits the formation of a soliton.

To be more specific, note in (5.2) that the "χ-terms" act as a source for the longitudinal sound. Once generated, this sound energy acts in (5.1) as a "potential well" for the bond energy which prevents its dispersion.

Some information on the level of anharmonicity in the hydrogen bonds of α-helix protein is available in [Schuster et al, 1976]. For the purpose of our numerical studies, we have assumed

$$\chi_1 = \chi_2 \equiv \chi \tag{4}$$

and allowed χ to be an adjustable parameter. Rough estimates, outlined in Appendix B, however, indicate that

$$\chi \sim 2 - 6 \times 10^{-11} \text{ newtons.}$$

6) Sound energy, W. The total longitudinal sound energy is defined as W in (5.3) and enters as an additional energetic term in (5.1). Including it to the approximation indicated in (5.1') indicates that its effect is merely to speed the rate of phase advance by a small amount. The numerical effect of this term in our results is negligible.

C. Numerical Observations

Equations (5.1) - (5.3) contain too many physical constants for a convenient numerical study. Thus the equations we have actually computed are written in the normalized form

$$\begin{aligned} i\frac{dA_{n\alpha}}{d\tau} = {} & 1.41\ A_{n\alpha} \sum_{n\alpha} \left[\left(\frac{dB_{n\alpha}}{d\tau}\right)^2 + (B_{n\alpha} - B_{n-1,\alpha})^2 \right] \\ & - .058(A_{n-1,\alpha} + A_{n+1,\alpha}) + .092(A_{n,\alpha+1} + A_{n,\alpha-1}) \\ & + .372 \times (10^{10}\chi) \Big[(B_{n+1,\alpha} - B_{n-1,\alpha})An_{\alpha} \\ & + B_{n+1,\alpha}A_{n+1,\alpha} - B_{n-1,\alpha}A_{n-1,\alpha} - B_{n\alpha}(A_{n+1,\alpha} - A_{n-1,\alpha}) \Big] \end{aligned} \tag{5.1'''}$$

$$\frac{d^2B_{n\alpha}}{d\tau^2} - (B_{n+1,\alpha} - 2B_{n\alpha} + B_{n-1,\alpha}) =$$

$$.132(10^{10}\chi)[|A_{n+1,\alpha}|^2 - |A_{n-1,\alpha}|^2]$$

$$+ A^*_{n\alpha}(A_{n+1,\alpha} - A_{n-1,\alpha}) + (A^*_{n+1,\alpha} - A^*_{n-1,\alpha})A_{n\alpha} \quad . \qquad (5.2'')$$

In these equations:

$$a_{n\alpha} \equiv A_{n\alpha} \exp(-i\, E_0 t/\hbar) \quad , \qquad (5.5a)$$

$$\beta_{n\alpha} \equiv B_{n\alpha} \cdot 10^{-11} \text{ meters} \quad , \qquad (5.5b)$$

$$\tau \equiv t\sqrt{w/M} \quad , \qquad (5.5c)$$

where we note that (5.5a) absorbs the fast phase advance in amide-I bond amplitude, (5.5b) measures displacement in units of 0.1 Å, and (5.5c) measures time in units of

$$\sqrt{\frac{M}{w}} = 3.92 \times 10^{-14} \text{seconds.}$$

This is the natural period of longitudinal sound vibration (divided by 2π). From (5.2") it can be seen that the longitudinal sound velocity is unity in these units.

We take the total number of unit cells

$$n_{max} = 200.$$

Since, as previously mentioned, the length of a single unit cell along a spine is 4.5 Å, this corresponds to a total length of 900 Å: about the length of a typical myosin molecule in a thick fiber of striated muscle.

As initial conditions we begin with

$$A_{n\alpha} = \begin{cases} 1 & \text{for } n = 1 \;, \\ 0 & \text{for } n \neq 1 \;, \end{cases}$$

and

$$B_{n\alpha} = 0 \text{ for all } n$$

at $\tau = 0$. Physically this corresponds to the introduction of one quantum of amide-I bond energy onto each of the three spines to excite what Davydov has called the symmetric mode.

The only item remaining to be specified in (5.1''') and (5.2'') is χ, the level of anharmonicity in the hydrogen bond. Guided by our previously mentioned estimates (see Appendix B), we choose $\chi = 10^{-11}$ newtons. The calculations are displayed in Figure 11 where

$$U(1) \equiv A_{n1}^2 + A_{n2}^2 + A_{n3}^2$$

and

$$U(2) \equiv \sum_{\alpha} \left[\left(\frac{dB_{n\alpha}}{d\tau} \right)^2 + (B_{n+1,\alpha} - B_{n,\alpha})^2 \right].$$

Consider first the result for $\tau = 100$ which is shown in Figure 11a. The bond energy, U(1), has dispersed somewhat and is moving away from the point of initiation at a normalized speed of about 0.1. The longitudinal sound energy, U(2), consists of two distinct components: a "fast" component traveling at the limiting sound speed and therefore found at n = 100, and a "slow" component which is locked to the bond energy. Does the interaction of bond energy and slow sound constitute a soliton? Flipping through Figures 11b ($\tau = 600$) and 11c ($\tau = 900$), we see that the bond energy does not settle into the hyperbolic secant shape that characterizes a soliton [Davydov, 1976, 1977, to appear]. On the contrary, it continues to disperse until at $\tau = 1500$ it has spread itself over half of the molecule.

To see how the bond energy dispersion at $\chi = 10^{-11}$ newtons differs from linear dispersion, turn to Figure 12 where the computation is repeated for the case $\chi = 0$. Note that the linear bond dispersions at $\tau = 600$ (Fig. 12a) and at $\tau = 900$ (Fig. 12b) are identical to those previously displayed in Figs. 11b and 11c. Thus we must conclude that nonlinear coupling between amide-I energy and sound energy plays no role in the computations of Fig. 11. The threshold level has not been attained; solitons have not formed.

If the nonlinearity parameter is raised an order of magnitude, to the level $\chi = 10^{-10}$ newtons, the dynamic behavior of the bond energy is strikingly different. As Fig. 13 clearly shows, it no longer disperses but propagates along the helix with a fixed shape and a normalized velocity of 0.132. In this case the level of amide-I bond excitation and the nonlinearity are great enough to permit the sound energy to form a "potential well" which holds the bond energy together. The threshold level is exceeded and Davydov's soliton is observed.

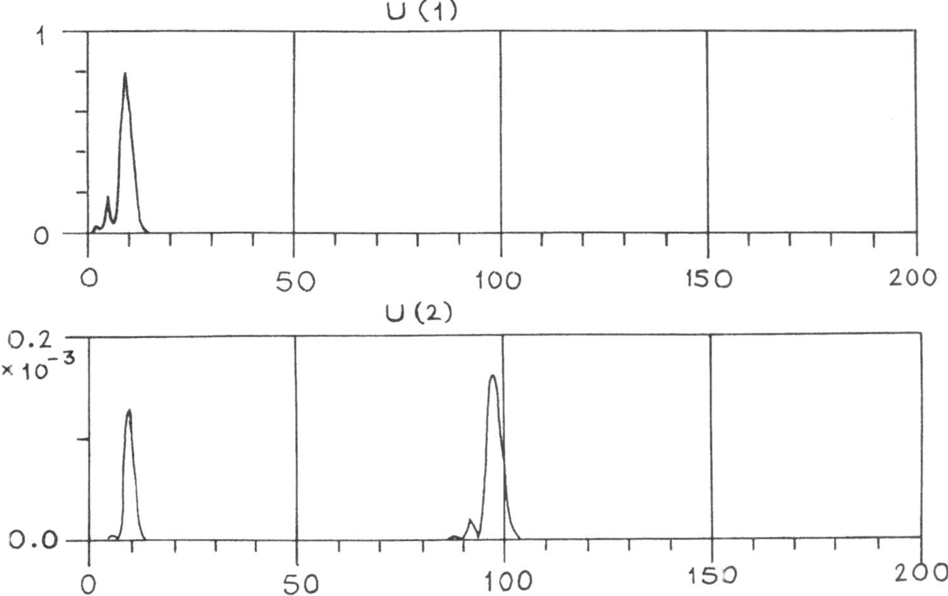

Fig. 11a. Symmetrical 3 spine excitation at $\chi = 10^{-11}$ newtons, $\tau = 100$.

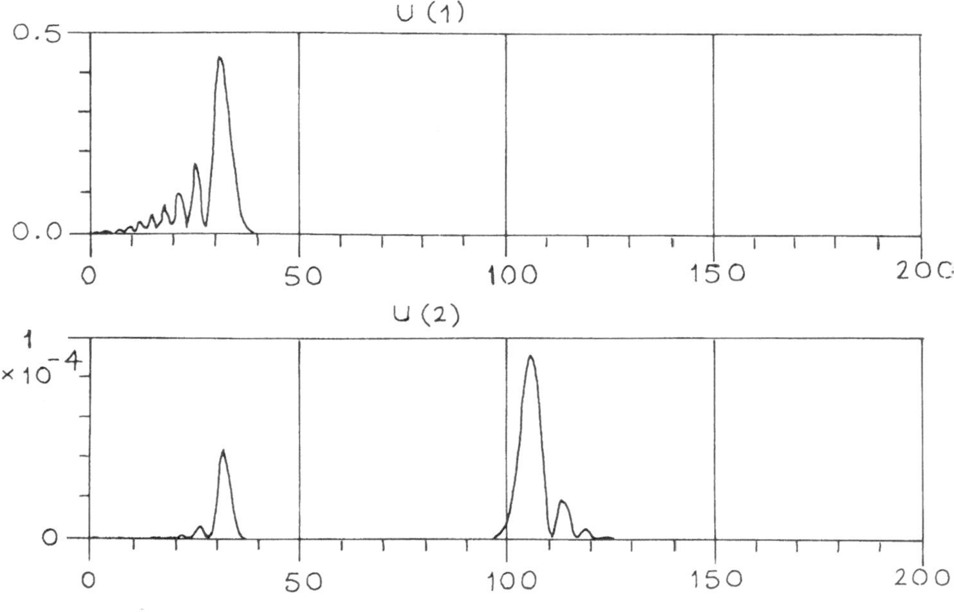

Fig. 11b. Symmetrical 3 spine excitation at $\chi = 10^{-11}$ newtons, $\tau = 600$.

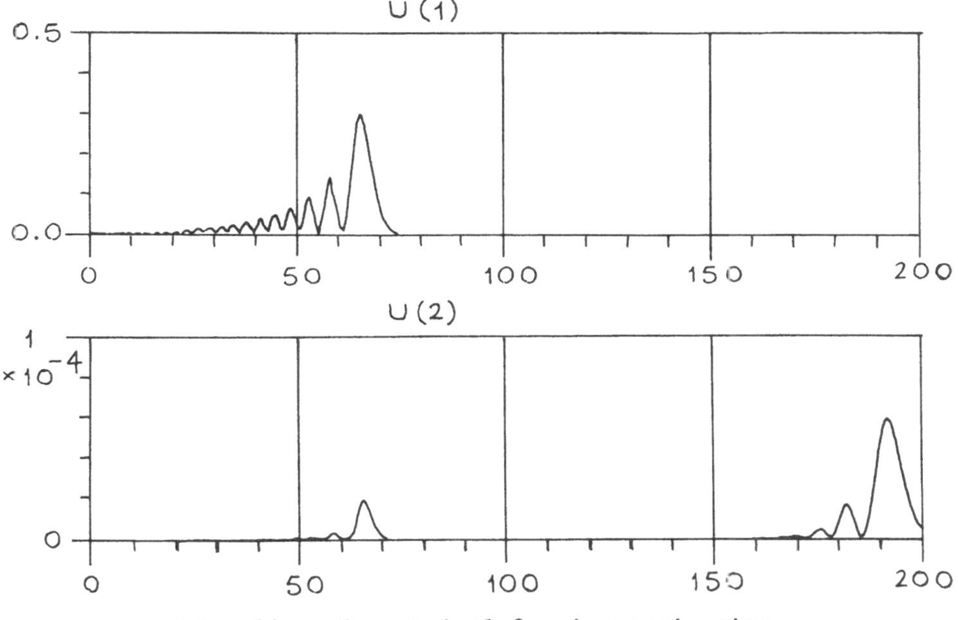

Fig. 11c. Symmetrical 3 spine excitation at $\chi = 10^{-11}$ newtons, $\tau = 900$.

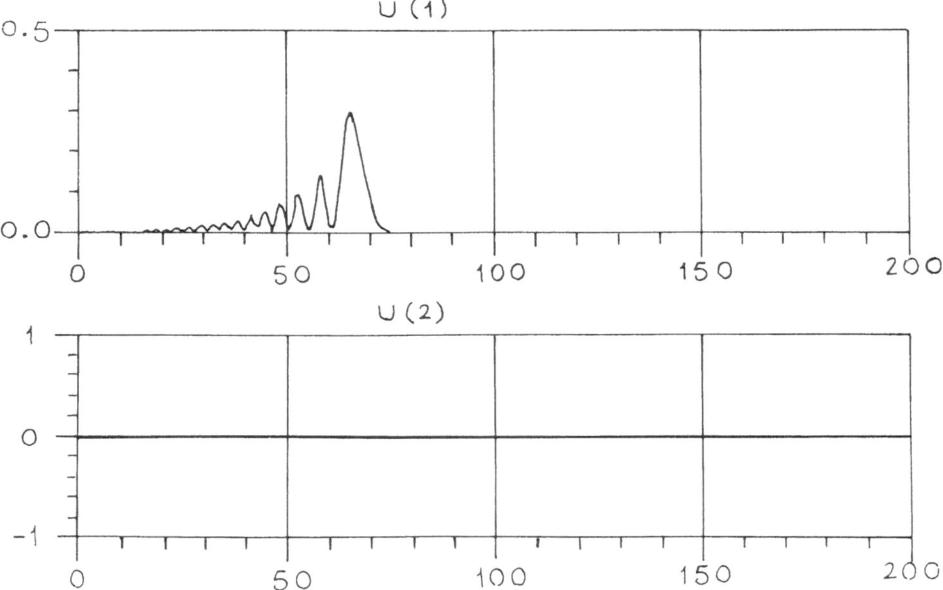

Fig. 12a. Symmetrical 3 spine excitation at $\chi = 0$, $\tau = 600$.

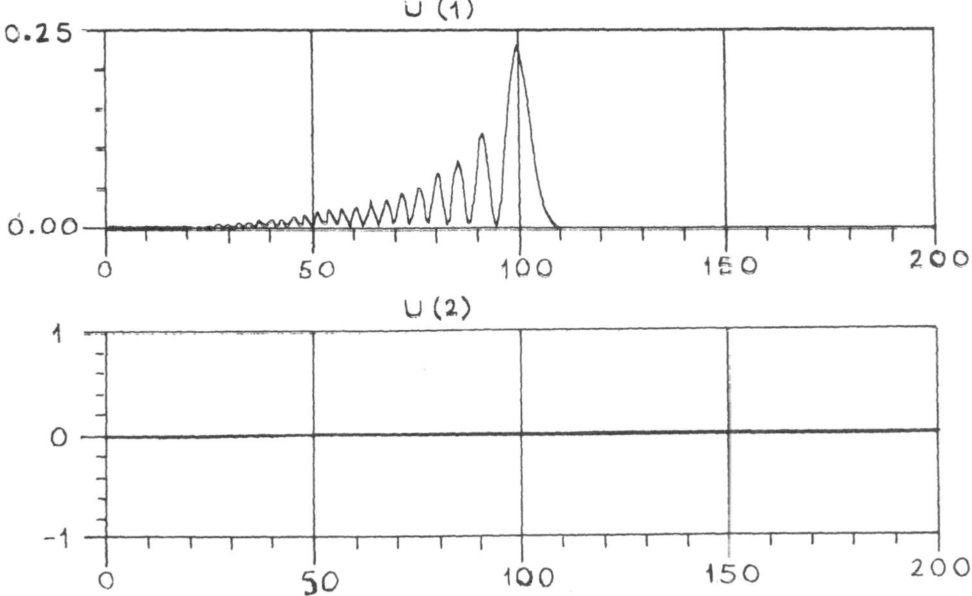

Fig. 12b. Symmetrical 3 spine excitation at $\chi = 0$, $\tau = 900$.

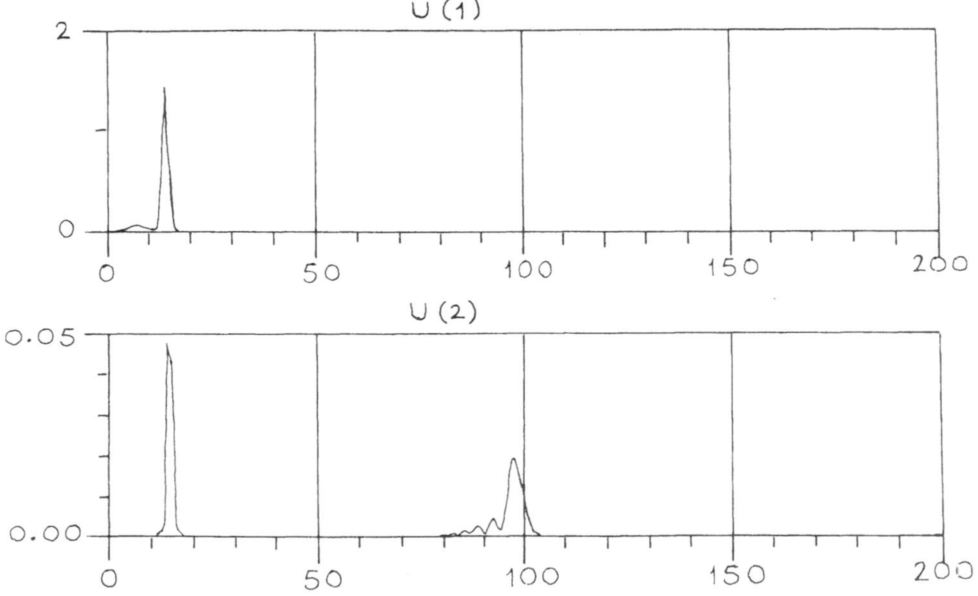

Fig. 13a. Symmetrical 3 spine excitation at $\chi = 10^{-10}$ newtons, $\tau = 100$.

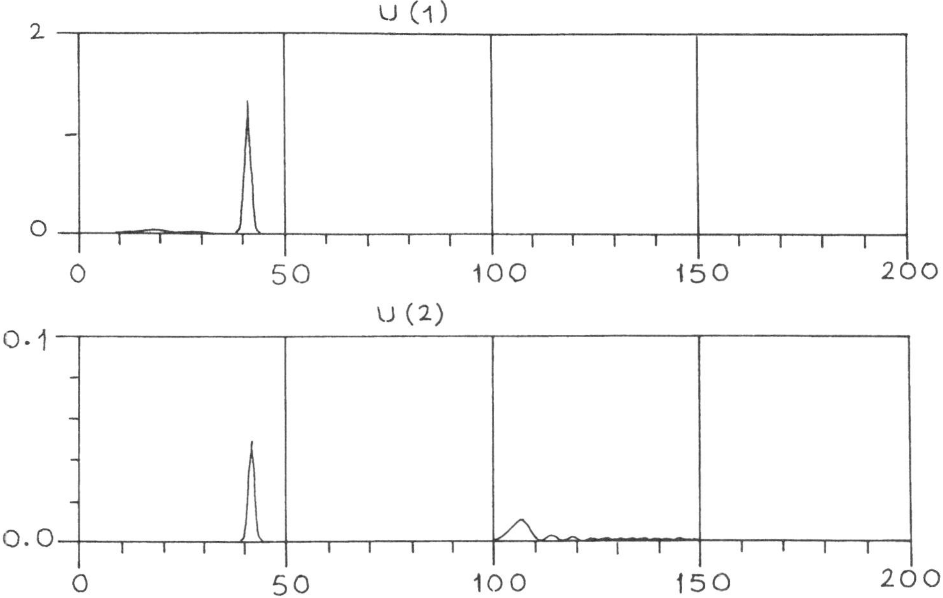

Fig. 13b. Symmetrical 3 spine excitation at $\chi = 10^{-10}$ newtons, $\tau = 300$.

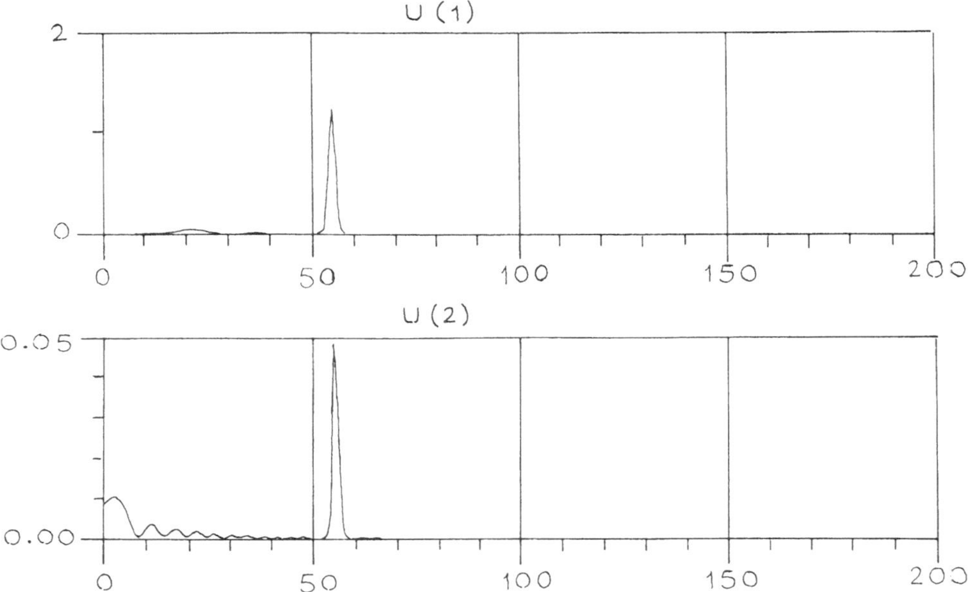

Fig. 13c. Symmetrical 3 spine excitation at $\chi = 10^{-10}$ newtons, $\tau = 400$.

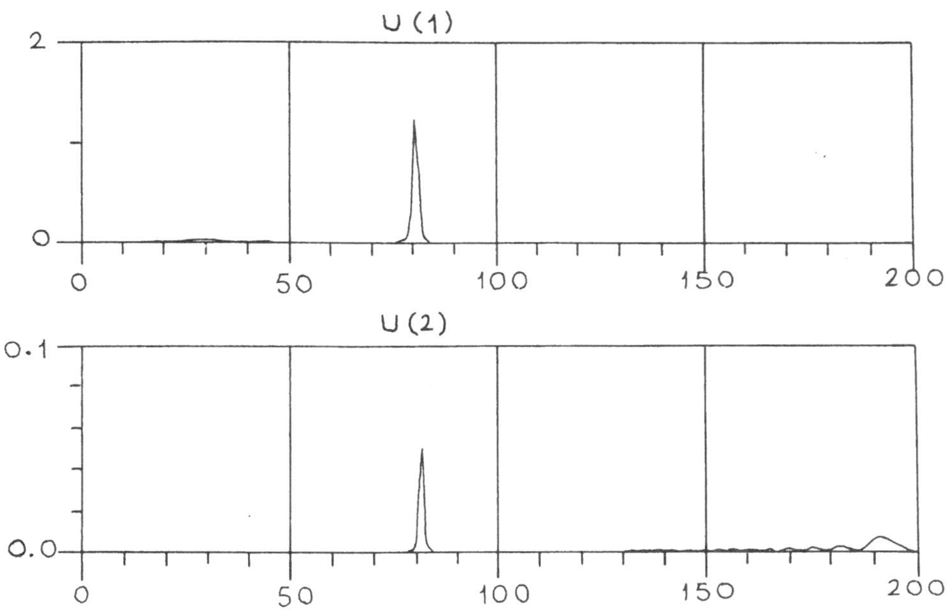

Fig. 13d. Symmetrical 3 spine excitation at $\chi = 10^{-10}$ newtons, $\tau = 600$.

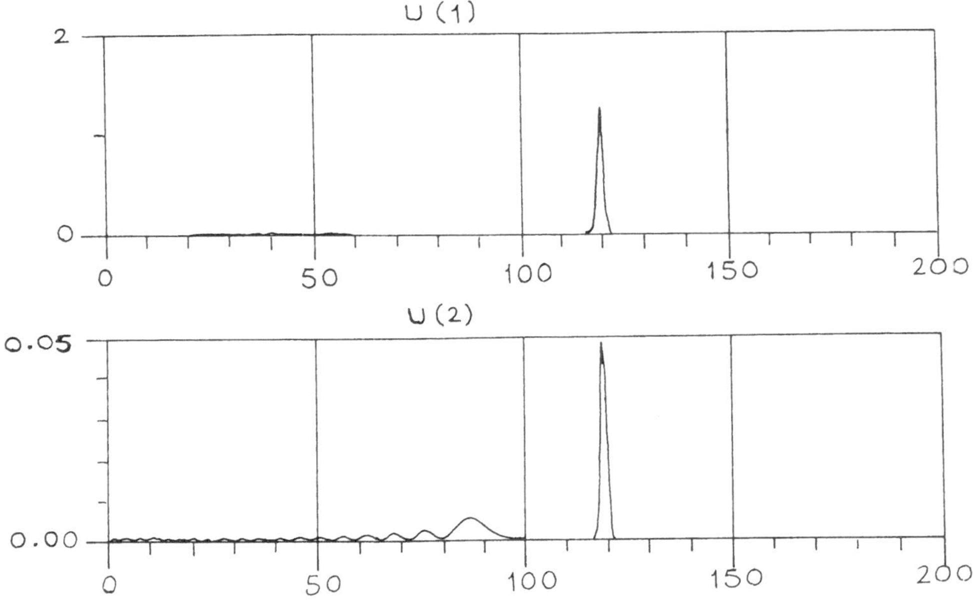

Fig. 13e. Symmetrical 3 spine excitation at $\chi = 10^{-10}$ newtons, $\tau = 900$.

How sharp is this threshold? To answer this we obtained expanded plots of bond energy for $\chi = 3 \times 10^{-11}$ newtons and $\chi = 5 \times 10^{-11}$ newtons. From these we observed that the soliton is just beginning to form at the lower level and is well developed at the upper level. The detailed computations in Appendix A indicate that the threshold is inversely proportional to both χ and the energy of bond excitation. Thus if N is the number of amide-I quanta introduced onto a single spine

$$\boxed{\chi N > 3 \times 10^{-11} \text{ newtons}} \tag{5.6}$$

should be a useful threshold condition for the formation of Davydov's solitons.

D. Summary of Results

1) Threshold for soliton formation. Our numerical and analytical studies show that with symmetrical (3 spine) excitation the threshold level of nonlinearity for soliton formation is

$$\chi > \frac{3}{N} \times 10^{-11} \text{ newtons}$$

where N is the number of amide-I quanta introduced onto each spine. Comparing this result with our order estimates

$$\chi \sim 2 - 6 \times 10^{-11} \text{ newtons}$$

for the nonlinear parameter, we see a possibility of soliton formation with a single quantum on each spine. As the number of quanta introduced becomes larger, the likelihood of soliton formation increases. In this connection it is important to note that the 0.5 electron volts released in each event of ATP hydrolysis is more than enough to introduce two quanta into an amide-I bond.

2) Soliton speed. From both numerical computations and analytical calculations we find the soliton speed near threshold to be almost equal to the group velocity of a linear pulse below threshold. From our numerical computations it is .11 of the longitudinal sound speed or

soliton speed $\sim 1.26 \times 10^3$ meters/second.

Thus the time required for a soliton to traverse 1000 Å (about the length of a typical myosin molecule in striated muscle) is about 80 picoseconds.

3) Mechanical bending. The mechanical bending of the α-helix under symmetrical excitation is zero, but if the soliton is antisymmetric (in the sense defined by Davydov) or if all the bond energy is confined to a single spine, the bend angle is approximately $\theta \doteq .45\ N(10^{10}\chi)$. This could be a significant effect for several amide-I quanta in the soliton and a nonlinear parameter somewhat larger than the values estimated in Appendix B.

E. Open Questions

In an exploratory study such as this it is as important to indicate what we have not shown as to itemize our results. Davydov has clearly changed the question posed at the beginning of this report to: "Is biological energy transmitted by solitons?", but a definitive answer is not yet available. Indeed it is rather exciting to await future scientific developments that should indicate whether the real level of nonlinearity in an α-helix is approximately that estimated in Appendix B (indicating that solitons should easily form at the single quantum level), or substantially smaller (indicating that several quanta must participate in order to form a soliton). We feel that the following questions should be given high priority.

1) Additional numerical studies. The numerical studies presented here are not complete, and additional investigations should include the following: a) A more careful study of relaxation from single spine excitation, b) Initial conditions for exciting Davydov's antisymmetric mode, c) Inclusion of additional dipole-dipole coupling terms from [Nevskaya and Chirgadze, 1976], d) Augmentation of Davydov's equations to include additional degrees of freedom, and e) Study of soliton propagation through a nonuniform α-helix.

2) The level of anharmonicity. Every effort should be given to obtain better experimental measurements and theoretical estimates of the anharmonicity (χ) in the hydrogen bonds of α-helix protein. We have not found really satisfactory estimates from the literature [Schuster et al, 1976], and present the order estimates of Appendix B as a rough guide. But we are not biochemists (nor chemists, even) so relevant information may be lying about. The level of anharmonicity is the most important fact in nonlinear biomolecular dynamics.

APPENDIX A: THE INITIAL VALUE PROBLEM IN LINEAR AND NONLINEAR LIMITS

In this appendix the original difference-differential equations, (5.1''') and (.2''), are approximated as partial differential equations, and studied analytically. In the linear limit, a Fourier transform (FT) solution is discussed. In the nonlinear limit, the inverse scattering transform (IST) [Ablowitz et al, 1974] is used to find the threshold for soliton creation and soliton speed.

From the numerical results presented above, it is evident that several unit cells participate in the structure of each soliton. Thus, as Davydov has shown [Davydov et al, 1978], (5.1'''), and (5.2'') can be approximated as

$$i\,\frac{\partial A}{\partial \tau} + .058\,\frac{\partial^2 A}{\partial \xi^2} - F(\tau)A \doteq .744(10^{10}\chi)\rho A\ , \qquad \text{(A.1)}$$

$$\frac{\partial^2 \rho}{\partial \tau^2} - \frac{\partial^2 \rho}{\partial \xi^2} \doteq -.264(10^{10}\chi)\,\frac{\partial^2}{\partial \xi^2}\,|A|^2\ , \qquad \text{(A.2)}$$

where ξ is a continuous variable approximating the longitudinal index n.

Also

$$\rho \equiv -\partial B/\partial \xi$$

and

$$F(\tau) \equiv .068 + 1.41\,\sum_{\alpha}\int\left[\left(\frac{\partial}{\partial \tau}\right)^2 + \left(\frac{\partial B}{\partial \xi}\right)^2\right]d\xi\ .$$

The term $F(\tau)A$ can be eliminated from (A.1) by adjusting the phase of A as

$$A = \mathcal{A}\,\exp\left[-i\int^{\tau} F(\tau')d\tau'\right]\ .$$

The numerical results also show that the soliton speed is slow compared with the sound speed. Thus (A.2) becomes approximately

$$\rho \doteq \frac{.264(10^{10}\chi)}{1-s^2}\,|\mathcal{A}|^2 \qquad \text{(A.3)}$$

where s is the wave speed. With these approximations, (A.1) takes the form

$$i \frac{\partial A}{\partial \tau} + .058 \frac{\partial^2 A}{\partial \xi^2} \doteq - \frac{.196(10^{10}\chi)^2}{1 - s^2} |A|^2 A \ . \qquad \text{(A.4)}$$

This is the "nonlinear Schroedinger equation" which has been exactly solved by Zakharov and Shabat [1972] for arbitrary initial conditions.

We are interested in the initial conditions (see inset)

$$A = \begin{cases} \frac{N}{p} \text{ for } 0 < \xi < p \\ -\frac{N}{p} \text{ for } -p < \xi < 0 \\ 0 \text{ for } |\xi| > p \ . \end{cases}$$

These initial conditions deserve a word of explanation. In our numerical computations a certain number (n) of amide-I quanta were put onto the n = 0 bond at time $\tau = 0$. All other amide-I bonds were without energy at $\tau = 0$ and the energy at n = -1 was maintained at the value zero throughout the computations. An antisymmetrical form is chosen for the full line ($-\infty < \xi < +\infty$) in order to maintain the boundary condition of zero at the origin (n = -1). We don't have a precise value for p, but it should be approximately equal to two.

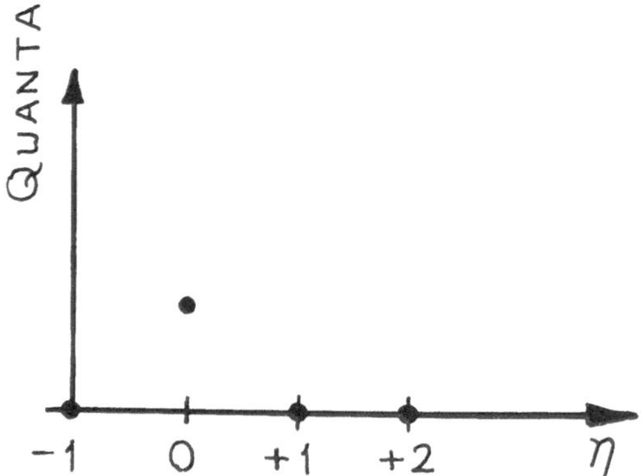

For analysis it is convenient to normalize (A.4) by writing

$$A = 3.19 \frac{\sqrt{1 - s^2}}{(10^{10}\chi)} \phi ,$$

$$\xi = .241\ x ,$$

$$\tau = t .$$

Then (A.4) takes the form

$$i\,\phi_t + \phi_{xx} = -2|\phi|^2\phi \tag{A.4'}$$

with the initial conditions

$$\phi = \begin{cases} \dfrac{(10^{10}\chi)N}{3.19\ p\sqrt{1-s^2}} & \text{for } 0 < x < \dfrac{p}{.241} , \\ -\dfrac{(10^{10}\chi)N}{3.19\ p\sqrt{1-s^2}} & \text{for } -\dfrac{p}{.241} < x < 0 , \\ 0 \text{ for } |x| > \dfrac{p}{.241} . & \end{cases}$$

1. Linear Limit

When the amplitude in (A.4') is small enough so the nonlinear term ($|\phi|^2\phi$) can be neglected, it becomes simply $i\phi_t + \phi_{xx} = 0$ with the Fourier transform solution

$$\phi(x,t) \propto \int_{-\infty}^{\infty} \frac{\sin^2(k\tilde{p}/2)}{(k\tilde{p}/2)} \exp[i(kx - k^2t)]dk \tag{A.5}$$

where

$$\tilde{p} \equiv p/.241.$$

The integrand in (A.5) takes its maximum value when k is the root of $\tan(k\tilde{p}/2) = k\tilde{p}$ or

$$k_{max} = \frac{2.331...}{\tilde{p}}$$

and the corresponding group velocity (where the phase $(kx - k^2 t)$ is stationary) is $2k_{max}$. Thus the linear pulse velocity (in unnormalized units which corresponds to the numerical computations) is

$$\text{linear pulse velocity} = \frac{.271}{p} \quad . \tag{A.6}$$

For p = 2 this implies a linear pulse velocity of 0.135 while from the numerical computations displayed in Figure 12 we find a velocity of 0.11.

The amplitude in (A.5) should fall asymptotically as $t^{-\frac{1}{2}}$ indicating that the maximum bond energy in the linear limit should fall as $1/\tau$ [Whitham, 1974]. This in turn implies that the bond energy must "spread out" over a length of the α-helix that is proportional to τ. Such an effect is observed, eg., in the data of Figures 11.

2. Soliton Limit

We now use the analytical tools of the inverse scattering transform method to find the threshold for soliton formation and its corresponding velocity.

For the Zakharov and Shabat linear scattering operator [Zakharov and Shabat, 1972]

$$i \begin{bmatrix} \partial_x & -\phi \\ -\phi^+ & -\partial_x \end{bmatrix} \begin{bmatrix} \psi_1 \\ \psi_2 \end{bmatrix} = \gamma \begin{bmatrix} \psi_1 \\ \psi_2 \end{bmatrix}$$

and the initial conditions listed below (A.4'), we assume asymptotic scattering amplitudes at t = 0 to be

$$\begin{bmatrix} \psi_1 \\ \psi_2 \end{bmatrix} = \begin{bmatrix} 1 \\ 0 \end{bmatrix} \exp(-i\gamma x) \quad \text{for} \quad x < -p$$

and

$$\begin{bmatrix} \psi_1 \\ \psi_2 \end{bmatrix} = a(\gamma) \begin{bmatrix} 1 \\ 0 \end{bmatrix} \exp(-i\gamma x) + b(\gamma) \begin{bmatrix} 0 \\ 1 \end{bmatrix} \exp(+i\gamma x)$$

for x > p. Then we find

$$a(\gamma) = \exp(2i\gamma\tilde{p})\left[\left(\cos m\tilde{p} - \frac{i\gamma}{m}\sin m\tilde{p}\right)^2 + \frac{K^2}{(m\tilde{p})^2}\sin^2 m\tilde{p}\right] \tag{A.7}$$

where

$$m^2 \equiv \gamma^2 + K^2/\tilde{p}^2 ,$$

$$\tilde{p} \equiv p/.241 ,$$

and

$$K \equiv \frac{(10^{10}\chi)N}{.241 \times 3.19\sqrt{1-s^2}} .$$

Solitons correspond to zeros of (A.7) which lie in the upper half of the γ-plane. For such a zero at $\gamma = \gamma_r + i\gamma_i$, the corresponding soliton has [Ablowitz et al, 1974]

$$\text{speed} = 4\gamma_r ,$$

$$\text{amplitude} = 2\gamma_i .$$

For K small (i.e. as $\chi \to 0$), $a(\gamma)$ has no UHP zeros. Thus the threshold for soliton formation occurs when the first zero of (A.7) crosses the real axis of the γ-plane. Since a zero of (A.7) implies

$$\cot(m\tilde{p}) = i\left(\frac{\gamma\tilde{p} \pm K}{m\tilde{p}}\right) ,$$

a real axis zero can only occur where $\cot(m\tilde{p}) = 0$ or at

$$\pm\gamma_r\tilde{p} = K = \frac{\pi}{2\sqrt{2}} .$$

In units that correspond to the numerical computations, the threshold condition is (for s = 0.11)

$$\boxed{\chi N > 7.64 \times 10^{-11} \text{ newtons.}} \tag{A.8}$$

This threshold level is higher than that in Hamilton and Ibers [1968].

The soliton velocity at threshold (again in units that correspond to the numerical computations) is

$$\text{soliton velocity} = \frac{.258}{p} \quad . \tag{A.9}$$

Comparison with (A.6) shows that the soliton velocity at threshold should be quite close (i.e. within 5%) of the linear pulse velocity. This is confirmed by the numerical results.

We note finally that solitary wave solutions of the set (A.1) and (A.2) have been studied rather extensively in plasma physics where they are called "Langmuir solitons". The paper by Gibbons et al [1977] is a particularly lucid introduction to this work.

APPENDIX B: ORDER ESTIMATES OF HYDROGEN BOND NONLINEARITY

The purpose of this appendix is to obtain order of magnitude estimates for the level of anharmonicity to be expected in longitudinal vibrations of α-helix protein. To this end we view the structural relation between the amide-I and hydrogen bonds as

$$\underbrace{C = 0}_{\text{amide-I}} \;\; \overbrace{\underbrace{------}_{\text{hydrogen bond}}}^{|\; x + x_0 \;|} \; N$$

The hydrogen bond is assumed to have the anharmonic potential (about the minimum at x_0)

$$U(x) = \frac{1}{2}wx^2 + ax^3 . \tag{B.1}$$

Nonlinearity enters Davydov's Hamiltonian formalism as an "interaction" term

$$H_{int} = \chi B^{\dagger}Bx \tag{B.2}$$

where $B^{\dagger}B$ gives the number of quanta in the amide-I bond. If the restoring force of this bond is taken to be K newtons per meter, the bond extension is

$$B^{\dagger}B\hbar\,\omega_0 = \frac{1}{2}K\,x^2 \tag{B.3}$$

where

$$\hbar\omega_0 = E_0 = .328 \times 10^{-19} \text{ joule ,}$$

the quantum energy of an amide-I vibration. From (B.1), (B.2), and (B.3)

$$\chi = \frac{2\hbar^2 a}{E_0 M_r} \tag{B.4}$$

where the "reduced mass" of amide-I is

$$M_r = \frac{48}{7} \times \text{mass of proton}$$

$$= 1.15 \times 10^{-26} \text{ kgm .}$$

To estimate the parameter a in (B.1) we note, from Pauling et al [1951], that the binding energy of a hydrogen bond in α-helix protein is about 8 kcal/mole or

$$\Delta U = 5.5 \times 10^{-20} \text{ joules.}$$

But from (B.1) we find (see inset) that $\Delta U = w^3/54\ a^2$, so

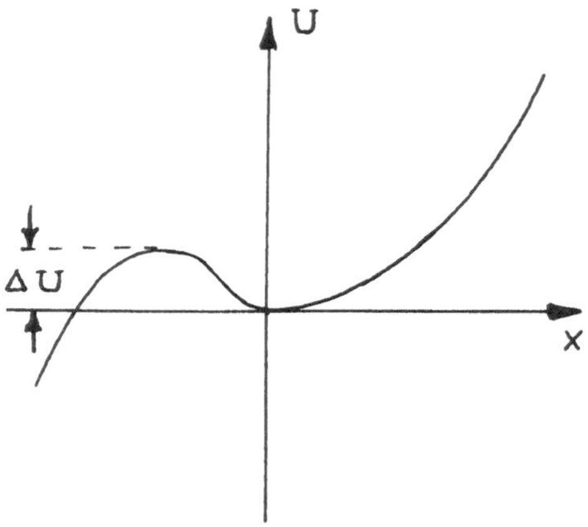

$$a = \frac{w^{3/2}}{\sqrt{54 \Delta U}}$$

$$= 38.3 \times 10^{10} \text{ newtons per meter}^2 \ .$$

Thus from (B.4) we obtain the order estimate

$$\chi \sim 2 \times 10^{-11} \text{ newtons.}$$

We expect the reader to be as suspicious of this estimate as we are; thus we turn to reference [Schuster et al, 1976]. There the potential (B.1) is written in the form

$$U = \frac{1}{2} kq^2 + k_3 q^3 \qquad \text{(B.1')}$$

where

$$q \equiv x/\lambda$$

is a normalized space variable and

$$\lambda \equiv [\hbar/\omega\mu]^{\frac{1}{2}}$$

in which ω is the radian frequency and μ the reduced mass of the resulting oscillation. Evidently

$$w = k/\lambda^2 \ ,$$

$$a = k_3/\lambda^3 .$$

Sandorfy [Schuster, Zundel and Sandorfy, 1976, p. 617] states:

"Model calculations ... have shown that k_3 must not exceed 7 or 8% of k ... for [second-order perturbation theory] to be valid. According to our experience this is probably fulfilled for weak hydrogen bonds with ΔU values not higher than about 3 or 4 kcal/mole." Since our ΔU is taken to be 8 kcal/mole, it seems reasonable to assume $k_3/k = 0.15$. and calculate

$$a = \frac{k_3}{k} \frac{w^{5/4} \mu^{1/4}}{\hbar^{1/2}} .$$

With μ the reduced mass of an O --- N,

$$\mu = \frac{112}{15} \times \text{mass of proton} ,$$

we find

$$a = 10.95 \times 10^{11} \text{ newtons per meter}^2$$

and therefore

$$\chi \sim 6.3 \times 10^{-11} \text{ newtons.}$$

REFERENCES

Ablowitz, M.J., Kaup, D.J., Newell, A.C., and Segur, H., 1973, Method for solving the sine-Gordon equation, Phys. Rev. Lett., 30:1262.

Ablowitz, M.J., Kaup, D.J., Newell, A.C., and Segur, H., 1974, The inverse scattering transform-Fourier analysis for nonlinear problems, Stud. in Appl. Math., 53:249.

Ann. N.Y. Acad. Sci., 1974, (Annual Meeting of the New York Academy of Sciences, 1974).

Barbashov, B.M., and Chernikov, N.A., 1967, Solution of the two plane wave scattering problem in a nonlinear scalar field theory of the Born-Infeld type, Sov. Phys. JETP, 24:437.

Barone, A., Esposito, F., Magee, C.J., and Scott, A.C., 1971, Theory and applications of the sine-Gordon equation, Riv. Nuovo Cimento, 1:227.

Barut, A.O., 1978, "Nonlinear Evolution Equations in Physics and Mathematics," Reidel.

Benney, D.J., and Newell, A.C., 1967, The propagation of nonlinear wave envelopes, J. Math. and Phys., 46:133.

Bespalov, V.I., and Talanov, V.I., 1966, Filamentary structure of light beams in nonlinear liquids, JETP Lett., 3:307.

Bishop, A.R., and Schneider, T., 1979, "Solitons and Condensed Matter Physics," Springer, Berlin.

Born, M., 1934, On the quantum theory of the electromagnetic field, Proc. Roy. Soc. A, 143:410.

Born, M., and Infeld, L., 1934a, Foundations of the new field theory, Proc. Roy. Soc. A, 144:425.

Born, M., and Infeld, L., 1934b, Quantization of the new field equations. Part I, Proc. Roy. Soc. A, 147:522.

Born, M., and Infeld, L., 1935, Quantization of the new field theory. Part II, Proc. Roy. Soc. A, 150:141.

Calogero, F., 1978, "Nonlinear Evolution Equations Solvable by the Spectral Transform," Pitman, London.

Caudrey, P.J., and Bullough, R.K., to appear, "Solitons," Springer, Berlin.

Chu, F.Y.F., and Scott, A.C., 1974, Bäcklund transformations and the inverse method, Phys. Lett, A 47:303.

Dashen, R.F., Hasslacher, B., and Neveu, A., 1974, Nonperturbative methods and extended-hadron models in field theory. I. Semi-classical functional methods, Phys. Rev., D 10:4114.

Dashen, R.F., Hasslacher, B., and Neveu, A., 1975, Particle spectrum in model field theories from semiclassical functional integral techniques, Phys. Rev., D 11:3424.

Davydov, A.S., and Kislukha, N.I., 1976, Solitons in one-dimensional molecular chains, Phys. Stat. Sol. (b), 75:735.

Davydov, A.S., 1977, Solitons as energy carriers in biological systems, Stud. Biophys., 62:1.

Davydov, A.S., Eremko, A.A., and Sergienko, A.I., 1978, Solitonui b α-spiralnui belkovuix molekulax, Ukraniskii Fizicheskii Zhurnal, 23:983.

Davydov, A.S., to appear a, Solitons in molecular systems with applications in biology, Physica Scripta.

Davydov, A.S., to appear b, Solitons and energy transfer along protein molecules, J. Theor. Biol.

de Broglie, L., 1960, "Nonlinear Wave Mechanics," Elsevier, Amsterdam.

de Broglie, L., 1963, "Introduction to the Vigier Theory of Elementary Particles," Elsevier, Amsterdam.

Derrick, G.H., 1964, Comments on nonlinear wave equations as models for elementary particles, J. Math. Phys., 5:1252.

Einstein, A., 1954, "Ideas and Opinions," Crown, New York.

Enz, U., 1963, Discrete mass, elementary length, and a topological invariant as a consequence of a relativistic invariant variational principle, Phys. Rev., 131:1392.

Faddeev, L.D., 1975, Hadrons from leptons?, JETP Lett., 21:64.

Faraday, M., 1910, "The Chemical History of a Candle," The Harvard Classics 30, P.F. Collier and Son, New York.

Feenberg, E., 1935, Born-Infeld field theory of the electron, Phys. Rev., 47:148.

Fermi, E., Pasta, J.R., and Ulam, S.M., 1955, "Studies of Nonlinear Problems," LASL Rept. No. LA-1940.

Fermi, E., 1965, "Collected Works of Enrico Fermi, Vol. II," U. of Chicago Press.

Fife, P.C., 1979, "Mathematical Aspects of Reacting and Diffusing Systems," Springer Lecture Notes in Biomathematics, Berlin.

Fisher, R.A., 1937, The waves of advance of advantageous genes, Ann. Eugenics (now Ann. Human Genetics), 7:355.

Flashka, H., 1974, The Toda lattice. II. Existence of integrals, Phys. Rev., B 9:1924.

Frenkel, J., 1934, Born's theory of the electron, Proc. Roy. Soc. A, 146:930.

Frenkel, J., and Kontorova, T., 1939, On the theory of plastic deformation and twinning, J. Phys. (USSR), 1:137.

Gardner, C.S., Green, J.M., Kruskal, M.D., and Miura, R.M., 1967, Method for solving the Korteweg-deVries equation, Phys. Rev. Lett., 19:1095.

Gibbons, J., Thornhill, S.G., Wardrop, M.J., and ter Haar, D., 1977, On the theory of Langmuir solitons, J. Plasma Physics, 17:153.

Goldstone, J., and Jakiw, R., 1975, Quantization of nonlinear waves, Phys. Rev., D 11:1486.

Green, D.E., 1973, Mechanism of energy transduction in biological systems, Science, 181:583.

Hamilton, W.C., and Ibers, J.A., 1968, "Hydrogen Bonding in Solids," W.A. Benjamin, New York.

Heisenberg, W., 1966, "Introduction to the Unified Field Theory of Elementary Particles," Interscience, N.Y.

Helmholtz, H., 1850, Messungen über den zeitlichen Verlauf der Zuchung animalischer Muskeln und die Fortplanzungsgeschwindigkeit der Reizung in der Nerven, Arch. Anat. Physiol., 276.

Helmholtz, H., 1896, Zwei hydrodynamische Abhandhingen: I. Ueber Wirbelbewegungen (1858), II. Ueber discontinuirliche Flüssigkeits bewegungen (1868), Ostwald's Klassiker, 79.

Hénon, M., 1974, Integrals of the Toda lattice, Phys. Rev., B 9:1921.

Hermann, R., 1977a, "The Geometric Theory of Nonlinear Waves," Math. Sci. Press, Brookline, N.Y.

Hermann, R., 1977b, "Toda Lattices, Cosymplectic Manifolds, Bäcklund Transformations and Solitons," Math. Sci. Press, Brookline, N.Y.

Hermann, R., 1977c, "Geometric Theory of Nonlinear Differential Equations, Bäcklund Transformations and Solitons," Math. Sci. Press, Brookline, N.Y.

Hobart, R.H., 1963, On the instability of a class of unified field models, Proc. Phys. Soc., 82:201.

Hodgkin, A.L., and Huxley, A.F., 1952, A quantitative description of membrane current and its application to conduction and excitation in nerve, J. Physiol., 117:500.

Ichikawa, Y.H., Imamura, T., and Taniuti, T., 1972, Nonlinear wave modulation in collisionless plasmas, J. Phys. Soc. Jap., 33:189.

Jakiw, R., 1977, Quantum meaning of classical field theory, Rev. Mod. Phys., 49:681.

Kelley, P.L., 1965, Self-focussing of optical beams, Phys. Rev. Lett., 15:1005.

Khodorov, B.I., 1974, "The Problem of Excitability," Plenum, New York.

Kolmogoroff, A., Petrovsky, I., and Piscounoff, N., 1937, Étude de l'équation de la diffusion avec croissance de la quantité de matière et son application a un problème biologique, Bull. Univ. Moscow, Série Int., A 1:1.

Korteweg, D.J., and de Vries, G., 1895, On the change of form of long waves advancing in a rectangular canal, and on a new type of long stationary waves, Phil. Mag., 39:422.

Lamb, H., 1932, "Hydrodynamics," 6'th edition, Dover, New York.

Lamb, Jr., G.L., 1971, Analytical descriptions of ultrashort optical pulse propagation in a resonant medium, Rev. Mod. Phys., 43:99.

Lamb, Jr., G.L., 1976, Bäcklund transformations at the turn of the century, in: "Bäcklund Transformations," R.M. Miura, ed., Springer Math. Series #515.

Lamb, Jr., G.L., to appear, "Elements of Soliton Theory," Wiley-Interscience, New York.

Lax, P.D., 1968, Integrals of nonlinear equations of evolution and solitary waves, Comm. Pure Appl. Math., 21:467.

Lonngren, K., and Scott, A., 1978, "Solitons in Action," Academic Press, New York.

Makhankov, V.G., 1978, Dynamics of classical solitons (in non-integrable systems), Phys. Repts., 35:1.

McLaughlin, D.W., and Scott, A.C., 1978, Perturbation analysis of fluxon dynamics, Phys. Rev., A 18:1652.

Mie, G., 1912a, Theory of matter, Ann. d. Physik, 37:511.

Mie, G., 1912b, Foundation of a theory of matter, Ann. d. Physik, 39:1.

Mie, G., 1912c, Basis of a theory of matter, Ann. d. Physik, 40:1.

Miles, J.W., 1980, Solitary waves, Annual Rev. of Fluid Mechanics, 12:11.

Miura, R.M., 1968a, Korteweg-de Vries equation and generalizations. I. A remarkable explicit nonlinear transformation, J. Math. Phys., 9:1202.

Miura, R.M., 1968b, Korteweg-de Vries equation and generalizations. II. Existence of conservation laws and constants of motion, J. Math. Phys., 9:1204.

Miura, R.M., 1976, The Koretweg-de Vries equation: A survey of results, SIAM Rev., 18:412.

Nagumo, J., Arimoto, S., and Yoshizawa, S., 1962, An active pulse transmission line simulating nerve axon, Proc. IRE, 50:2061.

Nagumo, J., Yoshizawa, S., Arimoto, S., 1965, Bistable transmission lines, Trans. IEEE on Circuit Theory, CT-12:400.

Nevskaya, N.A., and Chirgadze, Yu. N., 1976, Infrared spectra and resonance interactions of amide-I and II vibrations of α-helix, Biopolymers, 15:637.

Newell, A.C., 1974, "Nonlinear Wave Motion," AMS Lectures in Appl. Math. 15, (Conference held at Clarkson College, Potsdam, N.Y. 1972.).

Offner, F., Weinberg, A., and Young, G., 1940, Nerve conduction theory: Some mathematical consequences of Bernstein's model, Bull. Math. Biophys., 2:89.

Parsa, Z., 1979, Topological solitons in physics, Am. J. Phys., 47:56.

Pauling, L., Corey, R.B., and Branson, H.R., 1951, The structure of proteins: Two hydrogen-bonded helical configurations of the polypeptide chain, Proc. N.A.S., 37:205.

Perring, J.K., and Skyrme, T.H.R., 1962, A model unified field equation, Nucl. Phys., 31:550.

Rebbi, C., 1979, Solitons, Scientific American, 240:92.

Robinson, J., and Russell, J.S., 1838, 7th Meeting of the British Assoc. Adv. Sci.

Rosen, G., 1965, Particle-like solutions to nonlinear scalar wave theories, J. Math. Phys., 6:1269.

Rubinstein, J., 1970, Sine-Gordon equation, J. Math. Phys., 11:258.

Russell, J.S., 1844, Report on waves, 14th Meeting of the British Assoc. Adv. Sci.

Russell, J.S., 1885, The waves of translation in the oceans of water, air and ether, London.

Schuster, P., Zundel, G., and Sandorfy, C., eds., 1976, "The Hydrogen Bond," (3 volumes) North-Holland Publ. Co., Amsterdam.

Scott, A.C., 1962, Analysis of nonlinear distributed systems, Trans. IRE, CT-9:192.

Scott, A.C., 1975, The electrophysics of a nerve fibre, Rev. Mod. Phys., 47:487.

Scott, A.C., 1977, "Neurophysics," Wiley-Interscience, N.Y.

Scott, A.C., Chu, F.Y.F., and Reible, S.A., 1976, Magnetic flux propagation on a Josephson transmission line, J. Appl. Phys., 47:3272.

Scott, A.C., 1979, Sine-Gordon breather dynamics, Physica Scripta, 20:509.

Scott, A.C., and Luzader, S.D., 1979, Coupled solitary waves in neurophysics, Physica Scripta, 20:395.

Seeger, A., Donth, H., and Kochendörfer, 1951, Theorie der Versetzungen in eindimensionalen Atomreihen, II., Z. Phys., 130:321.

Seeger, A., Donth, H., and Kochendörfer, 1953, Theorie der Versetzungen in eindimensionalen Atomreihen, III., Z. Phys., 134:173.

Skyrme, T.H.R., 1971, Kinks and Dirac equation, J. Math. Phys., 12:1735.

Takhtadzhyan, L.A., and Faddeev, L.D., 1974, Essentially nonlinear one-dimensional model of classical field theory, Theor. and Math. Phys., 21:1046.

Tappert, F., and Varma, C.M., 1970, Asymptotic theory of self-trapping of heat pulses in solids, Phys. Rev. Lett., 25:1108.

Toda, M., 1967a, Vibration of a chain with nonlinear interaction, J. Phys. Soc. Japan, 22:431.

Toda, M., 1967b, Wave propagation in anharmonic lattices, J. Phys. Soc. Japan, 23:501.

Toda, M., 1969, Mechanical and statistical mechanics of nonlinear chains, J. Phys. Soc. Japan, 26 sup.:235.

Toda, M., and Wadati, M., 1973, A soliton and two solitons in an exponential lattice and related equations, J. Phys. Soc. Japan, 34:18.

Toda, M., 1975, Studies of a nonlinear lattice, Phys. Reports, 18:1.

Wahlquist, H.D., and Estabrook, F.B., 1973, Bäcklund transformation for solution of the Korteweg-de Vries equation, Phys. Rev. Lett., 31:1386.

Whitham, G.B., 1965a, A general approach to linear and nonlinear dispersive waves using a Lagrangian, J. Fluid Mech., 22:273.

Whitham, G.B., 1965b, Nonlinear dispersive waves, Proc. Roy. Soc. A, 283:238.

Whitham, G.B., 1967a, Nonlinear dispersion of water waves, J. Fluid Mech., 27:399.

Whitham, G.B., 1967b, Variational methods and applications to water waves, Proc. Roy. Soc., 299:6.

Whitham, G.B., 1970, Two-timing, variational principles and waves, J. Fluid Mech., 44:373.

Whitham, G.B., 1974, "Linear and Nonlinear Waves," Wiley, N.Y.

Zabusky, N.J., and Kruskal, M.D., 1965, Intèraction of solitons in a collisionless plasma and the recurrence of initial states, Phys. Rev. Lett., 15:240.

Zakharov, V.E., and Shabat, A.B., 1972, Exact theory of two-dimensional self-focussing and one-dimensional self-modulation of waves in nonlinear media, Sov. Phys. - JETP, 34:62.

REMARKS ON NONLINEAR EVOLUTION EQUATIONS AND THE INVERSE SCATTERING TRANSFORM*

Mark J. Ablowitz

Department of Mathematics and Computer Science
Clarkson College of Technology
Potsdam, N.Y. 13676

* This work was partially supported by the Air Force Office of Scientific Research, and the office of Naval Research.

TABLE OF CONTENTS

REMARKS ON NONLINEAR EVOLUTION EQUATIONS AND THE INVERSE SCATTERING TRANSFORM

Mark J. Ablowitz

Department of Mathematics and Computer Science
Clarkson College of Technology
Potsdam, N.Y. 13676

I. INTRODUCTION

In recent years there has been considerable attention devoted to a new and rapidly developing area of mathematical physics, namely the Inverse Scattering Transform (I.S.T. for short). This method has allowed us to solve certain physically interesting nonlinear evolution equations. By now there are a number of review articles [for example, see Scott et al, 1973; Miura, 1976; Ablowitz, 1978] on this subject as well as some new books [for example, see Zakharov et al, 1980; Ablowitz and Segur, to appear], all of which contain numerous references.

At this time there are quite a few areas of current interest. Some of these are the following:

"Classical" results and methods: Namely, solitons; direct and inverse scattering; direct methods for finding solutions via Hirota's bilinear forms or direct operations on Gel'fand-Levitan-Marchenko integral equations; asymptotic solutions as $|t| \to \infty$; prolongation structures; periodic and other types of boundary conditions (not the doubly infinite line); perturbations; discretizations; physical applications; numerical simulations; etc.

Alternative formulations of inverse scattering via Riemann-Hilbert Problems [see Zakharov et al, 1980; Zakharov and Shabat, 1979; Zakharov and Mikhailov, 1978].

Multidimensional problems: Scattering and inverse scattering problems; some multidimensional nonlinear evolution equations solvable by I.S.T. are the Kadomstev-Petviashvilli equation, Davey-Stewartson equation, three wave equations, self-dual Yang-Mills equation and others [for example, see Zakharov et al, 1980; Ablowitz and Segur, to appear; Zakharov and Manakov, 1979; Kaup, 1980; Proc. of Joint U.S.-U.S.S.R. Conf. on Soliton Theory, 1979; Zakharov and Manakov (Crete Conference)].

Ordinary differential equations of Painlevé type; monodromy preserving deformations [for example, see Ablowitz et al, 1978; Sato et al, 1977; Flashka and Newell, to appear].

Direct and inverse scattering problems associated with certain nonlinear singular integro-differential equations, namely the Intermediate Long-Wave equation and Benjamin-Ono equation, [Satsuma et al, 1979; Satsuma and Ablowitz, 1980; Kodama et al, to be published; Nakamura, 1979; Bock and Kruskal, 1979].

In this lecture I intend to review some of the results associated with certain continuous and discrete (differential-difference, partial difference) nonlinear evolution equations solvable by inverse scattering.

II. EXAMPLES USING THE NONLINEAR SCHRÖDINGER EQUATION

The prototype equations we will study are for the nonlinear Schrödinger (NLS) equation.

(a) Partial differential equation,

$$iU_t = U_{xx} \pm 2U^2U^* \tag{2.1}$$

(U^* is the complex conjugate of U).

(b) Differential-difference equations,

$$iU_{nt} = \frac{1}{(\Delta x)^2}(U_{n+1}+U_{n-1}-2U_n) \pm U_nU_n^*(U_{n+1}+U_{n-1}) \ . \tag{2.2}$$

(c) Partial-difference equations,

$$\frac{i\Delta^m U_n^m}{\Delta t} = \frac{1}{2(\Delta x)^2}\left[\left(U_{n+1}^m + U_{n-1}^m \prod_{-\infty}^{n-1} \Lambda_k^m - 2U_n^m\right) + \left(U_{n-1}^{m+1} \prod_{-\infty}^{n} \Lambda_k^m - 2U_n^{m+1} + U_{n-1}^{m+1}\right)\right]$$

$$\pm \frac{1}{4}\left[2U_n^m U_n^{m*} U_{n+1}^{m+1} \prod_{-\infty}^{n} \Lambda_k^m + 2U_{n-1}^m U_n^{m+1} U_n^{m+1*} \prod_{-\infty} \Lambda_k^m\right.$$

$$+ U_n^m (U_n^{m*} U_{n+1}^m + U_n^{m+1*} U_{n+1}^{m+1}) + U_n^{m+1} (U_{n-1}^m U_n^{m*} + U_{n-1}^{m+1} U_n^{m+1*})$$

$$\left. - U_n^m \sum_{-\infty}^{n+1} \Delta^m S_k^m - U_n^{m+1} \sum_{-\infty}^{n} \Delta^m S_k^{m*}\right], \qquad (2.3)$$

where

$$\Lambda_k^m = \frac{1 \pm (\Delta x)^2 U_k^{m+1} U_k^{m+1*}}{1 \pm (\Delta x)^2 U_k^m U_k^{m*}},$$

$$S_k = U_{k-1}^m U_{k-2}^{m*} + U_k^m U_{k-1}^{m*} .$$

Each of these equations has solitons and an infinite number of conserved quantities, and is solvable by linear integral (for continuous cases) or summation (for discrete cases) equations. Each of the results for the discrete equation relax properly to the continuum limit as $\Delta x \to 0$ and $\Delta t \to 0$ (in the partial difference case). It should be stressed that these are only prototype results. In principle one may get analogous results associated with any continuous equation (e.g., Korteweg-deVries (KdV), modified Korteweg-deVries, etc.). We also remark that recently we have carried out some numerical simulations using Eq. (2.3) as a difference approximation of Eq. (2.1) and have found it to be very accurate and efficient [Taha and Ablowitz, 1980].

The basic point of view is to work with the associated linear scattering problem and its respective discretizations. Namely, for some partial differential equations, such as (2.1), the associated scattering problem is:

$$\begin{aligned} V_{1x} &= -i\zeta V_1 + qV_2 \\ V_{2x} &= i\zeta V_2 + rV_1 \end{aligned} \qquad (2.4)$$

and for differential-difference and partial-difference equations, such as (2.2), (2.3):

$$\begin{aligned} V_{1n+1} &= zV_{1n}+Q_nV_{2n} \ . \\ V_{2n+1} &= \frac{1}{z}V_{2n}+R_nV_{1n} \ . \end{aligned} \tag{2.5}$$

$(Q_n = \Delta x q_n,\ R_n = \Delta x r_n)$.

Following the methods outlined in Ablowitz [1978] and Ablowitz and Segur [to appear] one can find classes of nonlinear evolution equations (both continuous associated with (2.4) and discrete associated with (2.5) where each equation has solitons, an infinite number of conserved quantities, and is solvable by I.S.T.

Briefly, the main results may be summarized as follows:

(a) For the continuous problem, from initial conditions we may calculate:

$$F(x) = \frac{1}{2\pi}\int_{-\infty}^{\infty} b(k,0)e^{-i(kx-\omega(2k)t)}dk - i\sum_{1}^{N} C_j(0)e^{-i(\zeta_j x-\omega(2\zeta_j)t)} \ , \tag{2.6a}$$

where $\omega(k)$ is the linearized dispersion relation, obtained by seeking solutions of the form

$$u \sim e^{i(kx-\omega t)} \ .$$

For example, in the NLS equation $\omega(k) = -k^2$, and from (2.4),

$$\{b(k,0),\ \{C_j(0),\ \zeta_j\}_{j=1}^{N}\}$$

are the required scattering data [for example, see Ablowitz, 1978; Ablowitz and Segur, to appear]. If we can solve the linear integral equation $(y > x)$

$$K(x,y)-F(x+y) \pm \int_x^{\infty}\int_x^{\infty} K(x,s)F^*(s+z)\,F(z+y)dzds = 0 \ , \tag{2.6b}$$

then $U \equiv r = \pm q^*$ is given by

$$U(x,t) = \pm 2K^*(x,x) \ . \tag{2.6c}$$

(b) For the discrete problem, the results may be summarized as follows. From the initial data we must calculate:

$$F(n) = \frac{1}{2\pi i}\oint b(z;t/m)z^{n-1}dz - \sum C_j(t/m)z_j^{n-1} , \tag{2.7a}$$

where

$$b(z;t/m) = b(z;0) \cdot \begin{Bmatrix} e^{-i\omega(z^2)t} & \text{differential-difference} \\ (\omega(z^2))^m & \text{partial-difference} \end{Bmatrix} ,$$

$$C_j(t/m) = C_j(0) \cdot \begin{Bmatrix} e^{-i\omega(z_j^2)t} \\ (\omega(z_j^2))^m \end{Bmatrix} .$$

From (2.5) the functions $\{b(z,0), \{C_j(0), z_j\}_{j=1}^{N}\}$ are the required scattering data. Then we must solve the linear summation equation ($\ell > n$)

$$\kappa(n,\ell) - F(n+\ell) \pm \sum_{n',n''=n+1}\sum \kappa(n,n'')F^*(n''+n')F(n'+\ell) = 0 \tag{2.7b}$$

and find the solution ($U_n \equiv R_n/\Delta x = \mp Q_n^*$)

$$U_n = \pm \kappa^*(n,n+1)/\Delta x . \tag{2.7c}$$

Since the details of soliton calculations, conserved quantities, direct and inverse scattering are worked out in Ablowitz [1978], Ablowitz and Segur [to appear] and Zakharov et al [1980], I will not pursue these matters further here.

III. THE INTERMEDIATE LONG-WAVE EQUATION

In the remaining portion of these lectures I wish to make some comments on a problem I mentioned earlier. This concerns very recent work we have been doing on the intermediate long wave (I.L.W.) equation, namely,

$$U_t + 2UU_x + U_x/\delta + \hat{T}(U_{xx}) = 0 , \tag{3.1}$$

where

$$\hat{T}U = \frac{1}{2\delta} \int_{-\infty}^{\infty} \coth\frac{\pi}{2\delta}(\xi - x)U(\xi)d\xi \tag{3.1a}$$

and $\displaystyle\fint_{-\infty}^{\infty}$ represents the Cauchy principal value integral. Equation (3.1) is a nonlinear singular integro-differential equation. As $\delta \to 0$ we have the KdV equation,

$$U_t + 2UU_x + \frac{\delta}{3} U_{xxx} = 0 \tag{3.2}$$

and as $\delta \to \infty$ we have the Benjamin-Ono equation

$$U_t + 2UU_x + H(U_{xx}) = 0 , \tag{3.3}$$

where

$$H(U) = \frac{1}{\pi} \fint_{-\infty}^{\infty} \frac{U(\xi)}{\xi - x} d\xi$$

is the usual Hilbert transform. Thus it is intermediate between two important equations. Physically speaking, it has been derived in the context of long internal waves in a stratified fluid [see Joseph, 1977; Kubota et al, 1978].

Recently it has been shown that there are solitons, an infinite number of conserved quantities, a Bäcklund transform, an associated novel type of linear scattering problem, and solutions via linear Gel'fand-Levitan-Marchenko type integral equations (see Satsuma et al, 1979; Satsuma and Ablowitz, 1980; Kodama et al, to be published Nakamura, 1979; Bock and Kruskal, 1979; and associated references). I shall only mention the basic results here.

To appreciate the novelty of the scattering problem, we should recall the following Plemelj formulae associated with the operator $\hat{T}$. Namely let $\psi^{\pm}(x)$ be the boundary values of certain functions $\psi^{\pm}(z)$ analytic in horizontal strips of width 2δ ($\psi^{+}(z)$ analytic for $0 < \text{Im} z < 2\delta$, $\psi^{-}(z)$ analytic for $-2\delta < \text{Im} z < 0$) and periodically extended, namely,

$$\psi^{+}(x) = \lim_{\text{Im} z \downarrow 0} \frac{1}{2\delta} \int_{-\infty}^{\infty} \coth \frac{\pi}{2\delta} (\xi - z)\, u(\xi) d\xi = (i + \hat{T})\, u(x) , \tag{3.4a}$$

$$\psi^{-}(x) = \lim_{\text{Im} z \uparrow 0} \frac{1}{2\delta} \int_{-\infty}^{\infty} \coth \frac{\pi}{2\delta} (\xi - z)\, u(\xi) d\xi = (-i + \hat{T})\, u(x) . \tag{3.4b}$$

By periodicity $\psi^{-}(x) = \psi^{+}(x + 2i\delta)$. The scattering problem and associated time dependence is given by

$$i\psi_x^+ + (u-\lambda)\psi^+ = \mu\psi^- , \tag{3.5}$$

$$i\psi_t + 2i(\lambda+\frac{1}{2\delta})\psi_x^\pm + \psi_{xx}^\pm + [(\mp i - \hat{T})\, U_x + \nu]\psi^\pm = 0 \tag{3.6}$$

where λ,μ,ν are constants appropriately defined by the Jost functions.

Some of the main points to note are the following:

(a) Equation (3.5) is interpreted as a differential Riemann-Hilbert problem. For finite δ it also is a differential-difference (for complex x) equation.

(b) Compatibility of (3.5) and (3.6) yields the I.L.W. equation (3.1).

(c) As $\delta \to 0$ all results associated with KdV and the Schrödinger scattering problem are recovered.

(d) The following Gel'fand-Levitan equation produces the solution of the I.L.W. equation

$$K(x,y)+F(x,y) + \int_x^\infty K(x,s)F(s,y)ds = 0, \quad (y > x)$$

where,

$$u(x) = i(K^+(x,x)-K^-(x,x)) ,$$

$$\hat{T}u(x) = -(K^+(x,x)+K^-(x,x)) ,$$

$$K^-(x,y) = K^+(x+2i\delta,y+2i\delta) , \tag{3.7}$$

and $F(x,y)$ satisfies,

$$L_1F = (i\partial_x + \frac{1}{2\delta})F^+(x,y) + (i\partial_y - \frac{1}{2\delta})F^-(x,y) = 0 , \tag{3.8a}$$

$$L_2F = (i\partial_t + \partial_x^2 - \partial_y^2)F(x,y) = 0 . \tag{3.8b}$$

N-soliton solutions can be constructed by assuming exponential solutions for F, i.e.,

$$F(x,y) = \sum_{\ell=1}^{N} C_\ell(t)\exp(i\zeta_{-\ell}x+i\zeta_{+\ell}y) , \tag{3.9}$$

where

$$\zeta_{\pm\ell} = i\kappa_\ell \pm (\kappa_\ell \cot 2\kappa_\ell \delta - \frac{1}{2\delta}) ,$$

$$C_\ell(t) = C_\ell(0)\exp(-4\kappa_\ell(\kappa_\ell \cot 2\kappa_\ell \delta - \frac{1}{2\delta})t) .$$

A one-soliton solution is given by

$$u = \frac{2\kappa_1 \sin 2\kappa_1 \delta}{\cosh 2\kappa_1 (x - x_o(t)) + \cos 2\kappa_1 \delta} , \tag{3.10}$$

where

$$x_o(t) = (2\kappa_1)^{-1} \ln(C_1(t)/2\kappa_1) .$$

It should be remarked that the analytical scattering and inverse scattering analysis yields an explicit representation for $F(x,y)$ [see Kodama et al, to be published];

$$F(x,y) = \frac{1}{2\pi}\int_{-1/2\delta}^{\infty} \frac{b(k,t)}{a(k,t)} e^{i\zeta_- x + i\zeta_+ y}$$

$$+ \sum_{\ell=1}^{N} C_\ell(t)\exp(i\zeta_{-\ell}x + i\zeta_{+\ell}y) \tag{3.11}$$

where

$$b(k,t) = b(k,0)\exp(-4ik(k\coth 2k\delta + \frac{1}{2\delta})t) ,$$

$$a(k,t) = a(k,0) ,$$

$$C_\ell(t) = -ib(k_\ell = i\kappa_\ell, t)/a'(k = i\kappa_\ell) .$$

These results are valid for given δ and $\max|u(x,o)|$ chosen small enough. When $\delta \to \infty$ (the Benjamin-Ono limit) new singularities may appear. We are presently investigating this situation.

REFERENCES

Ablowitz, M.J., 1978, Lectures on the Inverse Scattering Transform, Stud. Appl. Math., 58:17.

Ablowitz, M.J., Ramani, A., and Segur, H., 1978, Nonlinear evolution equations and ordinary differential equations of Painlevé type, Lett. Nuovo Cimento, 23:333.

Ablowitz, M.J., and Segur, H., to appear, Solitons and the Inverse Scattering Transform, SIAM Monograph, Stud. in Appl. Math.

Bock, T.L., and Kruskal, M.D., 1979, A two-parameter Miura transformation of the Benjamin-Ono equation, Phys. Lett., 74A:173.

Flaschka, H., and Newell, A.C., to appear, Monodromy and spectrum preserving deformations, I. Comm. Math Phys.

Joseph, R.I., 1977, Solitary waves in a finite depth fluid, J. Phys., 10A:L225.

Kaup, D.J., 1980, The inverse scattering solution for the full three-dimensional three-wave resonant interaction, Physica D, 1:45.

Kodama, Y., Satsuma, J., and Ablowitz, M.J., to be published, The nonlinear intermediate long wave equation: Analysis and method of solution.

Kubota, T., Ko, D.R.S., and Dobbs, L., 1978, Propagation of weakly nonlinear internal waves in a stratified fluid of finite depth, AIAA J. of Hydronautics, 12:157.

Miura, R.M., 1976, The Korteweg-deVries equation: A survey of results, SIAM Rev., 18:412.

Nakamura, A., 1979, Bäcklund transform and conservation laws of the Benjamin-Ono equation, J. Phys. Soc. Jap., 47:1335.

Proceedings of Joint U.S.-U.S.S.R. Conference on Soliton Theory, Kiev, U.S.S.R., 1979, to be published.

Sato, M., Miwa, T., and Jimbo, M., 1977, A series of papers entitled "Holonomic quantum fields, I-V", Pub. RIMS, I is in 14:223; Also a series of papers entitled "Studies on holonomic quantum fields, I-SVI", Proc. Japan Acad., I is in 53A:6.

Satsuma, J., Ablowitz, M.J., and Kodama, Y., 1979, On an internal wave equation describing a stratified fluid with finite depth, Phys. Lett., 73A:283.

Satsuma, J., and Ablowitz, M.J., 1980, Solutions of an internal wave equation describing a stratified fluid with finite depth, in: "Nonlinear Partial Differential Equations in Engineering and Applied Science," R.L. Sternberg, A.J. Kalinowki and J.S. Papadakis, eds., Marcel Dekker, New York.

Scott, A.C., Chu, F.Y.F., and McLaughlin, D.W., 1973, The soliton, a new concept in applied science, Proc. IEEE, 61:1443.

Taha, T., and Ablowitz, M.J., 1980, Numerical calculations, Clarkson College.

Zakharov, V.E., and Mikhailov, A.V., 1978, Relativistically invariant two-dimensional models of field theory which are integrable by menas of the inverse scattering problem method, Sov. Phys. J.E.T.P., 47:1017.

Zakharov, V.E., and Shabat, A.B., 1979, Integration of the nonlinear equations of mathematical physics by the method of the inverse scattering problem, II, Funkts. anal. i ego prilozh, 13:13.

Zakharov, V.E., and Manakov, S.V., 1979, Soliton theory, (Sov. Sci. Rev. Sec. A) Physics Reviews, 1:133.

Zakharov, V.E., Manakov, S.V., Novikov, S.P., and Pitayevsky, L.P., to appear 1980, Theory of solitons, the method of the inverse problem.

Zakharov, V.E., and Manakov, S., 1980, Lecture by S. Manakov "On the self-dual Yang-Mills equation in 2n + 1 dimensions." Conference on Solitons, Crete, Greece.

THE LINEARITY OF NONLINEAR SOLITON EQUATIONS AND THE THREE WAVE RESONANCE INTERACTION

D.J. Kaup

Physics Department
Clarkson College of Technology
Potsdam, N.Y. 13676

ABSTRACT

Based on a representation found by Professor Newell, it is illustrated how the general solution of nonlinear soliton equations are really simple linear sums. Then the general three-wave resonant interaction is discussed, what the resonance is, and the three-dimensional scattering problem which is used to solve it. In particular, we contrast the standard one-dimensional inverse scattering with this form of three-dimensional inverse scattering. The principle difference is that no bound states occur in this form of three-dimensional inverse scattering, and thus no solitons exist in three-dimensions. However, it is shown that from a Bäcklund transformation, one can construct localized solutions, called "lump" solutions, and their properties are discussed.

TABLE OF CONTENTS

THE LINEARITY OF NONLINEAR SOLITON EQUATIONS AND THE THREE WAVE RESONANCE INTERACTION

D.J. Kaup

Physics Department
Clarkson College of Technology
Potsdam, N.Y. 13676

I. INTRODUCTION

The purpose of a school such as this is to provide a background of information for those interested in a given particular subject. The basic fundamentals of solitons have already been well expanded on by the previous talks, and in general, the most that I could add without excessive duplication is simply additional references [Kaup, 1977; Kaup and Newell, 1978a, b; Kaup et al, 1979], which describe my viewpoint on these basics. However, there is one basic point that still has not been emphasized here, and which I have always considered to be striking and important. That is the fact that although these systems are indeed nonlinear, their behavior so closely mocks or imitates linear systems, that one is frequently ahead if he simply forgets that it is nonlinear, and looks upon the system as being essentially linear. For example, Professor Newell [Kaup and Newell, 1978b] has demonstrated a striking representation of the general solution for q in terms of the squared eigenstates. This expansion, for the Zakharov-Shabat (ZS) case with $r = -q^*$, is

$$q = -\frac{1}{\pi}\int_{-\infty}^{\infty}\left(\frac{b}{a}\psi_1^2 + \frac{b^*}{a^*}\psi_2^{*2}\right)d\zeta + 2i\sum_k\left(\gamma_k\psi_{1k}^2 - \gamma_k^*\psi_{2k}^{*2}\right), \quad (1.1)$$

with the notation as in Kaup and Newell [1978b]. Note that the solution, as given in (1.1), is an <u>exact linear sum</u> (integral) over a radiation (continuous spectrum) part, and a <u>linear sum</u> over each and every bound state (soliton). There is absolutely no difficulty in identifying what part of the solution is radiation, and what part belongs to the kth soliton. It's right there in the expansion given by (1.1). The only way that the nonlinearity enters at all in (1.1) is in the squared eigenstates. In the radiation part, the eigenstate is a plane wave only as $x \to \pm\infty$, and for x finite, undergoes phase shifts and a WKB swelling of its amplitude as it passes over the position of any bound state. But still, qualitatively these continuous eigenstates are very much like plane waves, and the radiation part of (1.1) is simply a "nonlinear Fourier transform", [Ablowitz et al, 1974], where b/a is the continuous part of the nonlinear Fourier transform. Similarly, since the soliton part of (1.1) is given in terms of the bound states, it follows that each soliton solution is localized. And again, the only way the nonlinearity enters in the soliton part of (1.1) is through the squared bound states, where the radiation will only, at most, minorly influence the shape of these bound states.

Thus these exactly integrable systems exhibit the feature that although they are nonlinear, their behavior is very remarkably linear. Mathematically, this is demonstrated by the system being exactly integrable [Ablowitz et al, 1974], by (1.1), and even may be visually seen in almost any computer plot of such solutions. As is well known, by a simple cursory visual examination of almost any such solution, one can pick out all the solitons and identify the radiation part, due to the usual striking contrast between the bound states and the continuous eigenstates. And, except for phase shifts, one can actually visually see that these states are noninteracting.

So, what I want to leave with you at this point, is the empirical observation, that if you consider these nonlinear systems to be "almost" linear, many of their "strange" features become understandable, and one can be lead to predictions [Kaup, 1977; Kaup and Newell, 1978 a,b; Kaup et al, 1979] concerning further features of such systems.

For the remainder of this lecture, I want to concentrate on some of the basics associated with three-dimensional inverse scattering, such as that which is used in solving the three dimensional three-wave resonant interaction (3WRI). But, before we even look at that, we first should describe this 3WRI, and explain what it is, because a basic understanding of it is very useful in understanding the associated 3D inverse scattering.

There are many examples that one can use to illustrate the 3WRI, one of which dealing with electromagnetic interactions, was given in our general review [Kaup et al, 1979] of the one-dimensional case. Instead of duplicating that one here, or something similar, I shall go almost to the other extreme, and instead consider only a very simple mathematical model. We could actually choose any hyperbolic system possessing a quadratic nonlinearity as its lowest order nonlinearity. But to keep things simple, we shall choose one of the simplest of all such models, namely the wave-equation with a nonlinear term,

$$\partial_t^2\phi - \nabla^2\phi = \varepsilon\phi^2 \tag{1.2}$$

where ϕ is a real field. Again, I emphasize that (1.2) is simply a mathematical model, which contains the essential features which we require: propagating waves and a quadratic nonlinearity. Almost any such system with propagating waves and a quadratic nonlinearity will give rise to the 3WRI in the weakly nonlinear limits. We shall now illustrate this with the model (1.2). With $\varepsilon = 0$, (1.2) is exactly linear, and one can create solutions of (1.2) using linear superposition. And, since such a linear solution remains linear for all time when $\varepsilon = 0$, there is nothing more that needs to be said about such a solution. But, when we "turn-on" the interaction and allow ε to be nonzero, sooner or later something is going to happen to destroy this linear superposition. Letting ϕ be a general linear solution becomes too difficult to attempt when $\varepsilon \neq 0$. So we simplify and the easiest simplification is to let ϕ be a sum of almost planar waves, that is let

$$\phi = \sum_{j=1}^{N} e^{i\theta_j} u_j(\vec{x},t) + \text{c.c.} \tag{1.3}$$

where

$$\theta_j = \vec{k}_j\cdot\vec{x} - \omega_j t , \tag{1.4}$$

c.c. means the complex conjugate, and where the envelopes, u_j, are allowed to be complex and are assumed to be slowly varying with respect to the phase of the plane waves. In other words, any envelope will contain many, many oscillations or wavelengths under it. Furthermore, we assume each $\vec{k}_j$ to be well separated from the others.

Under these conditions, the zeroth-order solution (1.3) will appear to be composed of N distinct envelopes. Each one of these envelopes will contain many, many oscillations, and each of these oscillations will be described by a wave-vector, $\vec{k}_j$, and a frequency, ω_j. And as one looks at any specific wave or oscillation, inside of any envelope, and considers only its nearest neighboring waves, the solution does appear to be a plane wave. It is only when one considers the solution over many oscillations that one does then note the eventual decay of the wave, due to the envelope. The purpose of these envelopes are two-fold. First, they serve to keep the total energy and spacial extent finite, and second, there are no real physical plane-wave solutions. All physical solutions are really bounded, and so using an envelope gives us a physical solution.

When $\varepsilon = 0$, (1.3) is an exact solution, providing we restrict our choices of ω_j to

$$\omega_j^2 = \omega^2(\vec{k}_j) = k_j^2 \quad . \tag{1.5}$$

This is just the linear dispersion relation. Now, to understand what can happen to the solution when $\varepsilon \neq 0$, we shall first consider the $\varepsilon\phi^2$ term to be an inhomogeneous term, and thereby driving ϕ, in (1.3), like a source term would. So, we use the linear Green's function, and have that

$$\phi = \phi_{hom} + \varepsilon \int_{-\infty}^{t} dt' \int d^3x' \; G(\vec{x}-\vec{x}', t-t')\phi^2(\vec{x}', t') \ , \tag{1.6}$$

where

$$G(\vec{x},t) = \frac{1}{(2\pi)^3} \int \frac{d^3k}{\omega(\vec{k})} e^{i\vec{k}\cdot\vec{x}} \sin[t\omega(\vec{k})] \ , \tag{1.7}$$

and

$$\omega(\vec{k}) = (k^2)^{\frac{1}{2}} \quad . \tag{1.8}$$

In (1.6), ϕ_{hom} is the homogeneous solution of (1.3). Now, I want to remind you that one interpretation of (1.6) is in terms of Huygen's principle. This interpretation is as follows. At $\vec{x}'$ and t', an amount of ϕ is created by the ϕ^2 term, of amplitude and phase ϕ^2, which is then propagated to $\vec{x}$ and t by G, and its final amplitude and phase at $\vec{x}$ and t is simply $\varepsilon\phi^2$. Now, one simply integrates over all possible values of $\vec{x}'$ and t' for the final result, (1.6). Well, let's simply consider what terms might be present in ϕ^2, and take the simple case where we have only two envelopes. Then

$$\phi^2 = u_1^2 e^{2i\theta_1} + u_2^2 e^{2i\theta_2} + u_1^* u_1 + u_2^* u_2$$

$$+ \; 2u_1 u_2^* \, e^{i(\theta_1-\theta_2)} + 2u_1 u_2 e^{i(\theta_1+\theta_2)} + \text{c.c.} \tag{1.9}$$

Now, take (1.6) and (1.7) and rearrange it so that we do the integral over $\vec{x}'$ first, over $\vec{k}$ second, and over t' last. The first four items in (1.9) correspond to simple harmonic generation [Kaup, 1978] and side band modulations, which are only special cases of the 3WRI. So, we shall ignore these terms here, and put our attention on the 5th and 6th terms, which represent an interaction between the two envelopes. Then in (1.6), the integral over $\vec{x}'$ will give its largest contribution when $\vec{k} = \vec{k}_1 \pm \vec{k}_2$, and also when $\vec{k} = -(\vec{k}_1 \pm \vec{k}_2)$.

And, at these values of $\vec{k}$, the integral over $\vec{x}'$ also gives very strongly peaked functions of $\vec{k}$, which are almost delta functions in the above variables. Thus, in $\vec{k}$-space, the interaction separates into distinct regions, centered at each of the above values of $\vec{k}$. The width of each region in $\vec{k}$-space is, of course, inversely proportional to the width of the associated envelope in $\vec{x}$-space. Now each of these regions in $\vec{k}$-space will have a time dependent phase associated with it. At $\vec{k} = \vec{k}_1 + \vec{k}_2$, this phase is $-(\omega_1 + \omega_2)t$, at $\vec{k} = \vec{k}_1 - \vec{k}_2$, this phase is $-(\omega_1 - \omega_2)t$, etc., all of which follows from (1.4), (1.6), and (1.9).

Next, we consider what happens when we do the d^3k integral in (1.7). And since the d^3x' integral gave essentially delta functions in $\vec{k}$-space, the d^3k integral simply gives us a sum, one term for each region in $\vec{k}$-space. But this just, in effect, gives us back what we started with, except that the amplitude and phase factors in (1.7) have been inserted. For example, about $\vec{k} = \vec{k}_1 + \vec{k}_2$, we obtain

$$2\varepsilon \frac{u_1 u_2}{\omega(\vec{k}_1+\vec{k}_2)} e^{i(\vec{k}_1\cdot\vec{x}+\vec{k}_2\cdot\vec{x}-\omega_1 t'-\omega_2 t')} \sin[(t-t')\omega(\vec{k}_1+\vec{k}_2)], \quad (1.10a)$$

and about the region $\vec{k} = \vec{k}_1 - \vec{k}_2$, we similarly have

$$2 \frac{u_1 u_2^*}{\omega(\vec{k}_1-\vec{k}_2)} e^{i(\vec{k}_1\cdot\vec{x}-\vec{k}_2\cdot\vec{x}-\omega_1 t'+\omega_2 t')} \sin[(t-t')\omega(\vec{k}_1-\vec{k}_2)], \quad (1.10b)$$

and similarly for the other terms.

The result exemplified by (1.10), interpreted in terms of Huygen's principle, is simply the net result of all wavelets produced by $\varepsilon\phi^2$ at $\vec{x}'$ and at t', upon summing over all possible values of $\vec{x}'$, chosen such that all wavelets end up at $\vec{x}$, and at the time t.

We now consider the last integral over dt'. Note the time dependencies in (1.10a) and (1.10b). We shall simply state the result. If in (1.10a), $\omega \neq \omega_1 + \omega_2$, then this term will remain bounded by the order of ε for all time. Thus the effect of this term is simply to produce a small perturbation on the linear solution. But, if $\omega = \omega_1 + \omega_2$, then this term will grow linearly in t, and cannot be considered as a small perturbation. In fact, it can grow so large that it starts to deplete the initial envelopes. The interaction has now become fully nonlinear. This term does not then produce just a small perturbation on the linear solution. This term will drastically alter these initially linear solutions, and will cause them to interact. In terms of Huygen's principle, the wavelets produced by the $\vec{k}_1 + \vec{k}_2$ term, (1.10a), at t', reach t with the phase factors $(\omega_1 + \omega_2 \pm \omega)t$, since the sine term in (1.7) contains both signs for ωt. Thus, in general, these wavelets will reach t with various and almost random phases, such that when one adds over all t', much cancellation occurs. But, if $\omega = \omega_1 + \omega_2$, every one of these wavelets reaches t with exactly the same phase as every other wavelet. The wavelets are now all coherent and in phase, so that when one adds over all t', one obtains a linear growth in (t-t'), which can only be limited in its growth by the actual finite extent of the initial interacting envelopes, which up until now, we have considered to be absent, by using only plane waves. This is a resonance phenomena, when $\omega = \omega_1 + \omega_2$. When this resonance phenomena occurs, a small perturbation, as in (1.2), can drastically affect the linear solution, so much so, that one must now consider the nonlinear solution, and recognize that (1.6) is not valid when resonance occurs. And similarly, for those terms which do not contain resonance, (1.6) is still a valid method of treatment, since they always remain small.

Now, what happens at such a resonance? As one can see from (1.10a), a new wave, of wavevector $\vec{k}_1 + \vec{k}_2$, is being created. It has a frequency $\omega = \omega(\vec{k}_1 + \vec{k}_2)$, and thus is a linear solution of (1.2), due to (1.5). It grows linear in t (at least initially), and when the interaction turns off (by the two initial envelopes eventually separating) it propagates as a free wave. So, two waves have interacted resonantly, producing a third, wave, whence the name "three-wave resonant interaction".

Of course, this interaction is not restricted to the model (1.2), but will occur in any system which has envelope waves and a quadratic nonlinear interaction, and where the resonance conditions

$$\vec{k}_1 + \vec{k}_2 + \vec{k}_3 = 0, \tag{1.11a}$$

$$\omega_1 + \omega_2 + \omega_3 = 0, \tag{1.11b}$$

$$\omega_i = \omega(\vec{k}_i), \tag{1.11c}$$

can be satisfied. In (1.11), we are introducing a convention where the sum of all $\vec{k}$'s and ω's are zero, which can always be done for a specific set of $\vec{k}$'s and ω's, by simply redefining them with appropriate signs. This form is chosen, since the resulting 3WRI equations will be highly symmetric.

So, when the resonance condition (1.11) is satisfied, we cannot obtain a solution by simple perturbation techniques, as in (1.6), but must consider the weakly nonlinear ($\varepsilon \to 0$) case. Since two waves are interacting to produce a third, we start with three waves

$$\phi = e^{i\theta_1} u_1 + e^{i\theta_2} u_2 + e^{i\theta_3} u_3 + \text{c.c.} \tag{1.12}$$

and due to (1.11), the three phases are not independent, but satisfy

$$\theta_1 + \theta_2 + \theta_3 = 0 \quad . \tag{1.13}$$

Now let's insert (1.12) into (1.2), retaining only the lowest order terms. This gives

$$- 2i \sum_{j=1}^{3} \omega_j e^{i\theta_j} (\partial_t - \frac{1}{\omega_j} \vec{k}_j \cdot \vec{\nabla}) u_j$$

$$= \varepsilon \begin{bmatrix} 2u_1^* u_2^* e^{-i(\theta_1+\theta_2)} + 2u_1^* u_2 e^{-i(\theta_1-\theta_2)} \\ \\ + u_1^* u_1 + \ldots \end{bmatrix} , \tag{1.14}$$

and where we shall still ignore all effects of harmonic generation[1]. Upon balancing terms of like phase in (1.14), one obtains

$$(\partial_t - \vec{v}_j \cdot \vec{\nabla}) u_j = \frac{i\varepsilon}{\omega_j} u_k^* u_\ell^* \quad , \tag{1.15}$$

where $\vec{v}_j = \vec{k}_j/\omega_j$ is the group velocity (since dispersion is absent) and (j,k,ℓ) in (1.15) is cyclic in (1,2,3). There are thus three equations contained in (1.15). Upon scaling the envelope amplitudes, one can obtain (1.15) in the canonical form [Kaup et al, 1979],

$$(\partial_t - \vec{v}_j \cdot \vec{\nabla}) q_j = i\gamma_j q_k^* q_\ell^* \quad , \tag{1.16a}$$

where the q's are the scaled amplitudes, and

1 Strictly speaking, for the model (1.2), one cannot ignore harmonic generation, since given any $\vec{k}$, there exists a solution for $\omega(2\vec{k}) = 2\omega(\vec{k})$. Thus due to the absence of dispersion in (1.2), harmonic generation is always resonant, and can never be ignored. In more physical models which have dispersion, solutions will exist to $\omega(2\vec{k}) = 2\omega(\vec{k})$, only for special values of $\vec{k}$. Thus, in general, harmonic generation is <u>not</u> resonant, and indeed may be ignored, unless one of the $\vec{k}$'s happen to be one of these special values.

$$\gamma_j = \operatorname{sgn}(\omega_j E_j) \quad , \tag{1.16b}$$

where E_j is the energy of the jth envelope.

Once given (1.16), the method of solution depends first on the number of independent dimensions, or directions, appearing in (1.16a). For example, if you were given the q's as being independent of $\vec{x}$, then you have only a one-dimensional problem (only a t-dependence) regardless whether or not the group velocities are in independent directions. Another way to have a one-dimensional problem is to have all three group velocities equal. Now, we have only one characteristic direction, with the different streamlines never crossing or mixing. So, we simply solve the one-dimensional problem along each and every streamline.

In each case when the 3WRI reduces to a one-dimensional problem, one can always give the general solution in terms of elliptic functions [Armstrong et al, 1962]. When only two of the derivatives in (1.16) are independent, one has what is commonly referred to as the "one-dimensional" case, by which one means one spatial dimension (plus one time dimension, giving a total of two dimensions). Solutions for this case were given first by Zakharov and Manakov [1973], and the recent paper [Kaup et al, 1979] in Reviews of Modern Physics gives an up-to-date account of what is known about this two-dimensional case, both in terms of inverse scattering and numerical calculations, and how they both compare.

So, this only leaves the most general case, namely when (1.16) has all three derivatives as being independent. This is the full three-dimensional (3D) case, and is the case I shall discuss here. I shall not attempt to explain all the "ins-and-outs" of 3D inverse scattering here, since first there is not sufficient time, and second, this audience is not primarily interested in these ins-and-outs, but rather in the general principles involved. So, here I shall seek to emphasize how this 3D problem differs from 1D inverse scattering, and how it is still similar. As for references on the 3D-3WRI inverse scattering, Cornille [1979] and Kaup [1980a] give the first successful attempt to create a workable, albeit clumsy, inverse scattering. The papers by Kaup [1980b; 1980c] are concerned with the "global scattering problem", which could nevertheless solve a special class of initial value problems. Kaup [1979] gives the final and complete solution of the general initial value problem, and with remarkable hindsight, the paper by Kaup [1981] then allows one to completely bypass inverse scattering, and construct physical solutions with a technique very similar to, and undoubtedly related to Bäcklund transformations. What I shall do next is to discuss the principle ideas involved in Kaup [1979], and conclude by discussing the three principle types of solutions given in Kaup [1981].

II. THE SCATTERING PROBLEM FOR THE THREE-DIMENSIONAL THREE-WAVE RESONANT INTERACTION

Before discussing the scattering problem for this system, I first want to ensure that the shape and nature of the solutions of the 3D-3WRI are understood, because centering on this, we can simplify the method of solution. First, the interaction region in space-time is often finite, and is defined to be that region where two or more of the envelopes overlap. If you are in a region of space-time where there is no overlap of the envelopes, there is no interaction, since $q_j^* q_k^* = 0$, and by (1.16), each envelope then propagates as a free envelope. To know what happens in the interaction region, we must solve the full nonlinear equations. But if we block this region off, and consider only all other parts of space-time, then we are considering only the noninteracting region, where the envelopes are free. And since there is no dispersion, in this region, these free envelopes simply undergo a simple translation, as shown by the free solution of (1.16). Thus in the noninteraction region, we know exactly what the solution is. And, usually this region is much larger than the interaction region. Contrast this with other cases, where dispersion is present. Then, one only knows the solution at infinity. Almost all of space-time is now the interaction region, and you only know the solution on the boundary. But due to the absence of dispersion in the 3WRI equations, the opposite occurs here, in that one knows the solution in almost all of space-time, while it is usually unknown in only a finite region of space-time.

So, let's now consider the solution in the noninteracting region. By (1.16), each envelope here satisfies

$$(\partial_t + \vec{v}_j \cdot \vec{\nabla})q_j = 0, \tag{2.1}$$

which therefore moves with the velocity $\vec{v}_j$, without any distortion. Now, note that in total, we only have three operators, $(\partial_t + \vec{v}_j \cdot \vec{\nabla})$, while there are four space-time coordinates. This means that there is at least one linear combination of the space-time coordinates which is independent of these three operators. We shall call this combination χ_4, and define it by

$$(\partial_t + \vec{v}_j \cdot \vec{\nabla})\chi_4 = 0, \tag{2.2}$$

for j = 1,2, and 3. For example, to find χ_4, go to the rest frame of one envelope, say q_1, then $\vec{v}_1 = 0$. Now rotate your coordinate system such that $\vec{v}_2$ and $\vec{v}_3$ lie in the x-y plane. Now, it is very obvious from (2.2) that $\chi_4 = z$. As a further point, consider (1.16) in this situation. Then no derivatives with respect to z ever occur in (1.16). Thus z (or rather, χ_4 in general) appears in (1.16) only as a parameter. This means that you may solve (1.16) by fixing χ_4, obtain a solution, then fix χ_4 to be another value, etc. In other words, solutions at different values of χ_4 never mix or affect one another.

So, with χ_4 being defined by (2.2), there remains three other independent coordinates, which we shall define by

$$\frac{\partial}{\partial \chi_j} = \partial_j = -\partial_t - \vec{v}_j \cdot \vec{\nabla} \quad , \tag{2.3}$$

for j = 1,2,3. These coordinates we shall call "characteristic coordinates". As an example, if $\vec{v}_1 = 0$, $\vec{v}_2 = \hat{i}$, and $\vec{v}_3 = \hat{j}$, then a solution of (2.2) and (2.3) is

$$t = -\chi_1 - \chi_2 - \chi_3 \quad , \tag{2.4a}$$

$$x = -\chi_2 \quad , \tag{2.4b}$$

$$y = -\chi_3 \quad , \tag{2.4c}$$

$$z = \chi_4 \quad . \tag{2.4d}$$

Of course, other different solutions will exist for other orientations of the velocities.

From (2.1) and (2.3), we have for the free solutions that

$$\partial_j q_j = 0, \tag{2.1'}$$

or

$$q_j = q_j(\chi_k, \chi_\ell, \chi_4) \quad , \tag{2.5}$$

with j,k,ℓ cyclic in 1,2,3. Explicitly in terms of the very special and model situation represented by (2.4), the three equations contained in (2.5) are

$$q_1 = q_1(-x,-y,z), \tag{2.6a}$$

$$q_2 = q_2(-y, t-x-y,z), \tag{2.6b}$$

$$q_3 = q_3(t-x-y,-x,z). \tag{2.6c}$$

As one will note, q_1 satisfies $\partial_t q_1 = 0$, and q_2 satisfies $(\partial_t+\partial_x)q_2 = 0$, and q_3 satisfies $(\partial_t+\partial_y)q_3 = 0$.

So, what the above illustrates is that these characteristic coordinates are very special, are determined by the velocities, and are natural coordinates to use in this problem. When we use these coordinates, the free solutions take on a very special form, as given by (2.5).

Since χ_4 will only enter our equations as a parameter, we shall hereafter ignore it.

Note that at constant χ_k and χ_ℓ, the free solution for q_j does not change as χ_j is changed (provided we remain in the noninteracting region, where (2.1) is valid). Thus q_j can be considered to propagate along the $(\pm)\chi_j$-axis.

Usually, the initial envelopes are localized and could be considered to be bounded in space. As seen in the model (2.6), if bounded in x,y, and z, the solution is also bounded in characteristic coordinates. Since characteristic coordinates are natural coordinates, we shall use these coordinates instead of space-time coordinates. Thus our results shall be independent of the actual group velocities, and shall be quite general.

Consider a localized free solution for q_j as in (2.5), and as shown in Figure 1. In Figure 1, I have attempted to pictorially represent what a typical solution of the 3D-3WRI (for some value of χ_4) might look like, using characteristic coordinates, and not space-time coordinates. (I want to emphasize that the angles between the χ-axes in Fig. 1 need not be 90° as shown. What these angles would be, would depend on the three group velocities, and their directions. However, since no result depends on the angles between these characteristic coordinates, we may arbitrarily stretch these angles out to be 90°. Thus, in Fig. 1, we show the angles to be 90°.) As it is drawn, we have two envelopes q_i and q_j coming together and interacting, creating some q_k, and then everything separates as $t \to +\infty$, with different final envelopes. We note that, as in (2.4a), the relation

$$t = -\chi_i - \chi_j - \chi_k \quad , \tag{2.7}$$

is a consequence of (2.3), and does not depend on how the velocities are directed. It is from (2.7) that one can understand how time is directed in Fig. 1, since (2.7) defines a plane with the directional derivative (1,1,1). Imagine such a plane in Fig. 1. As this plane would be moved toward the right in Fig. 1, t would become more negative. At any time t, the intersection of this plane with q_i, gives the profile of q_i at that value of t. Note how if this plane would be moved toward the origin, q_i and q_j would then approach each other. This would correspond to two pulses (envelopes) approaching each other. At the origin, they overlap and interact.

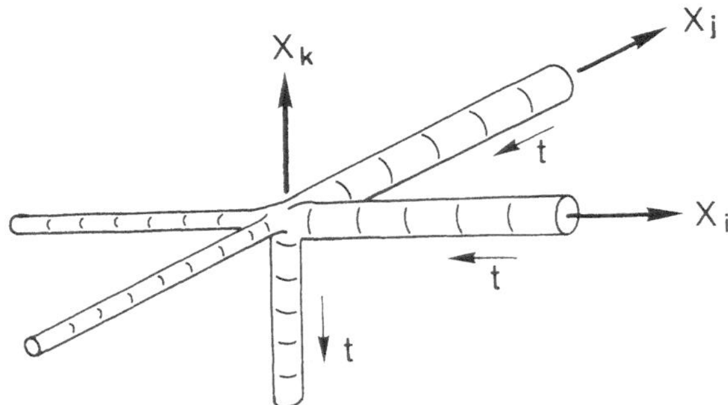

Fig. 1. A graphic representation of a solution of the 3D-3WRI in characteristic coordinates. This solution represents an initial q_i envelope and q_j envelope (each at the right) colliding, and then emerging with a reduced energy. In the process of interacting, an amount of q_k has been produced, and is seen traveling away from the collision center along its own characteristic.

Then as t becomes positive, they both come out the other side, with their original profiles changed and with q_k now being nonzero. Once t has become sufficiently large so that the envelopes no longer overlap, then the envelopes once again become free envelopes, and propagate according to (2.1).

Of course, there can be initial conditions where the envelopes would never separate, and there can be initial conditions whereby the solutions become singular. However those are special cases, and before one attempts to understand them, he had best understand this simple case as shown in Fig. 1. All it is, is two envelopes coming together, colliding and interacting, then separating. And in the process of interacting, a third envelope has been produced or created, which moves away along its own characteristic direction.

Let us now turn to the scattering problem itself. The original scattering problem was given by Zakharov and Shabat [1974]. However, it was not in a form that one could introduce characteristic coordinates into, which caused it to appear to be more complicated than it really was. Ablowitz and Haberman [1975] found another scattering problem which was in characteristic coordinate form, and is the one which we shall use. It is

$$\partial_k \psi_i = \gamma_k q_j^* \psi_k, \tag{2.8a}$$

$$\partial_i \psi_k = \gamma_i q_j \psi_i, \tag{2.8b}$$

where i,j, and k are cyclic, so that there are a total of six equations in (2.8). If one considers the integrability conditions for these six equations, one finds that the q's must satisfy

$$\partial_i q_i = \gamma_i q_j^* q_k^* , \tag{2.9}$$

where again, (i,j,k) are cyclic, so that (2.9) contains three equations, which if we interpret ∂_i as $\partial/\partial\chi_i$, are just the 3D-3WRI equations, (1.16).

Now in the initial value problem, one is given all three q's at some value of t, which we shall take to be at t = 0 for convenience. From (2.7), in terms of characteristic coordinates, we therefore initially know the q's only along the plane

$$\chi_j = -\chi_i - \chi_k . \tag{2.10}$$

So, we want to solve (2.8), only knowing this information. When one considers (2.8) relative to the plane defined by (2.10), one finds that three of the equations in (2.8) describe how the three

ψ's will evolve off of the plane (analogous to the t-evolution equations in 1D-IST) and the other three describe how the three ψ's propagate in this plane. Of course, since the plane is two-dimensional, having only three-first-order partial differential equations for three functions leaves the solution underdetermined, and it is exactly due to this underdeterminancy that this problem, even without any eigenvalue in (2.8), is still a scattering problem. It is not an eigenvalue problem as in 1D-IST, but it is still a scattering problem.

To restrict ourselves to the plane given by (2.10), we arbitrarily pick χ_i and χ_k to be our independent coordinates in this plane. Then provided we are restricted to be in this plane, the derivatives with respect to our two independent characteristic coordinates are related to the $\vec{\nabla}$ operator by

$$\partial_k = (\vec{v}_j - \vec{v}_k)\cdot\vec{\nabla} \quad , \tag{2.11a}$$

$$\partial_i = (\vec{v}_j - \vec{v}_i)\cdot\vec{\nabla} \quad , \tag{2.11b}$$

and the three equations from (2.8) which correspond to the scattering problem are

$$\partial_k\psi_i = \partial_k q_k^*\psi_k - \gamma_j q_k\psi_j \quad , \tag{2.12a}$$

$$(\partial_k - \partial_i)\psi_j = \gamma_k q_i\psi_k - \gamma_i q_k^*\psi_i \quad , \tag{2.12b}$$

$$\partial_i\psi_k = \gamma_i q_j\psi_i - \gamma_j q_i^*\psi_j \quad , \tag{2.12c}$$

and the reason this is a scattering problem is simply because $\partial_i\psi_i$, $(\partial_k + \partial_i)\psi_j$, and $\partial_k\psi_k$ are unspecified in (2.12). What (2.12a) does specify is how ψ_i changes as one moves in the χ_k direction, and nothing is said about how ψ_i must change as one moves in the χ_i direction. Similarly for (2.12b) and (2.12c), with different directions being involved. Due to this underspecification of these functions, integration of (2.12) gives

$$\psi_i = g_i(\chi_i) + \int_{\chi_k}^{\infty} \{\ \}(\chi_i, u)\,du \quad , \tag{2.13a}$$

$$\psi_j = g_j(\chi_i + \chi_k) + \int_{\chi_k}^{\infty} \{\ \}\,(\chi_i + \chi_k - v, v)\,dv \quad , \tag{2.13b}$$

$$\psi_k = g_k(\chi_k) + \int_{\chi_i}^{\infty} \{\ \}(w, \chi_k)\,dw \quad , \tag{2.13c}$$

where the brackets in (2.13) simply stand for the right-hand sides of (2.12), the exact form of which is unimportant for (2.13) and the following arguments. The three g's in (2.13) are arbitrary functions, and their specification will uniquely determine the solution. We emphasize that they are arbitrary, and they are arbitrary simply because (2.12) did underdetermine the solution for the ψ's.

Now, (2.12) are linear equations. So, the general solution may be given as a sum over some selected complete set of fundamental solutions. Furthermore, we can replace the arbitrary g's with the Fourier components, $e^{i\zeta\chi_i}$, etc., since from these plane waves, one can reconstruct any arbitrary function. So, for the fundamental solution we call ψ^i, we take

$$g_i(\chi_i) = e^{i\zeta\chi_i} , \tag{2.14a}$$

$$g_j = g_k = 0 , \tag{2.14b}$$

while for the fundamental solution ψ^j, we take

$$g_i = g_k = 0 , \tag{2.15a}$$

$$g_j(\chi_i + \chi_k) = e^{-i\zeta(\chi_i+\chi_k)} , \tag{2.15b}$$

and for ψ^k, we take

$$g_i = g_j = 0 , \tag{2.16a}$$

$$g_k(\chi_k) = e^{i\zeta\chi_k} . \tag{2.16b}$$

Now, let us see what the solution ψ^i, (2.14), corresponds to. In Fig. 2, we have pictorially represented this solution in the t = 0 plane. We assume the initial data to be on compact support, whence we can assume that the potentials (the q's) are only nonzero inside the central circle, as indicated. From (2.14) and (2.13a), we have for any χ_k above the circle, that $\psi^i_i = e^{i\zeta\chi_i}$ exactly. Also note that if χ_i lies to the left or the right of the circular region, then again $\psi^i_i = e^{i\zeta\chi_i}$. Thus ψ^i_i can only differ from $e^{i\zeta\chi_i}$ either inside the circular region, or directly underneath it. Similarly, one can argue that ψ^i_k is only nonzero in the circular region or directly to its left, and ψ^i_j is only nonzero in the circular region or to the lower right, as indicated in Fig. 2.

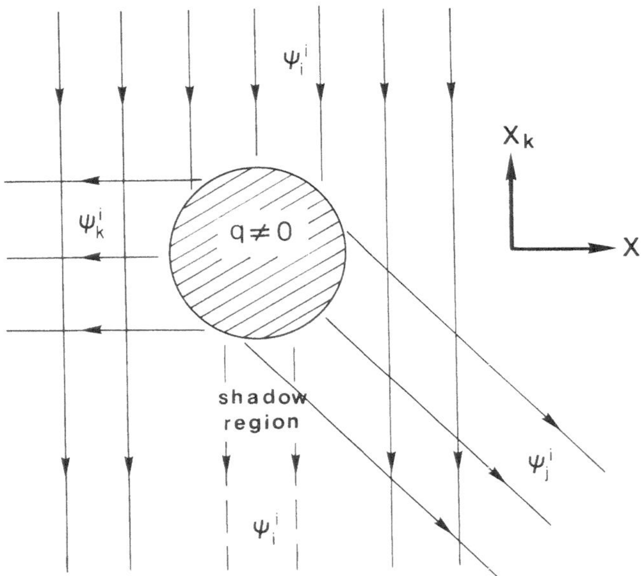

Fig. 2. The ψ^i solution in the χ_i-χ_k plane. The incident wave, ψ_i^i, is at the top, the transmitted wave, ψ_i^i, is the shadow region below, and ψ_k^i and ψ_j^i are the two reflected waves.

Now, note that each integral in (2.13) serves to propagate the solution of a particular component in a particular direction. These directions are indicated by the arrows in Fig. 2. We can now interpret Fig. 2 as follows. In Fig. 2, we see an incident wave, ψ_i^i entering at the top, which has transverse oscillations, $e^{i\zeta\chi_i}$. This wave interacts with the q's and creates out of this interaction two different waves, ψ_k^i and ψ_j^i, each of which propagates off in its own characteristic direction. So (2.12) is indeed a scattering problem, in that one wave is partially converted (scattered) into two others.

But what about the final profiles of $\psi_k^i(\chi_k)$ and $\psi_j^i(\chi_i+\chi_k)$? Well, as one would suspect, each of these final profiles, when Fourier transformed, defines a reflection coefficient, which depends on the exact structure of the potentials inside the circular region. Similarly, the final profile of $\psi_i^i(\chi_i)$ defines a transmission coefficient. So, we do have the potentials being mapped into reflection coefficients, as in 1D-IST. From the three solutions, ψ^i, ψ^j, and ψ^k, defined by (2.14)-(2.16), we therefore have two reflection coefficients per solution, for a total of six reflection coefficients. But, we have only three profiles, so that three of these six reflection coefficients are not independent, but are dependent on the other three.

Furthermore, (2.13) is not the only choice for a solution to (2.12), since one could choose different and the opposite limits for the integrals in (2.13). For example, changing the integral in (2.13a) to be from $-\infty$ to χ_k, corresponds in Fig. 2 to reversing the direction of the corresponding reflected wave. In all, since there are 8 independent ways of choosing the limits on the integrals in (2.13), and since there are three independent solutions per choice, we have a total of 24 independent fundamental solutions, and therefore 48 different reflection coefficients, of which only three are independent. Contrast this with the 1D-IST case, where we have only 4 fundamental solutions [Ablowitz et al, 1974], $\psi, \overline{\psi}, \phi$, and $\overline{\phi}$, and only four different reflection coefficients [Ablowitz et al, 1974], b/a, $b/\overline{a}$, $\overline{b}/a$, and $\overline{b}/\overline{a}$, of which at most, two of the latter are independent.

So, 3D-IST has a much larger redundancy in it than does 1D-IST, with the reason for this arising from the larger number of directions which one could use for the incident beam, and the reflected beams. Furthermore, when one starts considering the number of possible forms for the inverse scattering equations themselves, we again find a similar large number, whereas in 1D-IST, there are only two possible forms: inversion about $+\infty$ or about $-\infty$. The methods required to obtain the 3D inverse scattering equations are detailed in Kaup [1979] and due to the complexity of the algebra involved, we shall not reproduce those results here. Instead we shall simply outline them, and also contrast them with 1D-IST.

First, we note that the t-dependence of the reflection coefficient can be determined in the standard manner. We use the three equations in (2.8) which corresponds to evolving the ψ's off of the $t = 0$ plane, and simply evaluate them for $\chi_j = \pm\infty$ and/or $\chi_k = \pm\infty$, whence the t-evolution of the reflection coefficients will follow, since the potentials are assumed to vanish at these limits. Second, one finds that each fundamental solution of (2.12), where the incident wave is represented as in (2.14)-(2.16), has a particular analytic property with respect to ζ. For example, the fundamental solution ψ^i, defined by (2.13) and (2.14) is in general only bounded on the real ζ-axis, whereas $\psi^j(\psi^k)$ is analytic in the lower-(upper-)half ζ-plane. Similarly, each of the 24 fundamental solutions has a definite analytical property, with respect to ζ. Furthermore, each fundamental solution can be related to almost any other set of three fundamental solutions, via the various transmission and reflection coefficients, since only three of the 24 fundamental solutions are linearly independent. From this, one can construct "linear integral dispersion relations" whereby one fundamental solution is given in terms of an inhomogeneous part and an integral in the complex ζ-plane, over products of reflection coefficients and other fundamental solutions. Once these are constructed, one has, in effect, solved the inverse scattering problem, because all that is left is simply "doing the algebra".

So far, everything which has been described above, has very closely paralleled 1D-IST, and has differed really only in the dimensionality, larger number of solutions, etc. But, it is at the point of constructing the linear integral dispersion relations, that one will note a distinct difference between 3D-IST and 1D-IST.

In 1D-IST, the poles of the transmission coefficient, $a^{-1}(\zeta)$, in the upper half ζ-plane, define the bound-state eigenvalues [Ablowitz et al, 1974], each one of which gives rise to a soliton. But in this 3D-IST, every transmission coefficient <u>is bounded, and therefore no poles occur</u>, and therefore no solitons exist, as we know them from 1D-IST. The inverse scattering is done completely in terms of the continuous spectrum, and only in terms of the continuous spectrum. No bound states ever occur.

But, in 1D-IST, one property of the inverse scattering equations is that when the continuous spectrum is absent, closed form solutions exist, which are the N-soliton solutions. This occurs because the kernels in the integral equations (which are the inverse scattering equations) become what is called "separable". So, in 1D-IST, one could equivalently define solitons as being the separable solutions of the IST equations. So, when one considers the 3D-IST equations, quite surprising, we find that these equations can possess separable kernels, and the corresponding solutions are localized! And, one may also construct the analogy of the 1D N-soliton solution, except that these 3D objects are not quite solitons. And to distinguish them, we shall call these 3D objects "lumps". These 3D "lump solutions" we shall now discuss from a different point of view in the next section.

III. THE THREE DIMENSIONAL BÄCKLUND TRANSFORMATION

As one will recall, one of the more general means of constructing N-soliton solutions in one-dimension, is to use the Bäcklund transformation, whereby one starts with a trivial zero solution, and generates nontrivial solutions. Following a suggestion by Jim Corones [1976], simply applying a general technique, I have found that a Bäcklund transformation does exist for the 3D-3WRI, which is completely integrable, and is

$$\tilde{q}_j = q_j + \psi_i^* \psi_k / D \quad , \tag{3.1a}$$

where

$$\partial_i D = -\gamma_i \psi_i^* \psi_i \quad , \tag{3.1b}$$

and each above equation is cyclic in (i,j,k). In (3.1a), q_j is an initial known solution, and ψ_i is a solution of (2.8) for the known q's. By direct substitution, one may verify that if (2.8), (2.9) and (3.1b) is true, then $\tilde{q}_j$ will satisfy (2.9), being therefore a new solution. From (2.8), one may easily show that (3.1b) is integrable.

To see what kind of solutions we can generate with (3.1), just set $q_i = 0$, whence it follows from (2.8) that $\psi_j = g_j(\chi_j)$ and

$$\tilde{q}_j = \frac{1}{D} g_i^*(\chi_i) g_k(\chi_k) \quad , \tag{3.2a}$$

where

$$D = 1 + \gamma_1 G_1 + \gamma_2 G_2 + \gamma_3 G_3 \quad , \tag{3.2b}$$

$$G_i(\chi_i) = \int_{\chi_i}^{\infty} g_i^* g_i(u)\, du \quad . \tag{3.3}$$

The solution (3.2) is called a "1-lump" solution.

A trivial 1-lump solution is when one of the g's is chosen to be zero. We take $g_1 = 0$, then it follows that $q_2 = 0 = q_3$, while

$$q_1 = \frac{g_3^* g_2}{1+\gamma_2 G_2+\gamma_3 G_3} , \qquad (3.4)$$

where in (3.4), g_i is a function only of χ_i. Now, (3.4) is not a soliton solution as we know from 1D-IST. First, we note that its shape and amplitude is quite arbitrary. Except for the denominator, it can be any product of any function of the other two characteristic coordinates. Note that it is independent of χ_1, which must be so, for a free envelope to exist. Thus in characteristic coordinate space, (3.4) describes a "tube" parallel to the χ_1-axis, when g_2 and g_3 are localized.

Now, these N-lump solutions can be constructed from a Bäcklund transformation, as we have seen here for the 1-lump solution. But, in 1D-IST, the Bäcklund transformation is a well-known and well-used procedure for generating 1D N-soliton solutions. So, one could say that these lump solutions are the extension of 1D solitons, and at the same time one could say that they are not. The choice that one takes will depend on one's point of view. But, the main point to be recognized is that these lump solutions <u>are similar to and also are different from solitons</u>. Thus they are indeed a different quantity, and to emphasize this, we have chosen to use the word "lump" to describe them.

Now, the general 1-lump solution, (3.2) although simple, is found to contain in it a breadth of information and examples, which is remarkable. With just this one solution, we can generate and exhibit such effects as pulse decay, upconversion, and explosive instabilities, all with one simple mathematical solution [Kaup, 1981]. The detailed analysis of (3.2) is carried out in Kaup [1981], so what I shall do here is to present the results, and hopefully enough of the mathematics to convince you that these results are true and reasonable. The first thing which must be done is to parameterize the solutions (3.2) in some manner. We choose to use the standard concept of "action" and define the initial and final action (a) of an envelope by

$$a_j = \int_{-\infty}^{\infty} d\chi_i \int_{-\infty}^{\infty} d\chi_k \; q_j^* q_j \quad . \qquad (3.5)$$

The relation that the above integral has to the corresponding integral in space-time, is discussed in Kaup [1981]. Here, we shall simply comment that for the initial and final pulses (but not necessarily for in-between times) the above integral does reduce to the usual definition of action.

Now, with the definition of initial and final actions by (3.5), one then finds that the specification of the three initial actions either i) uniquely determine all constants in (3.2), except for trivial phase factors, or ii) no solution of (3.2) exists for those initial actions. In the latter case, usually the 1-lump solution was just not sufficiently general to represent such a solution. So, there are some interaction regimes which are inaccessible to the 1-lump solution. But at the same time, we find that a very large interaction regime is accessible to this solution.

Of course, if one can determine an initial solution from the three initial actions, it follows that the final actions can be given as functions of the initial actions. This is the manner in which we shall discuss our results. We input a certain amount of action into each of the three "channels", or initial envelopes. Note that we say nothing about the shape of the profile. All we do is to specify a global quantity, the "action". Then after the interaction, the initial action will be redistributed among the three envelopes. We shall usually graph one of these final actions vs. two of the initial actions.

The results that one finds are as follows. For the positive energy case, one of the γ's must differ in sign from the other two, which we shall choose to be $\gamma_1 = -\gamma_2 = -\gamma_3$, where ω_1 is then the largest frequency. The decay case is one of the possible solutions in this case, and is pictorially represented in Fig. 3 as a function of space and time. The large bottom cylindrically shaped object corresponds to an initially intense high frequency envelope, and the lower dashed lines correspond to small amounts of the other waves, on a collision course with the high frequency envelope. These three waves collide and interact in the middle of the graph, and during the interaction, action is lost by the high frequency envelope (it decays in strength), and reappears in the low frequency waves. Thus, by this interaction, high frequency waves can decay into lower frequency waves, with the exact amount of loss of action by the high frequency envelope depending on the initial actions. In Fig. 4, I have graphed the final action of the high frequency envelope vs. the initial actions for $a_{20} = a_{30}$. One should note that, for a_{20} fixed, as a_{10} is increased from zero, at first nothing happens since $a_{1f} = a_{10}$. But soon, a critical threshold is reached after which a_{1f} remains at some constant value, with all action beyond this value being lost into the two lower frequency daughter waves.

The other positive energy solution is just the time reversal of the above decay interaction. In this case, simply reverse the direction of time in Fig. 3, and you will see two low frequency waves coming together on a collision course (also with a small amount of the high-frequency wave present), which then interact, surrendering action to the high frequency wave, which emerges

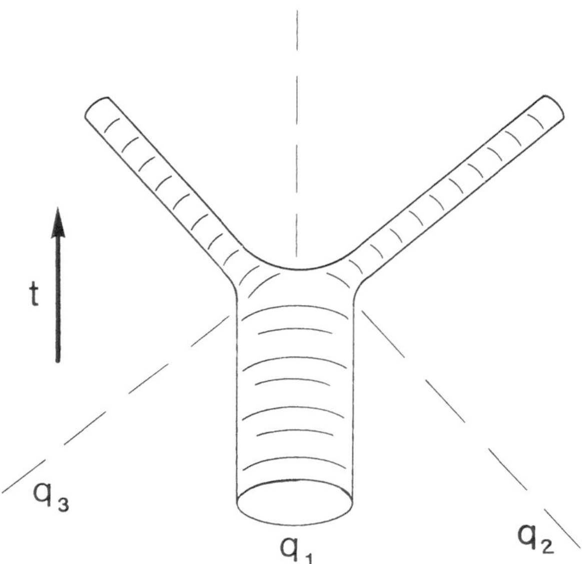

Fig. 3. The decay solution of the 3D-3WRI 1-lump solution. As shown, q_1 is the high frequency envelope, and is very intense. In the center region, a very small amount of q_2 and q_3 collides with q_1, causing a decay of q_1 into q_2 and q_3.

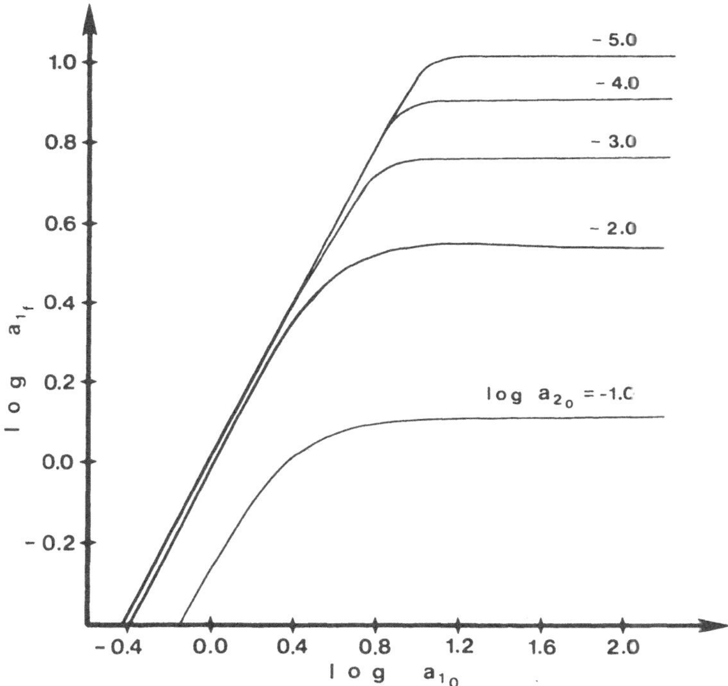

Fig. 4. The decay case. An initially intense high frequency pump of action a_{10} will decay to a final action a_{1f}, depending on the initial actions ($a_{20} = a_{30}$) in the daughter waves.

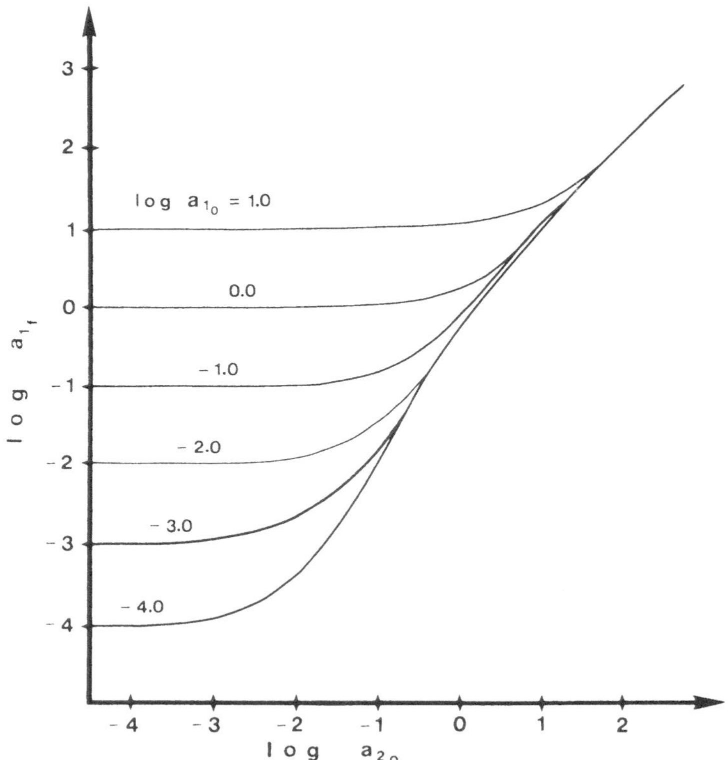

Fig. 5. The upconversion case. Two interacting daughter waves of action $a_{20} = a_{30}$ interact and pump energy into the high frequency wave, creating a pump wave of final action a_{1f}.

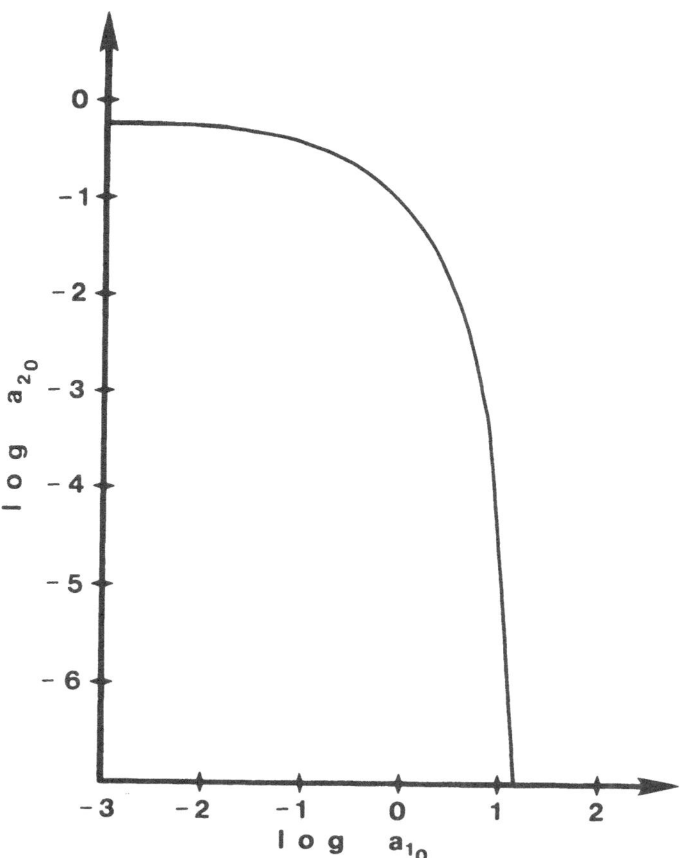

Fig. 6. The negative energy case for $a_{20} = a_{30}$. When the initial actions lie to the right and above the curve, the solution will always become singular in a finite time. All solutions under the curve are stable.

enhanced. This process is called upconversion, whereby energy passes from low frequency waves up into a higher frequency wave. The general results in this case for the 1-lump solution is demonstrated in Fig. 5. Here one should note again the threshold phenomena, whereby for fixed a_{10}, a_{1f} does not change until a_{20} passes a critical value, beyond which a_{1f} increases linearly with a_{20}.

The last case of a 1-lump solution is what is known as an explosive instability, which involves negative energy waves. In this case, $\gamma_1 = \gamma_2 = \gamma_3 = -1$, and as one can readily see, the denominator in (3.2a) may now become zero. When this happens, the solution becomes singular in a finite time, with each envelope having a singularity. The range of initial actions, for $a_{20} = a_{30}$, in which this instability occurs is shown in Fig. 6. Briefly, Fig. 6 simply shows that if a negative energy solution is to remain nonsingular, one must keep the initial actions within certain limits.

Of course, one is not assured that these one-lump solutions are typical of 3D interactions since the solution is, in a sense, quite special. However, comparing these solutions with 1D solutions, with the linear limit, and with what one expects to occur in 3D, one is lead to strongly suspect that these 1-lump solutions do indeed at least qualitatively, represent the typical 3D-3WRI. However, at the moment, this is only a conjecture, which shall have to wait for further results for verification.

REFERENCES

Ablowitz, M.J., Kaup, D.J., Newell, A.C., and Segur, H., 1974, The inverse scattering transform - Fourier analysis for nonlinear problems, Stud. Appl. Math., 53:249.

Ablowitz, M.J., and Haberman, R., 1975, Resonantly coupled nonlinear evolution equations, J. Math. Phys., 16:2301.

Armstrong, J.A., Bloembergen, N., Ducuing, J., and Pershan, P.S., 1962, Interactions between light waves in a nonlinear dielectric Phys. Rev., 127:1918.

Cornille, H., 1979, Solutions of the nonlinear 3-wave equations in three spatial dimensions, J. Math. Phys., 20:1653.

Corones, J., 1976, Solitons and simple pseudopotentials, J. Math. Phys., 17:756.

Kaup, D.J., 1977, Coherent pulse propagation: a comparison of the complete solution with the McCall-Hahn theory and others, Phys. Rev., A16:704.

Kaup, D.J., and Newell, A.C., 1978a, Theory of nonlinear oscillating dipolar excitations in one-dimensional condensates, Phys. Rev., B18:5162.

Kaup, D.J., and Newell, A.C., 1978b, Solitons as particles, oscillators, and in slowly changing media: a singular perturbation theory, Proc. Roy. Soc. London, A361:413.

Kaup, D.J., 1978, Simple harmonic generation: an exact method of solution, Stud. Appl. Math., 59:25.

Kaup, D.J., Rieman, A., and Bers, A., 1979, Space-time evolution of nonlinear three-wave interactions. I. Interaction in a homogeneous medium, Rev. Mod. Phys., 51:275.

Kaup, D.J., 1979, to appear in the Proceedings of the Soviet-American Soliton Symposium, Kiev, USSR, 2-15 September 1979, (Physica D, 1981).

Kaup, D.J., 1980a, A method for solving the separable initial-value problem of the full three-dimensional three-wave interaction, Stud. Appl. Math., 62:75.

Kaup, D.J., 1980b, The inverse scattering solution for the full three-dimensional three-wave resonant interaction, Physica D, 1:45.

Kaup, D.J., 1980c, Determining the final profiles from the initial profiles for the full three-dimensional three-wave resonant interaction, in: "Mathematical Methods and Applications of Scattering Theory," DeSanto, Sáenz and Zachary, eds., Lecture Notes in Physics 130, Springer-Verlag, New York.

Kaup, D.J., 1981, to appear, The Backlünd transformation and N-lump solutions of the three-dimensional three-wave resonant interaction, J. Math. Phys.

Zakharov, V.E., and Manakov, S.V., 1973, Resonant interaction of wave packets in nonlinear media, ZhETF Pis. Red., 18:413. [JETP Letts., 18:243.]

Zakharov, V.E., and Shabat, A.B., 1974, A scheme for integrating the nonlinear equations of mathematical physics by the method of inverse scattering problem. I., Funkts. Anal. Prilozh., 8:43.

Zakharov, V.E., 1976, Exact solutions to the problem of the parametric interaction of three-dimensional wave packets, Dokl. Akad. Nauk. SSSR, 228:1314. [Sov. Phys. Dokl., 21:322.]

CONTOUR DYNAMICS: A BOUNDARY INTEGRAL EVOLUTIONARY METHOD FOR INVISCID INCOMPRESSIBLE FLOWS*

Norman J. Zabusky

Department of Mathematics and Statistics
University of Pittsburgh
Pittsburgh, Pennsylvania 15260

* The work was accomplished in collaboration with Mr. M. Landau (stationary V-states), Professor E.A. Overman, II (regularization and ionospheric plasma cloud evolution), and Professor A. Schwartz (V-state interactions). For the plasma cloud work, we acknowledge helpful conversations with S.L. Ossakow, B.E. McDonald, and I.B. Bernstein. The work was supported by the Office of Naval Research under contracts N0014-77-C-0520, Task No. 062-583 (Euler equations), and N00013-78-C-0074 (plasma cloud). This paper was typed while the author visited the Laboratory of Applied Mathematical Physics at the Technical University of Denmark, Lyngby, Denmark.

TABLE OF CONTENTS

CONTOUR DYNAMICS: A BOUNDARY INTEGRAL EVOLUTIONARY METHOD FOR INVISCID INCOMPRESSIBLE FLOWS

Norman J. Zabusky

Department of Mathematics and Statistics
University of Pittsburgh
Pittsburgh, Pennsylvania 15260

I. INTRODUCTION

The method of contour dynamics, a generalization of the water-bag model, is ideally suited to study the evolution of incompressible and nondissipative fluids in two dimensions. For example, the method is applicable to the Euler equations, in homogeneous and stratified media and the equations for a "deformable" dielectric (or an ionospheric plasma cloud). In essence the "sources" of motion are singular points and/or piecewise-constant regions of "density" whose boundaries are advected with the local flow velocity. This velocity is derived from a stream-function that is obtained by solving an elliptic equation. In all cases to date the inviscid flows are area-preserving mappings of the regions.

Longuet-Higgins and Cokelet [1976, 1978] have studied incompressible shallow and deep water waves on boundaries where the mass density is piecewise-constant. Strong nonlinear waves and plunging breakers have been investigated. Baker, Meiron and Orszag [1980] have investigated the Rayleigh-Taylor problem of a heavy fluid over a light fluid. Zabusky and co-workers [Zabusky et al, 1979; Deem and Zabusky, 1979; see also Deem and Zabusky, 1978] have investigated the Euler equations with piecewise-constant finite-area-vortex-regions (FAVR's). Finally, Overman and Zabusky [1980] have developed a new algorithm and validated it against a linear stability analysis for the evolution of piecewise-constant deformable dielectrics (or ionospheric plasma ion density regions).

In Sec. II we review recent progress with the Euler equations, including examples of a variety of new singly and multiply connected vortex stationary states, or "V-states"; the merger of two isolated vortex domains and the interaction of translating dipolar "V-states". In Sec. III we discuss regularization procedures that will allow long-time calculations. In Sec. IV we present a nonlinear algorithm for a deformable dielectric or plasma cloud. We validate its performance against a linear stability analysis.

II. EULER EQUATIONS

A. Equations of Motion

The Euler equations in two space dimensions can be written in vorticity - stream function form as

$$\omega_t + u\omega_x + v\omega_y \equiv d_t\omega = 0 , \tag{2.1a}$$

$$\Delta\psi = -\omega , \tag{2.1b}$$

where

$$u = \psi_y \text{ and } v = -\psi_x , \tag{2.1c}$$

where (2.1a) is the inviscid advection equation which indicates that all vortex points or constant vorticity contours are advected with the flow and (2.1b) is the Poisson equation. There are no near boundaries and we consider here the sources of the flow as point (singular) vortices, and piecewise-constant FAVR's all interacting in a self-consistent manner.

The deformable FAVR may be considered a desingularized source of flow with internal degrees of freedom resulting from waves and deformations on the bounding contour. One could also include sheets of vorticity as sources of the flow.

For FAVR's in an infinite domain

$$\omega(x,y,t) = \sum_{i=1}^{N} \omega_i \chi_i(x,y,t) \tag{2.2}$$

where χ_i are a set of characteristic functions in the plane with boundary ∂D_i and ω_i is the strength. Thus the stream function of the flow

$$\psi(x,y,t) = -(2\pi)^{-1} \sum_{i=1}^{N} \iint_{IR^2} \omega_i \chi_i \, G(x-\xi,y-\eta)d\xi \, d\eta , \tag{2.3}$$

where

$$G = \log|\zeta - z| = \tfrac{1}{2}\log r^2 = \tfrac{1}{2}\log\left[(x-\xi)^2 + (y-\eta)^2\right] , \quad (2.4)$$

is the two dimensional Green's function for flow in an unbounded domain and $\zeta = \xi + i\eta$ and $z = x + iy$. Since

$$u + iv = -(i/2)\overline{(\partial_z \psi)} , \quad (2.5)$$

then using Green's theorem in the form

$$\int_{D_i} (\partial_z f)\, d\xi\, d\eta = i \oint_{\partial D_i} f(d\xi - i d\eta) , \quad (2.6)$$

we obtain

$$(u + iv) = \sum_i (\omega_i/2\pi) \oint_{\partial D_i} f(d\xi - i d\eta) \quad (2.7)$$

where $\partial_z = (\tfrac{1}{2})(\partial_x - i\partial_y)$. Thus we have reduced the dimensionality by one. That is, the problem is now reduced to the interaction of contours or "strings" that bound regions of constant area.

For the present numerical algorithm, we discretize each contour ∂D_i with M_i nodes and assume that they are joined by <u>linear</u> segments. Thus, the contour integral in (2.6) is represented by a sum over segments, where the integration over segment j, namely $(\Delta u_{\ell j}, \Delta v_{\ell j})$, can be obtained exactly [Zabusky et al, 1979]. Thus, we have

$$2\left(\sum_i M_i\right)$$

ordinary differential equations

$$(\dot{x}_\ell, \dot{y}_\ell) = \left(\sum_{i=1}^{N} \sum_{j=1}^{M_i} (\Delta u_{\ell j}) , \sum_{i=1}^{N} \sum_{j=1}^{M_i} (\Delta v_{\ell j})\right) , \quad (2.8)$$

which we solve by a two-step predictor-corrector or a leap-frog algorithm.

B. Stationary Vortex States ("V-States") of the Euler Equations

If we add a small harmonic perturbation to a single circular FAVR of radius a and density ω_1, we obtain a unidirectional wave on the surface [Lamb, 1932]. That is, if

$$r = a[1 + \varepsilon \cos(m\theta - \Omega_m t)] , \tag{2.9}$$

then the linear dispersion relation is

$$\Omega_m = (\omega_1/2)(m - 1)/m . \tag{2.10}$$

In the late-nineteenth century, Kirchoff found that an elliptical FAVR is stationary in a frame of reference rotating with angular velocity $\Omega^{(2)} = \omega_1 ab/(a + b)^2$, where a and b are the semi-major and semi-minor axes respectively. Note

$$\Omega^{(2)}\Big|_{a=b} = \Omega_2 ,$$

and one may ask if there exists states of higher symmetry associated with the bifurcation parameter $\Omega^{(m)}$ which approaches Ω_m as the states become circular.

Deem and I [1979] found several examples of rotating and translating stationary solutions that we called "V-states". The set of uniformly rotating V-states $V(m,\Omega^{(m)})$, are single FAVR's of constant vorticity density, having m-fold symmetry and angular velocity $\Omega^{(m)}$. These states may also be characterized by the ratio of the minimum-to-maximum radius, α. The set of uniformly translating V-states are "desingularized" representations of two oppositely-signed point vortices and are the two-dimensional analog of axisymmetric vortex rings in three dimensions.

The equations for the "free boundaries" are obtained from the boundary relation for normal components of velocity

$$\underset{\sim}{n} \cdot \underset{\sim}{v}_{particle} = \underset{\sim}{n} \cdot \underset{\sim}{v}_{boundary} . \tag{2.11}$$

For a single rotating state this is

$$\partial_s \psi + \Omega r(dr/ds) = 0 , \quad (x,y) \in \partial D . \tag{2.12}$$

Integrating we obtain the nonlinear intergrodifferential equation

$$\psi + (\tfrac{1}{2})\Omega r^2 = c , \quad (x,y) \in \partial D . \tag{2.13}$$

Numerical results have been obtained using a first-order relaxation algorithm [Landau and Zabusky, 1981], where the boundary is parameterized by $R(\theta_j)$ at discrete nodes (θ_j) and for each j

$$R^{(n+1)} = R^{(n)} + (\Delta R)^{(n)} \tag{2.14}$$

and

$$\Delta R^{(n)} = \frac{\psi^{(n)}(R^{(n)}) + \frac{1}{2}\Omega^{(n)} R^{(n)2} - c^{(n)}}{\psi_r^{(n)} + \Omega^{(n)} R^{(n)}} , \qquad (2.15)$$

where $\psi^{(n)}$ and $\psi_r^{(n)} \equiv (\partial_r\psi)^{(n)}$ can be written as line integrals. $\Omega^{(n)}$ and $c^{(n)}$ are weakly dependent on the detailed contour and are computed at the end of an iteration cycle from (2.13) and from the fact that ψ is constant on a contour and the contour is assigned a priori the boundary points, $r = 1$ and $r = \alpha$. Numerical results for $m = 3$ and $m = 6$ are shown in Fig. 1, and many further details are given in Landau and Zabusky [1981]. The numerical evidence is strong that for each m there exists a band

$$\Omega_L^{(m)} \equiv (m-2)/(m-1) \leq \Omega^{(m)} \leq (m-1)/m \equiv \Omega_U^{(m)}, \; m > 2 \qquad (2.16)$$

where V-states exist. The upper termination point Ω_U is the circle where the band originates and the lower termination point is where we observe that the boundaries of the V-states become nonanalytic, that is they develop corners or cusps.

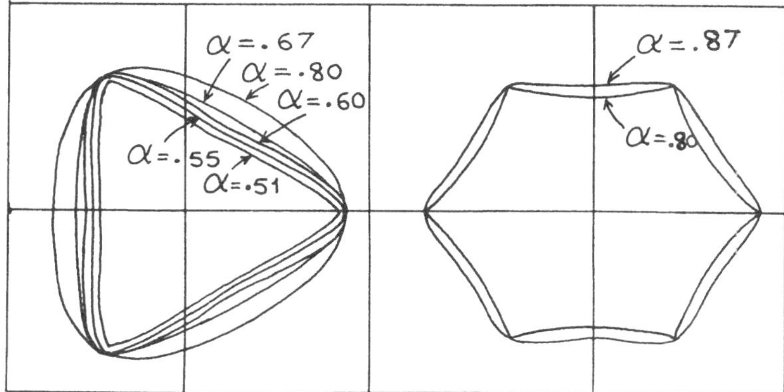

Fig. 1. Rotating singly-connected V-state $V(3,\Omega^{(3)})$ and $V(6,\Omega^{(6)})$ parameterized by their aspect ratio, α = minR/maxR.

Note also

$$\Omega_L^{(m+1)} = \Omega_U^{(m)} \ . \tag{2.17}$$

Recent analytic work by J. Burbea [to be published] and D.J. Hebert [private communication] using series expansions have validated the above results. The nature of the nonanalytic behavior of the V-states is still under investigation.

Figs. 2, 3, and 4 show doubly-connected regions which may be considered as "desingularized" versions of corresponding pair of point vortices. Fig. 2 results ($\omega_1 = \omega_2 = 1$) are for uniform rotation about the origin. Fig. 3 ($\omega_1 = -0.8$ (left) and $\omega_2 = +1.0$ (right)) are for uniform rotation about a center to the right and Fig. 4 ($\omega_1 = -1.0$ and $\omega_2 = +1.0$) results are for uniform translation U in a downward direction. The last results are obtained from the boundary relation

$$\psi(x,y) + Ux = c_i, \ (x,y) \in D_i, \quad (i = 1,2) \tag{2.18}$$

which is solved by a relaxation procedure similar to (2.14) and (2.15). The results given in Fig. 2 agree with Saffman and Szeto [1980]. The results given in Fig. 4 confirm the main body of the calculations of Pierrehumbert [1980]. The stability of these stationary states is an open problem except for the ellipse, $V(2,\Omega^{(2)})$ which was considered by Love [1893]. He showed that if $(a/b) > 3.0$, the $m = 3$ mode is unstable, and if (a/b) is increased, more of the low modes become unstable. In the next section we present numerical results for the evolution of an unstable ellipse which shows vorticity filamentation and "breaking". From this consideration it seems evident that all "elongated" stationary states are unstable,

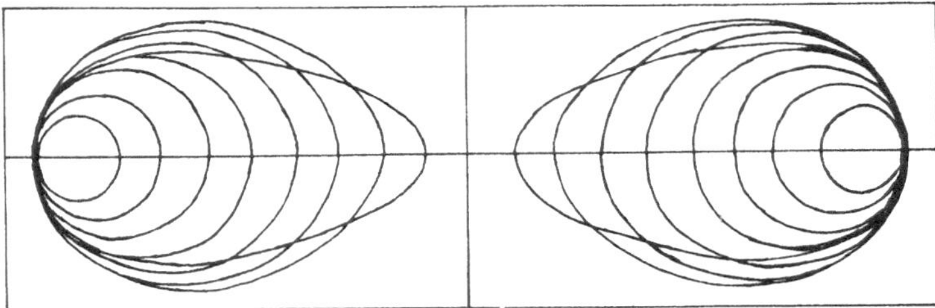

Fig. 2. Rotating doubly-connected V-states with $\omega_1 = \omega_2 = 1$. The rotation is counter clockwise with the center at the origin.

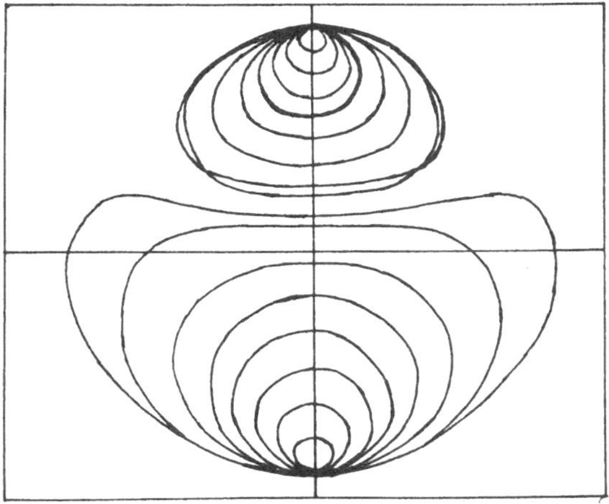

Fig. 3. Rotating doubly-connected V-states with $\omega_1 = -0.8$ (left) and $\omega_2 = 1.0$ (right). The rotation is counter-clockwise with the centre to the right.

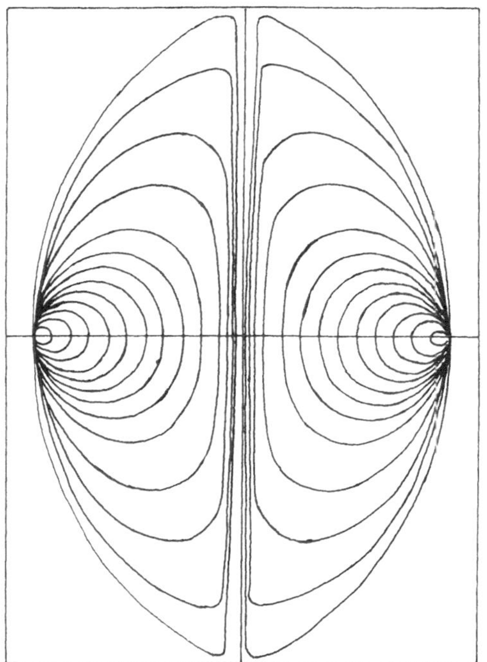

Fig. 4. Downward translating doubly-connected V-states with $-\omega_1 = \omega_2 = 1.0$.

e.g. the elongated members of the set of like-signed doubly connected V-states of Fig. 2. An appropriate perturbation will drive an instability that will lead to partial or complete merger; that is, the vortex regions will attract one another and fill the "intermediate" region. Simultaneously they may eject filamentary structures from the contour regions furthest from the center of rotation.

C. Instability, Breaking, and Interaction of FAVR's

We now apply the time-dependent contour dynamical algorithm to study dynamical processes. We will show examples of the breaking of an unstable ellipse and the merger of FAVR's and a FAVR-point vortex.

Fig. 5a shows a perturbed 3.5:1 ellipse at t = 0 represented by

$$x + iy = c \cos(\xi + i\eta)$$

$$\xi = \xi_o - \varepsilon \cos(m\eta)[(a \sin\eta)^2 + (b \cos\eta)^2], \quad 0 \leq \eta \leq 2\pi \quad (2.19)$$

$$c^2 = a^2 - b^2, \quad \xi_o = \cosh^{-1}(a/c)$$

with $\varepsilon = 0.05$, $m = 3$, $N = 90$ modes, area = 0.898, and perimenter = 4.37. The perturbation was chosen to be unsymmetrical so as to favor breaking at the right, the point of largest curvature. Figs. 5b and 5c show one large filament emerge and pinch at two points. The furthest tip shows signs of "roll-up". We used a primitive node-insertion algorithm (this item is discussed in Section III), that maintained a fixed distance between nodes in the filaments. Thus, although the results are to be taken as heuristic, the following observed features will probably persist: the vorticity in the filamentary structure quickly approaches a constant value that is dependent on (a/b) and the amplitude of the perturbation, ε, provided $\varepsilon \ll b$; the remaining contour will have maximum-to-minimum amplitude ratio < 3.0; and the vortex filament will begin a slow roll-up.

Fig. 6 shows an example of "merger" of a circular FAVR and a point vortex. The ratio of circulation is FAVR: point = 10:1. In Fig. 6a, their centers are sufficiently far apart so that the self-consistent motion is a rotation about the origin plus waves excited in the FAVR contour. As they are moved closer together, these waves become larger and finally "break" and are drawn around the point vortex. Fig. 7 shos the merger of two like-signed FAVR's with a 10:1 circulation ratio. Fig. 8 shows a comparable run with a point vortex replacing the smaller FAVR, but having the same circulation. Although these runs were made with an early code without a node-insertion process, we expect that the qualitative

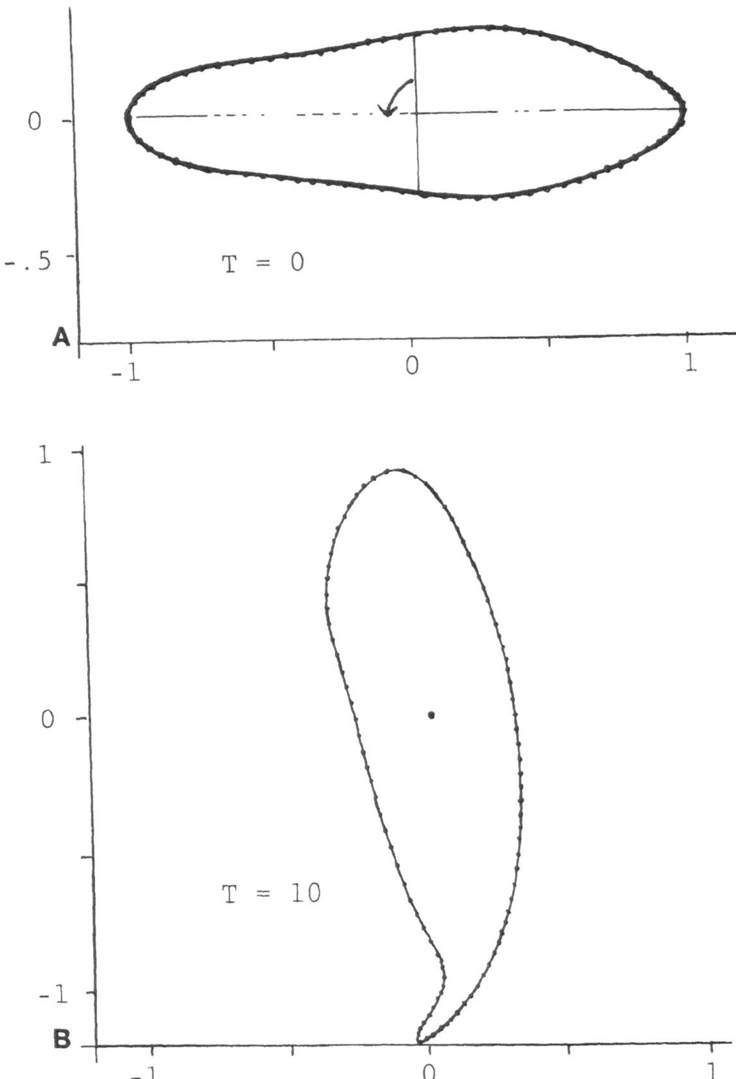

Fig. 5.(a,b). Evolution of a 3.5:1 ellipse with ω_1 = 1.0 and initial area 0.898, (a) t = 0; (b) t = 10.0.

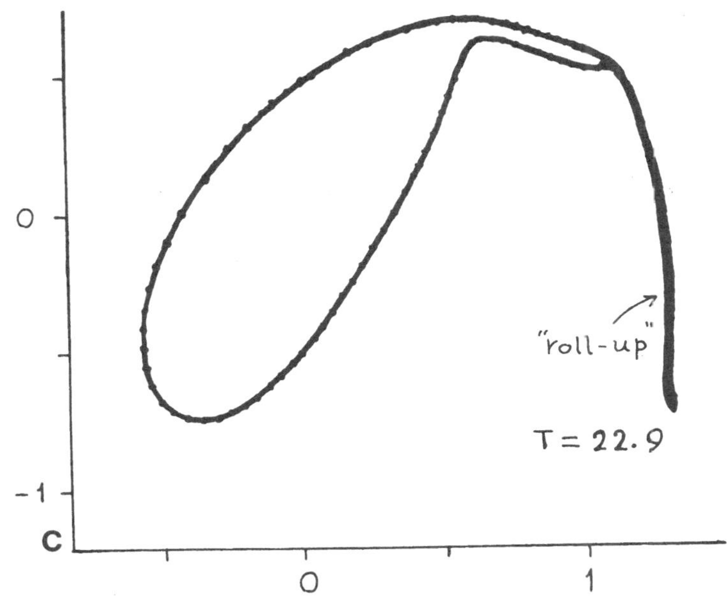

Fig. 5.(c). Evolution of a 3.5:1 ellipse with ω_1 = 1.0 and initial area 0.898, (c) t = 22.9.

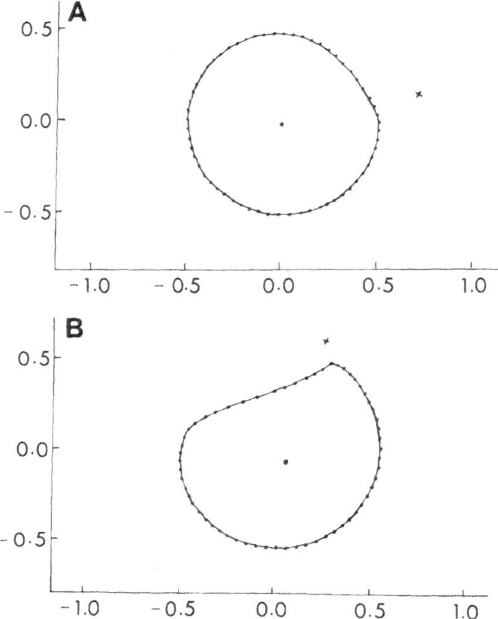

Fig. 6. Evolution of a FAVR and a point vortex with circulation ratio 10:1. (a) t = 1.0; (b) t = 5.0.

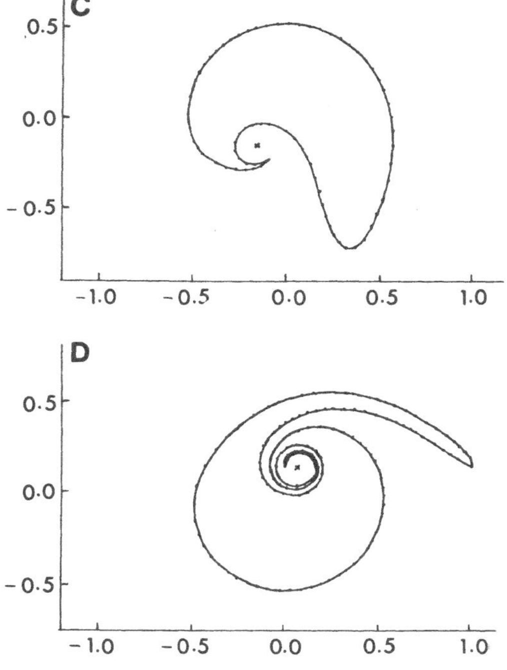

Fig. 6. (c) t = 12.0; (d) t = 19.0.

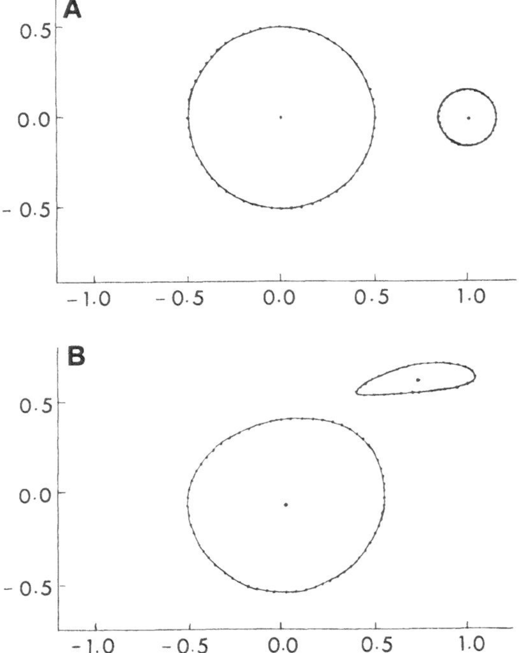

Fig. 7. Evolution of two merging FAVR's with circulation ratio 10:1. (a) t = 0; (b) t = 5.3.

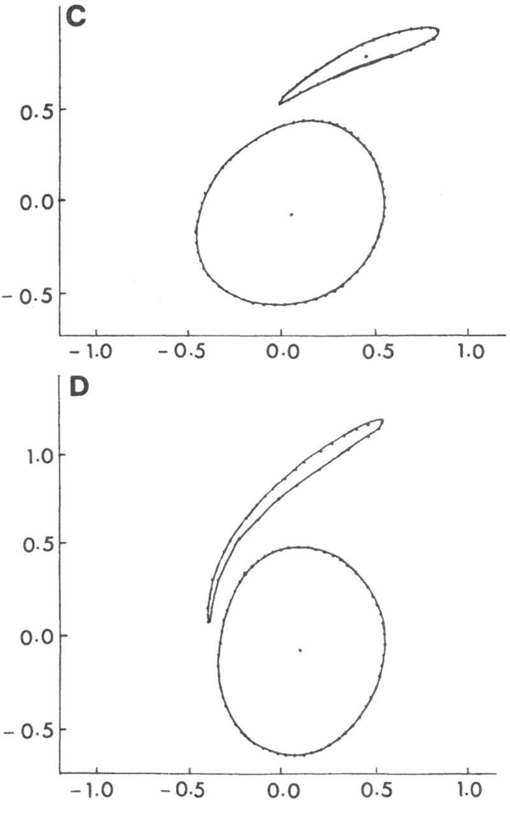

Fig. 7. (c) t = 8.0; (d) t = 11.0.

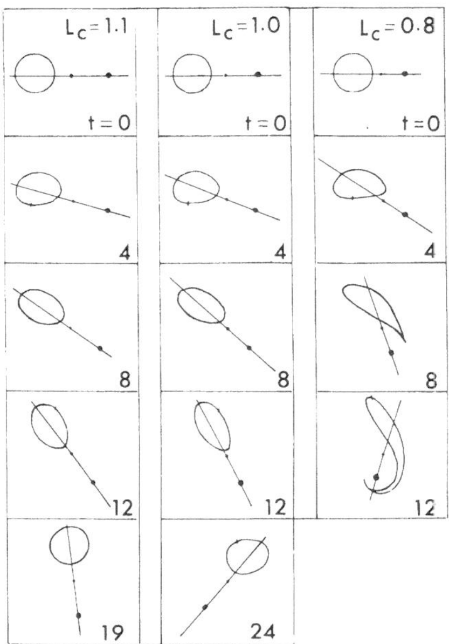

Fig. 8. Evolution of a FAVR and a point vortex, each with the same circulation. The light line joining their centres exhibits the gross rotation of the pair. L_c denotes the initial centre separation.

features will persist. That is, in the former case, Fig. 7, merger means "wrap-around", whereas in the latter, Fig. 8, it means "entrainment" and "strong winding". This difference in behavior arises because the velocities become singular in the vicinity of point vortices, and because they do not possess internal degrees of freedom. This strongly suggests the point-vortex algorithms are inadequate for determining fine-scale motion if there are many close encounters. In fact, a proper data summary of simulations with point-vortex algorithms should indicate the number of close encounters as defined by the total time interval where a particle exceeds a prescribed initial velocity (e.g. typical of a local sound speed).

Fig. 9 shows the head-on interaction of a travelling V-state moving to the right and a pair of point vortices moving to the left, both with the same axis of symmetry, $y = 0$. (Only the upper-half plane is displayed.) The parameters are chosen so that Fig. 9a shows a weak interaction, where the FAVR suffers a phase shift and its perimeter deviation

$$([P(t) - P(0)]/P(0), \; (P(t) = \oint_{\partial D} ds))$$

shows an oscillatory behavior after the interaction, i.e. $t > 8.0$. The parameters of Fig. 9b are chosen so that a strong interaction takes place, and the vortex breaks and ejects a filament.

III. REGULARIZATION OF CONTOUR DYNAMICAL ALGORITHMS

Figs. 5 and 9b shows examples of motion which indicate that long-time calculations will require improvements, including: node-insertion and removal; topology-alteration, e.g. "pinching" or the creation of multiply-connected regions and "merging" of regions with common boundaries, and finally regularization or smoothing of contours to avoid excessive perimeter growth and the development of corners or cusps. The last is very important in highly unstable problems like the deformable dielectric where growth rates are proportional to the wave number of the perturbation.

Two types of regions require the addition of nodes to avoid the growth of local truncation errors. First, when the local curvature $\kappa \equiv d\theta/ds$ (where $\tan\theta \equiv dy/dx$) is large, we insert nodes so that their density is proportional to κ or

$$(\Delta s_k) \;=\; c\,/\max(|\kappa|) \;. \qquad (3.1)$$

where the following constraints are satisfied

$$(\Delta s)_{max} \geq (\Delta s_k) \geq (\Delta s)_{min}$$

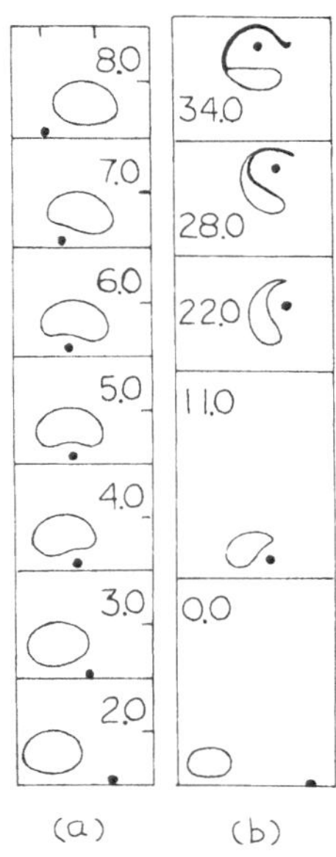

Fig. 9. Head-on (approaching) and strong head-on interaction of a translating V-state with a pair of point vortices. (a) weak interaction; (b) strong interaction.

and

$$(1 - r)(\Delta s_{k-1}) \leq (\Delta s_k) \leq (1 + r)(\Delta s_{k-1}), \quad (r \leq 0.3) \;.$$

Second, when non-adjacent nodes approach closer than adjacent nodes, as can happen when filaments evolve. Here, one examines

$$\kappa_{na} = d\theta_{na}/ds \tag{3.2}$$

(where $\tan\theta_{na} = (\eta - y)/(\xi - x)$), that is the rate of change of the line joining two non-adjacent nodes and inserts nodes in both regions proportional to κ_{na}. The first algorithm has been implemented, and the second is under development. If the distance between non-adjacent nodes falls below a prescribed amount, one must cut-and-rejoin the contour, i.e. create two singly connected regions. Some work in this direction has been described by Berk and Roberts [1970]. Also, if contours bounding different regions of vorticity approach, as they do in Fig. 7d, one must delete the common boundary.

To avoid the excessive growth of perimeter and the tendency toward the formation of corners or cusps one introduces a physically-motivated smoothing or regularization [Marchuk, 1975]. If the contour is parameterized by σ, then

$$s_\sigma^2 = x_\sigma^2 + y_\sigma^2 \quad \text{and} \quad \tan\theta = y_\sigma/x_\sigma \;, \tag{3.3}$$

and we find that we can write the evolution

$$(\dot{s}_\sigma, \dot{\theta}_\sigma) = (\hat{\dot{s}}_\sigma, \hat{\dot{\theta}}_\sigma) + \mu(\partial_s[\kappa], -s_\sigma[\kappa^2]), \; (x,y) \in \partial D \;, \tag{3.4}$$

or in Cartesian form

$$(\dot{x}, \dot{y}) = (\hat{\dot{x}}, \hat{\dot{y}}) + \mu(x_{ss}, y_{ss}), \; (x,y) \in \partial D \;, \tag{3.5}$$

where the circumflexed quantities are obtained from the advective part of the mapping and may involve first derivatives with respect to s. The second partial derivatives with respect to s indicate that this is a fully nonlinear algorithm, and the resulting partial differential equations must be solved by implicit methods.

This is one of many possible algorithms and is consistent to first order with the inclusion of a linear viscosity, when one examines the evolution of small-scale structures. Other nonlinear viscosities could be obtained by inserting terms like κ^{2P} under the bracket of (3.4). Dispersive smoothing is accomplished by increasing the order of differentiation by one of the bracketed terms [Zabusky and Overman, manuscript in preparation].

With (3.4) or (3.5) one finds that the area and perimeter obey

$$A_t = \hat{A}_t - \mu t \text{ and } P_t = \hat{P}_t - \mu\oint\kappa^2 \, ds \tag{3.6}$$

where $\hat{A}_t$ and $\hat{P}_t$ are the area and perimeter change respectively due to the advective terms $(\hat{\dot{x}},\hat{\dot{y}})$. The decrease in area in (3.6) contradicts the physically expected slow increase one expects in "diffusive" spreading. To obtain global spreading we are presently implementing a large-scale regularization procedure for increasing A and decreasing the magnitude $\Omega = \Omega(t)$ of piecewise-constant functions. These are based on the integral expressions

$$\partial_t \iint_{|R^2} \omega \, da = 0 \quad \Rightarrow \quad d_t(\Omega A) = 0 \,, \tag{3.7}$$

$$\partial_t \iint_{|R^2} \omega^2 \, da = -2\nu \iint_{|R^2} |\underset{\sim}{\nabla}\omega|^2 \, da \quad \Rightarrow \quad d_t(\Omega^2 A) = -\nu\,\Omega^2\, P/[(2\pi)^{\frac{1}{2}}(\delta_o^2 + 4\nu t)^{\frac{1}{2}}] \,, \tag{3.8}$$

where P is the instantaneous perimeter and δ_o is a length scale longer than any wavelength on the contour.

IV. INSTABILITY AND NONLINEAR EVOLUTION OF A PLASMA CLOUD (DEFORMABLE DIELECTRIC) VIA REGULARIZED CONTOUR DYNAMICS

Ionospheric, collision-dominated, low-β plasma clouds, driven by ambient, uniform electric fields or winds are being studied for two main reasons. First, they are probes of properties of the natural and disturbed ionosphere. Second, they evolve nonlinearly and yield magnetic field-aligned fine-scale structures ("irregularities" or "striations") that can degrade radio wave propagation [Ossakow, 1979].

Past analytic studies of the evolution of ion densities have been almost entirely confined to the linear stability of one-dimensional (1D) stationary states. Linson and Workman [1970], Shiau and Simon [1972], and Völk and Haerendel [1971] attributed the cause of striations to the $\underset{\sim}{E}\times\underset{\sim}{B}$ "gradient-drift" instability [Simon, 1963]. Presently there are no 2D stationary states of continuous density variation and no 2D stability analyses exist.

In recent computational studies, finite-difference algorithms were used to study the evolution of 1D and 2D ion density clouds with small-amplitude 2D perturbations [Zabusky et al, 1973; Doles et al, 1976; Scannapieco et al, 1974; Scannapieco et al, 1976]. It has been found that the growth of sinusoidal perturbations on 1D clouds agrees with linear theories [Zabusky et al, 1973; Doles et al, 1976]. These perturbations evolve into finger-like striations that emanate from the cloud's "backside" (the direction opposite to the drift velocity $\underset{\sim}{E}\mathrm{x}\underset{\sim}{B}/|\underset{\sim}{B}|^2$). The results obtained are in qualitative agreement with field experiments [Rosenberg, 1971; Davis et al, 1974; Baker and Ulwick, 1978].

We now investigate the linear stability of a two-dimensional model of an idealized cloud, namely a piecewise-constant distribution of ions, N,

$$N(x,y,t) = \begin{cases} N_- & \text{for } (x,y) \in D \\ N_+ & \text{for } (x,y) \notin D \end{cases} , \tag{4.1}$$

where D is a simply-connected, bounded region in R^2 with boundary Γ. The contour, Γ, deforms with a velocity $\underset{\sim}{V}_d = \underset{\sim}{E}_- \mathrm{x}\underset{\sim}{B}/|\underset{\sim}{B}|^2$, where $\underset{\sim}{E}_-$ is the self-consistent electric field on the "inside" of Γ and $\underset{\sim}{B} = B_0\underset{\sim}{e}_z$ is the earth's magnetic field, assumed to be constant. In the rest of this paper we set $B_0 = 1$. Our contributions are two-fold: (1) We introduce a contour dynamical model of the piecewise-constant cloud which generalizes the "waterbag" method. The evolution equations include a physically motivated diffusive regularization procedure [Overman and Zabusky, submitted for publication] which inhibits the formation of contour singularities and makes the system well-posed. (2) We analyze the linear stability of a circular region and demonstrate a new linear phenomenon, "downward-cascade", namely the wavenumber of maximum amplitude decreases with time. Our work is a combined analytical-numerical study. We also use our results to validate a numerical algorithm which solves the nonlinear contour evolution model.

The equations of motion of the continuum ionospheric plasma system of Fig. 10 (inset) have been given as [Zabusky et al, 1973; Doles et al, 1976]

$$\underset{\sim}{\nabla} \cdot (N\underset{\sim}{\nabla}\Phi) = 0 , \tag{4.2}$$

$$\partial_t N + \underset{\sim}{v} \cdot \underset{\sim}{\nabla} N = \nu\nabla^2 N , \tag{4.3}$$

$$\underset{\sim}{v} = -\underset{\sim}{e}_x \partial_y \Phi + \underset{\sim}{e}_y \partial_x \Phi , \tag{4.4}$$

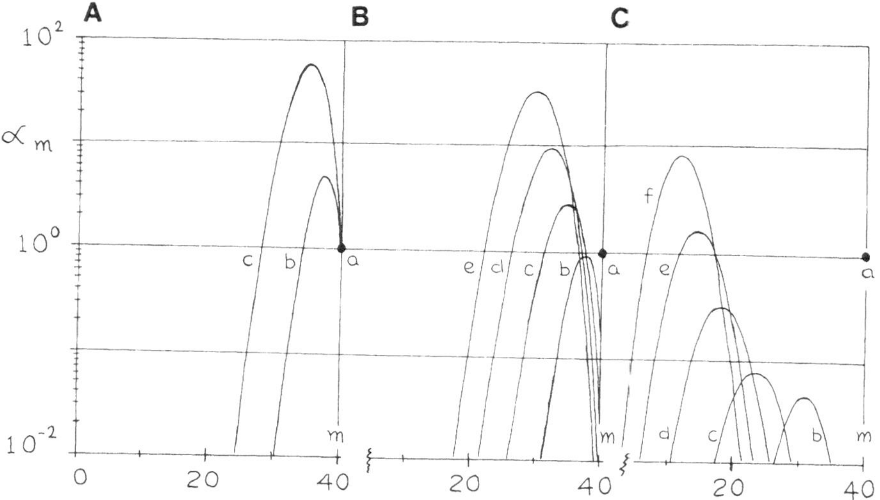

Fig. 10 (A),(B),(C). Solutions of (4.15) ($\alpha_m(\tau)$ vs. m) with $r_o = 1$, $m_o = 40$, $\alpha_{40}(0) = 1$, $\lambda = 2$, and (A): Q = 0.0 at (a) $\tau = 0.0$, (b) $\tau = 0.08$, (c) $\tau = 0.16$; (B): Q = 0.0135 at (a) $\tau = 0.0$, (b) $\tau = 0.08$, (c) $\tau = 0.16$, (d) $\tau = 0.24$, (e) $\tau = 0.32$; (C): Q = 0.0315 at (a) $\tau = 0.0$, (b) $\tau = 0.24$, (c) $\tau = 0.48$, (d) $\tau = 0.72$, (e) $\tau = 0.96$, (f) $\tau = 1.20$.

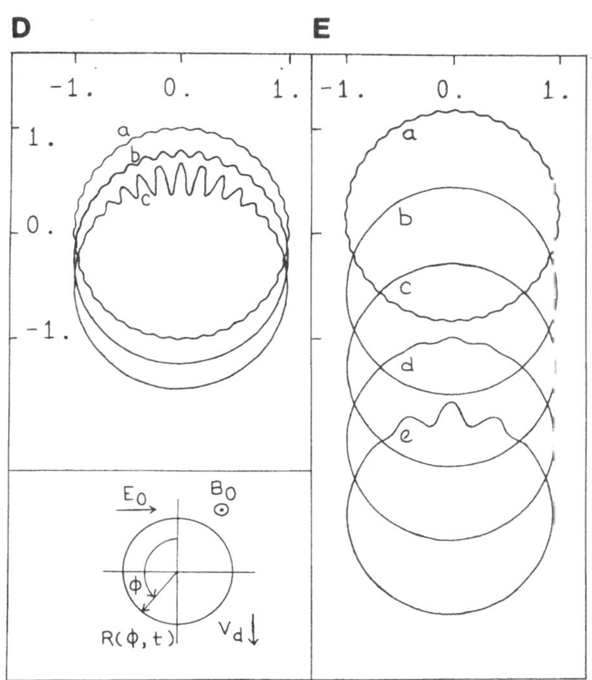

Fig. 10 (D),(E). $R(\varphi,t)$ in the laboratory frame. Solutions of (4.15) with $\varepsilon = 0.01$ and (D): at (a) $\tau = 0.0$, (b) $\tau = 0.08$, (c) $\tau = 0.16$ for conditions identical to Fig. 10 (B); (E): at (a) $\tau = 0.0$, (b) $\tau = 0.24$, (c) $\tau = 0.48$, (d) $\tau = 0.72$, (e) $\tau = 0.96$ for conditions identical to Fig. 10 (C). (Inset: Schematic of circular cloud in the (ζ,η) coordinate system drifting downward with velocity $V_d = 2E_0/(1+\lambda)$, $\lambda = N_-/N_+$.)

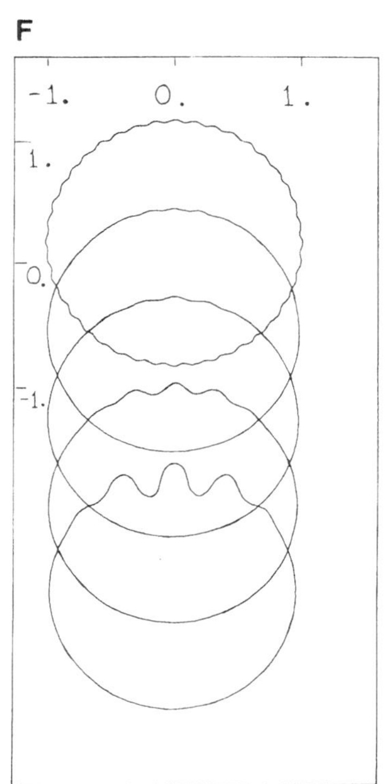

Fig. 10 (F). Solution $R(\varphi,t)$ of the nonlinear equations (4.5), (4.6), (4.7) for conditions and times identical to Fig. 10 (E).

where $\Phi \to -E_0 x$ as $|(x,y)| \to \infty$. Here ν is the dissipation parameter (typically very small), E_0 is the ambient electric field, and $\underset{\sim}{E} = -\underset{\sim}{\nabla}\Phi$, where Φ is the potential.

For piecewise-constant N, (4.1), Eq. (4.2) is equivalent to

$$\underset{\sim}{\nabla}^2\Phi_- = 0 \text{ for } (x,y) \in D, \quad \underset{\sim}{\nabla}^2\Phi_+ = 0 \text{ for } (x,y) \notin D \tag{4.5}$$

with the boundary conditions

$$\Phi_+ \to -E_0 x \text{ as } |(x,y)| \to \infty \quad , \tag{4.6a}$$

$$\Phi_+ = \Phi_- \text{ , and } \partial_n\Phi_+ = \lambda\, \partial_n\Phi_- \text{ on } \Gamma \quad , \tag{4.6b}$$

where ∂_n is the directional derivative normal to the boundary Γ and $\lambda = N_-/N_+$. These are equivalent to the equations for a dielectric in a uniform electric field. Eqs. (4.3) and (4.4) are equivalent to

$$\partial_t(x,y) = (-\partial_y\Phi_-, \partial_x\Phi_-) + \nu\partial_s^2(x,y), \text{ for } (x,y) \in \Gamma \quad , \tag{4.7}$$

where s is the arc length on Γ. If $\nu = 0$, (4.7) agrees identically with (4.3) and (4.4) for piecewise-constant distributions. For $0 < \nu << r_0^2\, \gamma$ (see (4.14)) the regularization term, $\nu\, \partial_s^2(x,y)$ in (4.7) is consistent with the dissipative term, $\nu\, \nabla^2 N$, in (4.3) for small scale structures to lowest order in ν. Thus, we propose (4.5), (4.6), and (4.7) as a short-time model of a cloud with steep sides.

We now analyze the linear stability of a circular region. It translates downward with a drift velocity $V_d = 2E_0/(\lambda+1)$ associated with the lowest order internal electric field, E_-. We translate to the (ξ,η) coordinate system, $\xi = x$ and $\eta = y + V_d t$, and then convert to polar coordinates (r,ϕ) where ϕ is the angle measured counterclockwise from the positive η-axis. We will solve (4.5), (4.6), and (4.7) on the perturbed boundary

$$\begin{aligned} R(\phi,t) &= \rho(t) + \varepsilon\rho^{(1)}(t), \\ &= \rho(t) + \varepsilon(\alpha_0^{(1)}(t) + \sum_{m=1}^{\infty} \alpha_m^{(1)}(t) \cos m\phi) \quad , \end{aligned} \tag{4.8}$$

where, for convenience, we have assumed the perturbation is symmetric about $\phi = 0$. From (4.5) and (4.6a) we have

$$\Phi_{-}(r,\phi) = A_0 + \sum_{m=1}^{\infty} r^m B_m \sin m\phi \quad , \tag{4.9}$$

$$\Phi_{+}(r,\phi) = E_0 r \sin\phi + C_0 + \sum_{m=1}^{\infty} r^{-m} D_m \sin m\phi \quad . \tag{4.10}$$

Substituting (4.9) and (4.10) into (4.6b) we obtain, to first-order,

$$B_m^{(1)} = -2[(\lambda-1)/(\lambda+1)^2]E_0\rho^{-m}\alpha_{m+1}^{(1)} \text{ , for } m \geq 1 \quad . \tag{4.11}$$

Note that the induction of a dipole field in the first-order solution causes a downshift by one between the Fourier coefficients of the potential $B_m^{(1)}$ and the boundary perturbation, $\alpha_{m+1}^{(1)}$.

To calculate the time evolution of $\alpha_m^{(1)}$ we write (4.7) in polar coordinates in the translating frame of reference

$$\begin{aligned} \partial_t R = & -R^{-1}(\partial_\phi \Phi_{-} + R'\partial_r \phi_{-}) + V_d(\cos\phi + (R'/R)\sin\phi) \\ & -\nu[R^2 + 2(R')^2 - RR'']/[R(R^2 + (R')^2)] \text{ ,} \end{aligned} \tag{4.12}$$

where $R' = \partial R/\partial\phi$. Substituting (4.8) into (4.12), we obtain, to zeroth order,

$$\partial_t \rho = -\nu\rho^{-1}$$

so

$$\rho = (r_0^2 - 2\nu t)^{\frac{1}{2}} \quad ,$$

where r_0 is the radius of the circle at $t = 0$. To first-order,

$$\partial_\tau \alpha_m = m(r_0/\rho)\alpha_{m+1} - Q(m^2 - 1)(r_0/\rho)^2\alpha_m \quad , \tag{4.13}$$

where we have suppressed superscripts and where

$$\gamma = 2r_0^{-1} E_0(\lambda-1)/(\lambda+1)^2 \text{ , } \tau = \gamma t \text{, and } Q = \nu/(r_0^2\gamma) . \tag{4.14}$$

Thus, <u>strong</u> (large λ) <u>clouds evolve more slowly and appear more dissipative</u>. We simplify (4.13) with two nonessential assumptions: First, replace ρ by r_0 since the zeroth-order radius change only weakly affects stability. Secondly, replace $m^2 - 1$ by m^2. We obtain

$$\partial_\tau \alpha_m = m\alpha_{m+1} - Qm^2\alpha_m \quad . \tag{4.15}$$

If $Q = 0$ and

$$\alpha_m(0) = \delta_{m,m_0}, \ (\text{so } R(\phi,0) = r_0(1+\varepsilon \cos m_0\phi)) \ , \tag{4.16}$$

then the solution of (4.15) is

$$\alpha_m(\tau) = \binom{m_0 - 1}{m_0 - m} \tau^{(m_0 - m)} \ , \quad 1 \le m \le m_0 \quad . \tag{4.17}$$

This has the "downward-cascade" property with the wavenumber of maximum amplitude being

$$m_{max} \approx m_0/(1+ \tau) \ , \tag{4.18}$$

at time τ.

Fig. 10A shows the graph $\alpha_m(\tau)$ vs. m of (4.17) with $m_0 = 40$, and the times indicated. We also find the "halfwidth" (width in m at half-amplitude) of (4.17) increases as

$$\Delta_{\frac{1}{2}} \approx 2(2 \ln 2)^{\frac{1}{2}} [m_{max}(m_0-m_{max})/m_0]^{\frac{1}{2}} \quad . \tag{4.19}$$

$R(\phi,\tau)$ attains its maximum value R_{max} when $\phi = 0$, or

$$R_{max} = r_0 + \varepsilon \sum_{m=1}^{m_0} \binom{m_0 - 1}{m_0 - m} \tau^{(m_0 - m)} = r_0 + \varepsilon(1+\tau)^{(m_0 - 1)}$$

$$\approx r_0 + \varepsilon\, e^{(m_0 - 1)\tau} \quad . \tag{4.20}$$

Since α_m is strongly peaked around $m_{max}(\tau)$, as indicated by (4.19), the minimum value of R, R_{min}, is attained when $\phi \approx \pi/m_{max}$, or

$$R_{min} \approx r_0 - \varepsilon\, e^{(m_0 - 1)\tau} \quad .$$

Thus, the maximum amplitude of the perturbation, $S(\tau)$, is

$$S(\tau) \approx \varepsilon \sum_{m=1}^{m_0} \alpha_m(\tau) \approx \varepsilon\, e^{(m_0 - 1)\tau} \quad .$$

Thus, the dissipationless problem is ill-posed.

For $Q > 0$, the solution of (4.15) with the initial condition (4.16) is

$$\alpha_m(\tau) = [(m_0 - 1)!/(m-1)!] \sum_{\ell=m}^{m_0} \left\{ e^{-Qk^2\tau} / \left[\prod_{\ell=k}^{m_0} Q(\ell^2 - m^2) \right] \right\} . \tag{4.21}$$

In Fig. (10.B) we show $\alpha_m(\tau)$ with $m_0 = 40$ and $Q = 0.0135$ and in Fig. (10.D) we show the corresponding $R(\phi,t)$ in the (x,y) laboratory coordinates with $r_0 = 1$ and $\varepsilon = .01$. In Figs. (10.C) and (10.E) we do the same but for $Q = 0.0315$. Note in Fig. (10.B) how α_m increases in time more slowly than in Fig. (10.A). In Fig. (10.C) we see an initial decrease followed by an increase at a much lower wavenumber.

Since (4.21) is presently rather impenetrable, we convert (4.15) to a partial differential equation in $\alpha(m,\tau)$ which is easier to analyze. We replace $\alpha_{m+1}(\tau)$ by $\alpha(m,\tau) + \partial_m\alpha(m,\tau)$ and obtain

$$\partial_\tau \alpha = m(\alpha + \partial_m \alpha) - Q\, m^2 \alpha . \tag{4.22}$$

With the initial condition $\alpha(m,0) = \exp[-(m - m_0)^2/2\ell^2]$, we find the maximum growth in amplitude $S \propto \exp[(2Q)^{-1} - 1]\tau$ when

$$m_0 = (2Q)^{-1} . \tag{4.23}$$

Fig. 10.F presents a solution of the full nonlinear equations (4.5), (4.6), and (4.7) with initial conditions as in Fig. 10.E. This was obtained with a regularized contour dynamical algorithm with a node insertion-and-removal algorithm [Overman and Zabusky, manuscript in preparation]. The close agreement between the linear and nonlinear solutions validates the numerical algorithm.

REFERENCES

Baker, K.D., and Ulwick, J.C., 1978, Measurement of electron density structure in barium clouds, Geophys. Res. Lett., 5:723.

Baker, G.R., Meiron, D.I., and Orszag, S.A., 1980, Vortex simulations of the Rayleigh-Taylor instability, Phys. Fluids, 23:1485.

Berk, H., and Roberts, K.V., 1970, The "water-bag" model, Methods in Comput. Phys., 9:87.

Burbea, J., to be published, J. Fluid Mech.

Davis, T.N., Romick, G.J., Wescott, E.M., Jeffries, R.A., Kerr, D.M., and Peek, H.M., 1974, Observations of the development of striations in large barium ion clouds, Planet. Space Sci., 22:67.

Deem, G.S., and Zabusky, N.J., 1979, Vortex waves: Stationary "V-states," - interactions, recurrence, and breaking, Phys. Rev. Lett., 40:859.

Deem, G.S. and Zabusky, N.J., 1978, Stationary V-states, introduction, recurrence, and breaking, in: "Solitons in Action," K. Lonngren and A. Scott, eds., Academic Press, N.Y.

Doles, III, J.H., Zabusky, N.J., and Perkins, F.W., 1976, Deformation and striation of plasma clouds in the ionosphere. 3. Numerical simulations of a multilevel model with recombination chemistry, J. Geophys. Res., 81:5987.

Hebert, D.J., private communication.

Lamb, H., 1932, "Hydrodynamics," 6th edition, Dover, N.Y.

Landau, M., and Zabusky, N.J., 1981, submitted for publication, Stationary solutions of the Euler equations in two dimensions. Singly doubly-connected "V-states" .

Linson, L.M., and Workman, J.B., 1970, Formation of striations in ionospheric plasma clouds, J. Geophys. Res., 75:3211.

Longuet-Higgins, M.S., and Cokelet, E.D., 1976, The deformation of steep surface waves on water. I. A numerical method of computation, Proc. Roy. Soc. London A, 350:1.

Longuet-Higgins, M.S., and Cokelet, E.D., 1978, The deformation of steep surface waves on water. II. Growth of normal mode instabilities, Proc. Roy. Soc. London A, 364:1.

Love, A.E.H., 1893, On the stability of certain vortex motions, Proc. Lond. Math. Soc. (1), 35:18.

Marchuk, G.I., 1975, "Methods of Numerical Mathematics," Springer-Verlag.

Ossakow, S.L., 1979, Ionospheric irregularities, Rev. of Geophys. and Space Phys., 17:521.

Overman, II, E.A., and Zabusky, N.J., 1980, Instability and nonlinear evolution of plasma clouds via regularized contour dynamics, Phys. Rev. Lett., 45:1693.

Overman, II, E.A., and Zabusky, N.J., submitted for publication, Regularization of contour dynamical algorithms.

Pierrehumbert, R.T., 1980, A family of steady, translating vortex pairs with distributed vorticity, J. Fluid Mech., 99:129.

Rosenberg, N.W., 1971, Observations of striation formation in a barium ion cloud, J. Geophys. Res., 76:6856.

Saffman, P.G., and Szeto, R., 1980, Equilibrium shapes of a pair of equal uniform vortices, Phys. Fluids, 23:2339.

Scannapieco, A.J., Ossakow, S.L., Book, D.L. McDonald, B.E., and Goldman, S.R., 1974, Conductivity ratio effects on the drift and deformation of F region barium clouds coupled to the E region ionosphere, J. Geophys. Res., 79:2913.

Scannapieco, A.J., Ossakow, S.L., Goldman, S.R., and Pierre, J.M., 1976, Plasma cloud late time striation spectra, J. Geophys. Res., 81:6037.

Shiau, J.N., and Simon, A., 1974, Barium cloud growth and striation in a conducting background, J. Geophys. Res., 79:1895.

Simon, A., 1963, Instability of a partially ionized plasma in crossed electric and magnetic fields, _Phys. Fluids_, 6:382.

Völk, H.J., and Haerendel, G., 1971, Striations in ionospheric ion clouds, I, _J. Geophys. Res._, 76:4541.

Zabusky, N.J., Doles, III, J.H., and Perkins, F.W., 1973, Deformation and striation of plasma clouds in the ionosphere, 2. Numerical simulation of a nonlinear two-dimensional model, _J. Geophys. Res._, 78:711.

Zabusky, N.J., Hughes, M.H., and Roberts, K.V., 1979, Contour dynamics for the Euler equations in two dimensions, _J. Comp. Phys._, 30:96.

Zabusky, N.J., and Overman, II, E.A., manuscript in preparation, Regularization of contour dynamical algorithms.

NUMERICAL COMPUTATION OF NONLINEAR WAVES*

Bengt Fornberg

Department of Applied Mathematics
California Institute of Technology
Pasadena, California 91125

*The author is supported by Control Data Corporation and by D.O.E. (Office of Basic Energy Sciences).

TABLE OF CONTENTS

NUMERICAL COMPUTATION OF NONLINEAR WAVES

Bengt Fornberg

Department of Applied Mathematics
California Institute of Technology
Pasadena, California 91125

I. INTRODUCTION

Equations that can be described as wave equations arise in a large number of physical situations. Most often, they take a form which includes partial derivatives. In contrast to the case of ordinary differential equations, it tends to be true that each equation involving partial derivatives requires a special solution technique. There are of course some very general approaches, like finite differences, finite elements, spectral decomposition etc. but there are likely to be lots of details that differ from case to case. Program packets which more or less automatically handle a wide range of different problems have so far had little or no impact on wave calculations and I do not think any major changes in that respect are in sight.

This makes it unavoidable for me to limit this discussion to a few special equations. A reasonable choice seems to be the following two equations:

Korteweg - de Vries:

$$u_t + uu_x + u_{xxx} = 0 \tag{1.1}$$

Deep water waves:

The velocity potential ϕ satisfies

$$\Delta\phi = 0 \qquad \left(\Delta = \frac{\partial^2}{\partial x^2} + \frac{\partial^2}{\partial y^2}\right) \tag{1.2}$$

inside the water, with

$$\eta_t = \phi_y - \phi_x \eta_x \tag{1.3}$$

$$\phi_t = -g\eta - \tfrac{1}{2}\left(\phi_x^{\,2} + \phi_y^{\,2}\right) \tag{1.4}$$

as boundary conditions on the free surface $y = \eta(x,t)$.

There is a very rich literature, both mathematical and numerical, for each of these equations. The KdV equation arises as a first approximation to a large number of phenomena in fields like shallow water waves, plasma physics, lattices, elastic rods and bubble-liquid mixtures. We will try to give brief reviews of the main numerical calculations on these equations. The lists will be very incomplete but we hope they will at least give an impression of the directions in which progress has been made.

II. MAIN DIFFERENCE SCHEMES FOR HYPERBOLIC EQUATIONS

The following list summarizes very briefly, with the relevant stencils sketched, the most basic finite difference approaches to hyperbolic problems. We assume for simplicity that we have one space and one time variable.

Explicit, 2 levels.

```
            *             ← new level
- - - * * * * * - - -     ← old level
```

Leap-Frog. Special case of explicit scheme, 3 levels.

```
              *             ← new level
-- * * * * * * * - -        ← old level
              *             ← old level
```

Implicit, 2 or more levels.

```
  - - - * * * - - -        ← new level
- - - * * * * * - - -      ← old level
- - - * * * * * - - -      ← old level
- - - - - - - - - - -        ........
```

Hopscotch. Every second new point is determined explicitly, every other second point is filled in by implicit approximations. The complete scheme may turn out explicit or implicit.

```
* + * + * + * + * + *  ← new level
* * * * * * * * * * *  ← old level
```

The whole process is shifted one step sideways for each new time step.

Method of lines. The equation is discretized initially only in space. The resulting system of ordinary differential equations, one for each meshpoint in space, is solved by any standard ODE-solver.

Spectral method. Same as method of lines but the equation is first transformed to Fourier space. Each equation in the system of ODEs describes how the coefficient for one Fourier mode changes in time.

Pseudospectral method. All spatial derivatives are approximated from the interpolating trigonometric polynomial at the latest time level with use of Fast Four Transforms (FFTs). The method of lines or Leap-Frog are most often used to advance in time.

Finite elements. A wide class of different techniques. At each time level the solution is represented as a smooth combination of basis functions, normally with local support. This representation can be used to find derivatives or to directly solve an optimization problem equivalent to a given differential equation.

For particular equations, less general methods may be natural. The "Sine-Gordon" equation $u_{xx}-u_{tt} = \sin u$ can, for example, be solved very accurately by exploiting the fact that the characteristics are the straight lines $x + t = \text{const.}$ and $x - t = \text{const.}$

The following five concepts are fundamental in describing a difference approximation:

Consistency: Expresses the notion that the difference scheme "locally resembles" the differential equation.

Accuracy: Let h and k denote the discretization steps in space and time respectively. If the local truncation error is $O(h^r) + O(k^s)$, the scheme is said to be accurate of order (r,s).

Dissipation: A measure of how fast different Fourier modes are damped with increasing time.

Stability: A scheme is stable if no solution can tend to infinity at a finite time as h and k $\to 0$. Stability for explicit schemes often require conditions on k like $k/h <$ const., $k/h^3 <$ const. or something similar. Linear stability for initial value problems can easily be checked with a Fourier transform method. Initial-boundary value stability is much more involved but can also be analyzed with a modal decomposition [Gustafsson et al., 1972]. Special kinds of nonlinear instabilities are possible in linearly stable schemes [Fornberg, 1973].

Convergence: The approximate solution tends to the analytic one as h and k tend to zero. For a consistent scheme to a well posed linear initial value problem, stability is necessary and sufficient for convergence. (Lax Equivalence theorem.)

III. APPLICATIONS TO THE KdV EQUATION

A. Finite Difference Methods

Most of the finite difference methods we have sketched above have been applied to the KdV equation. Some of the calculations are:

Zabusky - Kruskal [1965]:

```
    *      ← n+1
* * * * *  ← n
    *      ← n-1
```

$$u_j^{n+1} = u_j^{n-1} - \frac{k}{h}\frac{\left(u_{j+1}^n + u_j^n + u_{j-1}^n\right)}{3}\left(u_{j+1}^n - u_{j-1}^n\right) - \frac{k}{h^3}\left(u_{j+2}^n - 2\,u_{j+1}^n + 2\,u_{j-1}^n - u_{j-2}^n\right) \tag{3.1}$$

Stability requires $k/h^3 < 2/3\sqrt{3} = 0.3849$. In this work, clean interactions between solitary waves were observed and the term 'soliton' was introduced.

Zabusky [1968]:

$$\begin{matrix} . & * & . & * & \leftarrow n+1 \\ * & . & * & . & \leftarrow n \end{matrix}$$

$$\frac{1}{h^3}\left(u_{j+2}^{n+1} - 3\,u_{j+1}^{n} + 3\,u_{j}^{n+1} - u_{j-1}^{n}\right) +$$

$$+ \frac{1}{8h}\left(u_{j+2}^{n+1} + u_{j-1}^{n}\right)\left(u_{j+2}^{n+1} + u_{j+1}^{n} - u_{j}^{n+1} - u_{j-1}^{n}\right) = 0 \tag{3.2}$$

This is a very compact first order scheme on a staggered grid. Consistency requires $k/h^3 = 1/4$. The solution is obtained explicitly by one-sided numerical sweeps from a boundary. This scheme was originally used to study how general initial states for the KdV equation developed into solitons and an oscillatory tail.

Vliegenthart [1971]:

This paper contains a survey of finite difference schemes for the KdV equation. Dissipative, second order two-level explicit schemes were introduced. The numerical calculations focus on how step functions develop oscillations which tend towards a train of solitary waves. The schemes introduced here seem not to have been used in later work since the alternative of non-dissipative schemes of high order together with artificial dissipation, if desired, appears to be more flexible.

Greig - Morris [1976]:

With $f = u^2/2$,

$$\begin{aligned} u_j^{n+1} &= u_j^n - \frac{1}{2}\frac{k}{h}\left(f_{j+1}^n - f_{j-1}^n\right) - \\ &\quad - \frac{1}{2}\frac{k}{h^3}\left(u_{j+2}^n - 2\,u_{j+1}^n + 2\,u_{j-1}^n - u_{j-2}^n\right), \quad j+n \text{ even} \\ u_j^{n+1} &= u_j^n - \frac{1}{2}\frac{k}{h}\left(f_{j+1}^{n+1} - f_{j-1}^{n+1}\right) - \\ &\quad - \frac{1}{2}\frac{k}{h^3}\left(u_{j+2}^{n+1} - 2\,u_{j+1}^{n+1} + 2\,u_{j-1}^{n+1} - u_{j-2}^{n+1}\right), \quad j+n \text{ odd} \end{aligned} \tag{3.3}$$

This is a Hopscotch scheme with conditional stability $k/h^3 < 1/2$. (A similar condition is also required for consistency.) The stencil at every second mesh point looks like

```
  *
* * * * *
```

and a similar stencil upside down is used at every other second point. A tridiagonal system is solved at each time level. The method is second order accurate.

Winther [1980]:

The KdV equation is rewritten in the form

$$u_t - w_x = 0 \tag{3.4}$$

$$w = -u_{xx} - u^2/2 \tag{3.5}$$

The first equation is made to be satisfied pointwise while the second one is approximated by a Galerkin finite element method. Unconditional stability is demonstrated for a method which turns out to be equivalent to a finite difference scheme with stencil

```
* * * *  ← new level
* * * *  ← old level
* * * *  ← old level
```

Another scheme with much simpler coefficients but still the same stencil and second order accuracy, not derived via finite elements, is also quoted to be unconditionally stable. No numerical tests were performed.

B. Pseudospectral Method

The spectral or pseudospectral approaches were considered by Gazdag [1973] and Tappert [1974]. We will here describe the pseudospectral method following Fornberg [1975] and Fornberg - Whitham [1978].

Let us consider the problem of accurately approximating $\partial/\partial x$ on a periodic, equidistant grid, for example with 32 points in the period:

```
→ * * * * * * * * * * * * * * * * * * * * * * * * * * * * * * * * ←
```

TABLE I

N	Order of Accuracy	Coefficients for difference approximations of ∂/∂x															Stability Limit
1	2	0.0	1.0000														1.0000
2	4	0.0	1.3333	-0.1667													0.7287
3	6	0.0	1.5000	-0.3000	0.0333												0.6305
4	8	0.0	1.6000	-0.4000	0.0762	-0.0071											0.5776
5	10	0.0	1.6667	-0.4762	0.1190	-0.0198	0.0016										0.5442
6	12	0.0	1.7143	-0.5357	0.1587	-0.0357	0.0052	-0.0004									0.5206
7	14	0.0	1.7500	-0.5833	0.1944	-0.0530	0.0106	-0.0014	0.0001								0.5029
8	16	0.0	1.7778	-0.6222	0.2263	-0.0707	0.0174	-0.0031	0.0004	-0.0000							0.4890
9	18	0.0	1.8000	-0.6545	0.2545	-0.0881	0.0252	-0.0056	0.0009	-0.0001	0.0000						0.4778
10	20	0.0	1.8182	-0.6818	0.2797	-0.1049	0.0336	-0.0087	0.0018	-0.0003	0.0000	-0.0000					0.4685
11	22	0.0	1.8333	-0.7051	0.3022	-0.1209	0.0423	-0.0124	0.0030	-0.0005	0.0001	-0.0000	0.0000				0.4606
12	24	0.0	1.8462	-0.7253	0.3223	-0.1360	0.0512	-0.0166	0.0045	-0.0010	0.0002	-0.0000	0.0000	-0.0000			0.4539
13	26	0.0	1.8571	-0.7429	0.3405	-0.1502	0.0601	-0.0211	0.0063	-0.0016	0.0003	-0.0000	0.0000	-0.0000	0.0000		0.4480
⋮																	
60	120	0.0	1.9672	-0.9360	0.5745	-0.3837	0.2645	-0.1837	0.1269	-0.0865	0.0580	-0.0380	0.0243	-0.0152	0.0092	...	0.3764
⋮																	
Limit		0.0	2.0000	-1.0000	0.6667	-0.5000	0.4000	-0.3333	0.2857	-0.2500	0.2222	-0.2000	0.1818	-0.1667	0.1538	...	0.3183

Coefficients for the classical difference approximation of ∂/∂x

At each of these grid points, $\partial/\partial x$ can be approximated to second order as

$$\left(-1\ u_{j-1} + 0\ u_j + 1\ u_{j+1}\right)\Big/ 2\ h$$

and to fourth order as

$$\left(\frac{1}{6}\ u_{j-2} - \frac{4}{3}\ u_{j-1} + 0\ u_j + \frac{4}{3}\ u_{j+1} - \frac{1}{6}\ u_{j+2}\right)\Big/ 2\ h$$

etc. Table 1 shows the right half of these antisymmetric sets of coefficients for methods of increasing orders. We notice that approximations of increasing accuracies tend to a simple limit approximation with coefficients $(-1)^i \times 2/i$, $i = 1, 2, 3, \ldots$. A similar limit can be found for finite periodic grids of any size. Corresponding limit approximations are, of course, available for third derivatives as well. Applying such approximations to all the grid points becomes equivalent to cyclic convolutions, performed in $O(N \log N)$ operations with FFTs. One can show that a completely equivalent approach to the limit method is to transform the mesh data to the discrete Fourier space, then find the derivatives for the interpolating trigonometric polynomial (by multiplying the coefficients by the wave numbers) and finally transform back to physical space.

Writing F for the discrete Fourier transform operator (and F^{-1} for the inverse), the straightforward implementation with Leap-frog time differencing would be

$$u_j^{n+1} - u_j^{n-1} + 2\ iu_j^n\ k\ F^{-1}\{\nu\ Fu\} - 2\ i\ kF^{-1}\{\nu^3\ Fu\} = 0, \quad (3.6)$$

One can show that the simple modification

$$u_j^{n+1} - u_j^{n-1} + 2\ iu_j^n\ k\ F^{-1}\{\nu\ Fu\} - 2\ iF^{-1}\{\sin(\nu^3 k)\ Fu\} = 0, \quad (3.7)$$

keeps the formally infinite order spatial accuracy but vastly improves the accuracy in the Leap-Frog time step. This modification also improves the stability condition from $k/h^3 = 1/\pi^3 = 0.0323$ to $k/h^3 = 3/2\pi^2 = 0.1520$. Any linear dispersion relation, rather than the particular one corresponding to the u_{xxx} term, can be implemented with no increase in cost. Figures 1 to 5 show examples of different wave phenomena calculated with this method [from Fornberg and Whitham, 1978]. Figure 6 illustrates how the logarithm of the error varies as a function of the spatial resolution for methods of different accuracies. (A detailed description of this figure can be found in Fornberg and Whitham [1978].)

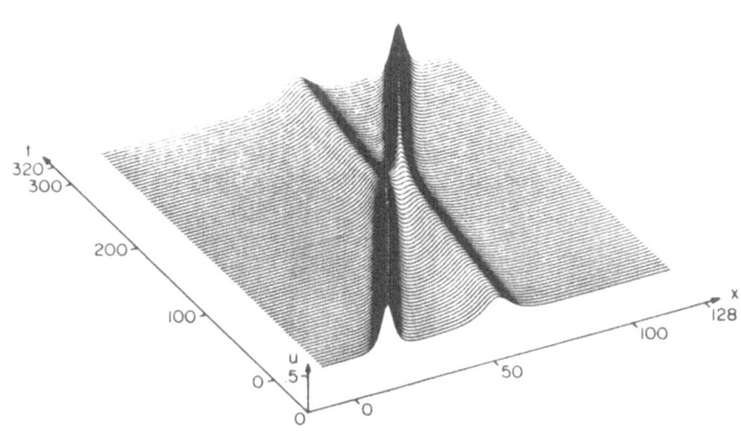

Fig. 1. Soliton interaction for $u_t + 3u^2 u_x + u_{xxx} = 0$.

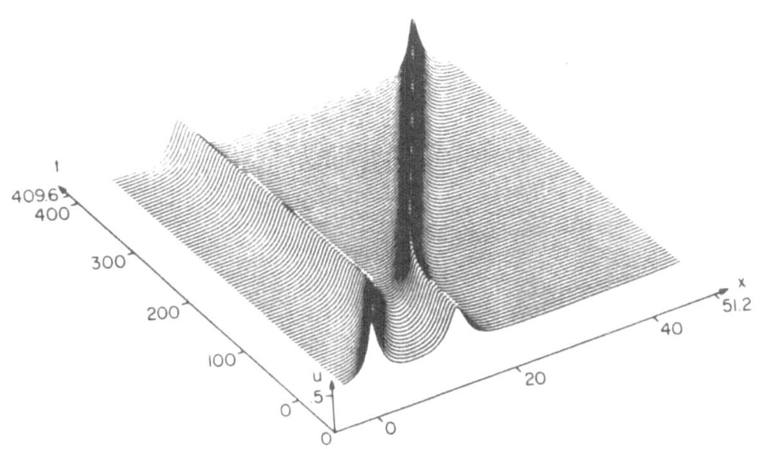

Fig. 2. Soliton interaction for $u_t + \frac{3}{2} uu_x - \frac{5}{4} u_x + \int K(x-\xi) u_\xi d\xi = 0$ with $K(x) = \frac{\nu}{2} e^{-\nu|x|}$, $\nu = \pi/2$.

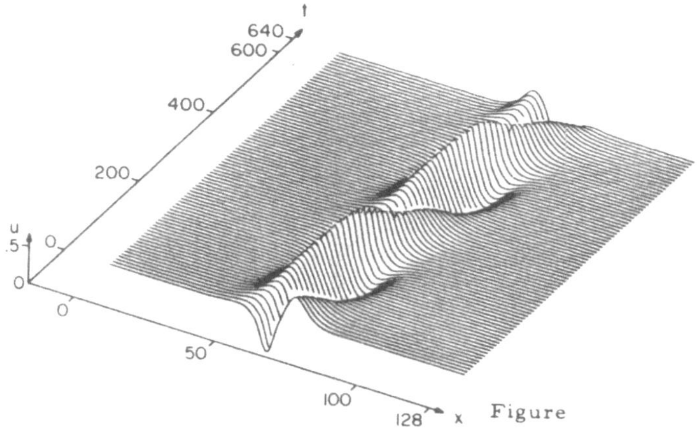

Fig. 3. Wave packet solution to $u_t + 3u^2u_x + u_{xxx} = 0$.

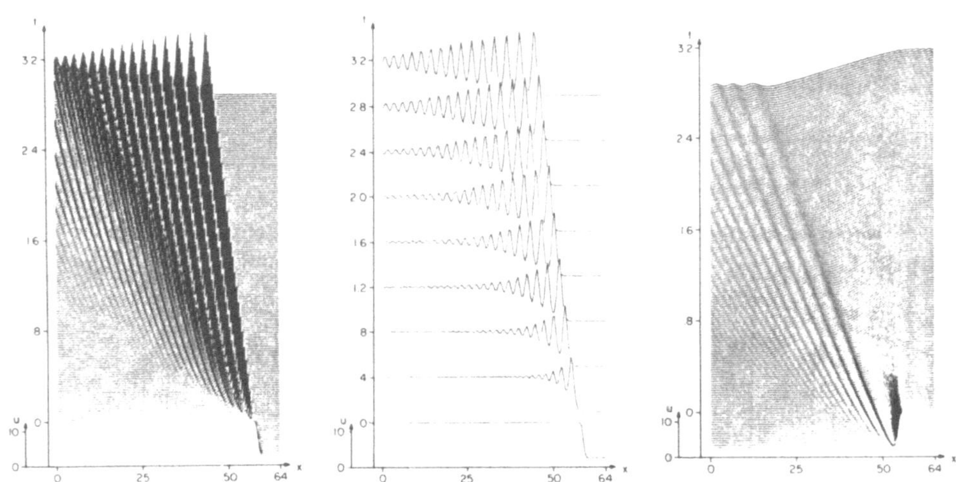

Fig. 4. Solution to $u_t + uu_x + u_{xxx} = 0$ for step function input.

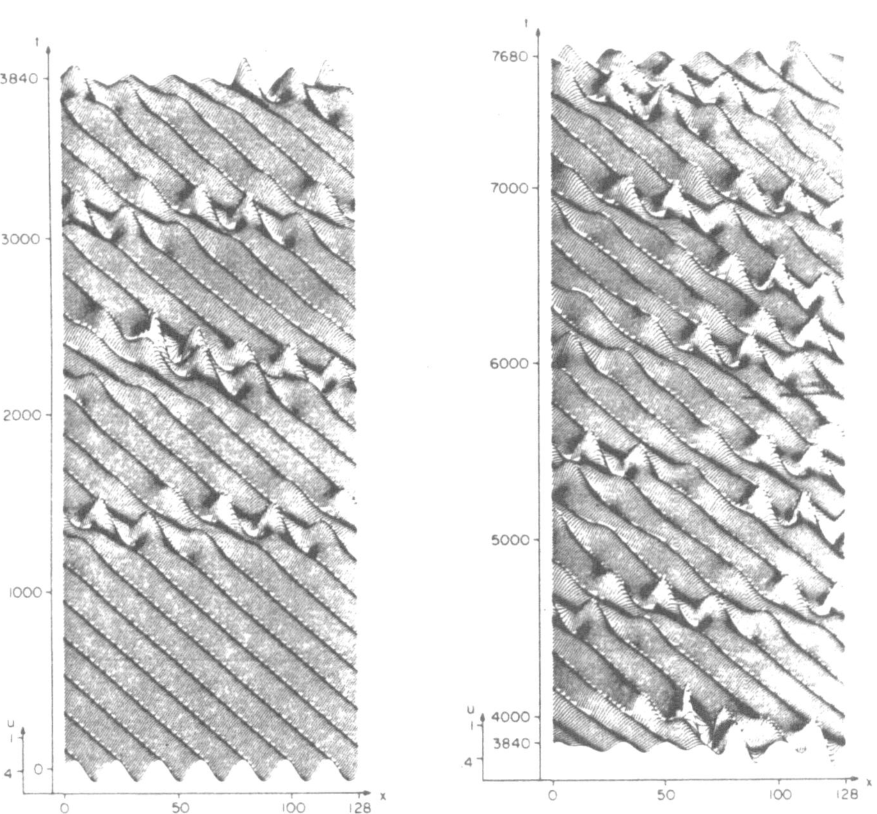

Fig. 5. Instabilities of a uniform wavetrain, $u_t + 3u^2u_x + u_{xxx} = 0$.

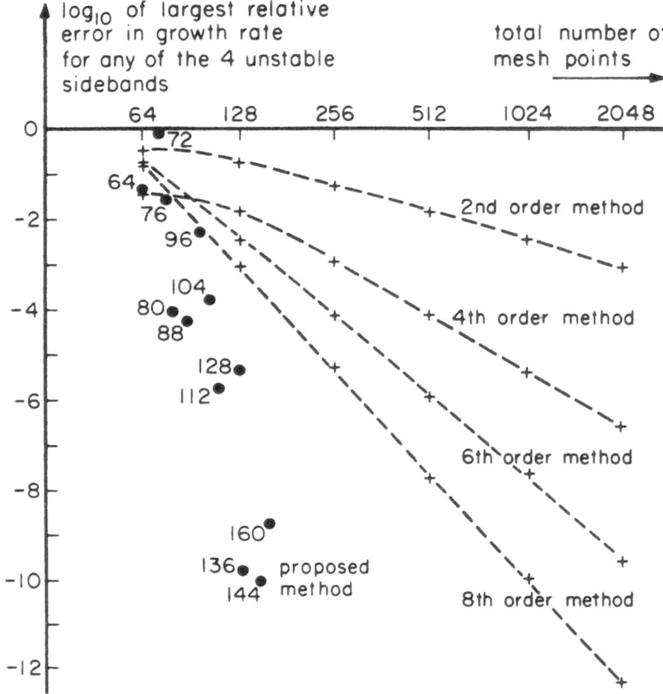

Figure 6. Comparison of accuracies for different approximations in space.

There are two recent papers which review different methods for the KdV equation. Eilbeck [1978] considers some of the schemes mentioned here together with some results for other wave equations. Abe and Inoue [1980] compare second order finite difference schemes (Zabusky - Kruskal's and Greig - Morris' schemes) with the method by Gazdag and a direct pseudospectral implementaion.

It is sometimes said that implicit schemes are more economical than explicit ones because of the stability restriction. For the KdV equation, explicit schemes will, in general, require $k/h^3 <$ const. This argument is fallacious. As we have just seen, it is very easy to make a scheme, say, second order in time and sixth order in space. Stability still requires $k/h^3 <$ const., but $k = O(h^3)$ is now exactly the right ratio to make the accuracies match. Compared to a scheme, second order in time and space and unconditionally stable, the only difference is that we have got away with a much larger space step. For the same accuracy, the time steps would be just the same.

IV. WAVES ON DEEP WATER

We assume there is only one horizontal space dimension, that the water is inviscid, incompressible and irrotational and that gravity is the only restoring force. With u and v denoting x - and y - velocities, the absence of vorticity implies the existence of a velocity potential ϕ satisfying

$$\phi_x = u \;, \quad \phi_y = v \;. \tag{4.1}$$

Incompressibility now gives

$$\Delta\phi = 0 \tag{4.2}$$

The two boundary conditions on the free top surface are:

i) Particles on the surface $y = \eta(x,t)$ remain on the surface

$$\eta_t = \phi_y - \phi_x \eta_x \tag{4.3}$$

ii) Bernoulli's equation for the pressure

$$\phi_t = -g\eta - \frac{1}{2}\left(\phi_x^2 + \phi_y^2\right) \tag{4.4}$$

In contrast to shallow water, it can be shown that no solitary waves of permanent shape exist. The simplest solutions are either steady uniform wavetrains which translate in time (called Stokes waves after pioneering work by G.G. Stokes at the end of the last century) and wave packets with a soliton type envelope. Quite naturally, virtually all calculations until very recently have been on Stokes waves since that is the only case which does not require a full time dependent code.

A. Some Calculations and Theoretical Results for Time Independent Waves

A summary of some calculations and theoretical results for time independent waves is as follows:

Stokes [1880]:

These papers contain a collection of his earlier works on series expansions for small amplitudes and gives also an argument that the highest Stokes wave (if it exists) must have a top angle of exactly 120°.

Michell [1893]:

A series expansion calculation for this highest wave suggests a height over wavelength ratio of 0.142. (Currently the most accurate estimate [Longuet-Higgins and Fox, 1978] is 0.14107.)

Nekrasov [1921,1922][1], Levi-Civita [1925] and Struik [1926]:

Proofs of existence and uniqueness for small amplitude waves on water with finite or infinite depth.

Schwartz [1974]:

With the use of Nekrasov's formulation and computer algebra, Taylor coefficients in the wave height h up to order h^{117} were obtained. Extrapolations with Padé approximants showed leading coefficients to be non-monotonic in h.

Longuet-Higgins and Fox [1978]:

The limit of Stokes waves of increasing heights was shown to be singular. Wave quantities like energy and speed exhibit an infinity of oscillations as the height increases.

Toland [1978]:

Existence of wave of maximal height as a limit of increasing waves.

Chen and Saffman [1980]:

New types of uniformly translating waves were found where all crests are not equally high. Bifurcations appear at large amplitudes.

Saffman and Yuen [1980a, 1980b]:

Bifurcations in two dimensions to new types of steady waves appear already at infinitesimal amplitudes.

In the next section, we will look in more detail at the calculation that led to the discovery of these new Stokes waves and in the remaining sections at two techniques for time dependent calculations.

---

[1]Translated summaries in [Lichtenstein, 1925/28] and [Milne-Thompson, 1960].

B. Time Independent Waves With All Crests Not Equal

We denote the period in space by L (which may contain one or several waves) and represent the surface in parameter form

$$x = x(\sigma) \quad , \quad y = y(\sigma) \quad , \quad z = x + i\,y \quad . \tag{4.5}$$

Let further g denote the acceleration due to gravity and c the wave speed. The governing equation can then be written [Longuet-Higgins and Cokelet, 1976]

$$\left(1 - \frac{gL}{\pi c^2}\,\mathrm{Im}\,z\right)\frac{d\bar{z}}{d\sigma} - 1 = -\frac{i}{2\pi}\,P\int_0^{2\pi}\cot\left[\frac{z(\sigma)-z(\sigma_1)}{2}\right]d\sigma_1 \tag{4.6}$$

(P denotes principal value integral. The trivial solution $z = \sigma$ corresponds to a flat surface.) Since the Stokes wave is known not to be monotonic in c, another free parameter b is introduced. The limits b = 0 and b = 1 correspond to a flat surface and maximal wave respectively. After minor modifications of the equation (to eliminate arbitrariness in the solution, for example with respect to translations) the equation was discretized. The resulting nonlinear system could have been solved at this stage by Newton's method. If the Jacobian matrix for the system is non-singular, quadratic convergence to a (locally) unique solution is obtained. Points where the matrix is singular correspond to turning points or bifurcation points. At a turning point, the system is singular only because of non-monotonicity in the particular continuation parameter. At bifurcation points, new branches of solutions appear. The technique of "arclength" continuation due to Keller [1977] eliminates turning points and leads to rules on how to catch the different branches at bifurcation points. If u represents a solution vector and b the original continuation parameter, the technique starts with the introduction of a new independent variable s

$$\begin{aligned} u &= u(s) \\ b &= b(s) \end{aligned} \tag{4.7}$$

and the addition of a new equation

$$\|u\|_s^{\,2} + b_s^2 = 1 \quad . \tag{4.8}$$

After these steps were taken, Newton's method was used to solve the nonlinear system. Figures 7 to 10 show bifurcation diagrams and wave forms for the cases of uniform Stokes waves turning into waves with every second crest peaked and every one or two out of three crests peaked.

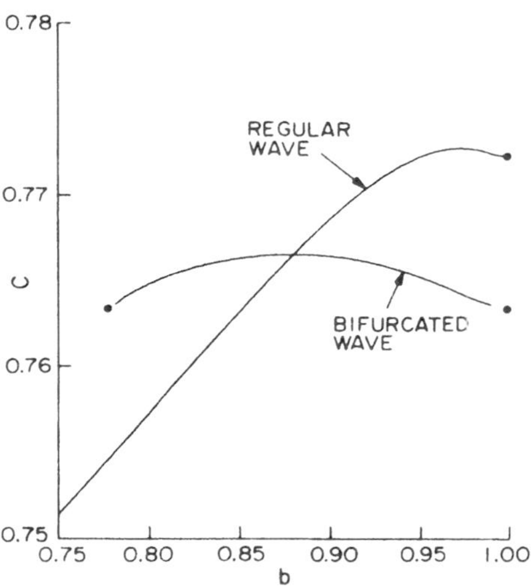

Fig. 7. Bifurcation diagram for waves with every second crest equal.

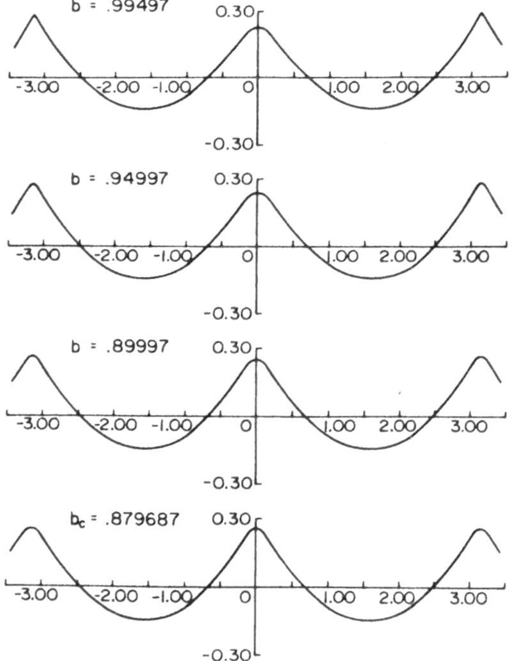

Fig. 8. Bifurcated waves with every second crest equal.

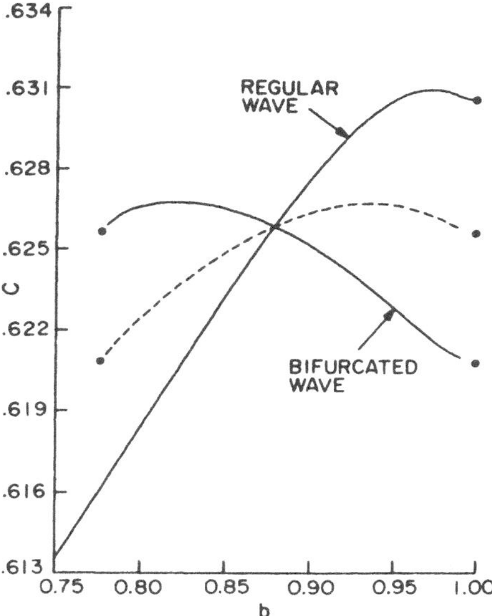

Figure 9. Bifurcation diagram for waves with two crests out of three equal.

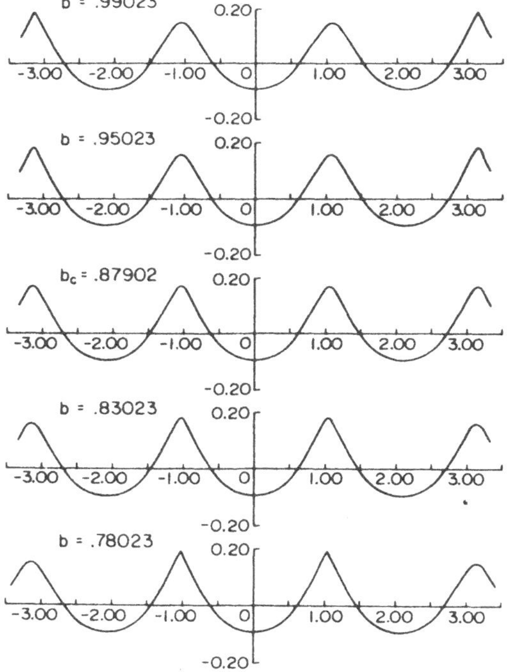

Figure 10. Bifurcated waves with two crests out of three equal.

C. Some Calculations and Theoretical Results for Time Dependent Waves

The following list summarizes some calculations and theoretical results for time dependent waves:

Benjamin [1967]:

Uniform wave trains are unstable with respect to long wave-length modulations. Known as "Benjamin-Feir instabilities".

Longuet-Higgins and Cokelet [1976,1978]:

Accurate time dependent calculations of a plunging breaker and of instabilities.

Longuet-Higgins [1978], Hasselmann [1979]:

More kinds of instabilities of Stokes waves are suggested including explosive instabilities at high amplitudes and superharmonic slow instabilities at low amplitudes.

There are two different approaches to reduce the full time dependent problem to one with only time and one space variable. Both depend, of course, on the fact that the governing equation inside the water is the simple linear relation $\Delta\phi = 0$. In the approach by Longuet-Higgins and Cokelet [1976,1978], a linear integral equation is solved at each time step. In another approach by the present author, a fast technique for conformal mappings reduces the problem at each time step to the trivial one of $\Delta\phi = 0$ on the unit circle. These two methods are described briefly below.

D. Integral Equation Method

The wave is described in an x,y coordinate system and $z = x + iy$. Assuming a spatial period of 2π, the transformation

$$\zeta = e^{-iz} \tag{4.9}$$

maps the region with water to a simply connected region with a removable cut to the origin. The tangential derivatives of ϕ can be calculated directly by interpolation. The problem is to find the normal derivatives. With the notation suggested in Figure 11, the normal derivative will satisfy

$$\int_c \frac{\partial\phi}{\partial n} \ln R \, ds = P \int_c \phi \, d\alpha - \pi\phi_o \tag{4.10}$$

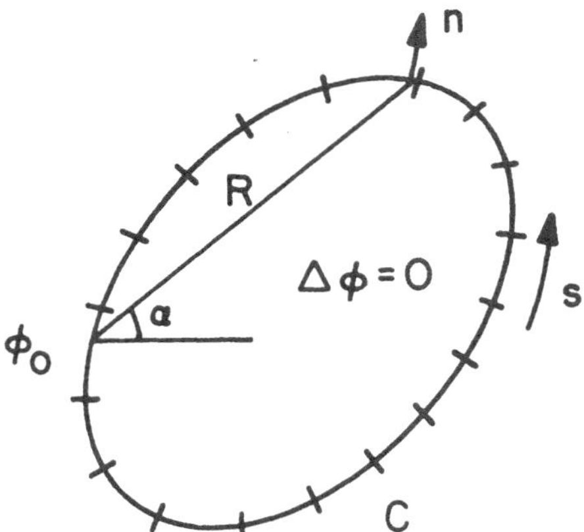

Figure 11. Notation for integral equation.

This linear system of equations (after discretization with N points) can be solved by Gaussian elimination in $O(N^3)$ operations. Longuet-Higgins and Cokelet used a fourth order Adams-Bashforth-Moulton scheme for the time stepping. Figures 12 to 14 show some results from Longuet-Higgins and Cokelet [1976].

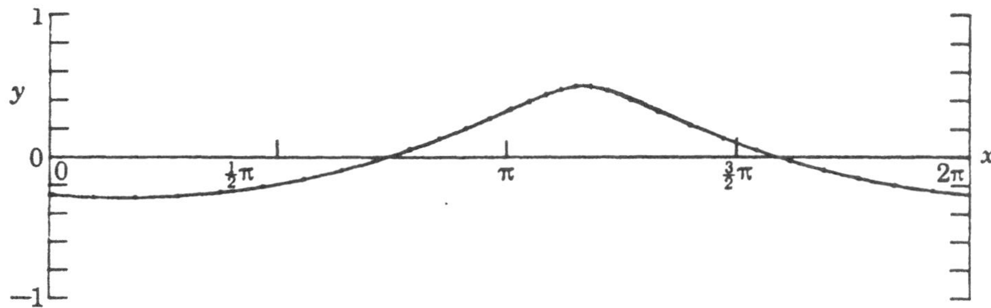

Figure 12. Comparison of exact and numerical solution for Stokes wave at time 2π. (The wave has travelled just over one period in space.) Spatial resolution 30 points.

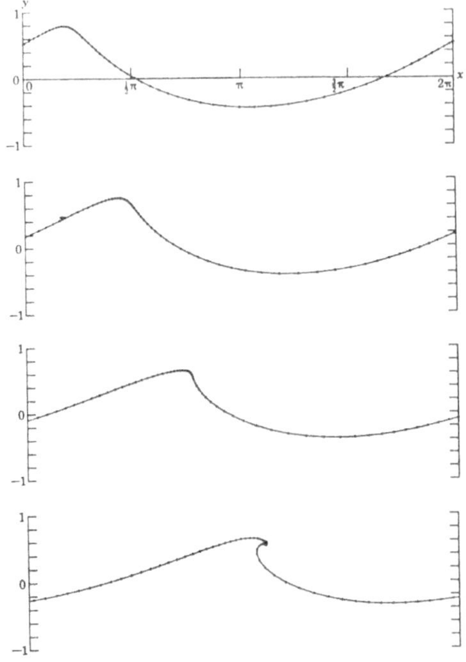

Fig. 13. Breaking waves.

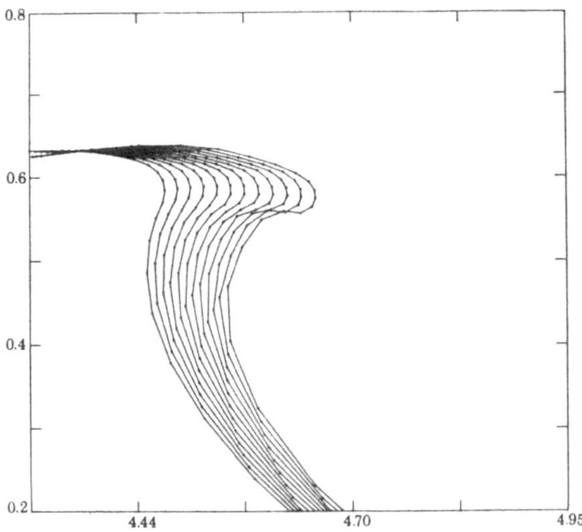

Fig. 14. Breaking waves.

E. Conformal Mapping Method

Figure 15 illustrates a series of mapping steps which all leave the equation $\Delta\phi = 0$ invariant. On either of the two rightmost regions, the necessary derivatives $\partial\phi/\partial s$ and $\partial\phi/\partial n$ can be calculated immediately. The corresponding derivatives at the original water

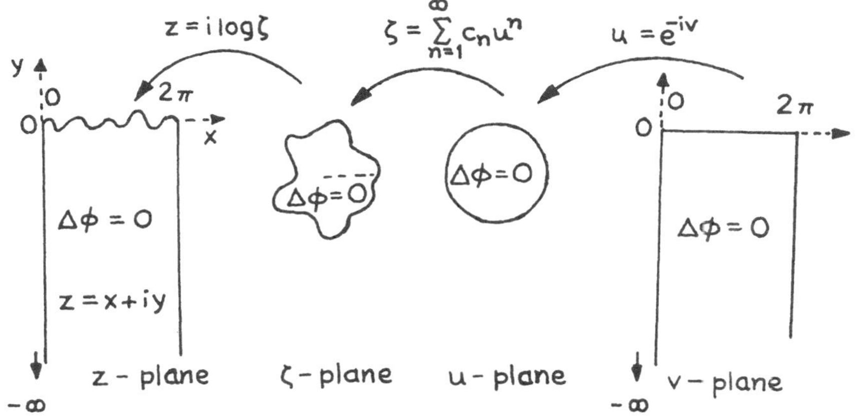

Fig. 15. Steps in conformal mapping method.

surface follow then directly. The crucial part of this procedure is to obtain cheaply the mapping between the unit circle and its wavy image. Table II summarizes the main available methods.

TABLE II

AUTHOR(S)	METHOD	COST "near-circular" regions	COST general regions
Theodorsen [1931, (see Henrici (1979))] Gutknecht [1979,1979]	Integral equation in polar coordinates. Direct iteration accelerated by SOR.	0(N log N)	-
Symm [1966]	Integral equation for inverse (linear) problem. Direct iteration on Fourier coefficients.	0(N log N)	-
Bauer et. al. [1975]	Integral equation in arc-length representation. Direct iteration.	0(N log N)	?
Hayes et. al. [1972]	Symm's equation solved by Gaussian elimination.		$0(N^3)$
Chakravarty and Anderson [1979]	Odd-even reduction on Cauchy-Riemann's eq.		$0(N^3)$
Fornberg [1980]	Nonlinear system for boundary correspondence solved by FFT, Newton's method and Conjugate gradients.		0(N log N)

The last method was developed especially for application to this problem of deep water waves. This application is in preliminary stages at this moment. Figure 16 illustrates the only test case completed so far. A second order centered Euler type scheme was used for the time stepping. With computational points placed uniformly on the unit circle, their images turn out to be concentrated in the wave troughs and distributed very sparsely at the crests (where, of course, the highest resolution is needed). For this reason, the main application of this technique seems to be the calculation of long wave trains, wavepackets etc. of medium amplitude. The cost of only O(N log N) operations per time step allows very long spatial intervals to be handled economically.

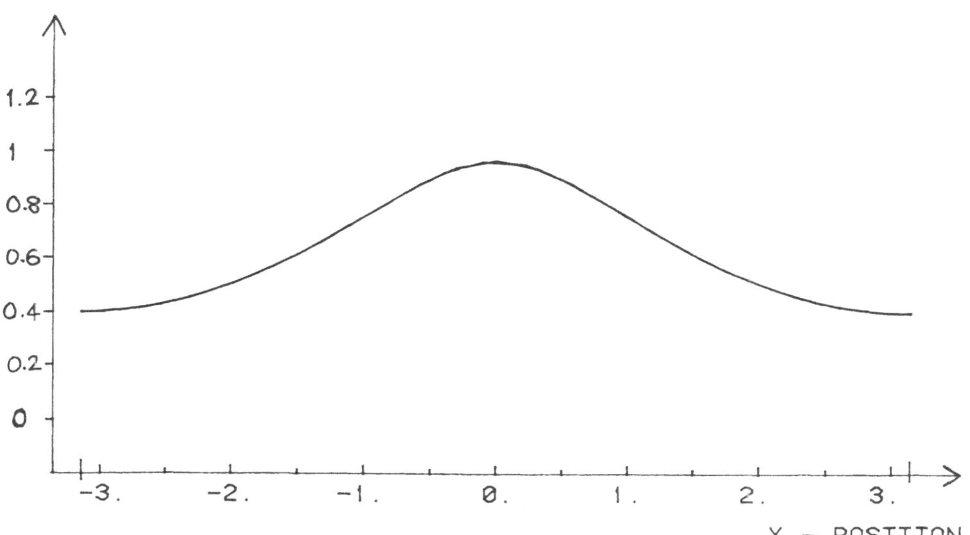

Figure 16. Comparison between a Stokes wave (with height over wavelength 0.09083) at times $t = 0$ and $t = 30.1628$, which corresponds to 5 periods in space. Spatial resolution 32 points. The discrepancy between the two curves is less than the thickness of the line.

REFERENCES

Abe, K., and Inoue, O., 1980, Fourier expansion solution of the Korteweg-de Vries equation, J. Comp. Phys., 34:202.

Bauer, F., Garabedian, P., Korn, D., and Jameson, A., 1975, "Supercritical Wing Sections II," Springer-Verlag.

Benjamin, T.B., 1967, Instability of periodic wavetrains in nonlinear dispersive systems, Proc. Roy. Soc. A, 299:59.

Chakravarthy, S., and Anderson, D., 1979, Numerical conformal mapping, Math. Comp., 33:953.

Chen, B., and Saffman, P.G., 1980, Numerical evidence for the existence of new types of gravity waves of permanent form on deep water, Stud. Appl. Math., 62:1.

Eilbeck, J.C., 1978, Numerical studies of solitons, in: "Proc. Symposium on Nonlinear (Soliton) Structure and Dynamics in Condensed Matter, Oxford 1978," A.R. Bishop and T. Schneider, eds., Springer-Verlag.

Fornberg, B., 1973, On the instability of leap-frog and Crank-Nicolson approximations of a nonlinear partial differential equation, Math. Comp., 27:45.

Fornberg, B., 1975, On a Fourier method for the integration of hyperbolic equations, SIAM J. Numer. Anal., 12:509.

Fornberg, B., 1980, A numerical method for conformal mappings, SIAM J. Sci. and Stat. Comput., 1:386.

Fornberg, B., and Whitham, G.B., 1978, A numerical and theoretical study of certain nonlinear wave phenomena, Phil. Trans. Roy. Soc. London A, 289:373.

Gazdag, J., 1973, Numerical convective schemes based on accurate computation of space derivatives, J. Comp. Phys., 13:100.

Greig, I.S., and Morris, L.L., 1976, A hopscotch method for the Korteweg-de Vries equation, J. Comp. Phys., 20:64.

Gustafsson, B., Kreiss, H.O., and Sundström, A., 1972, Stability theory of difference approximations for mixed initial boundary value problems, II, Math. Comp., 26:649.

Gutknecht, M.H., 1979, Solving Theodorsen's integral equation for conformal maps with the fast Fourier transform. Part I. Theory. Research report 79-02, Seminar für angewandte Mathematik, Eidgenössische Technische Hochschule, Zürich.

Gutknecht, M.H., 1979, Solving Theodorsen's integral equation for conformal maps with the fast Fourier transform. Part II. Practice. Research report 79-04, Seminar für angewandte Mathematik, Eidgenössische Technische Hochschule, Zürich.

Hasselmann, D., 1979, The high wavenumber instabilities of a Stokes wave, J. Fluid Mech., 93:491.

Hayes, J.K., Kahaner, D.K., and Kellner, R., 1972, An improved method for numerical conformal mapping, Math. Comp., 26:327.

Henrici, P., 1979, Fast Fourier methods in computational complex analysis, SIAM Review, 21:481.

Keller, H.B., 1977, Numerical solutions of bifurcation and nonlinear eigenvalue problems, in: "Applications of Bifurcation Theory," Academic Press, New York.

Levi-Civita, T., 1925, Détermination rigoureuse des ondes permanentes d'ampleur finie, Math. Ann., 93:264.

Lichtenstein, editor, 1925/28, "Jahrbuch über die Fortschritte der Mathematak 1921-1922," Walter de Gruyter & Co.

Longuet-Higgins, M.S., 1978, The instabilities of gravity waves of finite amplitude in deep water, Proc. Roy. Soc. London A, 360:471.

Longuet-Higgins, M.S., and Cokelet, E.D., 1976, The deformation of steep surface waves on water. I. A numerical method of computation, Proc. Roy. Soc. London A, 350:1.

Longuet-Higgins, M.S., and Cokelet, E.D., 1978, The deformation of steep surface waves on water. II. Growth of normal-mode instabilities, Proc. Roy. Soc. London A, 364:1.

Longuet-Higgins, M.S., and Fox, M.J.H., 1978, Theory of the almost-highest wave. Part 2. Matching and analytic extension, J. Fluid Mech., 85:769.

Michell, J.H., 1893, The highest waves in water, Phil. Mag., 36:430.

Milne-Thompson, L.M., 1960, "Theoretical Hydrodynamics," 4th ed., MacMillan, New York.

Nekrasov, A.I., 1921, On stationary waves (In Russian), Izv Ivanovo-Vosnosonk. Politehn Inst., 3:52.

Nekrasov, A.I., 1922, On stationary waves. Part 2. On nonlinear integral equations (In Russian), Izv Ivanovo-Vosnosonk. Politehn Inst., 6:155.

Saffman, P.G., and Yuen, H.C., 1980a, Bifurcation and symmetry breaking in nonlinear dispersive waves, Phys. Rev. Lett., 44:1097.

Saffman, P.G., and Yuen, H.C., 1980b, A new type of three-dimensional deep water gravity wave of permanent form, J. Fluid Mech., 101:797.

Schwartz, L.W., 1974, Computer extensions and analytic continuation of Stokes' expansion for gravity waves, J. Fluid Mech., 62:553.

Stokes, G.G., 1880, Considerations relative to the greatest height of oscillatory waves which can be propagated without change of form, in: "Math. and Phys. Papers," Cambridge Univ. Press 1.

Stokes, G.G., 1880, Supplement to a paper on the theory of oscillatory waves, in: "Math. and Phys. Papers," Cambridge Univ. Press 1.

Struik, D.J., 1926, Détermination rigoureuse des ondes irrotationelles periodiques dans un canal a profondeur finie, Math. Ann., 95:595.

Symm, G.T., 1966, An integral equation method in conformal mapping, Numer. Math., 9:250.

Tappert, F., 1974, Numerical solutions of the Korteweg-de Vries equation and its generalizations by the split-step Fourier method, in: "Lectures in Appl. Math., Amer. Math. Soc. 15."

Toland, J.F., 1978, On the existence of a wave of greatest height and Stoke's conjecture, Proc. Roy. Soc. London A, 363:469.
Vliegenthart, A.C., 1971, On finite-difference methods for the Korteweg-de Vries equation, J. Eng. Math., 5:137.
Winther, R.W., to appear, A conservative finite element method for the Korteweg-de Vries equation.
Zabusky, N.J., 1968, Solitons and bound states of the time-dependent Schrodinger equation, Phys. Rev., 168:124.
Zabusky, N.J., and Kruskal, M.D., 1965, Interactions of 'solitons' in a collisionless plasma and the recurrence of initial states, Phys. Rev. Lett., 15:240.

BIFURCATIONS, FLUCTUATIONS AND DISSIPATIVE STRUCTURES*

G. Nicolis

Department of Physics
Free University of Brussels
Brussels, Belgium

* Notes prepared by Y. Elskens for the lectures of G. Nicolis.

TABLE OF CONTENTS

BIFURCATIONS, FLUCTUATIONS AND DISSIPATIVE STRUCTURES

G. Nicolis

Department of Physics
Free University of Brussels
Brussels, Belgium

I. INTRODUCTION

The purpose of the present lectures is to discuss the emergence of structures from dissipative processes in macroscopic systems. This class of phenomena, generally studied by means of nonequilibrium thermodynamics, is quite different from solitons and other structures appearing in nonlinear Hamiltonian systems, which were covered by many lecturers at this school.

Indeed, Hamiltonian systems admit a variety of conservation laws, constraining their evolution very strictly; as a counterpart they present rather poor stability properties in the sense that their evolution remains strongly dependent on the initial conditions. Conversely, dissipation was long believed to lead all macroscopic systems to a final quiescent state of maximal disorder, known as the state of thermodynamic equilibrium: this evolution to thermal chaos was thought to follow from the second law of thermodynamics, which introduces irreversibility into physics.

The situation appears to be quite different today: new developments of thermodynamics [Glansdorff and Prigogine, 1971] showed that dissipation and irreversibility were not only compatible with order, but that they even could be the very source of order. The new forms of matter associated with such transitions to order are known as dissipative structures. Their existence is nowadays firmly established. Yet it should be fully realized that from the standpoint of classical dynamics there is something deeply shocking in them: in fact, dissipative structures arise thanks to the absence of stringent conservative laws - whereas such laws are precisely at the origin of solitons and other nonlinear phenomena in Hamiltonian systems.

One of our principal goals in these lectures is to describe the physical mechanisms by which dissipation can lead to a dynamical behaviour, on a macroscopic scale, which is very different from the equilibrium-like chaos. Nonequilibrium thermodynamics and statistical mechanics will often lay at the basis of our developments. We shall first illustrate the subject (Section II) with a few representative examples.

Then, in the second part of the lectures (Sections III to VI), we shall set up a purely phenomenological description of nonequilibrium transitions. We shall develop general classification schemes for the various behaviours and use them as an appropriate framework for interpreting the experimental observations. In Section III, the basic ideas of nonequilibrium thermodynamics and their application to general reaction-diffusion systems will be presented. The mathematical tools for the analysis of order in nonequilibrium systems are stability and bifurcation theory, which are discussed respectively in Section IV and Sections V - VI.

In the third part (Sections VII to IX), we will adopt a more fundamental approach and try to relate the behaviour observed at the macroscopic level to the dynamics at the microscopic level. This will lead us to a stochastic analysis of nonequilibrium systems (Section VII) and to the discussion of the effects of fluctuations on the solution of the phenomenological evolution equations, by means of general theorems (Section VIII) and of approximation schemes (Section IX).

In the conclusions (Section X), we shall summarize the results of these various approaches and emphasize some problems which still remain open.

II. ILLUSTRATIVE EXAMPLES

Transition phenomena leading to ordered patterns are very common in nonequilibrium systems. In the present section we give an overview of such phenomena in various branches of physical sciences and biology, and comment on their implications to the understanding of the complexity observed in nature at various scales.

In A. we present a "canonical" model illustrating the role of irreversibility in the emergence of order. In B. we describe various transition phenomena occurring in nature under nonequilibrium conditions. Some general comments are presented in C.

A. A "Canonical" Model

Mathematical examples of nonequilibrium transitions are numerous [see Nicolis and Prigogine, 1977, for a survey]. In this subsection we focus on a simple model describing a hypothetical system, which has the advantage of leading to exact solutions while being at the same time perfectly compatible with the constraints imposed by the laws of physics and chemistry. This model belongs to the class of reaction-diffusion systems (discussed in more detail in Section III), i.e. systems subject to fluxes of matter from the environment, in which chemical species diffuse and react. Although the vocabulary stems from chemistry, such examples are relevant to many other situations described by similar equations (ecology, solid-state physics, ...).

The particular model we choose to discuss was introduced by Schlögl [1972]. It involves three species (A,B,X) assumed to react without any release of heat:

$$\begin{aligned} B &\underset{k_1}{\overset{k_0}{\rightleftharpoons}} X \\ A + 2X &\underset{k_3}{\overset{k_2}{\rightleftharpoons}} 3X \end{aligned} \tag{2.1}$$

The kinetics of these reactions for an ideal mixture can be written as

$$w_A = -k_2 a x^2 + k_3 x^3 \tag{2.2a}$$

$$w_B = -k_0 b + k_1 x \tag{2.2b}$$

$$w_X = -w_A - w_B \tag{2.2c}$$

where

w_A is the rate of production of species A

a is the concentration of A

k_i is the rate constant of reaction i.

The rate constants are fixed by the conditions under which the reactions proceed (temperature, pressure, possible solvent,...). The evolution equations for the concentrations are:

$$\frac{da}{dt} = w_A + v_A \tag{2.3a}$$

$$\frac{db}{dt} = w_B + v_B \tag{2.3b}$$

$$\frac{dx}{dt} = w_X + v_X \tag{2.3c}$$

where the v denotes the flux associated with non-chemical phenomena. In an open system, v_A, v_B and v_X are related to the flows across the boundary of the reactor or to convection and diffusion. We first fix these "mechanical" contributions in such a way that:

$$v_X = 0 \tag{2.4a}$$

$$\frac{da}{dt} = 0 \tag{2.4b}$$

$$\frac{db}{dt} = 0 \tag{2.4c}$$

These concentrations a and b may be considered as constant parameters, and the state of the system is described by the single variable x:

$$\frac{dx}{dt} = -k_3x^3 + k_2ax^2 - k_1x + k_0b \tag{2.5}$$

<u>At equilibrium</u>, the system reaches a stationary state in which there is no net flow through the boundary:

$$v_A = v_B = 0 \tag{2.6}$$

The detailed balance law prevails for both reactions:

$$k_1x = k_0b \tag{2.7a}$$

$$k_3x^3 = k_2ax^2 \tag{2.7b}$$

These equations have two families of solutions

$$x_{eq} = 0 \quad , \quad b_{eq} = 0 \quad , \quad a_{eq} = \alpha \tag{2.8}$$

$$x_{eq} = \frac{k_2}{k_3}\alpha \quad , \quad b_{eq} = \frac{k_1k_2}{k_0k_3}\alpha \quad , \quad a_{eq} = \alpha \tag{2.9}$$

where α is a real positive parameter. The first solution reflects the need for X to be present initially to start its own synthesis from A, and the equilibrium between X and B. The second solution allows for a proportionality between the concentrations of X, A and B, in agreement with the law of mass action. Considering a and b as control parameters, we note that equilibrium can be fulfilled in only one non-trivial way:

$$\frac{b_{eq}}{a_{eq}} = \frac{k_1 k_2}{k_0 k_3} \quad . \tag{2.10}$$

Let now a and b be arbitrary, thus removing the system from equilibrium. We introduce the following parameters:

$$m = \frac{1}{3}\frac{k_2 a}{k_3} \geq 0 \tag{2.11a}$$

$$\delta = \frac{k_1}{k_3}\frac{1}{m^2} - 3 \quad , \quad (\geq -3) \tag{2.11b}$$

$$\delta' = \frac{k_0}{k_3}\frac{b}{m^3} - 1 \quad , \quad (\geq -1) \tag{2.11c}$$

and scaled variables

$$\tau = k_3 m^2 t \tag{2.12a}$$

$$\xi = \frac{x}{m} - 1 \quad , \quad (\geq -1) \ . \tag{2.12b}$$

In these variables the equilibrium states are found for the values of the parameters

$$\delta' = 3\delta + 8 \tag{2.13a}$$

and variable

$$\xi = 2 \ . \tag{2.13b}$$

The evolution equation then becomes

$$\frac{d\xi}{d\tau} = -\xi^3 - \delta\xi + \delta' - \delta \tag{2.14}$$

or

$$\frac{d\xi}{d\tau} = -\frac{\partial\Phi}{\partial\xi} \tag{2.15a}$$

with the "kinetic potential" Φ

$$\Phi = \frac{1}{4}\xi^4 + \frac{1}{2}\delta\xi^2 + (\delta - \delta')\xi + \Phi_0 \quad . \tag{2.15b}$$

The steady-state solutions are the extrema of the potential and satisfy the equation

$$\delta' = \xi^3 + (\xi + 1)\delta \quad . \tag{2.16}$$

In the three-dimensional (ξ,δ,δ') space, this is the equation of a manifold from which we derive the basic diagram which represents the surface by lines of equal ξ in the parameter space of δ and δ' (Fig. 2.1). When (2.16) has three real physical solutions, three steady states are accessible to the system: this is the "coexistence" region. The coexistence region is bordered by two curves along which two of these three solutions coalesce; they are obtained for

$$(\delta - \delta')^2 = -\frac{4}{27}\delta^3 \quad . \tag{2.17}$$

A special point corresponds to the <u>critical values</u>

$$\delta = \delta' = 0 \tag{2.18a}$$

or

$$a = \frac{\sqrt{3k_1k_3}}{k_2} \quad , \quad b = \frac{k_1k_2}{9k_0k_3}a \tag{2.18b}$$

for which all three stationary states coalesce. In terms of the solution ξ, this means that a <u>bifurcation</u> happens between completely different situations: around this point (clearly not accessible at equilibrium: see (2.13) or (2.10)), all extrapolations from an equilibrium theory are bound to fail.

The best way to study this so-called cusp bifurcation occurring at the critical point $\delta = \delta' = 0$ is to examine the variations of the steady states along the line

$$\delta = \delta' \tag{2.19}$$

which is the axis of the bifurcation (Fig. 2.2).

It is amazing to see that in such a simple model, the cubic non-linearity of the reaction kinetics leads already to the possibility of new kinds of behaviour, like regulating the performance of the

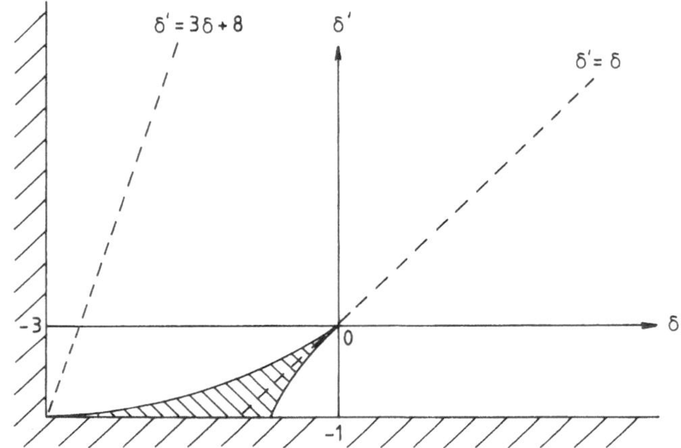

Fig. 2.1. The Schlögl model: diagram of states in parameters space. $\delta' = 3\delta + 8$: equilibrium line ($\xi = 2$); $\delta = \delta' = 0$: critical point ($\xi = 0$); dashed region: 3 steady states; white region: 1 steady state; $\delta \geq -3$, $\delta' \geq -1$: physical region.

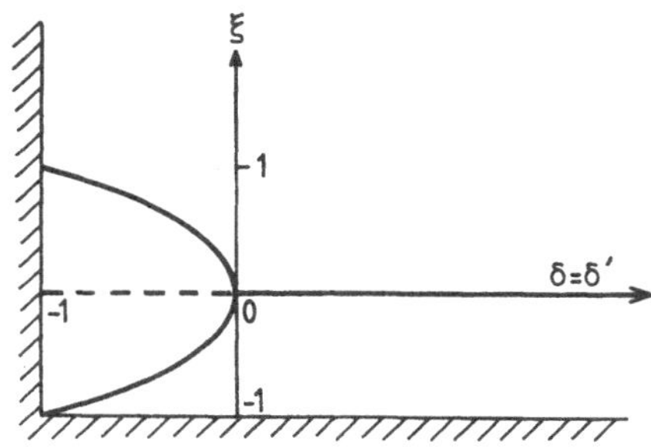

Fig. 2.2. The Schlögl model. Diagram of states along the axis $\delta = \delta'$.

system by conferring to it the possibility to be at one or another steady state in the coexistence region.

Many other models can be developed, with more intricate bifurcations, which allow for more elaborate behaviours. For example, time-periodic solutions, or limit cycles, can be generated in models like the following trimolecular chemical oscillator or "brusselator" [Prigogine and Lefever, 1968] involving two coupled reactants (intermediates X and Y):

$$\begin{aligned} A &\rightarrow X \\ B + X &\rightarrow Y + D \\ 2X + Y &\rightarrow 3X \\ X &\rightarrow E \end{aligned} \tag{2.20}$$

We come back to this model later on.

B. The Real World

Bifurcations in dissipative systems are observed in many fields of natural sciences. Without being exhaustive, we quote a number of characteristic examples.

(i) Fluid Mechanics

Bifurcations are well known in hydrodynamics, where they appear, among other cases, as convective instabilities [Verlarde and Normand, 1980; Chandrasekhar, 1961] in the Bénard circulation, with length scales ranging from a few millimeters (the drying of printed surfaces) to a few kilometers (the atmospheric currents). Interfaces too are notoriously rich in instabilities [Miller, 1978].

(ii) Geophysics

Other natural systems undergoing bifurcations are the terrestrial dynamo [Jacobs, 1976] and the earth-atmosphere system exchanging energy with the outer space [Fraedrich, 1978]. Plate tectonics can also be approached in relation with the convective instabilities [Richter, 1973].

(iii) Solid-State Physics

The lamellar growth of a eutectic alloy from its liquid phase [Langer, 1980] or simply the dendritic ramifications of snowflakes [Nakaya, 1954] show the building of solid-state patterns on scales by several orders larger than any crystallographic length: these beautiful structures are altogether very regular in each sample and very sensitive to the conditions of their formation.

In quantum physics, one can observe the formation of bands in constrained superconductors [Scalapino and Huberman, 1977].

(iv) Electric Systems

Engineering makes a large use of cooperative phenomena in nonlinear (active) devices, like the negative resistance of Gunn diodes [Haken, 1978; Nakamura, 1977] or the relaxation oscillations of a resonant circuit coupled to a triode as in van der Pol's model [van der Pol, 1930; Andronov, Witt and Khaikin, 1966].

(v) Quantum Optics

The generation of coherent light by a laser provides one of the most striking examples of nonequilibrium transitions [Haken, 1978], with a remarkable variety of operation regimes.

(vi) Chemical Reactions

In chemical systems, we note a flourishing variety of patterns arising from nonequilibrium constraints: the simple N_2O_4 decomposition under illumination [Creel and Ross, 1976] shows bistability and hysteresis. The Belousov-Zhabotinski reaction leads to a wave-like behaviour in a homogeneous environment [Zhabotinski, 1974]. Explosions and flames present highly organized spatial structures [Clavin and Guyon, 1978]. We saw some examples of chemical oscillations in Professor Howard's lectures. In fact, one of the oldest experimental reports (as early as 1828) on limit cycles describes oscillations in an electrochemical reaction [Degn, 1972].

(vii) Biological Systems

Nonequilibrium transitions in chemical systems already present striking analogies with some of the characteristic manifestations of biological organization. Living beings of course present a much greater variety of spatial and temporal patterns, and of adaptative cooperative behaviours [Nicolis and Prigogine, 1977].

In vitro, this is nicely illustrated by the chemical clock of the glycolytic oscillator [Boiteux, Goldbeter and Hess, 1975 ; Goldbeter, 1980] and the self-replication and chirality of various biopolymers for which at least qualitative models have been developed (see Professor Schuster's lectures). *In vivo*, the self-organizing trends range from the mitotic cycle of the cell [Kauffman and Wille, 1975] and the cAMP regulation associated with aggregation in the amoebe *Dictyostelium discoideum* [Gerisch and Hess, 1974] up to circadian and seasonal rhythms in most organisms, and to the process of morphogenesis during the embryonic development (see Professor Kauffman's lectures).

And note that ecosystems present even more complicated behaviours.

C. Discussion

Two essential features of transitions in nonequilibrium systems, justified in the models and supported by experiemntal observation of nature, deserve a special emphasis:

(i) The bifurcation clearly comes from the appearance of new "physical" solutions to the underlying equations of evolution when parameters vary. This would be impossible if one were limited to linear systems like

$$A \rightleftharpoons X \rightleftharpoons B \tag{2.21}$$

instead of (2.1). The nonlinearities are needed to develop qualitatively new behaviours in some (out-of-equilibrium) range. For one- and two- variable systems, this nonlinearity must be at least cubic for certain types of transition, but quadratic expressions are sufficient with several variables.

(ii) The bifurcation happens at a finite distance from equilibrium. This is necessary since, in an infinitesimal neighbourhood of equilibrium, a linear approximation is generally possible, dominating the nonlinear "corrections".

The combination of these two elements allows the competing dissipative processes to drive the system towards bifurcating states. In analogy with the equilibrium statistical mechanics, we shall call these bifurcations nonequilibrium phase transitions. However, a "nonequilibrium phase" of the system, also called a dissipative structure, is not a new phase of matter, but rather a new dynamical regime appearing under some constraints within a given phase of matter: for instance, in the Schlögl model the reacting species still form an ideal mixture (of gases or of solutes in solvent). That it must be so is easily understood by noting that the constraints maintaining (hence controlling) a dissipative structure are not the same as the parameters of a thermostatical phase, but are superimposed to the latter ones: in a system at given mean temperatures and pressure, we force a heat flow, or a flux of chemical species (v_A,v_B in Schlögl's model).

III. THERMODYNAMICAL ASPECTS

A. Introductory Remarks - Phenomenological Description

The examples discussed in the previous section illustrate the tendency to self-organization and complex behaviour in natural systems. The appearance of such new phenomena as nonequilibrium phase transitions leading to various dissipative structures, is a challenge to both intuition and mathematical ingenuity. Indeed, the equilibrium properties of matter can be understood, at least in principle, from the statistical mechanics of its constituents in terms of the short-range intermolecular interaction and the (random) thermal motion. In contrast, the appearance of long-range order out of equilibrium clearly calls for new approaches and techniques.

In this section, as well as Sections IV to VI, we make no attempt to derive *ab initio* the macroscopic properties of nonequilibrium systems from first principles. Rather, we set up a phenomenological description which is a natural extension of equilibrium considerations and which appears to be in agreement with experimental data on such systems as fluids or chemical reactions. We show how this phenomenological approach allows us to arrive at classification schemes and plausible interpretations of the non-equilibrium transition phenomena.

The basic approach, that of local thermodynamics, is the subject of this section. Sections IV to VI deal with linear stability analysis and bifurcation analysis.

B. Macroscopic Systems

Macroscopic systems by definition involve a large number (N) of degrees of freedom, the behaviour of which being, as a principle governed by well-known microscopic (classical or quantal) laws. On the other hand, these systems are practically described by only few relevant variables (X_i; $1 \leq i \leq n$) ($n << N$); obeying closed sets of evolution laws; the choice of these macrovariables and the derivation of their relationships (from experiment or theory) naturally include phenomenological hypotheses. We shall first present the classical version of this formulation [de Groot and Mazur, 1962; Glansdorff and Prigogine, 1971] and then specialize to the particular class of reaction-diffusion systems [Nicolis and Prigogine, 1977]. Some comments on the foundations of this approach will be presented in Section VII.

C. Local Thermodynamics

(i) Hypotheses

Consider a macrosystem Σ and its environment $\bar{\Sigma}$. The evolution of Σ in time is characterized by the rate of change (dx/dt) of an extensive macrovariable (x) and it reflects both the spontaneous evolution arising from the internal processes (d_ix) as well as the influence (d_ex) of external constraints associated to its coupling with the environment (Fig. 3.1)

$$\frac{dx}{dt} = \frac{d_ix}{dt} + \frac{d_ex}{dt} \quad . \tag{3.1}$$

This separation of the evolution in external and internal fluxes leads to the standard classification of the systems into

- isolated systems: $d_ex_k = 0 \quad \forall \quad k$ (3.2)

- closed systems: $d_ex_k = 0$ if x_k refers to the mass of a substance (3.3)

- open systems: $d_ex_k \neq 0$ in general . (3.4)

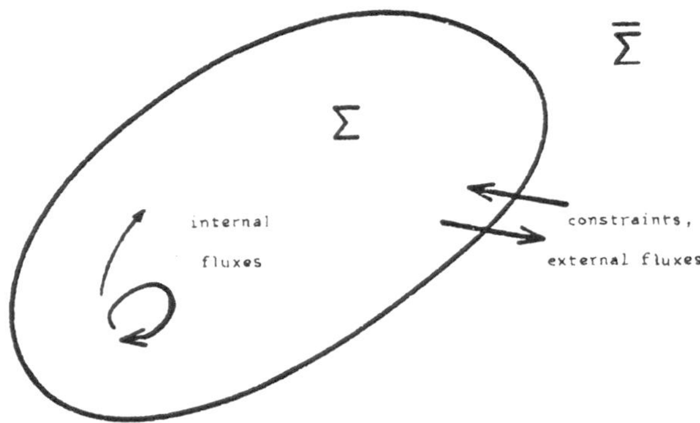

Fig. 3.1. The evolution of a macrosystem Σ is determined by its internal forces and fluxes, and by its coupling with the environment $\bar{\Sigma}$.

The fundamental hypotheses of local thermodynamics are:

1) a complete set of state variables is given by fields $\underset{\sim}{x}(\underline{r},t)$;

2) the state variables $\underset{\sim}{x}(\underline{r},t)$ for a homogeneous quiescent system reduce to the state variables of equilibrium thermodynamics;

3) the equations of state interrelating the fields $\underset{\sim}{x}(\underline{r},t)$ out of equilibrium are the same as in equilibrium (<u>local equilibrium</u> hypothesis).

For consistency, we assume that all external constraints act through the boundaries of the system (short-range constraints) and that the system is far from any thermostatical singularity. These simplifications single out the properly thermodynamical effects.

By virtue of the local equilibrium hypothesis, we may define an entropy field $s(\underline{r},t) = s[\underset{\sim}{x}]$ as in equilibrium and a global entropy

$$S(t) = \int_{\Sigma} s(\underline{r},t)d^3\underline{r} \quad . \tag{3.5}$$

We then state the second law of thermodynamics in the form

$$\frac{d_iS}{dt}(t) = \int_{\Sigma} \sigma(\underline{r},t)d^3\underline{r} \tag{3.6}$$

with

$$\frac{d_iS}{dt} = \sigma \geq 0 \quad . \tag{3.7}$$

This inequality leads to a classification of thermodynamical processes and states:

$$- \text{ if } \quad \frac{d_iS}{dt} = 0 \quad , \text{ the system is at equilibrium,} \tag{3.8}$$

$$- \text{ if } \quad \frac{d_iS}{dt} > 0 \quad , \text{ it is out of equilibrium} \quad . \tag{3.9}$$

A process evolving through equilibrium states is called reversible, otherwise it is irreversible or dissipative.

Moreover, the entropy production can be expressed in the bilinear form of Gibbs:

$$\frac{dS}{dt} = \sum_k X_k J_k \tag{3.10}$$

where $X_k(\underline{r},t) = X_k[\underset{\sim}{x}]$ is the "thermodynamic force" associated with the "thermodynamic flux"

$$J_k(\underline{r},t) = J_k\left[\frac{d_i}{dt}\underset{\sim}{x}\right] \quad . \tag{3.11}$$

Thermodynamic considerations also lead to other expressions

$$J_k = J_k[\underset{\sim}{X}] \tag{3.12}$$

on the form of which the second law (3.7) implies certain restrictions.

(ii) Stability Considerations

At equilibrium, all fluxes and forces vanish, and the system is homogeneous. Moreover, apart from singularities (like phase transitions) there is no qualitative change in the system when the state parameters (x_k) are varied within some range. In such a range, there is always a Lyapounov functional of the state variables to guarantee the global stability of the equilibrium state (e.g. the Gibbs free enthalpy at constant temperature and pressure for a simple fluid, or the entropy in isolated systems: see Fig. 3.2.a). In such a description, all nonlinearities become irrelevant after a sufficient lapse of time since there is no evolution away from the equilibrium state!

The near-equilibrium regimes are defined by the property that the constitutive relations between fluxes and forces are linear. In this range, and under a few additional technical hypotheses, Prigogine's theorem of minimum entropy production [Glansdorff and Prigogine, 1971] states that d_iS/dt is a Lyapounov functional (Fig. 3.2.b). Thus, nonequilibrium steady states are always globally stable. Because of this stabilizing effect of dissipation, all nonlinearities in the system's properties are ultimately wiped out. The striking similarity of this picture with that of equilibrium theory allows the transposition of several methods and concepts familiar from equilibrium theory.

In the nonlinear range, i.e. far from equilibrium, the qualitative properties of the (constrained) systems are very different. Spectacular phenomena like transitions between macroscopic patterns can take place, provided that the constitutive relations (3.12), relating fluxes and thermodynamic forces, present sufficiently strong nonlinearities.

As seen in Section II, such processes show a much greater variety than equilibrium transitions in the sense that they are not limited by the requirement that the system be uniform and time-independent.

D. Reaction-Diffusion Systems

A general dissipative system is described by an abstract vector field $\underset{\sim}{x}(\underline{r},t)$ and various "extrinsic" control parameters $\lambda_k(\underline{r},t)$. Its evolution equations may be written as

$$\frac{d}{dt}\underset{\sim}{x}(\underline{r},t) = \underset{\sim}{G}[\underset{\sim}{x}(r,t);\ \underset{\sim}{\lambda}(\underline{r},t)] \qquad (3.13)$$

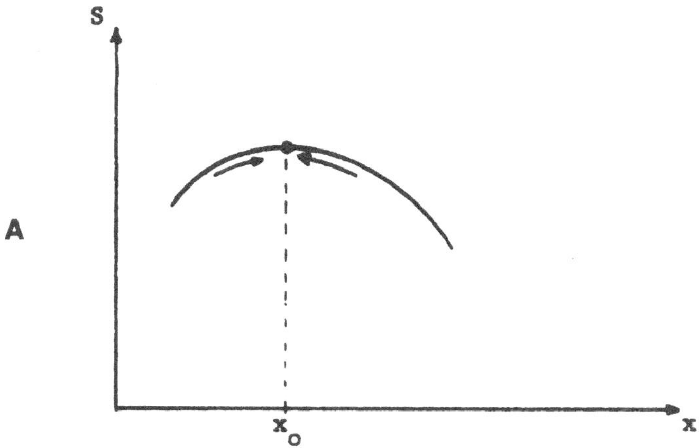

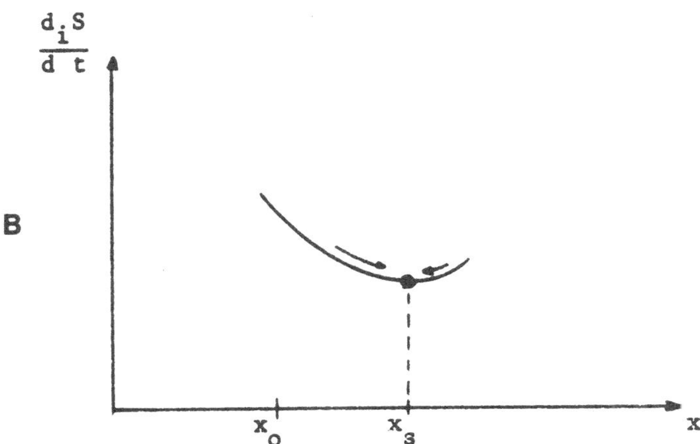

Fig. 3.2. Global stability.

(a) Equilibrium is the only stable steady state for an isolated system; the entropy is then maximal.

(b) A constrained system near equilibrium has a stable stationary state in the linear regime; then its entropy production is minimal.

All nonlinearities and symmetries are embedded in the explicit form of $\underset{\sim}{G}$, but a symmetry of the equation may be broken by its solution; e.g. it often happens that system (3.13) is autonomous

$$\frac{\partial}{\partial t}\underset{\sim}{\lambda} = 0 \tag{3.14}$$

and homogeneously constrained:

$$\nabla\underset{\sim}{\lambda} = 0 \tag{3.15}$$

and yet its solutions are time-dependent or nonuniform.

In general, the evolution functional $\underset{\sim}{G}$ is such that time-reversal symmetry is broken, in agreement with the observed irreversibility of the processes.

A typical example, with considerable physical relevance, is that of reaction-diffusion systems. Some of their most interesting properties stem from their coupling between scalar (chemical) and vector (diffusion) irreversible processes, which results in the emergence of spatial patterns. A reaction-diffusion system can be described by equations (3.13) with the following additional requirements:

- the state variables include the mass densities of the different chemical species and the energy density (which can be treated as an additional species for our purposes); all other variables will be treated as control parameters;
- the rate of change G_k of species k is separable into a chemical and a diffusive part:

$$G_k = F_k(\underset{\sim}{x}(\underline{r},t);\ \underset{\sim}{\lambda}(\underline{r},t)) + v_k(\underset{\sim}{x}(\underline{r},t);\ \underset{\sim}{\lambda}(\underline{r},t)) \tag{3.16}$$

- the diffusive part is assumed to obey the familiar Fick law, where, in addition, cross-effects are neglected:

$$v_k = \sum_{\ell} \operatorname{div}(D_{k\ell} \operatorname{grad} x_{\ell}) \tag{3.17}$$

with

$$D_{k\ell} = D_k(\underset{\sim}{x},\underset{\sim}{\lambda})\,\delta^{Kr}_{k\ell} \tag{3.18}$$

where $\delta^{Kr}_{k\ell}$ is the Kronecker delta.

- the reactive part reflects the creation and destruction of molecules of species k from the reactions:

$$F_k = \sum_{\rho} F_{k\rho} \tag{3.19}$$

$$F_{k\rho} = m_k \nu_{k\rho} w_\rho (\underset{\sim}{x}(\underline{r},t);\ \underset{\sim}{\lambda}(\underline{r},t)) \tag{3.20}$$

where

$\nu_{k\rho}$ is the (molar) stoichiometric coefficient of species k in reaction ρ

m_k is the mass of molecules of species k

w_ρ is the rate of reaction ρ

- each reaction satisfies the usual requirements imposed by thermodynamics, like for instance, the mass conservation laws in reactions:

$$\sum_k m_k \nu_{k\rho} = 0 . \tag{3.21}$$

The Schlögl model described in Section II.A is an example of such reaction-diffusion systems where a "good stirring" leads to a vanishing contribution from diffusion.

IV. CLASSICAL STABILITY ANALYSIS

The basic concepts in stability analysis can be traced to Lyapounov [Sattinger, 1973]. This section is intended to present the main ideas and methods which were developed along these lines.

We first classify the various types of stability (Section IV.A) for general nonlinear partial-differential systems. Then we discuss the stability of linear systems with constant coefficients and its relevance to nonlinear systems by means of linearization (Section IV.B).

We illustrate this approach on the homogeneous Schlögl model (Section IV.C) and finally discuss the general theory of stability of reaction-diffusion systems (Section IV.D).

A. Lyapounov's Stability

The basic concept in stability analysis is that of Lyapounov stability [Minorski, 1962], which merely expresses that solutions related to neighbouring initial conditions evolve in the neighbourhood of each other. To express this statement quantitatively, we make a few hypotheses on the system under consideration.

Consider a real vector function $\underset{\sim}{x}(\underline{r},t)$ defined on a spatial domain V in $\mathbb{R}^q (q \in \mathbb{N})$ for all times $t \in \mathbb{R}$. We restrict ourselves to continuous functions of time, square-integrable in space: such functions form a class $\mathcal{B} = \mathcal{C}(\mathbb{R}, M^n)$ where M^n is a Hilbert space with scalar product

$$(\underset{\sim}{x},\underset{\sim}{y}) = \sum_{k=1}^{n} \int_V y_k(\underline{r}) x_k(\underline{r}) d^q\underline{r} \quad . \tag{4.1a}$$

We denote by $\| \cdot \|$ the related norm:

$$\| x \|^2 = (\underset{\sim}{x},\underset{\sim}{x}) \quad . \tag{4.1b}$$

Each component x_k of $\underset{\sim}{x}$ belongs to the Hilbert space $M \equiv L^2(V,d^q\underline{r})$.

Consider the following evolution equations

$$\frac{\partial}{\partial t} \underset{\sim}{x}(\underline{r},t) = \underset{\sim}{G}[\underset{\sim}{x};\underset{\sim}{\lambda}] \tag{4.2}$$

with initial conditions

$$\forall \ \underline{r} \in V : \quad \underset{\sim}{x}(\underline{r},t_o) = \underset{\sim}{x}_o(\underline{r}) \tag{4.3}$$

and boundary conditions

- of Dirichlet type $\quad (\forall \ \underline{r} \in \partial V)(\forall \ t \in \mathbb{R}) \quad \underset{\sim}{x}(\underline{r},t) = \hat{\underset{\sim}{x}}_o \qquad (4.4a)$
- or of Neumann type $\quad (\forall \ \underline{r} \in \partial V)(\forall \ t \in \mathbb{R}) \quad n \cdot \nabla \underset{\sim}{x}(\underline{r},t) = \hat{\underset{\sim}{x}}'_o \qquad (4.4b)$

with constants $\hat{\underset{\sim}{x}}_o$ or $\hat{\underset{\sim}{x}}'_o$. Other boundary conditions, arising for instance in infinite systems, may lead to a redefinition of the scalar product (4.1) and accordingly to the consideration of other functions' spaces $\mathcal{B}$.

Definition 1

A solution $\underset{\sim}{x}(\underline{r},t)$ to problem (4.2-4) is Lyapounov-stable if

$$\forall \ \varepsilon > 0 \ , \ \forall \ t_o \in \mathbb{R} \ , \ \exists \eta > 0 :$$

$$\| \underset{\sim}{x}_o - \underset{\sim}{x}'_o \| < \eta \quad \Rightarrow \quad \forall t > t_o : \| \underset{\sim}{x} - \underset{\sim}{x}' \| < \varepsilon \tag{4.5}$$

where $\underset{\sim}{x}'(\underline{r},t)$ is a solution to problem (4.2-4) with initial condition $\underset{\sim}{x}'_o$.

Definition 2

A solution $\underset{\sim}{x}(\underline{r},t)$ to problem (4.2-4) is asymptotically stable if

$$\forall t_o \in \mathbb{R} \ , \ \exists \eta > 0 :$$

$$\| \underset{\sim}{x}_o - \underset{\sim}{x}'_o \| < \eta \ \Rightarrow \ \lim_{t \to +\infty} \| \underset{\sim}{x} - \underset{\sim}{x}' \| = 0 \qquad (4.6)$$

with the same abbreviations as in (4.5).

Definition 3

The system described by equations (4.2-4) is structurally stable if

$$\forall \varepsilon > 0 \ , \ \forall t_o \in \mathbb{R} \ , \exists \eta > 0 :$$

$$\| \underset{\sim}{\lambda} - \underset{\sim}{\lambda}' \| < \eta \ \Rightarrow \ \forall t > t_o : \| \underset{\sim}{x} - \underset{\sim}{x}'_s \| < \varepsilon \qquad (4.7)$$

where $\underset{\sim}{x}'(\underline{r},t)$ is a solution to problem (4.2-4) with parameter $\underset{\sim}{\lambda}'$.

$\| \cdot \|$ denotes a norm in the λ-space $\mathbb{R}^p$ too.

When a solution of (4.2-4) is not stable, we shall call it unstable. We may remark that asymptotic stability implies Lyapounov stability; stable solutions which are not asymptotically stable are marginally stable.

The notion of stability, explicitly refers to the limit $t \to +\infty$, implying the idea of a direction of time. This is most obvious in asymptotic stability, since it states that any finite perturbation will eventually disappear. An asymptotically stable solution is, therefore, an attractor in the space of evolutions $\mathcal{B} = \mathcal{C}(\mathbb{R}, M^n)$.

This viewpoint reduces the stability analysis to a problem of fixed point in $\mathcal{B}$. It also gives an indication on the methods which will be useful for theoretical analyses, such as the implicit functions' theorem or the center-manifold theorem. In short, we are interested in asymptotic solutions [Fife, 1978] of (4.2-4). Two particularly important classes of such solutions are:

- stationary solutions and

- time-periodic solutions.

More elaborate treatments would include multiperiodic and quasi-periodic solutions, or more general attractors [Lanford, 1973 ; Joseph, 1973]. As a matter of fact, a bifurcation analysis will partly indicate which kind of behaviours should be taken as references.

B. Linear Stability Analysis

The stability of a solution to the system (4.2-4) can, in principle, be analyzed by checking directly the definitions (4.5-7). However, this requires a general solution of the system, which is often too difficult. In a first step, one may limit the investigation to infinitesimal perturbations only around the asymptotic solution: this reduces a global, nonlinear problem to a local linear one. Lyapounov's theorem states the equivalence of both analyses for general systems.

We consider a reference solution $\underset{\sim}{x}_s$ to problem (4.2-4):

$$\frac{\partial}{\partial t}\underset{\sim}{x}_s = \underset{\sim}{G}[\underset{\sim}{x}_s;\underset{\sim}{\lambda}] \ . \tag{4.8}$$

Let $\underset{\sim}{x}$ be another solution of (4.2) satisfying the same boundary condition (4.4). The difference

$$\underset{\sim}{z} = \underset{\sim}{x} - \underset{\sim}{x}_s \tag{4.9}$$

satisfies the differential equation

$$\frac{\partial}{\partial t}\underset{\sim}{z} = \left(\frac{\delta \underset{\sim}{G}}{\delta \underset{\sim}{x}}\right)_{\underset{\sim}{x}_s,\underset{\sim}{\lambda}} \cdot \underset{\sim}{z} + 0(\|\underset{\sim}{z}\|) \tag{4.10}$$

provided that $\underset{\sim}{G}$ is Fréchet-differentiable around $(\underset{\sim}{x}_s,\underset{\sim}{\lambda})$; the boundary conditions for $\underset{\sim}{z}$ are homogeneous of the

- Dirichlet type: $\forall \underline{r} \in \partial V \ , \quad \underset{\sim}{z}(\underline{r},t) = 0$ (4.11)

- Neumann type: $\forall \underline{r} \in \partial V \ , \ \underline{n}\cdot\nabla\underset{\sim}{z}(\underline{r},t) = 0$ (4.12)

The null solution $\underset{\sim}{z} = 0$ stands for the reference solution $\underset{\sim}{x} = \underset{\sim}{x}_s$.

Defining the linear operator:

$$\underset{\approx}{L} = \underset{\approx}{L}(\underset{\sim}{\lambda}) = \frac{\delta \underset{\sim}{G}}{\delta \underset{\sim}{x}}[\underset{\sim}{x}_s;\underset{\sim}{\lambda}] \tag{4.13}$$

we find the first approximation to (4.10):

$$\frac{\partial}{\partial t} \underset{\sim}{z} = \underset{\approx}{L} \cdot \underset{\sim}{z} \quad . \tag{4.14}$$

We now state the relationship between the linear problem (4.12-14) and the nonlinear one (4.10-12):

<u>Theorem of linearized stability</u>

If the trivial solution of the problem (4.12-14) is asymptotically stable, then $\underset{\sim}{x}_s$ is an asymptotically stable solution of (4.2-4).

If the trivial solution of the problem (4.12-14) is unstable, then $\underset{\sim}{x}_s$ is an unstable solution of problem (4.2-4).

The main advantage of this theorem of linearized stability is that a stability analysis is generally easier for a linear system than for a nonlinear one. For <u>autonomous</u> systems, a bilateral Laplace transform on time

$$\underset{\sim}{z}(\underline{r},t) = \frac{1}{2\pi i} \int_{-i\infty}^{i\infty} \underset{\sim}{y}(\underline{r},\omega) e^{\omega t} \, d\omega \tag{4.15}$$

reduces the differential problem (4.12-14) to an algebraic problem with respect to ω:

$$\omega \, \underset{\sim}{y}(\underline{r},\omega) = \underset{\approx}{L}(\underset{\sim}{\lambda}) \cdot \underset{\sim}{y}(\underline{r},\omega) \tag{4.16}$$

which has nontrivial solutions only if the time constant ω is an eigenvalue of $\underset{\approx}{L}$:

$$\omega = \omega_m \quad . \tag{4.17}$$

The general solution to (4.12-14) is thus a superposition of eigenmodes of $\underset{\approx}{L}$:

$$\omega_m \underset{\sim}{\phi}_m(\underline{r}) = \underset{\approx}{L}(\underset{\sim}{\lambda}) \cdot \underset{\sim}{\phi}_m(\underline{r}) \tag{4.18}$$

$$\underset{\sim}{z}(\underline{r},t) = \sum_m c_m \underset{\sim}{\phi}_m(\underline{r}) \, e^{\omega_m t} \tag{4.19}$$

with coefficients c_m to be fixed by initial conditions. It is clear from the representation (4.19) that:

- If all the eigenvalues of $\underset{\approx}{L}$ have a strictly negative real part, the solution $\underset{\sim}{z} = 0$ is asymptotically stable.

- If an eigenvalue of $\underset{\approx}{L}$ has a strictly positive real part, the solution $\underset{\sim}{z} = 0$ is unstable.

Therefore it is even not necessary to calculate the whole spectrum of $\underset{\approx}{L}$; one has just to verify the sign of the real parts of the eigenvalues[1]. Still it is useful to determine explicitly the values of ω_m as well as the corresponding $\underset{\sim}{\phi}_m$ responsible for such instability: studying each eigenvalue as a function of $\underset{\sim}{\lambda}$, one can determine the range of the control parameters for which the corresponding mode is unstable (see Fig. 4.1).

The theorem of linearized stability permits one to discuss the stability of the reference solution $\underset{\sim}{x}_s$ in terms of a set of real (or complex conjugate) quantities, depending on a vector of control parameters $\underset{\sim}{\lambda}$. Thus, the linearized stability analysis directly leads to the structural stability properties of the reference solution (or branch) $\underset{\sim}{x}_s$.

However, the theorem does not give any indication on the critical case, i.e. when the linearized system is marginally stable. Such cases just mark the border between unstable and asymptotically stable regions of the parameter space for the branch of $\underset{\sim}{x}_s$. As the locus for the onset of instability, they also play an essential role in bifurcation phenomena, since the loss of stability for one solution $\underset{\sim}{x}_s$ is often accompanied by the gain of stability for another solution $\underset{\sim}{x}'_s$, qualitatively different from $\underset{\sim}{x}_s$. The properties of the eigenmodes $(\underset{\sim}{\phi}_m, \omega_m)$ responsible for the loss of stability usually prelude to the properties of the bifurcating branch $\underset{\sim}{x}'_s$.

A final warning is necessary: linear stability analysis refers to one reference state. It provides useful information on the transition from asymptotic stability of this state to another reference solution, i.e. for the first bifurcation from the studied state.

[1] When the system is invariant under time-reversal, the spectrum of $\underset{\approx}{L}$ is symmetrical with respect to zero: this makes a fundamental difference!

But as soon as this state is unstable, the analysis loses its physical relevance, and one has to switch to the study of other branches (which is possible e.g. by another linearization).

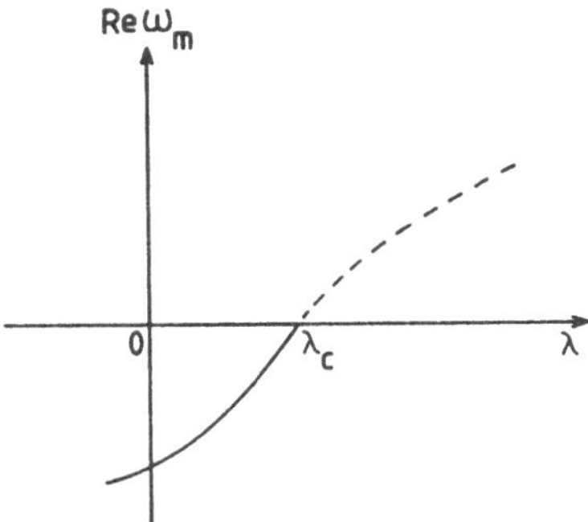

Fig. 4.1. The mode ϕ_m with eigenvalue ω_m is asymptotically stable for $\lambda < \lambda_c$ and unstable for $\lambda > \lambda_c$.

C. Illustration: The Schlögl Model

The homogeneous system limit of the Schlögl model is described by the evolution equations (2.5) with the assumption that all variables are uniform. This implies that the boundary conditions are symmetric and that all parameters are space-independent. To simplify further, we choose them to be time-independent as well. The (unique) state variable $x = x(t)$ obeys the equation

$$\frac{d}{dt} x = k_0 b - k_1 x + k_2 a x^2 - k_3 x^3 \quad . \tag{4.20}$$

We shall study the stability of the stationary solutions of (4.20) for the case:

$$b \neq 0 \quad . \tag{4.21}$$

With the scaling (2.11-12), the evolution equation becomes:

$$\frac{d\xi}{d\tau} = -\xi^3 - \delta\xi + (\delta' - \delta) \quad . \tag{4.22}$$

For simplicity, we further limit ourselves to a single control parameter $\delta = \delta'$: then,

$$\frac{d\xi}{d\tau} = -\xi(\delta + \xi^2) \quad . \tag{4.23}$$

The stationary states follow directly:

$$\xi_s = 0 \tag{4.24}$$

$$\xi_s = \pm\sqrt{-\delta} \quad . \tag{4.25}$$

The first solution exists for all parameter values; the two other solutions branch from the first one at the point $\delta = 0$; they are real for $\delta < 0$ and imaginary (thus non-physical!) for $\delta > 0$.

(i) Linear Analysis

The stability of these three solutions is discussed straightforwardly in terms of the linear operator

$$L = -3\xi_s^2 - \delta \quad . \tag{4.26}$$

For this simplistic case, the eigenvalue is

$$\omega = -3\xi_s^2 - \delta$$

and the (normalized) eigenvector is the unit. We see that for $\xi_s = 0$: $\omega = -\delta$.

Thus this branch is:

- asymptotically stable if $\delta > 0$
- at a critical point if $\delta = 0$
- unstable if $\delta < 0$.

For $\xi_s = \pm\sqrt{-\delta}$, one has $\omega = 2\delta$.

These branches are:

- asymptotically stable if $\delta < 0$
- at a critical point if $\delta = 0$
- nonexistent if $\delta > 0$.

At the critical point, there occurs an exchange of stability between the "reference state" $\xi = 0$ and the other branches, according to the linear analysis. But the behaviour of the system in this case cannot be discussed within the linear approach.

(ii) The Exact Solution

In the exceptionally simple case of systems involving one variable, a direct analysis can be performed using the exact solution of the equation (4.23).

For $\delta \neq 0$, the equation takes the form

$$\left(\frac{1}{\xi} - \frac{\xi}{\delta + \xi^2}\right) \frac{d\xi}{d\tau} = -\delta \tag{4.27}$$

which is integrated to yield,

$$\xi^2 = \frac{-\delta}{1 - A\,\exp(2\delta\tau')} \tag{4.28}$$

with

$$\tau' = \tau - \tau_0 \geq 0$$

$$A = \frac{\xi_0^{\;2}}{\delta + \xi_0^{\;2}}$$

and the initial condition $\xi = \xi_0$ at time $\tau = \tau_0$.

The stationary states are easily recovered (take A = 0 and $A \to \infty$), and we note that nonstationary states evolve towards them (with τ_0 finite)

$$- \text{ if } \quad \delta > 0 \ : \quad \lim_{\tau \to \infty} \xi^2 = 0 \tag{4.29a}$$

$$- \text{ if } \quad \delta < 0 \ : \quad \lim_{\tau \to \infty} \xi^2 = -\delta . \tag{4.29b}$$

For $\delta = 0$, the equation becomes

$$\frac{d\xi}{d\tau} = -\xi^3 \tag{4.30}$$

yielding

$$\xi^2 = \frac{\xi_0^{\,2}}{1+\xi_0^{\,2}\,\tau'} \tag{4.31}$$

with

$$\tau' = \tau - \tau_0 \geq 0$$

and initial condition $\xi(\tau_0) = \xi_0$. In the case $\xi_0 = 0$, we find the stationary solution $\xi = 0$ again. When $\xi_0 \neq 0$, the solution decays to this state in a non-exponential way.

It is easily seen that (4.31) is the limit of (4.27) for $\delta \to 0$. This analysis illustrates Lyapounov's theorem in showing the agreement between the statements on stability derived from a fully nonlinear analysis and from the linearization. We observe that in the noncritical cases ($\delta \neq 0$), the linearized theory correctly describes the asymptotic behaviour (in the limit $t \to \infty$) around the steady states. Moreover, it also shows the inadequacy of the linearized theory for the critical case: when $\delta = 0$, the convergence toward the steady state is only algebraic

$$\xi - \xi_s = 0(\tau^{-\frac{1}{2}}) \tag{4.32a}$$

instead of an exponential trend for $\delta \neq 0$

$$\xi - \xi_s = 0(e^{-|\delta\tau|}) . \tag{4.32b}$$

The solution (4.27) announces this phenomenon by a "critical slowing down" and a critical exponent for ξ equal to 1/2:

$$\tau = 0(|\delta|^{-1}) \tag{4.33a}$$

$$|\xi - \xi_s| = 0(|\delta|^{\frac{1}{2}}) \tag{4.33b}$$

D. General Reaction-Diffusion Systems

The reaction-diffusion equations were introduced in Sect. III.D as special cases of evolution equations like (4.2), of the form

$$\frac{\partial}{\partial t} \underset{\sim}{x}(\underline{r},t) = \underset{\sim}{F}(\underset{\sim}{x}(\underline{r},t);\ \underset{\sim}{\lambda}) + \underset{\approx}{D}\cdot\nabla^2\ \underset{\sim}{x}(\underline{r},t) \tag{4.34}$$

where $\underset{\sim}{F} = \underset{\sim}{F}(\underset{\sim}{x},\underset{\sim}{\lambda})$ is now a usual function of $\underset{\sim}{x}$, $\underset{\approx}{D} = \underset{\approx}{D}(\underset{\sim}{\lambda})$ is the matrix of diffusion coefficients $D_{jk} = D_k\delta^{Kr}_{jk}$. We consider a reference solution $\underset{\sim}{x}_s(\underline{r},t)$ of (4.34) and linearize the evolution equation around it:

$$\underset{\sim}{x} = \underset{\sim}{x}_s + \underset{\sim}{z} \tag{4.35}$$

$$\frac{\partial}{\partial t} \underset{\sim}{z}(\underline{r},t) = \underset{\approx}{F_x}(\underset{\sim}{x}_s,\underset{\sim}{\lambda})\cdot\underset{\sim}{z} + \underset{\approx}{D}\cdot\nabla^2\ \underset{\sim}{z} + 0(\|\underset{\sim}{z}\|) \tag{4.36}$$

with

$$\underset{\approx}{F_x} = \frac{\partial \underset{\sim}{F}}{\partial \underset{\sim}{x}}$$

and homogeneous boundary conditions. Neglecting all higher-order terms, we find that the linear operator related to (4.34) for the study of a reference solution $\underset{\sim}{x}_s$ has the matrix elements

$$L_{jk} = \left(\frac{\partial F_j}{\partial x_k}\right)_s + D_{jk}\nabla^2 \quad . \tag{4.37}$$

Since the spatial dependence only appears through the Laplacian operator, a first step towards the solution of the eigenvalue problem of (4.37) will be the determination of the diffusion eigenmodes (without reactions). This is done by solving the Laplace equation

$$\nabla^2\phi_m = -k_m{}^2\phi_m \tag{4.38}$$

for the given boundary conditions on the domain V, determining the eigenfunctions $\phi_m(\underline{r})$ and the eigenvalues k_m.

When the boundary conditions are either of the Dirichlet or of the Neumann type, the eigenvalues k_m^2 are real and positive, as it becomes obvious from the "variational" expression

$$k_m^2 = -(\phi_m, \nabla^2 \phi_m) = (\nabla\phi_m, \nabla\phi_m) \quad . \tag{4.39}$$

Expanding the disturbance $\underset{\sim}{z}(\underline{r},t)$ in the basis of eigenfunctions ϕ_m, we find solutions of the form

$$\underset{\sim}{z}(\underline{r},t) = \sum_m \underset{\sim}{c}_m \phi_m(\underline{r}) e^{\omega_m t} \tag{4.40}$$

where ω_m and $\underset{\sim}{c}_m$ are given by the linear equation

$$\underset{\approx}{L}_m \cdot \underset{\sim}{c}_m - \omega_m \underset{\sim}{c}_m = 0 \tag{4.41}$$

with

$$\underset{\approx}{L}_m = \underset{\approx}{L}(\underset{\sim}{\lambda}, k_m) = \underset{\approx}{L}_o - \underset{\approx}{D} k_m^2 \tag{4.42}$$

$$\underset{\approx}{L}_o = \left(\frac{\partial \underset{\sim}{F}}{\partial \underset{\sim}{x}} \right) (\underset{\sim}{x}_s; \underset{\sim}{\lambda}) \quad . \tag{4.43}$$

The homogeneous equation (4.41) for $\underset{\sim}{c}_m$ admits a nontrivial solution if and only if the corresponding characteristic equation is satisfied

$$\det(\underset{\approx}{L}_m - \omega_m \underset{\approx}{I}) = 0 \quad . \tag{4.44}$$

This fixes the time-constants ω_m corresponding to the eigenmodes $\underset{\sim}{c}_m \phi_m$ under consideration.

(i) <u>One Variable</u>

If there is only one state variable x, the linear analysis leads to

$$\omega_m = \left(\frac{\partial F}{\partial x} \right)_s - D k_m^2 \quad . \tag{4.45}$$

Since ω_m is real, there is no oscillation, and the only bifurcation branches from a steady state to another steady state.

Moreover, diffusion always stabilizes the system ($D > 0$, $k_m^2 \geq 0$), and this influence increases with the wavenumber k_m.

The first instability occurs for the lowest k_m, i.e. for the homogeneous system whenever possible. The Schlögl model is typical of such all-or-none transitions.

Infinite systems do not suffer from the restricting influence of boundaries; this additional freedom is manifested by the appearance of stable wave trains. We do not discuss this point here, as the theory of waves in dissipative systems has been covered in Professor Howard's lectures [see also Fife, 1978].

(ii) Two Variables

With two state variables, the characteristic equation becomes:

$$\omega_m^2 - T(k_m,\underset{\sim}{\lambda})\omega_m + \Delta(k_m,\underset{\sim}{\lambda}) = 0 \tag{4.46}$$

and therefore may admit complex roots:

$$\omega_m = \sigma_m \pm i\Omega_m \tag{4.47}$$

with

$$\sigma_m = \frac{1}{2} T(k_m,\underset{\sim}{\lambda}) \tag{4.47a}$$

$$\Omega_m^2 = \Delta(k_m,\underset{\sim}{\lambda}) - \sigma_m^2 \tag{4.47b}$$

$$\Delta(k,\underset{\sim}{\lambda}) = \det \underset{\approx}{L}(k,\underset{\sim}{\lambda}) \tag{4.48}$$

$$T(k,\underset{\sim}{\lambda}) = \mathrm{Tr}\, \underset{\approx}{L}(k,\underset{\sim}{\lambda}) = \left(\frac{\partial F_1}{\partial x_1}\right)_s + \left(\frac{\partial F_2}{\partial x_2}\right)_s - (D_{11}+D_{22})k^2 \tag{4.49}$$

When the two solutions (4.46) are complex ($\Omega_m^2 > 0$), the first transition to instability (cf. Fig. 4.1) occurs when Re $\omega_m = \sigma_m$ vanishes ($\Omega_m^2 > 0$):

$$(D_{11} + D_{22})k^2 = T(0,\underset{\sim}{\lambda}) \quad . \tag{4.50}$$

We expect that the transition should lead from a couple of damped oscillatory modes to finite-amplitude periodic solutions (Hopf bifurcation). As for one variable, diffusion plays a stabilizing role, since it always reduces σ_m.

Consider now the case where both solutions ω_m are real. The transition to instability then occurs via a state of non-oscillatory marginal stability where one root $\omega_m = 0$ and the second root remains negative. From (4.46) we verify that the condition for this is

$$\Delta(k,\underset{\sim}{\lambda}) = 0 \quad . \tag{4.51}$$

This equation is quadratic in k^2 and has two, one or no real positive solutions. When the parameters $\underset{\sim}{\lambda}$ are such that k^2 is one of the eigenvalues $k_m{}^2$ of the Laplace equation (4.38) as well as a solution of (4.51), the solution of the linearized equation (4.36) depends on space via the corresponding eigenfunction ϕ_m.

The first transition to instability thus should occur at a critical value $\underset{\sim}{\lambda} = \underset{\sim}{\lambda}_c$ that is the minimal value of $\underset{\sim}{\lambda}$ for which the requirement (4.51) can be satisfied. The corresponding eigenvalue $k^2 = k_c{}^2$ is thus fixed by the chemical kinetics and diffusion coefficients - but it is <u>not</u> related to the size, the geometry or even the dimensionality of the system. For this reason, it is said to be <u>intrinsic</u>.

However, only eigenvalues $k^2 = k_m{}^2$ of the Laplacian can be realized by the system. Therefore, the first instability actually occurs for the eigenvalue $k_1{}^2$, which gives the "effective minimum" of $\underset{\sim}{\lambda}$: this value is slightly different from $k_c{}^2$ in general. Successive instabilities in this region can be related to $k_1{}^2, k_2{}^2, \ldots$ gradually departing from $k_c{}^2$.

Finally, if $\Omega_m = 0$, there is one double, real eigenvalue. The transition then occurs through a degenerate state ($\omega_m = 0$, $\sigma_m = \Omega_m = 0$) and its parameters satisfy two equations:

$$\Delta(k,\underset{\sim}{\lambda}) = 0 \tag{4.52a}$$

$$T(k,\underset{\sim}{\lambda}) = 0 . \tag{4.52b}$$

We shall not discuss the case further.

A general diagram of the (λ, k^2) parameter space for the instabilities is presented in Fig. 4.2. The first transition to instability will occur either with $k^2 = 0$ or with $k^2 = k_1{}^2$, according to which λ is lowest (i.e. first encountered) in the diagram. In the first case, no spatial pattern formation is expected but the evolution will be periodic (complex ω_m). In the second case, a spatial pattern is expected to emerge, with a temporally steady behaviour (real ω_m).

Exceptional cases may be imagined like transitions with two diffusion eigenmodes ($k_1{}^2$ and $k_2{}^2$) or with limit cycles together with spatial patterns: such degenerate cases arise when the values of $\underset{\sim}{\lambda}$, corresponding to both instabilities, coincide. This is very improbable with a scalar parameter λ, but it can be envisaged when one analyzes the stability of the system with respect to one parameter λ_1, keeping a second parameter λ_2 constant. Different choices of λ_2 result in deformations of the qualitatively unchanged diagram (as Fig. 4.2) and a particular choice of λ_2 can lead to diagrams

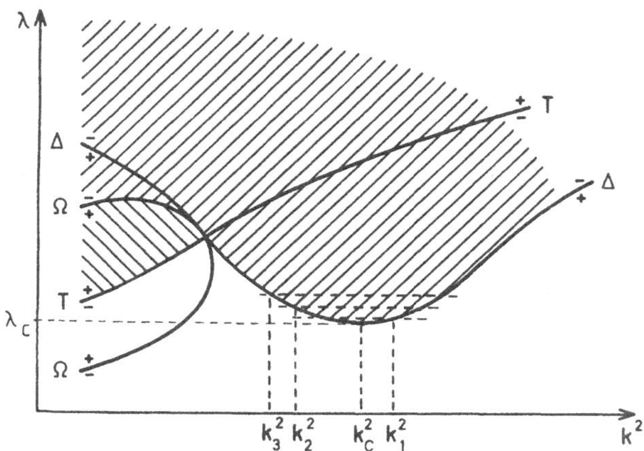

Fig. 4.2. Marginal stability diagram referring to complex conjugate ($\Omega^2 > 0$) and to real ($\Omega^2 < 0$) roots of the characteristic equation (4.46), for a system of two variables.

curve	property		equation
Δ	marginal stability	$\Delta(k,\lambda) = 0$	$\omega = 0$
T	opposite roots	$T(k,\lambda) = 0$	$\omega = \pm i\Omega$
Ω	double real root	$\Delta - \frac{1}{4}T^2 = 0$	$\Omega = 0$

The signs along the lines refer to their defining expressions in these domains. ▭ the undashed domain $\Delta > 0$. $T < 0$ is asymptotically stable. ▨ unstable domain with two comples conjugate roots. ▨ unstable domain with two real roots. λ_c: threshold for the inhomogeneous bifurcations. k_c: intrinsic (critical) wavenumber. k_1, k_2, k_3: successive unstable wavenumbers.

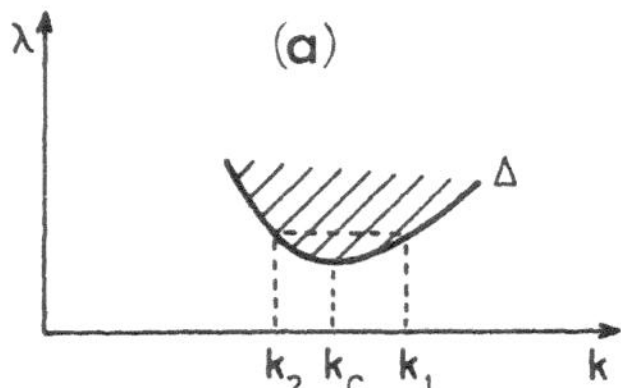

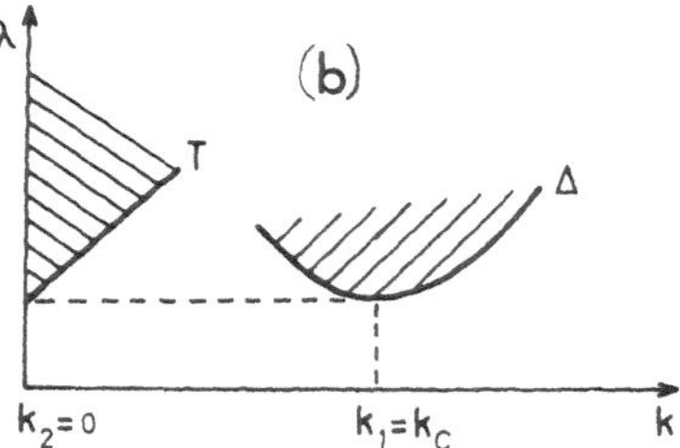

Fig. 4.3. Marginal stability curves allowing for a degenerate bifurcation in a system with two variables. (a) two stationary inhomogeneous modes; (b) one homogeneous time-periodic mode and one inhomogeneous stationary mode.

presented in Fig. 4.3. These phenomena related to secondary or degenerate bifurcations, obviously play an important role in the study of complex systems and will be discussed in Section VI.

In summary, the essential features of this two-variables case are that diffusion may play an active role, in leading to the formation of spatial reproducible inhomogeneities, and that periodic evolution can also be triggered by the chemical kinetics (at k = 0). In all cases, the structure arising from the transition is mainly intrinsic (with small adaptation to the boundary conditions). This is due to the coupling between scalar and vector phenomena, yielding natural length scales, and to the nonlinearities of the reaction kinetics, introducing natural time scales. This situation is very peculiar even in the context of dissipative systems (for example, the size of the Bénard convection cells is extrinsic, as it is proportional to the depth of the fluid layer [Chandrasekhar, 1961]).

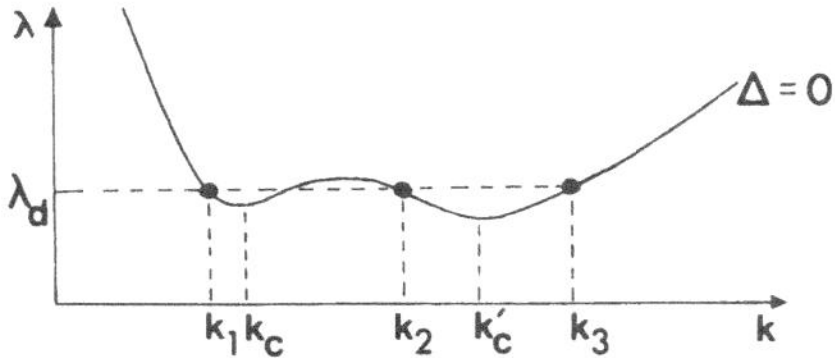

Fig. 4.4. Marginal stability curve for real roots of the characteristic equation, for a system of four variables. λ_d: value of λ corresponding to a triple degeneracy (k_1^2, k_2^2, k_3^2 are three eigenvalues of the Laplacian).

(iii) Three or More Variables

The cases in which more than two variables are coupled, may be discussed along the same lines. More complicated behaviours must be expected, since the degree of the characteristics equations is higher in ω as well as in k^2. For example, there may be several oscillatory modes of incommensurable frequencies, or interacting spatial modes, or combinations of such effects (Fig. 4.4). These situations may lead to chaotic evolutions.

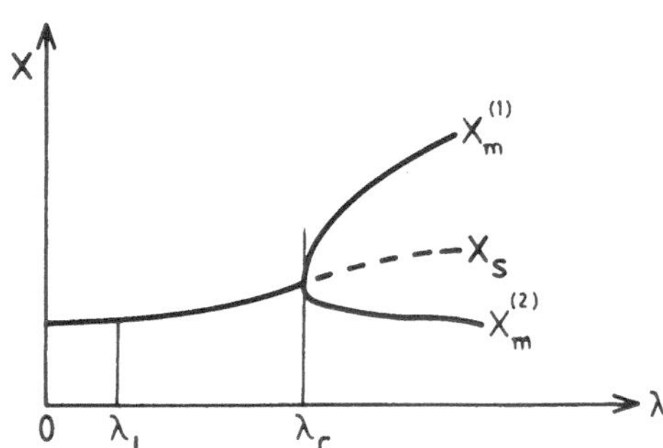

Fig. 4.5. The various ranges of thermodynamics.

parameter	state	Lyapounov function
$\lambda = 0$	equilibrium	entropy
$0 \leq \lambda < \lambda_\ell$	linear range	(quadratic) entropy production
$0 \leq \lambda < \lambda_c$	thermodynamic branch x_s	excess entropy production
$\lambda = \lambda_c$	critical threshold	
$\lambda > \lambda_c$	$\left\{\begin{array}{c}\text{new solutions: } x_m^{(1)}, x_m^{(2)} \\ \text{dissipative structures}\end{array}\right\}$	$\left\{\begin{array}{c}\text{the study of fluctuations} \\ \text{is necessary}\end{array}\right.$

The new solutions cannot be obtained by simple extrapolation from the range.

(iv) Conclusion of Physical Motivation

The example of reaction-diffusion systems indicates the vast potentialities of nonequilibrium systems. It also emphasizes some basic differences between equilibrium and nonequilibrium transition phenomena.

At equilibrium a well-known theorem of Duhem [Glansdorff-Prigogine, 1971, Section 4.5] states that stability with respect to diffusion ensures chemical stability. In our examples, diffusion is trivally stable (the diffusion coefficient matrix is positive definite), and yet we have a number of transition phenomena induced by chemical kinetics (e.g. limit cycles) or by the coupling of chemistry with diffusion (like pattern formation). The clue in this paradox can be found in more general thermodynamic stability criteria, valid far from equilibrium. Although no completely universal criterion has yet been demonstrated, it is striking to remark that when the transition occurs, the contribution of the critical mode $\underset{\sim}{\phi}_m$ to the excess entropy production just vanishes [Glansdorff-Prigogine, 1971 : Section 15.3; Nicolis-Prigogine, 1977: Section 8.11], which is impossible near the equilibrium state.

A typical thermodynamic diagram is sketched on Fig. 4.5.

V. BIFURCATION ANALYSIS

In the next two sections, we shall be concerned with the construction of the solutions of the nonlinear evolution equations (4.2-4) beyond a point of marginal stability. The most appropriate tool for this is bifurcation analysis. This method originates in Poincaré's works on classical mechanics and is currently an important chapter of modern mathematics [Sattinger, 1973; Sattinger, 1979; Thom, 1972].

Our main objective is to construct systematic perturbative expansions in relevant small parameters[2], resulting in converging iteration schemes in a neighbourhood of some reference state. When the state is noncritical, the method amounts to linearization, but at a critical point, one obtains reliable descriptions of these various branches near the point.

On the other hand, general theorems and criteria have been proved by topological arguments about the structure of the solutions; this approach, initiated by Leray and Schauder, gives results "in the large" and is not limited to neighbourhoods of the point under consideration.

---

[2] The small parameter is generally the amplitude of the destabilizing mode of linear analysis.

We shall first quote fundamental general theorems (V.A) and discuss the simple cases of primary bifurcations from a stationary state by expansion methods (V.B-D). These calculations lead us to a first interpretation of bifurcation phenomena (V.E), which is exploited in various examples (V.F).

A. General Statements

As in Section IV, we consider nonlinear equations of the type

$$\frac{\partial}{\partial t} \underset{\sim}{x}(\underline{r},t) = \underset{\sim}{G}[\underset{\sim}{x};\underset{\sim}{\lambda}]$$

and discuss the stability of a reference solution $\underset{\sim}{x}_s(\underline{r},t;\underset{\sim}{\lambda})$ as the control parameter $\underset{\sim}{\lambda}$ varies. Suppose that this reference solution is marginally stable when[3] $\underset{\sim}{\lambda} = \underset{\sim}{\lambda}_c$, because of some eigenvalue $\omega_c(\underset{\sim}{\lambda})$ crossing the imaginary axis at this critical point. The Leray-Schauder theory of topological degree leads to the following basic theorems, whose rigorous vocabulary and proof can be found in Sattinger's book [1973, Section 7.5].

In the statements of the theorems, we shall refer to a scalar parameter λ obtained by the choice of any line going through $\underset{\sim}{\lambda}_c$ in the $\underset{\sim}{\lambda}$-space; this is needed to define an order relation on this line.

Theorem B1

If an eigenvalue ω_c of the linearized equation associated to a branch $\underset{\sim}{x}_s$ has odd multiplicity for $\lambda < \lambda_c$, and becomes zero at $\lambda = \lambda_c$, then there is at least one branch outgoing from $(\underset{\sim}{x}_s,\lambda_c)$. Moreover, this "outgoing branch" either extends to infinity or meets another bifurcation point.

Theorem B2

Under the same hypotheses, if ω_c is a simple eigenvalue at $\lambda = \lambda_c$, then there are only two branches intersecting at (and only at) the bifurcation point $(\underset{\sim}{x}_s,\lambda_c)$.

These theorems are immediately extended to the case of a vector parameter $\underset{\sim}{\lambda}$, by considering a branch as a manifold of dimension p, and by constructing it from "generating lines" (i.e. sections by planes) with usual continuity assumptions.

---

[3] Note that if $\underset{\sim}{\lambda} \in \mathbb{R}^p$, the set of such points $\underset{\sim}{\lambda}_c$ has in general dimension p - 1.

Finally, the stability of the branches can be investigated; it depends on the way the branches cross each other on the critical manifold. So we introduce the following:

Definition 1

Consider two branches $(\underset{\sim}{x}_s,\lambda)$ and $(\underset{\sim}{x}'_s,\lambda)$, depending on a scalar parameter λ, crossing each other at a critical point $\lambda = \lambda_c(\underset{\sim}{x}_s = \underset{\sim}{x}'_s)$. Suppose that $\underset{\sim}{x}_s$ is asymptotically stable for $\lambda < \lambda_c$. A solution $\underset{\sim}{x}(\lambda)$ is said to be:

- subcritical if it has $\lambda < \lambda_c$
- supercritical if it has $\lambda > \lambda_c$

and the branch $(\underset{\sim}{x}'_s,\lambda)$ is said to be

- subcritical if all its elements $(\underset{\sim}{x}'_s,\lambda)$ are subcritical
- supercritical if all its elements are supercritical
- transcritical if some of its elements are subcritical and others are supercritical.

Definition 2

In the same context, if the branch $(\underset{\sim}{x}_s,\lambda)$ becomes unstable because of an eigenvalue $\omega_c(\lambda)$, this bifurcation is said to be transverse if

$$\mathrm{Re}\,\frac{d\omega_c}{d\lambda}\,(\lambda_c) \neq 0\ . \tag{5.1}$$

This vocabulary is easily extended to the general case of vector control parameters. We now state the theorem of stability for bifurcating branches [Sattinger, 1979, Sec. 3.2 and 4.4] illustrated by Fig. 5.1:

Theorem B3

If a branch $(\underset{\sim}{x}_s,\lambda)$, asymptotically stable for $\lambda < \lambda_c$ loses its stability at a simple critical point because of a transverse eigenvalue $\omega_c(\lambda)$, crossing the imaginary axis, then:

1) supercritical bifurcating branches are stable, and subcritical bifurcating branches are unstable;

2) bifurcating solutions from a stationary state will be
   sationary if $\mathrm{Im}\ \omega_c(\lambda_c) = 0$ and
   time periodic if $\mathrm{Im}\ \omega_c(\lambda_c) \neq 0$.

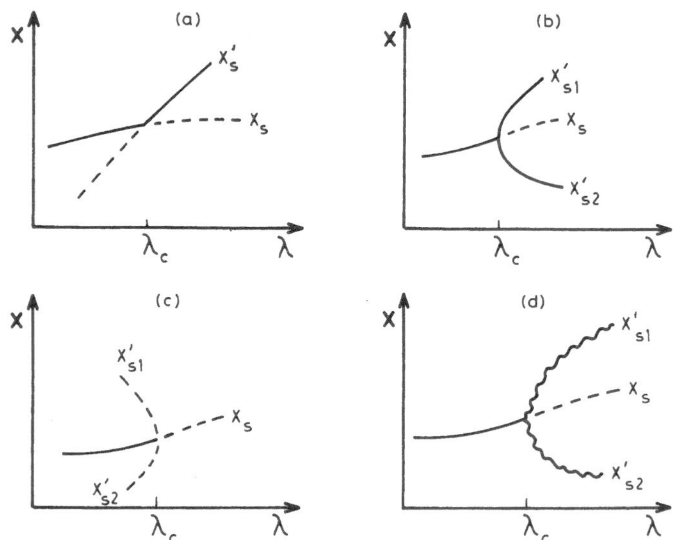

Fig. 5.1. Bifurcations by a simple (transversal) eigenvalue from a stationary solution x_s to:

(a) one stationary solution x_s' ;

(b) two stable stationary solutions x_{s1}' , x_{s2}' ;

(c) two unstable stationary solutions x_{s1}' , x_{s2}' ;

(d) one limit cycle x_s' ; the "branch" x_{s2}' differs from x_{s1}' only by its phase.

Conventions: —— denotes a stable solution;
--- denotes an unstable solution;
∿∿ denotes a periodic solution.

We also quote one version of the center manifold theorem [Lanford, 1973], which generalizes the fixed point theorem in the case of patterned attractors:

Theorem B4

Consider a branch $(\underset{\sim}{x}_s,\lambda)$, asymptotically stable for $\lambda < \lambda_c$, which loses its stability at $\lambda = \lambda_c$ because of d eigenvalues

$$\{\omega_k(\lambda)\}_{1\leq k\leq d} \,.$$

At the critical point, the reference solution $\underset{\sim}{x}_s$ may be unstable, but there is an invariant manifold, of dimension d, attracting all solutions within a neighbourhood of the reference solution $\underset{\sim}{x}_s$. Moreover, this manifold merges to the vector space generated by the destabilizing eigenmodes.

In other words, the center manifold here defined is a (locally) stable attractor for the critical state $\lambda = \lambda_c$; this manifold makes the connection between the various branches merging at the critical state.

B. Bifurcation Equations by the Poincaré-Lindstedt Series

The bifurcation equations arise from an attempt to describe together all the branches bifurcating from a critical manifold, without solving exactly the differential equation. To this end, we write the nonlinear equations (4.9) for the excess variable $\underset{\sim}{z}$ in the general form:

$$\forall\, \underline{r} \in V,\ \forall\, t \in \mathbb{R} :\ \underset{\approx}{L}(\lambda)\cdot\underset{\sim}{z} + \underset{\sim}{h}[\underset{\sim}{z};\underset{\sim}{\lambda}] \;=\; \frac{\partial}{\partial t}\,\underset{\sim}{z} \tag{5.2}$$

with the homogeneous boundary conditions (4.11-12):

either

$$\underset{\sim}{z}(\underline{r},t) \;=\; 0 \tag{5.2a}$$

or

$$\underline{n}\cdot\nabla\underset{\sim}{z}(\underline{r},t) \;=\; 0\,. \tag{5.2b}$$

Here we have introduced the definitions:

$$\underset{\sim}{z} \;=\; \underset{\sim}{x} - \underset{\sim}{x}_s(\underset{\sim}{\lambda}) \tag{5.3}$$

$$\underset{\approx}{L}(\underset{\sim}{\lambda}) = \underset{\sim}{G}_{\underset{\sim}{x}}[\underset{\sim}{x}_s(\lambda);\underset{\sim}{\lambda}] \tag{5.4a}$$

$$\underset{\sim}{h}[\underset{\sim}{z};\underset{\sim}{\lambda}] = \underset{\sim}{G}[\underset{\sim}{x};\underset{\sim}{\lambda}] - \underset{\sim}{G}[\underset{\sim}{x}_s;\underset{\sim}{\lambda}] - \underset{\approx}{L}(\underset{\sim}{\lambda})\cdot\underset{\sim}{z} \tag{5.4b}$$

$$\underset{\sim}{G}_{\underset{\sim}{x}} = \frac{\delta \underset{\sim}{G}}{\delta \underset{\sim}{x}}$$

The dot stands for the action of a (multi-) linear operator.

The reference state $\underset{\sim}{x}_s$ is taken on a branch which loses its stability at the critical point $\underset{\sim}{\lambda} = \underset{\sim}{\lambda}_c$; the choice of the boundary conditions (5.2a) or (5.2b) is related to the relevant boundary conditions (4.4a) or (4.4b).

We express the vicinity of the critical manifold by expanding the state function $\underset{\sim}{x}$ and the parameter $\underset{\sim}{\lambda}$ in powers of a (formal) small parameter ε through the so-called Poincaré-Lindstedt series:

$$\underset{\sim}{\gamma} = \underset{\sim}{\lambda} - \underset{\sim}{\lambda}_c = \varepsilon\underset{\sim}{\gamma}_1 + \varepsilon^2\underset{\sim}{\gamma}_2 + \ldots \tag{5.5a}$$

$$\underset{\sim}{z} = \underset{\sim}{x} - \underset{\sim}{x}_s(\underset{\sim}{\lambda}) = \varepsilon\underset{\sim}{z}_1 + \varepsilon^2\underset{\sim}{z}_2 + \ldots \tag{5.5b}$$

Note that ε has no physical meaning; its only use here is to display symmetrically the relation between $\underset{\sim}{x}$ and $\underset{\sim}{\lambda}$ near the critical manifold. This relation will lead to parametric equations for the branches.

To the zeroth order, insertion of (5.5) into (4.2) yields:

$$O(1): \quad \frac{\partial}{\partial t}\underset{\sim}{x}_s = \underset{\sim}{G}[\underset{\sim}{x}_s;\underset{\sim}{\lambda}] \tag{5.6a}$$

This is just the equation for the reference state. Of more interest for our purposes are the equations to the higher orders:

$$O(\varepsilon): \quad \frac{\partial}{\partial t}\underset{\sim}{z}_1 - \underset{\approx}{L}_c\cdot\underset{\sim}{z}_1 = 0 \tag{5.6b}$$

$$O(\varepsilon^2): \quad \frac{\partial}{\partial t}\underset{\sim}{z}_2 - \underset{\approx}{L}_c\cdot\underset{\sim}{z}_2 = \underset{\sim}{\gamma}_1\cdot\underset{\approx}{L}_{\underset{\sim}{\lambda}}\cdot\underset{\sim}{z}_1 + \tfrac{1}{2}\underset{\sim}{h}_{\underset{\sim}{x}\underset{\sim}{x}}:\underset{\sim}{z}_1\underset{\sim}{z}_1 \tag{5.6c}$$

$$\begin{aligned} O(\varepsilon^3): \quad \frac{\partial}{\partial t}\underset{\sim}{z}_3 - \underset{\approx}{L}_c\cdot\underset{\sim}{z}_3 = {} & \underset{\sim}{\gamma}_1\cdot\underset{\approx}{L}_{\underset{\sim}{\lambda}}\cdot\underset{\sim}{z}_2 + \underset{\sim}{\gamma}_2\cdot\underset{\approx}{L}_{\underset{\sim}{\lambda}}\cdot\underset{\sim}{z}_1 \\ & + \tfrac{1}{2}\underset{\sim}{\gamma}_1\underset{\sim}{\gamma}_1:\underset{\approx}{L}_{\underset{\sim}{\lambda}\underset{\sim}{\lambda}}\cdot\underset{\sim}{z}_1 \\ & + \underset{\sim}{h}_{\underset{\sim}{x}\underset{\sim}{x}}:\underset{\sim}{z}_1\underset{\sim}{z}_2 + \frac{1}{6}\underset{\sim}{h}_{\underset{\sim}{x}\underset{\sim}{x}\underset{\sim}{x}}\vdots\underset{\sim}{z}_1\underset{\sim}{z}_1\underset{\sim}{z}_1 \\ & + \tfrac{1}{2}\underset{\sim}{\gamma}_1\cdot\underset{\sim}{h}_{\underset{\sim}{x}\underset{\sim}{x}\underset{\sim}{\lambda}}:\underset{\sim}{z}_1\underset{\sim}{z}_1 \end{aligned} \tag{5.6d}$$

where all derivatives are taken at the critical solution:

$$\underset{\approx}{L}_{\lambda} = \underset{\approx}{L}_{\lambda}(\underset{\sim}{\lambda}_c) = \left(\frac{\partial}{\partial \underset{\sim}{\lambda}} \frac{\delta \underset{\sim}{G}}{\delta \underset{\sim}{x}}\right)[\underset{\sim}{x}_s ; \underset{\sim}{\lambda}_c]$$

$$\underset{\approx}{L}_c = \underset{\approx}{L}(\underset{\sim}{\lambda}_c) \ . \tag{5.7}$$

This Lyapounov-Schmidt hierarchy (5.6b-d) has a very simple structure, reflecting the complementarity between linarization and bifurcation analysis. The first-order equation (5.6b) just reduces to a linear homogeneous problem, identical to the linearized problem (4.11,12,14). The higher-order equations are inhomogeneous, with the same linear operator $\underset{\approx}{\mathcal{L}} = \underset{\approx}{L}_c - \partial/\partial t$ acting on the unknown function $\underset{\sim}{z}_k$; they are solvable only if their r.h.s. belong to the range of $\underset{\approx}{\mathcal{L}}$, i.e. if it is orthogonal[4] to the null-space of $\underset{\approx}{\mathcal{L}}$. This implies that the parameters γ_k satisfy a hierarchy of equations, the solution of which provides γ as a power series in ε. Then the solution of the linear equations in $\underset{\sim}{z}_k$ leads to $\underset{\sim}{x}$ as a joint series in ε. All these equations together thus describe the branches arising from the critical manifold near $(\underset{\sim}{\lambda}_c, \underset{\sim}{x}_s)$. Note that once the dominant order for $\underset{\sim}{z}$ has been solved, it is sufficient to solve the higher order equation within the range of $\underset{\approx}{\mathcal{L}}$ only; the arbitrary additional contribution from the null-space of $\underset{\approx}{\mathcal{L}}$ would be redundant with the dominant order and may always be omitted with no loss of generality.

Since the null-space of $\underset{\approx}{\mathcal{L}}$ will play a major role in this construction, we shall first consider the simplest case of a one-dimensional null-space. The multidimensional null-space will be investigated in Section VI.

Moreover, for the sake of clarity and for technical convenience, we shall not immediately discuss the solution of system (5.6) in general: we first limit ourselves to the case of a simple (real) eigenvalue with a scalar parameter λ (V.C). In Section V.D we shall indicate how these analyses can be extended to the emergence of periodicity, generated by simple complex eigenvalues.

C. Bifurcation at a Simple Eigenvalue: Stationary Solutions

In order to avoid technical complications, we first consider a bifurcation between stationary states:

[4] The appropriate definition of orthogonality depends on the form of $\underset{\approx}{\mathcal{L}}$ together with its boundary and initial condition.

$$\frac{\partial}{\partial t} \underset{\sim}{x} = 0 \ . \tag{5.8}$$

For reaction-diffusion systems, $\underset{\approx}{L}$ includes second-order spatial derivatives with constant coefficients (in our models); with our boundary conditions (5.2), the scalar product must be

$$(\underset{\sim}{y},\underset{\sim}{z}) = \int_V \sum_k y_k(\underline{r})\ z_k(\underline{r})\ d^q\underline{r} \ . \tag{5.9}$$

The null-space of $\underset{\approx}{L}_c$ is generated by the (normalized) eigenfunction $\underset{\sim}{\phi}_m$ associated to the vanishing eigenvalue ω_m. Therefore, the solution $\underset{\sim}{z}_1$ of (5.6c) is simply:

$$\underset{\sim}{z}_1 = A_1 \underset{\sim}{\phi}_m \tag{5.10}$$

with a free "amplitude" parameter $A \in \mathbb{R}$, and the second-order equatio reads:

$$\underset{\approx}{L}_c \cdot \underset{\sim}{z}_2 = -\gamma_1 A_1\ \underset{\approx}{L}_\lambda \cdot \underset{\sim}{\phi}_m - \tfrac{1}{2}\ A_1^2\ \underset{\sim}{\psi}_{mm} \tag{5.11}$$

where

$$\underset{\sim}{\psi}_{mm} = \underset{\sim}{h}_{xx} : \underset{\sim}{\phi}_m \underset{\sim}{\phi}_m \ . \tag{5.12}$$

Since the linear operator $\underset{\approx}{L}_c$ is singular, we impose the solvability condition discussed in the previous section. After a straightforward algebra one obtains:

$$-\gamma_1 A_1 P_1 + A_1{}^2 P_2 = 0 \tag{5.13}$$

where

$$P_1 = (\underset{\sim}{\phi}_m{}^*, \underset{\approx}{L}_\lambda \cdot \underset{\sim}{\phi}_m) = \frac{\partial\omega_m}{\partial\lambda}(\lambda_c) \tag{5.14a}$$

$$P_2 = -\tfrac{1}{2}(\underset{\sim}{\phi}_m{}^*, \underset{\sim}{\psi}_{mm}) \ . \tag{5.14b}$$

This solvability condition relates the first-order corrections γ_1 and $\underset{\sim}{z}_1$ near the critical point. However, it becomes trivial when P_1 or P_2 vanish, in which case we must continue the calculation to higher orders.

In order to keep the discussion short, we shall admit here the transversality hypothesis

$$\mathrm{Re}\ \frac{\partial\omega_m}{\partial\lambda}(\lambda_c) = 0 \ . \tag{5.15}$$

i) If $P_2 \neq 0$

If both coefficients in (5.13) are finite, we introduce the normalized amplitude

$$\beta = \varepsilon A_1 . \tag{5.16a}$$

Observing that to this order, $\varepsilon\gamma_1$ is simply

$$\varepsilon\gamma_1 = \gamma = \lambda - \lambda_c \tag{5.16b}$$

one obtains after multiplying (5.13) by ε^2

$$-\gamma P_1 \beta + P_2 \beta^2 = 0 \tag{5.17}$$

with

$$\underset{\sim}{z} = \beta \underset{\sim}{\phi}_m + 0(\beta) .$$

This bifurcation equation describes two branches emerging from the critical state:

- $\beta = 0$ the reference branch
- $\beta = \frac{P_1}{P_2}\gamma$ the bifurcating branch.

From this analysis, one can trace the graphic representation of the bifurcation to lowest order (Fig. 5.2a). Going back to the initial variable $\underset{\sim}{x}$ by determining all terms in the Poincaré-Lindstedt series, one gets the picture of the whole branches (Fig. 5.2b): the first approximation described by (5.17) replaces the branches $(\underset{\sim}{x}_s, \underset{\sim}{x}_s')$ by their tangents $(\underset{\sim}{x}_t, \underset{\sim}{x}_t')$ at the critical point:

$$\underset{\sim}{x}_s = \underset{\sim}{x}_s(\lambda)$$

$$\underset{\sim}{x}_s' = \underset{\sim}{x}_s'(\lambda) = \underset{\sim}{x}_s(\lambda) + \underset{\sim}{z}(\lambda)$$

$$\underset{\sim}{x}_t = \underset{\sim}{x}_s(\lambda_c) + (\lambda - \lambda_c)\left(\frac{\partial \underset{\sim}{x}_s}{\partial \lambda}\right)_{\lambda = \lambda_c}$$

$$\underset{\sim}{x}_t' = \underset{\sim}{x}_s'(\lambda_c) + (\lambda - \lambda_c)\left(\frac{\partial \underset{\sim}{x}_s'}{\partial \lambda}\right)_{\lambda = \lambda_c}$$

$$\underset{\sim}{x}_t' = \underset{\sim}{x}_t + \frac{P_1}{P_2}(\lambda - \lambda_c)\underset{\sim}{\phi}_m \tag{5.18}$$

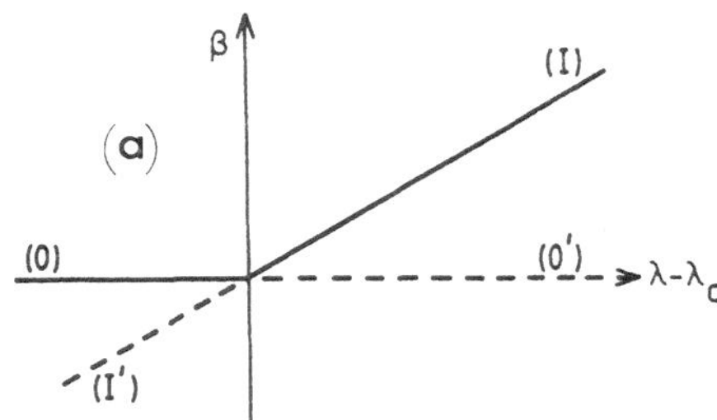

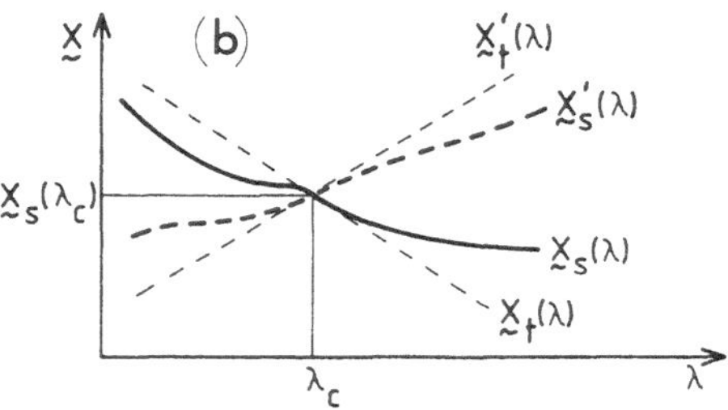

Fig. 5.2. All-or-none bifurcation between stationary states: (a) Near the critical point; (b) Global representation.

(0) reference solution, stable branch $\underset{\sim}{x}_s(\lambda)$

(0') reference solution, unstable branch $\underset{\sim}{x}_s(\lambda)$

(I) bifurcating solution, stable branch $\underset{\sim}{x}_s'(\lambda)$

(I') bifurcating solution, unstable branch $\underset{\sim}{x}_s'(\lambda)$

The branches intersect at the critical point: $\lambda = \lambda_c$, $\underset{\sim}{x} = \underset{\sim}{x}_s(\lambda_c)$. Near the bifurcation point, the branches are approximated by their tangents; in Fig. (a), β is the amplitude of the dominant mode.

ii) <u>If $P_2 = 0$</u>

If the coefficient P_2 vanishes, the bifurcating branch crosses the reference branch with a "vertical tangent"

$$\gamma_1 = 0 \quad , \quad A_1 \neq 0 \quad . \tag{5.19}$$

Then the function $\underset{\sim}{\psi}_{mm}$ belongs to the range of the linear operator $\underset{\approx}{L}_c$ and we can solve (5.12) in the form

$$\underset{\sim}{z}_2 = -\tfrac{1}{2} A_1^2 \underset{\sim}{\Gamma}_2 \tag{5.20}$$

where

$$\underset{\sim}{\Gamma}_2 = \underset{\approx}{K} \cdot \underset{\sim}{\psi}_{mm} \tag{5.21}$$

and $\underset{\approx}{K}$ is the regular operator $\underset{\approx}{L}_c^{-1}$ restricted to the range of $\underset{\approx}{L}_c$. The next equation is then (5.6d)

$$\underset{\approx}{L}_c \cdot \underset{\sim}{z}_3 = -\gamma_2 A_1 \underset{\approx}{L}_\lambda \cdot \underset{\sim}{\phi}_m - A_1^3 \underset{\sim}{\psi}_{mmm} \tag{5.22}$$

where

$$\underset{\sim}{\psi}_{mmm} = -\tfrac{1}{2} \underset{\approx}{h}_{xx} : \underset{\sim}{\phi}_m \underset{\sim}{\Gamma}_2 + \frac{1}{6} \underset{\approx}{h}_{xxx} \vdots\, \underset{\sim}{\phi}_m \underset{\sim}{\phi}_m \underset{\sim}{\phi}_m \quad . \tag{5.23}$$

The solvability condition for (5.22) finally yields:

$$-\gamma_2 A_1 P_1 + A_1^3 P_3 = 0 \tag{5.24}$$

where

$$P_3 = -(\underset{\sim}{\phi}_m^{*}, \underset{\sim}{\psi}_{mmm}) \quad . \tag{5.25}$$

As for equation (5.14), we must discuss (5.24) according to the value of P_3:

1) If $P_3 \neq 0$, to this order, the control parameter is

$$\lambda - \lambda_c = \gamma = \varepsilon^2 \gamma_2 + O(\varepsilon^2) \quad . \tag{5.26a}$$

Introducing the normalized amplitude

$$\beta = \varepsilon A_1 \tag{5.26b}$$

so that

$$\underset{\sim}{x} - \underset{\sim}{x}_s = \underset{\sim}{z} = \beta \underset{\sim}{\phi}_m + O(\beta) \tag{5.26c}$$

we rewrite (5.24) as

$$-(\lambda - \lambda_c)P_1\beta + P_3\beta^3 = 0 . \qquad (5.27)$$

This bifurcation equation describes again the reference branch $\beta = 0$ and the outcoming branches

$$\beta = \pm\left[\frac{P_1}{P_3}(\lambda - \lambda_c)\right]^{1/2} . \qquad (5.28)$$

The bifurcating branches are now approximated near the critical point by a parabola (Fig. 5.3a) with its axis on the line $\beta = 0$. The branches will be either subcritical or supercritical, according to the sign of P_3. Again, this picture is only an approximation to the true branches, which are found by resummation of the series (Fig. 5.3b).

2) If $P_3 = 0$, a similar development to higher orders will generally lead to bifurcation equations of the form

$$-\gamma_n A_1 P_1 + P_{n+1} A^{n+1} = 0 \qquad (n \in \mathbb{N}) \qquad (5.29)$$

with $P_k = 0$ for any $k(2 \leq k \leq n)$. They result, with the normalized quantities

$$\beta = \varepsilon A_1 \qquad (5.30a)$$

$$\lambda - \lambda_c = \gamma = \varepsilon^n \gamma_n$$

in the reference branch $\beta = 0$ and bifurcating branches

$$\beta^n = \frac{P_1}{P_{n+1}}\gamma \qquad (5.31)$$

to dominant order. For a real state variable $\underset{\sim}{x}$, this equation describes either one or two real branches for β.

3) Finally, if all coefficients P_n arising from the successive solvability conditions are zero, except for P_1, we meet a very singular case:

- either the reference branch crosses a "vertical branch", composed of a continuum of allowed states with the same value of the parameter $\lambda = \lambda_c$;

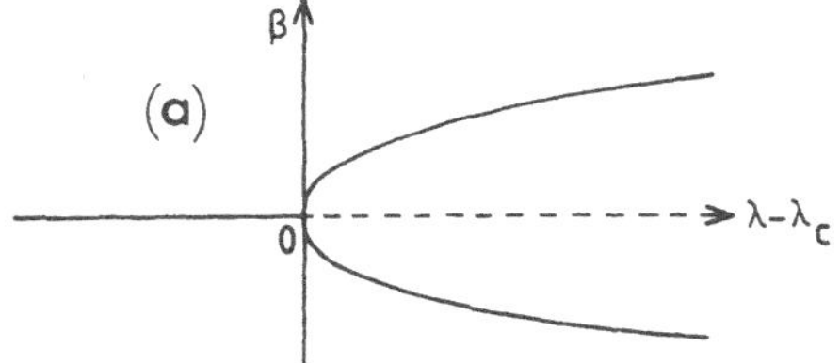

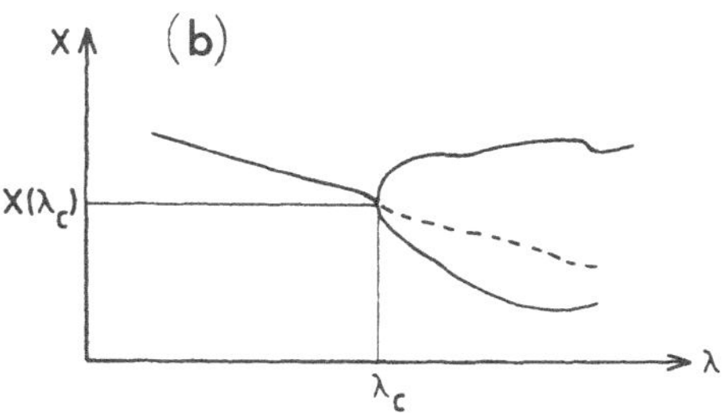

Fig. 5.3. Cusp bifurcation between stationary states. (a) Near the bifurcation point; (b) Global picture.

- or the reference branch loses its stability by crossing another branch whose equation is not analytic in λ and $\underset{\sim}{x}$ near the critical point;

- or the loss of stability comes from a more complicated process.

However, the last possibility has been excluded by our hypotheses in Section V.B.

iii) General Remarks

In this analysis, the small parameter was found to be actually the normalized amplitude β of the unstable mode; this follows from our transversality assumption (5.15) which is satisfied by physical systems. The multivaluedness of β as a function of λ, as it may arise from (5.31) as for (5.28) is not incompatible with the nondegeneracy of the eigenvalue ω_m: all branches here refer to the same (unique, real) eigenmode $\underset{\sim}{\phi}_m$.

In all cases, theorems B2 and B3 (see Section V.A) permit a determination of the stability of bifurcating branches.

The extension to a vector control parameter $\underset{\sim}{\lambda}$ is straightforward.

D. Bifurcation to Limit Cycles

(i) Preliminary Remarks

An equally important part of the classification of bifurcations are the transitions to periodic solutions, or Hopf bifurcations [Sattinger, 1973, 1979]. These are related to the instabilities of modes with complex eigenvalues

$$\begin{aligned} \omega_m &= \sigma_m + i\Omega_m \\ \bar{\omega}_m &= \sigma_m - i\Omega_m \end{aligned} \qquad (5.32)$$

occurring in the linear analysis of equation (5.6b)

$$\frac{\partial}{\partial t} \underset{\sim}{z}_1 = \underset{\approx}{L} \cdot \underset{\sim}{z}_1 = 0 \; . \qquad (5.33)$$

Here the time-dependence plays an essential role and cannot be removed as for a simple real eigenvalue. Also, the real combinations of the complex eigenfunctions form a two-parameter family:

$$\underset{\sim}{z}_1(\underline{r},t) = A_1 e^{\sigma_m t}\left(e^{i\Omega_m(t-t_o)} \underset{\sim}{\phi}(\underline{r}) + e^{-i\Omega_m(t-t_o)} \bar{\underset{\sim}{\phi}}(\underline{r})\right) \qquad (5.34)$$

where A_1 is the arbitrary amplitude and t_o is the arbitrary initial time. The overbar denotes complex conjucation.

It is convenient to scale the time-behaviour by introducing [Joseph and Sattinger, 1972]:

$$\tau = \Omega t \tag{5.35}$$

where $2\pi/\Omega$ is the exact period of the motion, such that the solution $\underset{\sim}{z}$ is 2π-periodic. The Lyapounov-Schmidt hierarchy is then generated by the nonlinear equation

$$\Omega \frac{\partial}{\partial\tau} \underset{\sim}{z} - \underset{\approx}{L}(\lambda) \cdot \underset{\sim}{z} = \underset{\sim}{h}[\underset{\sim}{z};\lambda] \tag{5.36}$$

with the periodicity condition in τ:

$$\underset{\sim}{z}(\underline{r},0) = \underset{\sim}{z}(\underline{r},2\pi) . \tag{5.37}$$

The boundary conditions (5.2) remain unchanged. The Poincaré-Lindstedt series (5.5)

$$\underset{\sim}{x} - \underset{\sim}{x}_s(\lambda) = \varepsilon \underset{\sim}{z}_1 + \varepsilon^2 \underset{\sim}{z}_2 + \ldots \tag{5.38a}$$

$$\lambda - \lambda_c = \varepsilon\gamma_1 + \varepsilon^2\gamma_2 + \ldots \tag{5.38b}$$

are not affected by this scaling, but an additional expansion is obviously needed:

$$\Omega - \Omega_c = \varepsilon\alpha_1 + \varepsilon^2\alpha_2 + \ldots \tag{5.38c}$$

in order to respect the dependence of the period on the control parameter. We have introduced the reference pulsation:

$$\Omega_c = \Omega_m(\lambda_c) = \text{Im}\ \omega_m(\lambda_c) . \tag{5.39}$$

The important point with the periodicity condition (5.37) is that we shall look directly for the time-periodic sclutions, in agreement with Theorem B3.

(ii) Bifurcation Analysis

The nonlinear equation (5.36) may be expanded into a hierarchy as (5.6) by insertion of the series (5.38). To first order in ε, one recovers the linear equation

$$\Omega_c \frac{\partial}{\partial\tau} \underset{\sim}{z}_1 - \underset{\approx}{L}_c \cdot \underset{\sim}{z}_1 = 0 \tag{5.40}$$

resulting in the general solution

$$\underset{\sim}{z}_1(\underline{r},\tau) = \underset{\sim}{z}_1(\underline{r},\tau;|A_1|,\tau_o) = A_1\underset{\sim}{\chi}_m(\underline{r},\tau) + \bar{A}_1\bar{\underset{\sim}{\chi}}_m(\underline{r},\tau) \tag{5.41}$$

with

$$\underset{\sim}{\chi}_m(\underline{r},\tau) = e^{i\tau}\,\underset{\sim}{\phi}_m(\underline{r}) \tag{5.42a}$$

and the phasor

$$A_1 = |A_1|e^{-i\tau_o} . \tag{5.42b}$$

The amplitude $|A_1|$ and the phase τ_o are unknown (real) parameters. The higher order equations read:

$$O(\varepsilon^2): \quad \left(\Omega_c \frac{\partial}{\partial\tau} - \underset{\approx}{L}_c\right)\cdot \underset{\sim}{z}_2 = \underset{\sim}{a}_2 \tag{5.43a}$$

$$O(\varepsilon^3): \quad \left(\Omega_c \frac{\partial}{\partial\tau} - \underset{\approx}{L}_c\right)\cdot \underset{\sim}{z}_3 = \underset{\sim}{a}_3 \tag{5.43b}$$

and so on, with

$$\underset{\sim}{a}_2 \equiv \gamma_1\underset{\approx}{L}_\lambda \cdot \underset{\sim}{z}_1 + \tfrac{1}{2}\underset{\approx}{h}_{xx} : \underset{\sim}{z}_1\underset{\sim}{z}_1 - \alpha_1 \frac{\partial}{\partial\tau}\underset{\sim}{z}_1 \tag{5.44a}$$

$$\begin{aligned}\underset{\sim}{a}_3 \equiv\ & \gamma_1\underset{\approx}{L}_\lambda \cdot \underset{\sim}{z}_2 + \gamma_2\underset{\approx}{L}_\lambda \cdot \underset{\sim}{z}_1 + \tfrac{1}{2}\gamma_1^{\,2}\underset{\approx}{L}_{\lambda\lambda} \cdot \underset{\sim}{z}_1 \\ & + \underset{\approx}{h}_{xx} : \underset{\sim}{z}_1\underset{\sim}{z}_2 + \frac{1}{6}\,\underset{\approx}{h}_{xxx} \vdots\, \underset{\sim}{z}_1\underset{\sim}{z}_1\underset{\sim}{z}_1 \\ & + \tfrac{1}{2}\gamma_1\underset{\approx}{h}_{xx\lambda} : \underset{\sim}{z}_1\underset{\sim}{z}_1 \\ & - \alpha_1\frac{\partial}{\partial\tau}\underset{\sim}{z}_2 - \alpha_2\frac{\partial}{\partial\tau}\underset{\sim}{z}_1 \quad .\end{aligned} \tag{5.44b}$$

Their solvability conditions are again orthogonality relations, but the consideration of periodic solutions has modified our function space. The natural scalar product is now:

$$(\underset{\sim}{y},\underset{\sim}{z}) = \int_o^{2\pi}\int_V \sum_k y_k(\underline{r},\tau)z_k(\underline{r},\tau)d^q\underline{r}\,d\tau \tag{5.45}$$

instead of (5.9).

The solvability conditions for the inhomogeneous linear equations (5.43a) are:

$$(\chi_m^*, a_2) = 0 \tag{5.46a}$$

$$(\bar{\chi}_m^*, a_2) = 0 \tag{5.46b}$$

where the asterisk denotes the adjoint eigenfunction

$$\chi_m^*(\underline{r},\tau) = e^{-i\tau} \phi_m^*(\underline{r}) \quad . \tag{5.47}$$

Simple calculations lead to the condition

$$(\gamma_1 P_1 - i\alpha_1)A_1 = 0 \tag{5.48}$$

with

$$P_1 = (\chi_m^*, L_\lambda \cdot \chi_m) = \frac{\partial\omega_m}{\partial\lambda}(\lambda_c) \quad . \tag{5.49}$$

Since we require A_1 to be finite, this condition may be split into:

$$\gamma_1 \operatorname{Re} P_1 = 0 \tag{5.50a}$$

$$\gamma_1 \operatorname{Im} P_1 = \alpha_1 \quad . \tag{5.50b}$$

With the transversality hypothesis (5.1)

$$\operatorname{Re} P_1 \neq 0 \tag{5.51}$$

the solvability conditions reduce to

$$\gamma_1 = 0 \tag{5.52a}$$

$$\alpha_1 = 0 \quad . \tag{5.52b}$$

Then the linear equation (5.43a) becomes

$$\left(\Omega_c \frac{\partial}{\partial\tau} - L_c\right)\cdot z_2 = \frac{1}{2}\left(A_1^2 \psi_{20} + 2A_1\bar{A}_1\psi_{11} + \bar{A}_1^2\bar{\psi}_{20}\right) \tag{5.53}$$

with

$$\psi_{20} = h_{xx} : \chi_m\chi_m = e^{2i\tau} h_{xx} : \phi_m\phi_m \tag{5.54a}$$

$$\psi_{11} = h_{xx} : \chi_m\bar{\chi}_m = h_{xx} : \phi_m\bar{\phi}_m \quad . \tag{5.54b}$$

This equation is easily integrated within the range of the linear operator, leading to an explicit solution in the form

$$\underset{\sim}{z}_2 = \tfrac{1}{2}(A_1{}^2\underset{\sim}{\Gamma}_{20}(\underline{r},\tau) + 2A_1\overline{A}_1\underset{\sim}{\Gamma}_{11}(\underline{r},\tau) + \overline{A}_1{}^2\overline{\underset{\sim}{\Gamma}}_{20}(\underline{r},\tau)) \tag{5.55}$$

where

$$\underset{\sim}{\Gamma}_{ij}(\underline{r},\tau) \equiv (\Omega_c \frac{\partial}{\partial\tau} - \underset{\approx}{L}_c)^{-1} \underset{\sim}{\psi}_{ij} \quad .$$

With the explicit solution for α_1, γ_1 and $\underset{\sim}{z}_2$ we can write the equation for the third order (5.43b) as:

$$\left(\Omega_c \frac{\partial}{\partial\tau} - \underset{\approx}{L}_c\right)\cdot \underset{\sim}{z}_3 = \left(\gamma_2 \underset{\approx}{L}_\lambda - \alpha_2 \frac{\partial}{\partial\tau}\right)\cdot \underset{\sim}{z}_1 + \underset{\sim}{h}_{\underset{\sim\sim}{xx}} : \underset{\sim}{z}_1\underset{\sim}{z}_2 + \frac{1}{6}\underset{\sim}{h}_{\underset{\sim\sim\sim}{xxx}} \vdots\, \underset{\sim}{z}_1\underset{\sim}{z}_1\underset{\sim}{z}_1 \quad . \tag{5.56}$$

The solvability conditions for this equation are obtained as for (5.43a):

$$(\gamma_2 P_1 - i\alpha_2)A_1 + P_3 A_1{}^2\overline{A}_1 = 0 \tag{5.57}$$

with

$$P_3 = \tfrac{1}{2}(\underset{\sim}{\chi}_m{}^*, \underset{\sim}{h}_{\underset{\sim\sim\sim}{xxx}} \vdots\, \overline{\underset{\sim}{\chi}}_m\underset{\sim}{\chi}_m\underset{\sim}{\chi}_m) + \tfrac{1}{2}(\underset{\sim}{\chi}_m{}^*, \underset{\sim}{h}_{\underset{\sim\sim}{xx}} : (\overline{\underset{\sim}{\chi}}_m\underset{\sim}{\Gamma}_{20} + 2\underset{\sim}{\chi}_m\underset{\sim}{\Gamma}_{11})) \quad . \tag{5.58}$$

Separating the real and imaginary parts in (5.57) we find two equations for three variables:

$$(\gamma_2 \text{ Re } P_1 + |A_1|^2 \text{ Re } P_3)A_1 = 0 \tag{5.59a}$$

$$(\gamma_2 \text{ Im } P_1 + |A_1|^2 \text{ Im } P_3 - \Omega_2)A_1 = 0 \quad . \tag{5.59b}$$

When Re $P_3 \neq 0$, we introduce the normalized phasor

$$\beta = \varepsilon A_1 \tag{5.60a}$$

along with

$$\gamma = \lambda - \lambda_c = \varepsilon^2\gamma_2 + 0(\varepsilon^2) \tag{5.60b}$$

$$\Omega - \Omega_c = \varepsilon^2\alpha_2 + 0(\varepsilon^2) \tag{5.60c}$$

$$\underset{\sim}{z} = \underset{\sim}{x} - \underset{\sim}{x}_s(\lambda) = (\beta\underset{\sim}{\chi}_m + \bar{\beta}\bar{\underset{\sim}{\chi}}_m) + O(|\beta|^2) \quad . \tag{5.60d}$$

The relations (5.59) become the bifurcation equations to dominant order, describing three branches (Fig. 5.4):

- the stationary reference branch:

$$\beta = 0 \tag{5.61}$$

- the periodic bifurcating branch

$$|\beta|^2 = P_3'(\lambda - \lambda_c) \tag{5.62a}$$

$$\Omega - \Omega_c = (\text{Im } P_1 - P_3' \text{ Im } P_3)(\lambda - \lambda_c) \tag{5.62b}$$

with

$$P_3' = \frac{\text{Re } P_1}{\text{Re } P_3} \quad .$$

The argument of β (i.e. the phase τ_o of the limit cycle) is completely arbitrary; this is consistent with the stability analysis of the limit cycle. Higher order corrections will, of course, modify these equations, but we only consider here the neighbourhood of the critical point. We remark that the limit cycle branches are either all subcritical (if $P_3' < 0$) and unstable (one speaks of an inverted bifurcation) or all supercritical (if $P_3' > 0$) and stable (one speaks of a normal bifurcation).

More complicated but similar behaviour occurs for the case Re $P_3 = 0$.

In conclusion, the analysis of the bifurcation to a limit cycle can be treated in a very similar way to the case of steady states. The two main differences are the existence of an additional parameter (the period $2\pi/\Omega$) of the cycle) and the presence of a "dummy" free parameter (the phase τ_o of the cycle).

A variant to the expansion of the pulsation Ω, is the method of matching asymptotic expansions using instead multiple time-scales, but its validity has been demonstrated under some restrictive conditions [Haberman, 1977; Kuramoto and Tzusuki, 1975].

E. Discussion

The perturbative bifurcation analysis, by the Poincaré-Linstedt series expansion, leads to a satisfactory description of the system to the dominant order in the vicinity of the bifurcation point. In special cases, exact solutions can be obtained over finite ranges

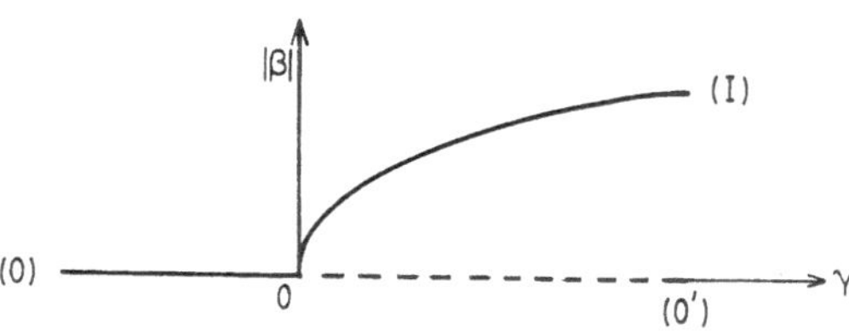

(a)

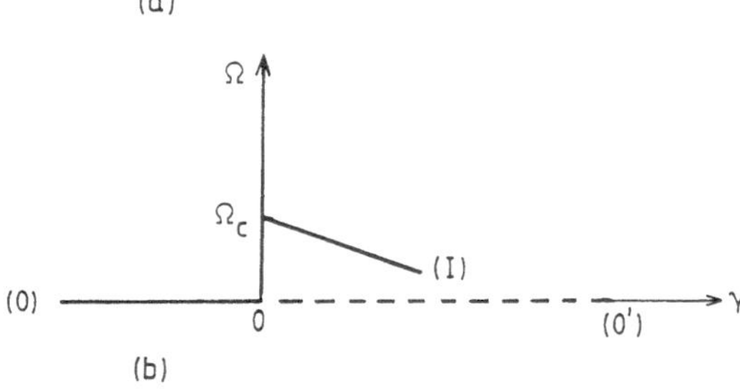

(b)

Fig. 5.4. Bifurcation from a stationary state to a limit cycle.

(0) : reference branch, stable
(0'): reference branch, unstable
(I) : bifurcating branch, stable

The bifurcation equations fix the amplitude $|\beta|$ and the period $2\pi/\Omega$, but the phase is arbitrary.

(a) : branching diagram
(b) : pulsations diagram

of the control parameter [Lefever, Herschkowitz-Kaufmann and Turner, 1977].

Through this analysis, we have been able to reduce the study of the system near the bifurcation point, to that of one (generally complex-valued) degree of freedom, the amplitude of the bifurcating solution β. This is analogous to the mean-field theories of phase transitions near the critical points, wherein β appears as the order parameter. This analogy is physical as much as mathematical, and one may also speak of "critical exponents" relating β and $\lambda - \lambda_c$.

From a more fundamental viewpoint, we may note that a bifurcation generally breaks a symmetry of the evolution equations:

- discrete (and abstract) symmetries for all-or-none transitions;

- temporal translation for transitions to limit cycles;

- translational or rotational symmetry for pattern formation.

All these symmetry-breaking phenomena become possible because of the breaking of the time-reversal symmetry in the underlying (dissipative) equations of evolution.

These considerations motivate the recent systematic approach to bifurcations by group theoretic methods [Sattinger, 1979].

In all this analysis, we limited our consideratons to autonomous systems, in which the control parameters are time-independent. Then we compared different solutions obtained for various choices of $\underset{\sim}{\lambda}$. This approach can be expected to describe also quite correctly the evolution of the system when the control parameters vary slowly with time (quasi-statically); explicit evaluations with a multiple time-scale formalism show the validity of this viewpoint, but also its limitation since the limit of "infinitely slow" variations of λ is ill-defined near singular points [Haberman, 1979].

F. The First Primary Branches

After discussing the various ways by which a system bifurcates beyond an instability, we want to investigate the various behaviours of the new solutions associated with bifurcation.

(i) Multiple Steady-States and Turning Points

The simplest kind of bifurcation is certainly the mere all-or-none transition between uniform stationary states. Already then, there is a possibility for unexpected behaviours since various branches may coexist for some range of the control parameter (see Fig. 5.3).

Note, however, that bifurcations are not the only way to generat coexisting solutions. A simple (and quite frequent) way is the turning point, occurring when the branch presents a local extremum of $\tilde{\lambda}$ as a function of $\tilde{x}$ (see Fig. 5.5a).

Although a turning point is not a bifurcation point, its existence is related to the effects of a bifurcation (like the cusp bifurcation of Sect. V.C) on the global structure of the space of sol-tions; it is sometimes referred to as a fold catastrophe [Thom, 1972] or "saddle-node bifurcation". For nondegenerate systems with a scalar parameter λ, there are thus two half-branches emerging from the turning point, on the same side of the critical (extremal) parameter value λ_c; at least one of them is unstable. In most cases, the following assertions are valid [Decker and Keller, 1980]:

- if one of the half-branches is stable, the other one is unstable;
- each half-branch either extends to infinity or undergoes other singularities.

A typical example is given in Fig. 5.5, in which we distinguish three (sub)-branches (b_1 and b_2 stable, u unstable) and three regions:

- for $\lambda < \lambda_D$: b_1 is the only branch
- for $\lambda > \lambda_C$: b_1 is the only branch
- for $\lambda_D < \lambda < \lambda_C$: three branches ($b_1, b_2$, u) coexist.

The system's choice between the branches for a given (fixed) λ depends on the initial condition; the size of the attraction domain (between the stable and unstable branches) indicates how easy a transition can be under the influence of disturbances. For instance, at $\lambda = \lambda_T$, a nucleation process is necessary, starting from one branch, before point T is reached and the transition continues spontaneously towards the other branch; whereas the transition to b_1 is spontaneous from the turning point D.

A remarkable property of such systems appears when the control parameter is varied slowly between $\lambda_D - \varepsilon$ and $\lambda_C + \varepsilon$ back and forth: the state then follows a cycle A'ACBB'BDAA' induced by the stability of the branches in the static limit (see the remark in Sect. V.E). This hysteresis phenomenon obviously results in a regulating process for the system under consideration.

Systems presenting multiple steady states for some values of their control parameters are very common, and they often associate all-or-none bifurcations with turning points.

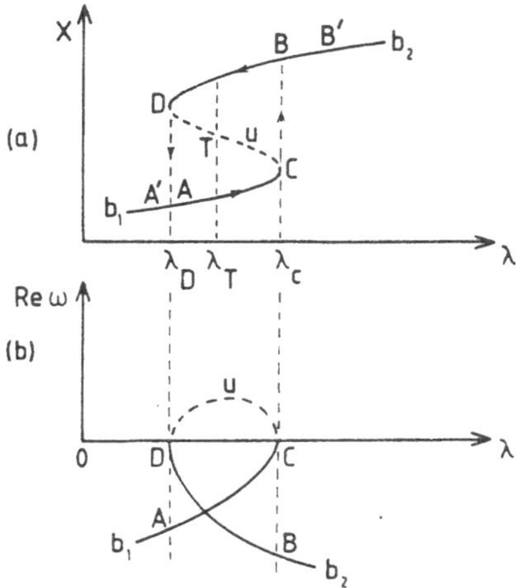

Fig. 5.5. Turning points. (a) The state diagram (x,λ). At $\lambda = \lambda_T$, two stable solutions coexist: if $x > x_T$, x is attracted by b_2; if $x < x_T$, x is attracted by b_1. At $\lambda = \lambda_C$, the transition to B is spontaneous. At $\lambda = \lambda_D$, the transition to A is spontaneous. The arrows along the lines ACBDA indicate the possible hysteretic behaviour. (b) The linear stability diagram (ω,λ).

The simple case pictured in Fig. 5.5 is found for the Schlögl model on the lines $\delta + \delta' = -1$ and $(\delta + 1)(\delta' + 1) = 1/4$. Such a situation was encountered by Ermentrout and Cowan [1978] in the modelling of neural networks; a system with more than two stable stationary states in some range was found by Mandel [1979] in the theory of lasers. Hysteresis cycles were observed indeed in lasers by Ruschin and Bauer [1979].

A typical chemical system displaying turning-point instabilities is the illuminated NO_2-N_2O_4 reactor according to Creel and Ross [1976]:

$$N_2O_4 \underset{k_2}{\overset{k_1}{\rightleftharpoons}} NO_2 \quad (5.63a$$

$$NO_2 + h\nu \longrightarrow NO_2^* \longrightarrow NO_2 + \Delta H \quad . \quad (5.63b$$

The exothermic disactivation of NO_2 releases heat which displaces the equilibrium of reaction (a) by raising the temperature. The evolution equations for the concentration x of NO_2 and the temperature T are thus coupled nonlinearly via the rate constants:

$$\frac{d}{dt} x = 2k_1(C - \tfrac{1}{2}x) - 2k_2x^2 \quad (5.64a$$

$$\frac{d}{dt} T = \alpha I_o x - \beta(T - T_o) - \lambda \frac{dx}{dt} \quad (5.64b$$

$$k_i = Z_i \exp\left[\frac{-E_i}{kT}\right] \quad (5.64c$$

where:

x	is the concentration of NO_2
C	is the initial concentration of N_2O_4 (constant)
I_o	is the intensity of illumination
T	is the temperature
T_o	is the external temperature
k_i	is the rate constant of reaction i

$\alpha, \beta, \lambda, k, Z_i, E_i$ are constants.

The diagram of stationary states has the form sketched in Fig. 5.5.

Similar equations with other nonlinear functions, are found by Lefever and Garay [1978] in a model for immune surveillance where the variable is the population of cancerous cells and the excitation-disactivation reaction (b) is realized by cytolysis of the cell. Other examples are found in the Eigen-Schuster theory of competition and selection (see Professor Schuster's lectures), as well as in electrothermal (negative resistivity) instabilities of circuits [Bedeaux et al, 1977]. Recall that bistable circuits are the basis of logical electronics!

(ii) Limit Cycle

Hopf bifurcations appear in the linear analysis as transitions from damped periodic modes to amplified periodic modes. In both cases the commanding time-constants occur in complex conjugate pairs $\omega = \sigma \pm i\Omega$. The two parameters fixing the evolution are the initial amplitude and phase of the perturbation.

The limit cycle emerging from the bifurcation is followed by the system in one way only: the system behaves like a clock recognizing a temporal orientation in its motion. On this periodic trajectory, there is a continuous family of solutions differing by the phase. The amplitude and period, however, are fixed by the bifurcation equations (5.59).

A good example of a limit cycle is provided by the brusselator model reaction (with or without diffusion). The evolution equations of the scaled variables are [Nicolis and Prigogine, 1977]:

$$\frac{\partial x}{\partial t} = a - (b + 1)x + x^2y + D_1\nabla^2x \tag{5.65a}$$

$$\frac{\partial y}{\partial t} = bx - x^2y + D_2\nabla^2y \quad . \tag{5.65b}$$

The linear stability analysis shows that the first bifurcation may lead either to a limit cycle or to a spatial pattern, according to the values of parameters (a,b,D_1,D_2). In particular, time-periodic solutions are favoured by the choice $D_1 = D_2$, and unlikely when D_1 and D_2 are very different.

Fig. 5.6 describes the bifurcation to a uniform limit cycle with Neumann boundary conditions and appropriate parameter values. The asymptotic stability of the cycle is illustrated by the attraction of 4 different initial conditions.

The corresponding stable periodic solution is

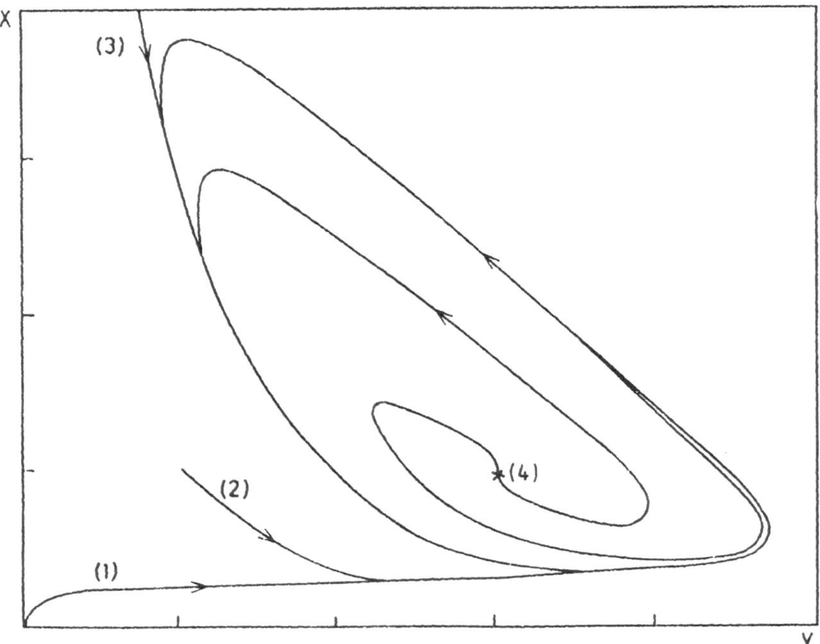

Fig. 5.6. Limit cycle solution associated with temporal symmetry-breaking. (1) to (4): different initial conditions lead to the same final periodic motion. This establishes the stability of the limit cycle.

$$\begin{pmatrix} x(t) \\ y(t) \end{pmatrix} = \begin{pmatrix} x_s \\ y_s \end{pmatrix} + \beta \begin{pmatrix} \cos \Omega t \\ \rho b_o \cos(\Omega t + \theta) \end{pmatrix}$$

$$+ \beta^2 \begin{pmatrix} u \cos(2\Omega t + \psi_1) \\ v_o + v \cos(2\Omega t + \varphi_1) \end{pmatrix} + O(\beta^3) \tag{5.66a}$$

with pulsation

$$\Omega(b) = a + \beta^2 \alpha_2 + O(\beta^4) \tag{5.66b}$$

where:

- b is the control parameter
- $\beta = \beta_o(b - 1 - a^2)$
- a is a fixed parameter
- $\beta_o, b_o, \rho, u, v, v_o, \alpha_2$ are known functions of a.

Further examples of systems exhibiting similar behaviour are:

- models of the Belousov-Zhabotinski reaction, like the oregonator, which was already described in Professor Howard's lectures;
- the glycolytic oscillator [Boiteux, Goldbeter and Hess, 1975], modelled by Goldbeter and Lefever as follows:

$$\frac{d\alpha}{dt} = \sigma_1 - \sigma_M \phi \tag{5.67a}$$

$$\frac{d\gamma}{dt} = -k_1 \gamma + q \sigma_M \phi \tag{5.67b}$$

$$\phi = \phi(\alpha, \gamma) = \frac{\alpha e(1+\alpha e)(1+\gamma)^2 + L\theta c e'\alpha(1+ce'\alpha)}{L(1+\alpha e)^2(1+\gamma)^2 + (1+ce'\alpha)} \tag{5.67c}$$

  with two variables $\alpha = \alpha(t)$, $\gamma = \gamma(t)$ and a nonlinear function ϕ; c, e, e', θ and L are constants;

- the cAMP oscillations in the metabolism of colonies of Dictyostelium Discoideum, causing aggregation and differentiation of the amoebae, well observed in vivo [Gerisch and Hess, 1974]; their dynamics can be understood on the basis of three normalized concentration variables $\alpha(t)$, $\beta(t)$ and $\gamma(t)$ obeying the kinetic equations

$$\frac{d\alpha}{dt} = \sigma_1 - f_1 - f_2 \tag{5.68a}$$

$$\frac{d\beta}{dt} = f_1 - k_1\beta + k_2\gamma \tag{5.68b}$$

$$\frac{d\gamma}{dt} = f_2 - k_2\gamma \tag{5.68c}$$

where

$$f_1 = f_1(\alpha,\gamma) = \sigma_{M1}\frac{\alpha(1+\alpha)(1+\gamma)^2}{L_1+(1+\alpha)^2(1+\gamma)^2} \tag{5.69a}$$

$$f_2 = f_2(\alpha,\beta) = \sigma_{M2}\frac{\alpha(1+\alpha)(1+\beta)^2}{L_2+(1+\alpha)^2(1+\beta)^2} \tag{5.69b}$$

correspond to enzymatic reaction rates; introducton of diffusion provides a realistic description of the waves propagating in the colonies;

- the Kauffman-Tyson mitotic oscillator [1975], arising is the study of the fusion of two cells; it is described by four time-dependent variables subject to cubic couplings;

- multimode lasers [Risken and Nummedal, 1968; Haken, 1978].

Oscillations of the limit cycle type are also encountered when the evolution is affected by time-delays, like in problems of population dynamics [May, 1976]. In this latter case, oscillations can arise even in systems involving a single variable.

(iii) Spatial Organization

We have already discussed some aspects of spatial pattern formation in the chapter on linear analysis. Here we stress some basic differences between this phenomenon and the transitions to limit cycles. Indeed, for a reaction-diffusion system free of external fields, the spatial derivatives appearing in the equations of evolution are of order two, whereas the temporal derivatives are of the first order. For this reason, no preferred spatial orientation can be induced by the transition from a homogeneous state to a pattern (except for asymmetric boundary conditions).

Fig. 5.7 shows a pattern obtained for a system of the brusselator type, in one space dimension, under no-flux boundary conditions: notice that both oriented solutions are equiprobable. The pattern's wavenumber is fixed by the control parameters (diffusion constants, precursors' concentrations, ...); on the other hand, the shape is affected by the need to match boundary conditions at the end points.

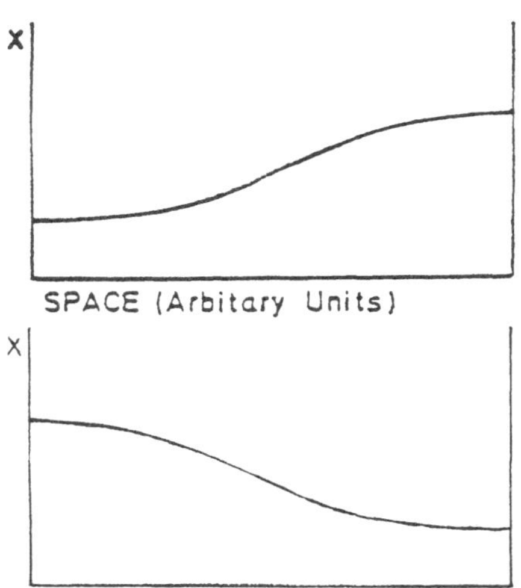

Fig. 5.7. Spatial symmetry-breaking associated with the first bifurcation: the new stable solutions display a polarity axis. The two orientations are equally probable.

Examples of such transitions induced by diffusion eigenmodes include:

- abstract reaction schemes, like the brusselator and the activator-inhibitor model for morphogenesis [Gierer and Meinhardt, 1972]:

$$\frac{\partial x}{\partial t} = k_1 a - k_3 x + \alpha \frac{x^2}{y} + D_1 \nabla^2 x \tag{5.70a}$$

$$\frac{\partial y}{\partial t} = k_2 C x^2 - k_4 y + D_2 \nabla^2 y \tag{5.70b}$$

- in physical problems, the kinetics of phase transitions [Langer, 1980; Lovett, Ortoleva and Ross, 1978] or the formation of Peierls state in insulators by illumination [Berggren and and Huberman, 1978].

Of course, the system need not be ideal as in our explicit models, and diffusion can present cross-couplings or non-Fickian effects.

VI. MORE COMPLEX BIFURCATIONS

A. Successive Bifurcations

From the preceding analysis of bifurcations, it was clear that the first bifurcation from the uniform stationary solution is dominated by a single spatial or temporal mode. As this is still a very simple structure, the question arises whether one can observe more complex solutions suitable for representing the behaviour of macroscopic systems.

A first step in the analysis of successive instabilities would be to explore the properties of solutions bifurcating from the reference state at other values of the control parameters; however, these primary branches generally emerge as unstable solutions and are therefore inaccessible to the system.

On the other hand, bifurcating branches may also lose their stability as a result of interaction with nearby solutions, and bifurcate into more intricate solutions: these secondary bifurcations(see Fig. 6.1) may be studied in much the same way as the primary one. But they are constituted from a superposition of various kinds of modes, and hence they lead to a much wider variety of forms than the primary bifurcations.

A central problem in the analysis of complex solutions is thus to investigate the mechanisms by which bifurcating branches may interact. A very elegant method reduces much of this question to perturbation theory, in a natural extension of the simple bifurcation formalism: this is the bifurcation analysis near multiple eigenvalues, discussed in Section VI.B.

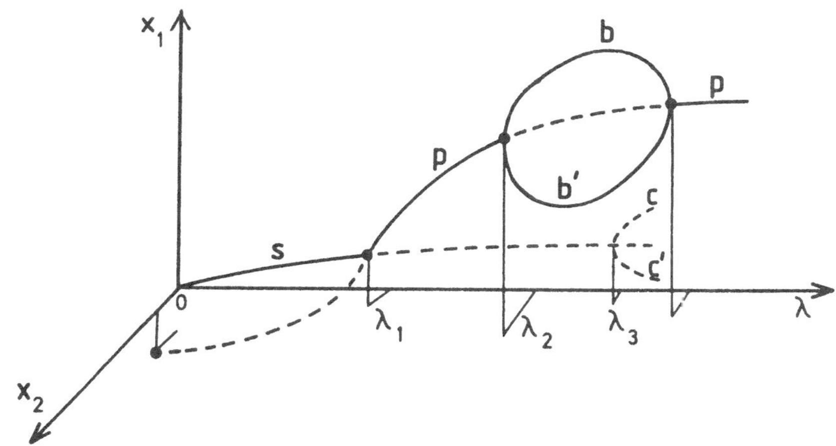

Fig. 6.1. Secondary bifurcations for two variables (x_1,x_2) and one parameter λ. s: reference branch; p: first primary branch; b,b': secondary branches; c,c': later primary branches; ——: stable; ----: unstable; λ_1,λ_2,λ_3: successive thresholds.

Beyond some degree of complexity, it may happen that the organization of the system becomes too intricate, for instance, the structure may be compared to a realization of some random process. This behaviour occurs in large classes of dynamical systems; the most famous example is certainly turbulence, which remains among the most challenging questions of physics. These questions are briefly discussed in Section IV.C.

We close the second part of these lectures with general remarks on phenomenological approaches (Section VI.C).

B. Bifurcations near Multiple Eigenvalues

Degenerate behaviour occurs whenever several eigenmodes of the linearized operator $\underset{\approx}{L}$ become unstable simultaneously. In most cases, these phenomena can be viewed as limiting cases of successive bifurcations for certain values of the control parameter $\underset{\sim}{\lambda}$. Actually, they are better pictured by varying two or more parameters instead of one, as in this way the distance between successive bifurcating branches can be controlled. In this subsection, we shall discuss the case of a double zero eigenvalue, by means of an extended Poincaré-Lindstedt expansion. More elaborate situations can also be discussed and classified systematically [Golubitsky and Schaeffer, 1978; Kopell and Howard, 1975].

We consider the nonlinear evolution equation

$$\frac{\partial}{\partial t}\underset{\sim}{x} = \underset{\sim}{G}[\underset{\sim}{x};\underset{\sim}{\lambda}] \tag{6.1}$$

with usual boundary conditions such as (4.4), where $\underset{\sim}{\lambda}$ represents a vector of two real parameters:

$$\underset{\sim}{\lambda} = \begin{pmatrix} \lambda \\ \mu \end{pmatrix}. \tag{6.2}$$

Suppose that:

1) the stationary solution $\underset{\sim}{x}_s(\underset{\sim}{\lambda})$ is stable for any $\underset{\sim}{\lambda}$ such that $\lambda < \lambda^*(\mu)$, $\mu < \mu_c$;

2) one simple real eigenvalue ω_1 of the linearized problem changes its sign at $\lambda = \Lambda_1(\mu)$ and another simple real eigenvalue ω_2 changes its sign for $\lambda = \Lambda_2(\mu)$;

3) the corresponding eigenmodes $\underset{\sim}{\phi}_m(\underline{r},t)$, $(m = 1,2)$, are distinct.

The critical values $\underset{\sim}{\lambda}_c = (\lambda_c, \mu_c)$ corresponding to $\Lambda_1(\mu_c) = \Lambda_2(\mu_c) = \lambda_c$ then determine a degenerate bifurcating point. As before, it will be convenient to define the deviation of the state variables $\underset{\sim}{x}$ as:

$$\underset{\sim}{z}(\underset{\sim}{\lambda}) = \underset{\sim}{x}(\underset{\sim}{\lambda}) - \underset{\sim}{x}_s(\underset{\sim}{\lambda}) \tag{6.3}$$

and rewrite the evolution equation (6.1) in terms of $\underset{\sim}{z}$:

$$\frac{\partial}{\partial t}\underset{\sim}{z} = \underset{\approx}{L}(\lambda) \cdot \underset{\sim}{z} + \underset{\sim}{h}[\underset{\sim}{z};\underset{\sim}{\lambda}] \tag{6.4}$$

where $\underset{\approx}{L}(\underset{\sim}{\lambda})$ and $\underset{\sim}{h}[\underset{\sim}{z};\underset{\sim}{\lambda}]$ correspond to the linear and nonlinear parts of $\underset{\sim}{G}[\underset{\sim}{x};\underset{\sim}{\lambda}]$ respectively. With our hypotheses, the null-space of $\underset{\approx}{L}(\underset{\sim}{\lambda}_c)$ is two-dimensional. We also assume that $\underset{\approx}{L}$ and $\underset{\sim}{h}$ admit the following representations near the critical point:

$$\underset{\approx}{L}(\underset{\sim}{\lambda}) = \underset{\approx}{L}_c + (\lambda - \lambda_c)\underset{\approx}{L}_\lambda + (\mu - \mu_c)\underset{\approx}{L}_\mu + \ldots \tag{6.5a}$$

$$\begin{aligned}\underset{\sim}{h}[\underset{\sim}{z};\underset{\sim}{\lambda}] &= \frac{1}{2}\underset{\approx}{h}_{xx}:\underset{\sim}{z}\underset{\sim}{z} + \frac{1}{6}\underset{\approx}{h}_{xxx}\vdots\,\underset{\sim}{z}\underset{\sim}{z}\underset{\sim}{z} + \\ &\quad + \frac{1}{2}\left((\lambda - \lambda_c)\underset{\approx}{h}_{xx\lambda} + (\mu - \mu_c)\underset{\approx}{h}_{xx\mu}\right):\underset{\sim}{z}\underset{\sim}{z} \\ &\quad + \ldots\end{aligned} \tag{6.5b}$$

when

$$\lambda - \lambda_c = O(\varepsilon)$$

$$\mu - \mu_c = O(\varepsilon)$$

$$\|\underset{\sim}{z}\| = O(\varepsilon)$$

and

$$\varepsilon \to 0 \,.$$

We propose the following perturbative expansions around the degenerate critical point:

$$\underset{\sim}{z} = \underset{\sim}{x} - \underset{\sim}{x}_s(\underset{\sim}{\lambda}) = \varepsilon \underset{\sim}{x}_1 + \varepsilon^2 \underset{\sim}{x}_2 + \ldots \tag{6.6a}$$

$$\lambda - \lambda_c = \varepsilon\lambda_1 + \varepsilon^2\lambda_2 + \ldots \tag{6.6b}$$

$$\mu - \mu_c = \varepsilon\mu_1 + \varepsilon^2\mu_2 + \ldots \tag{6.6c}$$

where the small parameter ε can be defined by either (6.6b) or (6.6c). After inserting these expressions in the evolution equation (6.4) and in the boundary conditions, and using (6.5), we equate to zero the coefficient of each power of ε separately. The first two equations are

$$O(\varepsilon): \quad \underset{\approx}{L}(\underset{\sim}{\lambda}_c) \cdot \underset{\sim}{x}_1 = 0 \tag{6.7a}$$

$$O(\varepsilon^2): \quad \underset{\approx}{L}_c \cdot \underset{\sim}{x}_2 = -(\lambda_1 \underset{\approx}{L}_\lambda + \mu_1 \underset{\approx}{L}_\mu) \cdot \underset{\sim}{x}_1 - \frac{1}{2} \underset{\approx}{h}_{xx} : \underset{\sim}{x}_1 \underset{\sim}{x}_1 \tag{6.7b}$$

with the same notations as in (5.6). The solution of the first equation is

$$\underset{\sim}{x}_1 = A_1 \underset{\sim}{\phi}_1 + A_2 \underset{\sim}{\phi}_2 \tag{6.8}$$

where $\underset{\sim}{\phi}_1$ and $\underset{\sim}{\phi}_2$ are two orthogonal eigenfunctions in the null-space of $\underset{\approx}{L}(\underset{\sim}{\lambda}_c)$. The two amplitudes A_1 and A_2 are still unknown. The equation for the second order (6.7b) can be solved only if the right-hand side satisfies certain solvability conditions, for the reasons explained in the preceding section. The latter lead (after some algebra) to the following equations

$$-(\lambda_1 P_{\lambda 1} + \mu_1 P_{\mu 1}) A_1 + Q_{111} A_1^2 + 2Q_{112} A_1 A_2 + Q_{122} A_2^2 = 0 \tag{6.9a}$$

$$-(\lambda_1 P_{\lambda 2} + \mu_1 P_{\mu 2}) A_2 + Q_{211} A_1^2 + 2Q_{212} A_1 A_2 + Q_{222} A_2^2 = 0 \tag{6.9b}$$

with the coefficients

$$P_{\lambda j} = (\underset{\sim}{\phi}_j^*, \underset{\approx}{L}_\lambda \cdot \underset{\sim}{\phi}_j) = \frac{\partial \omega_j}{\partial \lambda} (\underset{\sim}{\lambda}_c) \tag{6.10a}$$

$$P_{\mu j} = (\underset{\sim}{\phi}_j^*, \underset{\approx}{L}_\mu \cdot \underset{\sim}{\phi}_j) = \frac{\partial \omega_j}{\partial \mu} (\underset{\sim}{\lambda}_c) \tag{6.10b}$$

$$Q_{ijk} = -\frac{1}{2} (\underset{\sim}{\phi}_i^*, \underset{\approx}{h}_{xx} : \underset{\sim}{\phi}_j \underset{\sim}{\phi}_k) \; . \tag{6.10c}$$

In general, the expressions (6.9) can be reduced to canonical forms. For example, the interaction of two spatial modes (one even and one odd) in one space dimension with fixed boundary conditions leads to the following equations [Keener, 1976]:

$$-(\lambda_1 P_{\lambda 1} + \mu_1 P_{\mu 1}) A_1 + Q_{11} A_1^2 + Q_{22} A_2^2 = 0 \tag{6.11a}$$

$$-(\lambda_1 P_{\lambda 2} + \mu_1 P_{\mu 2}) A_2 + Q_{12} A_1 A_2 = 0 \tag{6.11b}$$

The system (6.9) describes the intersection of two quadrics in the $(\lambda_1,\mu_1;A_1,A_2)$ space. In order to maintain a dependence of (A_1,A_2) on (λ_1,μ_1), we assume the transversality of the eigenvalues ω_j:

$$\frac{\partial\omega_1}{\partial\lambda}\frac{\partial\omega_2}{\partial\mu} - \frac{\partial\omega_1}{\partial\mu}\frac{\partial\omega_2}{\partial\lambda} \neq 0 \ . \tag{6.12}$$

If the quadratic coefficients Q_{ij} are non-zero, (6.11) is compatible with a nontrivial pair (λ_1,μ_1). Then finite (λ_1,μ_1) yield at least one finite amplitude A_1 or A_2, and we may define the normalized amplitudes

$$\beta_1 = \varepsilon A_1 \tag{6.13a}$$

$$\beta_2 = \varepsilon A_2 \tag{6.13b}$$

which have order ε. The solution $\underset{\sim}{x}$ can then be written as

$$\underset{\sim}{x}(\underset{\sim}{\lambda}) = \underset{\sim}{x}_s(\underset{\sim}{\lambda}) + \beta_1\underset{\sim}{\phi}_1 + \beta_2\underset{\sim}{\phi}_2 + 0(\varepsilon) \tag{6.14a}$$

$$\underset{\sim}{\lambda} = \underset{\sim}{\lambda}_c + \varepsilon\underset{\sim}{\lambda}_1 + 0(\varepsilon) \ . \tag{6.14b}$$

The bifurcation equations (6.11) describe four branches of solutions:

- the reference solution (0)

$$\beta_1 = 0 \tag{6.15a}$$

$$\beta_2 = 0 \tag{6.15b}$$

- a bifurcating solution (I) dominated by $\underset{\sim}{\phi}_1$:

$$\beta_1 = \frac{1}{Q_{11}}\Big((\lambda - \lambda_c)P_{\lambda 1} + (\mu - \mu_c)P_{\mu 1}\Big) \tag{6.16a}$$

$$\beta_2 = 0 \tag{6.16b}$$

- two bifurcating solutions (II) and (II') combining $\underset{\sim}{\phi}_1$ and $\underset{\sim}{\phi}_2$:

$$\beta_1 = \frac{1}{Q_{12}}\Big((\lambda - \lambda_c)P_{\lambda 2} + (\mu - \mu_c)P_{\mu 2}\Big) \tag{6.17a}$$

$$\beta_2 = \pm\sqrt{\beta_1\frac{1}{Q_{22}}\Big((\lambda-\lambda_c)(P_{\lambda 1} - \frac{Q_{11}}{Q_{12}}P_{\lambda 2}) + (\mu-\mu_c)(P_{\mu 1} - \frac{Q_{11}}{Q_{12}}P_{\mu 2})\Big)} \tag{6.17b}$$

We see that near the degenerate critical point, three bifurcations occur, along three critical lines in the parameters' space (Fig. 6.3):

- a primary bifurcation from branch (0) to (I) when

$$(\lambda - \lambda_c)P_{\lambda 1} + (\mu - \mu_c)P_{\mu 1} = 0 \tag{6.18}$$

- a primary bifurcation from (0) to (II) or (II') when

$$(\lambda - \lambda_c)P_{\lambda 2} + (\mu - \mu_c)P_{\mu 2} = 0 \tag{6.19}$$

- a secondary bifurcation between the branches (I) and (II) or (II'), which is found as the locus of coalescence points of these branches. Comparing equations (6.16) and (6.17), we see that the condition for this bifurcation is (Fig. 6.2, point D):

$$Q_{12}(\lambda - \lambda_c)P_{\lambda 1} + Q_{12}(\mu - \mu_c)P_{\mu 1} =$$

$$= Q_{11}(\lambda - \lambda_c)P_{\lambda 2} + Q_{11}(\mu - \mu_c)P_{\mu 2} \quad . \tag{6.20}$$

The important point about the secondary bifurcation is to allow for an exchange of stability between the bifurcating branches (I), (II) and (II'). One representative case is depicted in Fig. 6.2.

If one or several quadratic coefficients in (6.9) vanish, the value $\lambda_1 = \mu_1 = 0$ would follow from (6.9). In this case, the analysis must go to higher orders (λ_k, μ_ℓ) and the perturbation expansions become more delicate to handle. However, the algebra can be carried out in a systematic way.

For example, when $\underset{\sim}{\phi}_1$ and $\underset{\sim}{\phi}_2$ represent respectively two modes with wavenumbers 1 and 2 in a one-dimensional system with zero-flux boundary conditions, the result reads [Erneux and Hiernaux, 1980]:

$$\left((\lambda - \lambda_c)P_{\lambda 1} + (\mu - \mu_c)P_{\mu 1}\right)\beta_1 = Q_1\beta_1\beta_2 \tag{6.21a}$$

$$\left((\lambda - \lambda_c)P_{\lambda 2} + (\mu - \mu_c)P_{\mu 2}\right)\beta_2 = Q_2\beta_1{}^2 + R\beta_2{}^3 \tag{6.21b}$$

Again a variety of secondary bifurcations will be generated for different values of the parameters.

More complex patterns of higher order bifurcations, involving, for instance, a time-periodic solution, can be treated in a similar way. These phenomena occur in many systems, e.g.:

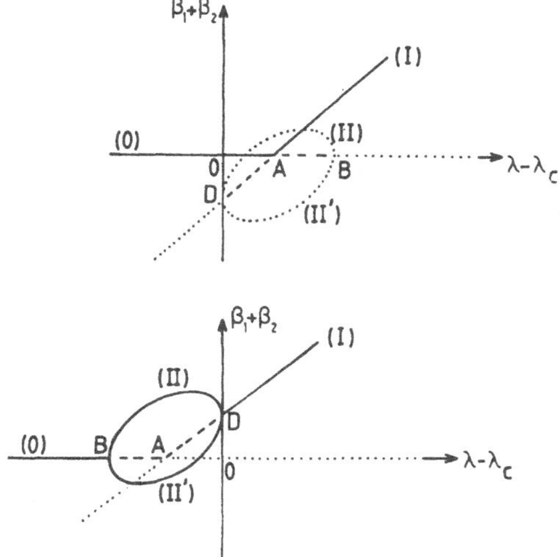

Fig. 6.2. Illustration of the secondary bifurcation described by (6.15-17). (0): reference solution; (I): bifurcating branch with $\beta_1 \neq 0$, $\beta_2 = 0$; (II)-(II'): bifurcating branches with $\beta_1 \neq 0$, $\beta_2 \neq 0$; ——— : stable branch; ---- : unstable with respect to one eigenmode; •••• : unstable with respect to two eigenmodes.

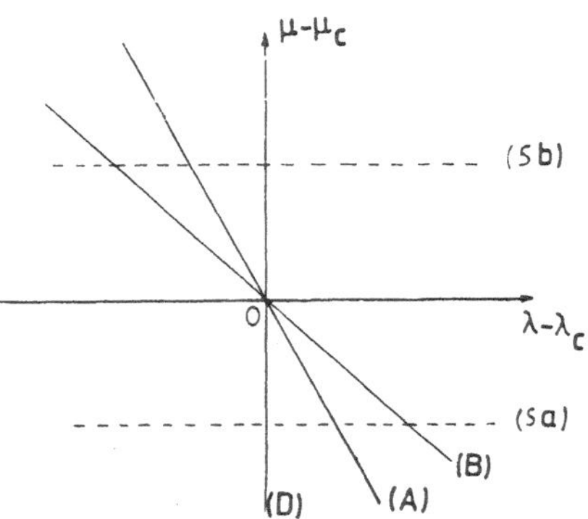

Fig. 6.3. The parameters' space near the degenerate bifurcation described by (6.18-20). (λ_c,μ_c) is the degenerate value of $\underset{\sim}{\lambda}$; (A): locus for the primary bifurcation (0)-(I); (B): locus for the primary bifurcation (0)-(II/II'); (D): locus for the secondary bifurcation (I)-(II/II'); (Sa): section for the bifurcation diagram 6.2a; (Sb): section for the bifurcation diagram 6.2b.

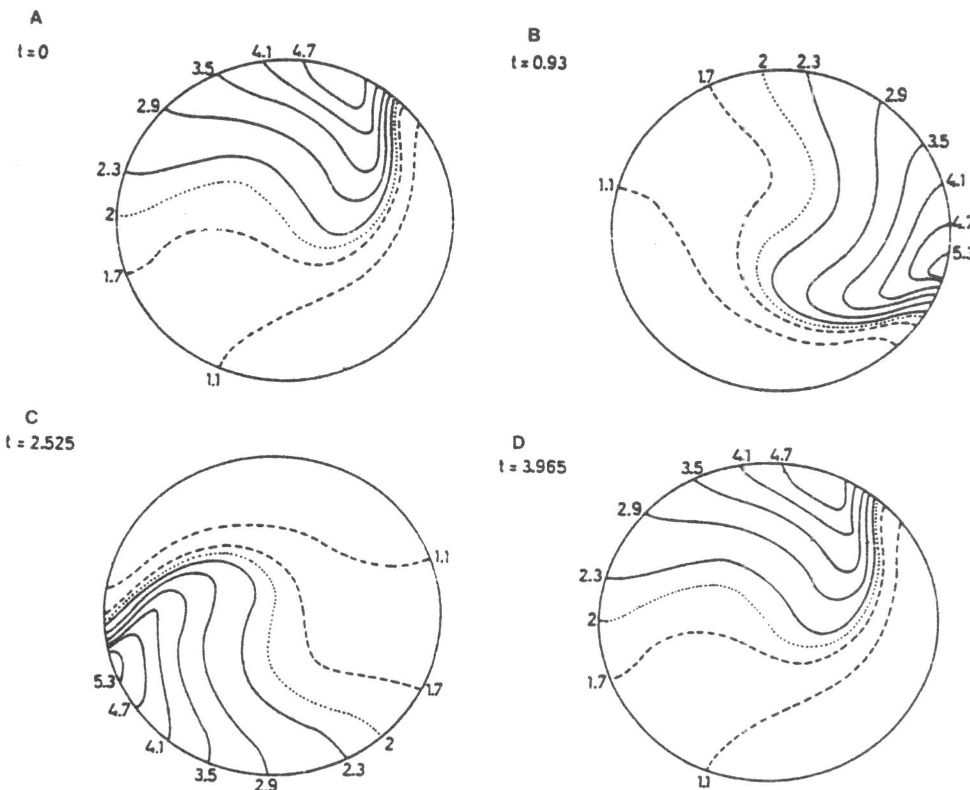

Fig. 6.4. Rotating waves for the brusselator in a disc with zero-flux boundary conditions. The first bifurcation led from a stationary homogeneous state to a periodic homogeneous state. An inhomogeneous time-periodic state is reached by a secondary bifurcation. (a)-(d): The four graphs represent the same system at four different times. Apart from small variations, they can be superposed by a simple rotation. The curves of equal concentration for X are represented by full or broken lines when the concentration is respectively larger or smaller than the value of the unstable homogeneous stationary state X = 2.

- the rotating waves in reaction-diffusion systems [Erneux and Herschkowitz-Kaufman, 1977] (Fig. 6.4) similar to those observed in the Belousov-Zhabotinski reaction in a Petri dish [Zhabotinski, 1974; Tyson, 1976];
- the Rayleigh-Bénard convection beyond the first instability [Dubois and Bergé, 1980];
- the formation of spatial concentration patterns determining the morphogeneis in various organisms like drosophila [see S. Kauffman's lectures in this school].

C. Transitions to Chaotic Behaviour

We have seen that bifurcations may increase the morphological complexity. It is therefore tempting to interpret the very distorted structures encountered in nonequilibrium systems as the result of successive bifurcations from a much simpler state.

Indeed, the interaction of two limit cycles leads to a biperiodic time-dependent regime. More generally, multiperiodic solutions with many (incommensurate) frequencies can easily be derived from a succession of Hopf bifurcations; these rather erratic solutions were long considered as the prototype of turbulence after Landau [Landau and Lifshitz, 1959]. But these solutions have poor structural stability properties and Ruelle and Takens [1971] proved that the interaction of three (or more) modes with nonlinear couplings could lead to very different types of asymptotic solutions, namely strange attractors (for an overview, see Chorin, Marsden and Smale [1977]).

Bifurcations leading to three coupled modes are, of course, very common (see Fig. 4.4). The transition to the strange attractor may therefore occur in systems obeying reaction-diffusion dynamics.

A remarkable property of strange attractors is their very high sensitivity to initial conditions: neighbouring initial values may repel each other in an exponential fashion; in the spectral analysis, this leads to a continuous band of frequencies in contrast with the peak spectrum of multiperiodic solutions.

A wide field of applications for strange attractors is the analysis of hydrodynamic turbulence. Experiments on the Rayleigh-Bénard convective instability [Ahlers and Walden, 1980; Bergé et al, 1980; see also Normand et al, 1977] allowed the determination of various successions of instabilities under reproducible constraints, finally leading to chaotic behaviour.

A favourite model describing these Rayleigh-Bénard instabilities is the well-known Lorenz model [Lorenz, 1963] leading to a strange attractor; it involves three variables and quadratic nonlinearities:

$$\frac{dx}{dt} = \sigma(y - x) \tag{6.22a}$$

$$\frac{dy}{dt} = -xz + rx - y \tag{6.22b}$$

$$\frac{dz}{dt} = xy - bz \tag{6.22c}$$

where x, y, z are the variables; σ, b are constants; and r is the control parameter. Physically, x is proportional to the convective velocity; y is proportional to the temperature difference between ascending and descending flows; z is proportional to the mean convective heat flow; σ is the Prandtl number; b is related to the wave-number of the structure; r is the reduced Rayleigh number

$$r = \frac{R_a}{R_{a_c}} .$$

The quiescent state (x,y,z) = (0,0,0) is obtained at equilibrium (r = 0).

A usual choice is

$$\sigma = 10 \tag{6.23a}$$

$$b = \frac{8}{3} \tag{6.23b}$$

which corresponds to the first destabilizing mode in a linear analysis of the fluid system; the bifurcation sequence of the model is not very sensitive to these choices. For these values, not only a strange attractor can be observed beyond a threshold value of the control parameter ($r \geq r_T = 24.7$) but it even degenerates into a limit cycle at a critical point (r = 148.4), which in turn bifurcates to another strange attractor at higher values of r [Manneville and Pomeau, 1980].

Among other examples directly relevant to our subject is also a model of the Belousov-Zhabotinski reaction known as the reversible oregonator [Field, 1975]. It involves the following sequence of reactions:

$$
\begin{aligned}
A + Y &\rightleftharpoons X \\
X + Y &\rightleftharpoons P \\
B + X &\rightleftharpoons Z + 2X \\
2X &\rightleftharpoons Q \\
Z &\rightleftharpoons fY
\end{aligned}
\tag{6.24}
$$

where f is a suitable stoichiometric coefficient. Using f and some kinetic rate constants as control parameters (with fixed quantities of A, B, P and Q), one finds the possibility for chaotic behaviours of the system (6.24), in agreement with recent experimental results [Vidal et al, in press]. Note that this possibility disappears when all reactons are taken as irreversible (from left to right) [Turner, 1980].

More complicated systems for which chaotic behaviours have been indicated are a form-variable model of the terrestrial dynamo [Jacobs, 1975] and various discrete-time mappings arising in the context of population dynamics [May, 1976].

D. Conclusions

Our initial interest was in the determination of the qualitative behaviour of nonlinear systems in the long-time limit when their parameters are varied.

Linear stability analysis led us to the discovery of instability thresholds, and bifurcation analysis allowed us to determine the merging between various kinds of behaviours at these critical points. The bifurcation here appears as the emergence of a new solution breaking some symmetry of the reference state: this is exactly what we mean by qualitative changes in the behaviour.

On the other hand, it is important to realize that despite those dramatic changes at the macroscopic level, nothing important seems to change on a microscopic scale: thermal (equilibrium-like) agitation remains chaotic, and the molecules still hardly recognize each other over distances larger than a few angströms. This process of successive bifurcations can be studied by appropriate local and global techniques; for the first few bifurcations, a convenient way is to introduce at each transition an "order variable" referring to the branch (the physical nonequilibrium phase) chosen by the system. Thus each transition enriches the set of relevant macrovariables by a new one, picked out from a "reservoir" of many[5] microscopic degrees

[5] In the thermodynamic limit!

of freedom. So the limiting case of "many" order variables, to be fixed all together by the control parameters and the constitutive relations, is likely to be undescriptible, as it may lead to chaos. This view, of chaotic behaviour arising brutally from more organized patterns in a few steps, can already be considered as a major progress in the understanding of nonlinear phenomena in complex systems. If chaos in dynamical systems turned out to be as common as turbulence in fluids dynamics, one would be led to the conclusion that macroscopic order is sandwiched between equilibrium thermal disorder and turbulent disorder, and it acquires its spectacular properties from this very peculiar situation.

VII. FLUCTUATIONS AND STOCHASTIC DESCRIPTION

The principal lesson to this point is that macroscopic systems, held out of equilibrium, are capable of undergoing transitions characterized by the appearance of coherence or order on a macroscopic scale. Using the methods of stability and bifurcation analyses, we have been able to classify the various kinds of behaviours in the vicinity of the bifurcation points.

Although not complete, our analysis showed that such phenomena are possible, indicating the enormous potentialities of nonequilibrium systems; but it did not justify the spontaneous choice of the system in favour of a particular branch, nor did it provide a mechanism for the transition. Noticing that the approach of the bifurcation point manifests itself by the loss of stability with respect to perturbations in some modes, we may suspect that triggering these modes could force the transition of the system onto a bifurcating branch.

Now, in all macroscopic systems, there is a spontaneous mechanism allowing them to test continuously stability with respect to many possible modes of evolution, namely the thermodynamic fluctuations. In the next three sections, we shall indicate how fluctuations can be incorporated in to our picture and lead us to a deeper understanding of nonequilibrium transition processes.

In this section, we discuss the physical meaning of fluctuations in macroscopic systems, focussing mainly on the formalism of master equations. General results are presented in Section VIII and asymptotic properties in Section IX.

A. The Role of Fluctuations

When fluctuations are incorporated in the description of a system, the state is no longer given by the macrovariables $\underset{\sim}{x}(\underline{r},t)$ but by the probability distribution $P(\underset{\sim}{x},\underline{r},t)$ of these macrovariables; the central question is therefore to find the evolution equation for P.

In the derivation of the phenomenological equations from statistical mechanics, the number of degrees of freedom is reduced severely. Such a procedure is only valid in some average sense, and it should therefore be completed by introducing a "noise" in the macrovariables' evolution, as a recollection of their coupling to the eliminated degrees of freedom. In some simple cases, this viewpoint led to useful results: for instance the theory of Brownian motion can be completely derived by these means [Résibois and De Leener, 1977].

Moreover, near equilibrium, one shows how irreversible macroscopic processes are related to reversible microscopic processes by the celebrated fluctuation-dissipation theorem [Kubo, 1966]. Unfortunately, no satisfactory theory has been developed yet for systems far from equilibrium, although some interesting steps were undertaken using formal projection methods [Mori, 1975; Zwanzig, 1972; Misra, Prigogine and Courbage, 1979].

For this reason we shall adopt here an alternative procedure which is "intermediate" between the phenomenological and the microscopic description: we shall assume that flutuations define a random process in an appropriate state space. Furthermore, the process will be assumed to be Markovian, that is, completely reducible to the conditional probability of occupying the state $\underset{\sim}{X}_1$ at time t_1 provided the state $\underset{\sim}{X}_2$ was occupied at t_2.

B. Master Equation for Birth-and-Death Processes

(i) The Breakdown of the Central Limit Theorem

In reaction-diffusion systems, the main phenomena going on a macroscopic scale are variations in the concentration of a species with time and space. Since the reacting system can be regarded as ideal (see Section II.A), we expect that if we divide the system into cells, with a size ΔV small by comparison with gradient and diffusion lengths, the distributions of two molecules of any species in the cell will be uncorrelated. The population of a cell will then be the sum of identically distributed independent random variables. If each cell is still big enough to contain a large number N of particles of the same species, we are in position to use the central limit theorem: as the size of the cell goes to infinity, the probability distribution for $x = X/\Delta V$ tends towards a Gaussian distribution with expectation $\overline{x} = \langle X\rangle/\Delta V$ which is an intensive quantity, and variance $\langle\delta X^2\rangle/\Delta V^2$ which depends on ΔV as ΔV^{-1}. It follows that in the limit of ΔV large there is a clear-cut separation between macroscopic behaviour and fluctuations.

Applying this description to the homogeneous Schlögl model (Section II.A), we guess that it fits perfectly in the subcritical domain, where the distribution function for X has a single peak corresponding to the thermodynamical branch. But in the super-critical range of the parameter, the distribution function for X must show two peaks (for the two stable branches) and a trough (the unstable branch). Since this phenomenon is macroscopic, a realistic approximation will represent X by the superposition of two unimodel distributions; for each branch the central limit theorem applies but it fails for the global distribution.

Example: if $P(X = X_1) = P(X = X_2) = \frac{1}{2}$

$$P(X \neq X_1 \text{ and } X \neq X_2) = 0$$

then

$$< X > = \frac{1}{2}(X_1 + X_2) = 0(V)$$

$$< \delta X^2 > = \frac{1}{4}(X_1 - X_2)^2 = 0(V^2)$$

$$\frac{\sqrt{< \delta X^2 >}}{< X >} = \left| \frac{X_1 - X_2}{X_1 + X_2} \right| = 0(1)$$

Obviously, a brute force application of the central limit theorem is not appropriate for separating the fluctuations out of the average evolution, even in the thermodynamic limit. The purpose of much of our discussion will be to develop the necessary tools for filling this gap.

As an additional physical motivation, we may recall that near equilibrium critical points the phenomenological theory has also to be corrected, in order to account for experimental data such as long-range correlations in space and time. Fluctuations play also an important role in determining the time-scales of the evolution. For instance, we expect that many transitions between coexisting stable branches of solution arising beyond bifurcations require a nucleation process, like for equilibrium phase transitions. Moreover, in the thermodynamic limit, the system cannot evolve at all, in finite times, without fluctuations in small volumes [Suzuki, 1976].

(ii) The Cell Picture

Having convinced ourselves on the necessity to incorporate fluctuations in the description of transition phenomena, we now discuss the basic assumptions invoked in their description.

The fundamental hypothesis in local thermodynamics (as presented in Section III.C), is that the relevant field variables are just the macroscopic observables of equilibrium statistical mechanics. In order to understand the origin of this hypothesis we divide the system into submacroscopic cells, the size of which is simultaneously much larger than the microscopic length scale (typically the dimensions of the constituting molecules) and much smaller than any macroscopic length scale (typically, the dimensions of the entire system) (Fig. 7.1). In a submacroscopic cell, the particles are so numerous, and the constraints applied by the environment vary so weakly, that the cell may be considered as an open ensemble in equilibrium: the macroscopic variables defined for the cell are then the thermodynamic state functions; on the macroscopic scale, the centres of the cells form a (quasi-) continuum and the interpolation of thermodynamic functions between these centres defines the thermodynamic fields. A similar "coarse-graining" is also performed on the time-scales.

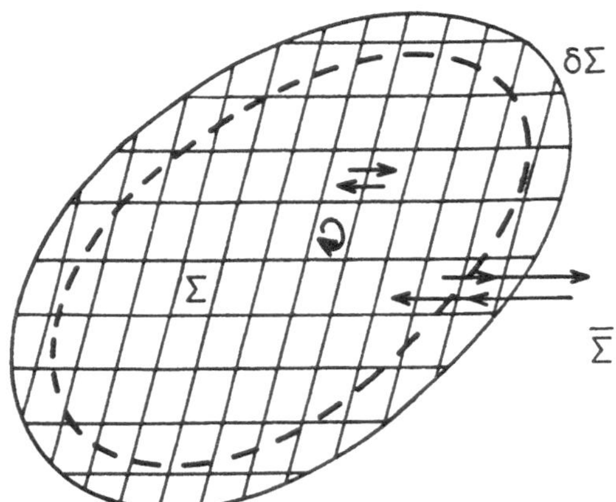

Fig. 7.1. The cell picture for a system Σ is valid only at some distance from the boundary δΣ. Three processes can occur: chemical reactions in one cell; diffusive exchanges between neighbouring cells; exchanges with the environment $\bar{\Sigma}$.

One restriction is obviously accepted in this picture: the "environmental" and "molecular" scales must be well decoupled. We do not expect it to be valid when these two characteristic lengths are comparable, like near the walls of a container, in a thin membrane or in interstellar space. In short, we see that the local equilibrium picture must be taken far from any kind of singularity.

(iii) The Multivariate Master Equation

Since we are mainly interested here in the coupling between chemical reactions and diffusion processes, the only variable we have to consider is the number $X_{i\underline{r}}$ of particles of a species i in a cell with centre $\underline{r}$; this is an extensive property. The state of the system is given by a probability distribution

$$P = P(\{X_{i\underline{r}}\},t) .$$

Within this framework, the most general Markovian equation of evolution for this function has the form

$$\frac{\partial}{\partial t} P(\{X_{i\underline{r}}\},t) = \sum_{\{X'_{j\underline{R}}\}} \Big(W(\{X'_{j\underline{R}}\}|\{X_{i\underline{r}}\}(P(\{X'_{j\underline{R}}\},t) - W(\{X_{i\underline{r}}\}|\{X'_{j\underline{R}}\})P(\{X_{i\underline{r}}\},t)\Big) \tag{7.1}$$

where we have introduced the rate

$$W(\{X'_{j\underline{R}}\}|\{X_{i\underline{r}}\})$$

for transitions from $\{X'_{j\underline{R}}\}$ to $\{X_{i\underline{r}}\}$. We assume that these rates result from the superposition of a chemical contribution W_{ch} and a diffusion contribution W_{df}:

$$W(\{X'_{j\underline{R}}\}|\{X_{i\underline{r}}\}) = \delta^{Kr}_{\underline{r}\,\underline{R}} W_{ch} + \delta^{Kr}_{ij} W_{df} . \tag{7.2}$$

The chemical evolution operator W_{ch} is a local operator. It is modelled by a birth-and-death process in each cell, with transition probabilities given by the kinetics of the reaction. For instance, the reaction $A + B \rightarrow D$ is described by the stochastic rate law:

$$W_{ch}(\{A+1,B+1,D-1\}|\{A,B,D\}) = K_2(A+1)(B+1) \tag{7.3}$$

where K_2 is the rate at which the reaction proceeds due to collisions in the cell:

$$K_2 = (\Delta V)^{-1} k_2 \quad , \tag{7.4}$$

k_2 being the usual rate constant of chemical kinetics.

The diffusive exchange of particles occurs by random walk between neighbouring cells:

$$W_{df}\left(\left\{X'_{i\underline{R}}\right\}\middle|\left\{X_{i\underline{r}}\right\}\right) = \sum_{\underline{L}} \delta^{Kr}_{\underline{R},\underline{r}+\underline{L}} \, W_d\left(\left\{X'_{i\underline{R}}\right\}\middle|\left\{X_{i\underline{r}}\right\}\right) \tag{7.5}$$

where $\underline{r}+\underline{L}$ is the location of any first neighbour of cell $\underline{r}$. For ideal mixtures and cubic cells, we specify

$$W_d\left(\left\{A_{\underline{r}+\underline{L}} + 1, A_{\underline{r}} - 1\right\}\middle|\left\{A_{\underline{r}+\underline{L}}, A_{\underline{r}}\right\}\right) = \frac{\mathcal{D}_A}{2d}\,(A_{\underline{r}+\underline{L}} + 1) \tag{7.6}$$

with a microscopic jump rate $\mathcal{D}_A$. The spatial dimensionality d was singled out for convenience.

In the following, we shall limit ourselves to ideal mixtures, for which all nonzero rates W are given by (7.2-6).

The connection with macrophysics implies that the size of the cells play no role in the field description arising from (7.1); this requires the extensivity of the transition rates in the thermodynamic limit $(\Delta V \to \infty)$

$$W\left(\left\{X'_{j\underline{R}}\right\}\middle|\left\{X_{i\underline{r}}\right\}\right) = w\left(\left\{x_{j\underline{R}}\right\},\left\{X_{i\underline{r}} - X'_{j\underline{R}}\right\}\right)\Delta V \tag{7.7}$$

when the probability distribution P for the population of each cell is replaced by the density p for the concentration in each cell:

$$p(\{x_{i\underline{r}}\},t) = \lim_{\Delta V \to \infty} \Delta V \, P(\{x_{i\underline{r}}\Delta V\},t) \quad . \tag{7.8}$$

The restriction (7.7) was already included in the form (7.4) of the chemical rate, and it is obviously satisfied by the diffusive rate (7.6) if $\mathcal{D}_A$ is independent of the population (or of the size ΔV).

However, the space variable in the phenomenological description of Section III.C is continuous; whereas the cell picture defines $\underline{r}$ only on a lattice with spacing

$$L = (\Delta V)^{1/d} \quad . \tag{7.9}$$

The transition from the discrete space to the continuous space requires a multiple-scale viewpoint, since it is performed via the continuum limit

$$\frac{\Delta V}{V} \to 0 \text{ , or } \frac{L}{\ell_h} \to 0$$

where V is the volume of the whole system, and ℓ_h is a typical macroscopic length. In order to recover Fick's diffusion law for the continuous system from the transition rate (7.6), we must also require the condition

$$\mathcal{D}_i = 2dL^{-2} D_i \tag{7.10}$$

where D_i is Fick's diffusion coefficient for the species i. The (local!) chemical evolution is not affected by this transformation.

(iv) Mean-Field Master Equation

The Markovian approach devoloped for reaction-diffusion systems applies equally well to homogeneous systems. Summing over all cells, we may define the global populations

$$X_i = \sum_{\underline{r}} X_{i\underline{r}} \tag{7.11}$$

the distribution of which obeys a mean-field master equation

$$\frac{\partial}{\partial t} P(\{X_i\},t) = \sum_{\{X_i'\}} \Big(W(\{X_i'\}|\{X_i\})P(\{X_i'\},t) - W(\{X_i\}|\{X_i'\})P(\{X_i\},t) \Big) \tag{7.12}$$

The theory of the master equations (7.1) and (7.12) was recently reviewed by Malek Mansour, Van den Broeck, Nicolis and Turner [1980].

C. Continuous Markov Processes

A second way to account for fluctuations in the macroscopic evolution is to introduce an additional random force in the phenomenological equations. Of course, the structure of the noise must reflect the specific features of the system under consideration.

A general phenomenological equation of the form

$$\frac{\partial x}{\partial t} = f[x,\lambda] - \text{div}\, \underline{j}[x,\lambda] \tag{7.13}$$

is replaced by a stochastic (Langevin) differential equation

$$\frac{\partial x}{\partial t} = f + q^f - \text{div}(\underline{j} + \underline{q}^j) \tag{7.14}$$

where q^f and $\underline{q}^j$ are distinct stochastic processes. In general, one chooses continuous Markov processes depending on the macrovariable:

$$q = q(x,\lambda,\underline{r},t) \quad . \tag{7.15}$$

Then:

$$< q^f > = 0 \tag{7.16a}$$

$$< \underline{q}^j > = \underline{0} \tag{7.16b}$$

$$< q^f(\underline{r},t)q^f(\underline{r}',t') > = Q^{(f)}\delta(\underline{r}-\underline{r}')\delta(t-t') \tag{7.17a}$$

$$< q^f(\underline{r},t)q^j_\alpha(\underline{r}',t') > = 0 \tag{7.17b}$$

$$< q^j_\alpha(\underline{r},t)q^j_\beta(\underline{r}',t') > = Q^{(j)}\delta^{Kr}_{\alpha\beta}\delta(\underline{r}-\underline{r}')\delta(t-t') \tag{7.17c}$$

where α,β are Cartesian indices and $Q^{(f)}$, $Q^{(j)}$ are some positive functionals of the macrovariable x (depending on the control parameters).

The importance of distinguishing between scalar and vector noise terms q^f and $\underline{q}^j$ was pointed out by Gardiner [1976] and Grossman [1976].

For these assumptions, one can generally convert the phenomenological equation (7.14) to a (multivariate) Fokker-Planck equation commanding the probability distribution of x, which is the fundamental quantity defining the state of the system [Arnold, 1973].

D. External Noise

So far we have limited ourselves to the internal fluctuations generated by the system's dynamics. However, in view of the important role of the control parameters in the evolution equations, their randomization can have tremendous consequences.

The evolution equation for the observable x reads in general

$$\frac{\partial x}{\partial t} = G(x,\lambda) \quad . \tag{7.18}$$

If λ is a stochastic process λ_t, this relation is conveniently rewritten in the standard form of stochastic differential equations

$$dx_t = G(x_t,\lambda_t)dt \tag{7.19}$$

for the stochastic process $x = x_t$. More assumptions are needed to solve Eq. (7.19), and a simple case is that of a Gaussian white noise λ_t:

$$\lambda_t = \bar{\lambda} + \xi_t \quad . \tag{7.20}$$

Here $\bar{\lambda}$ is the deterministic part of λ, whereas ξ_t is a Wiener process

$$\xi_t dt = dW_t \tag{7.21}$$

$$< \xi_t > = 0 \tag{7.21a}$$

$$< \xi_t \xi_{t'} > = \sigma^2 \delta(t - t') \tag{7.21b}$$

where σ is a real constant. The rate equation becomes:

$$dx_t = G(x_t,\bar{\lambda})dt + \sigma H(x_t)dW_t \tag{7.22}$$

where

$$H(x) = \frac{\partial G}{\partial \lambda}(x,\bar{\lambda}) \quad . \tag{7.23}$$

The system then evolves according to a Markovian process; the probability density p(x,t) satisfies the Fokker-Planck equation:

$$\frac{\partial}{\partial t} p(x,t) = - \frac{\partial}{\partial x}(G(x,\bar{\lambda})p(x,t)) + \frac{1}{2}\frac{\partial^2}{\partial x^2}\left(\sigma^2 H^2(x)p(x,t)\right). \tag{7.24}$$

Here G is called the drift coefficient and $\sigma^2 H^2$ is the diffusion coefficient for the stochastic process x_t.

E. Discussion

The relationships between tbe three approaches presented here are not completely elucidated. In the following sections, we shall focus on the multivariate master equation, which can be regarded as more "fundamental", in the sense that it incorporates, at the outset, the maximum amount of specific information pertaining to our system.

VIII. GENERAL PROPERTIES OF THE MASTER EQUATION

The solution of the linear equations (7.1) and (7.12) is hindered by several difficulties:

- a master equation refers to a probability distribution on an infinite number of states, leading to normalization problems in the large number limit and to a large number of coupled variables;
- the coefficients of this differential-difference equation are non-linear functions of the state variables;
- no obvious perturbation parameter is present in the equation;
- far from equilibrium, the behaviour of each system appears to be very specific.

So, before discussing the solution of these equations for reaction-diffusion processes, we would like to present some general results which are not based on approximation schemes.

A. The Competition of Reaction and Diffusion

The master equation (7.1) describes an irreversible Markovian evolution. In the absence of constraints, its stationary state is characterized by the usual rule of detailed balance, or microreversibility, from which the equilibrium distribution is easily deduced: this is a multinomial or a multipoissonian distribution, according to the boundary conditions [Malek Mansour and Nicolis, 1975; Moreau, 1979]. It has also been proved that the expressions (7.3-6) for the transition rates are necessary to obtain these equilibrium distributions. The contractions of those distributions to a few cells, are equivalent in the thermodynamic limit and present no spatial correlations:

$$< \delta X_{i\underline{r}} \delta X_{j\underline{R}} > \;=\; < \delta X_{i\underline{r}}^{\;2} > \delta^{Kr}_{ij}\, \delta^{Kr}_{\underline{r}\,\underline{R}} \qquad (8.1a)$$

with

$$< \delta X_{i\underline{r}}^{\;2} > \;=\; < X_{i\underline{r}} > \;. \qquad (8.1b)$$

Out of equilibrium, however, we have deviations from these laws. Chemical reactions modify the population of the cells and produce locally non-Poissonian behaviours if their kinetics are nonlinear. By tending to restore spatial homogeneity, diffusion correlates the distributions in neighbouring cells and propagates the non-Poissonian behaviour. As a result, the second factorial cumulant

$$\pi^{(2)}_{i\underline{r},j\underline{R}} = \langle \delta X_{i\underline{r}} \delta X_{j\underline{R}} \rangle - \langle \delta X_{i\underline{r}} \rangle \delta^{Kr}_{ij} \delta^{Kr}_{\underline{r}\,\underline{R}} \tag{8.2}$$

is no longer zero. Near equilibrium, it turns out to be extensive and proportional to the thermodynamic fluxes across the system.

The role of diffusion and of nonlinearity in the chemical kinetics is illustrated by the following model [Malek Mansour and Van den Broeck, 1980a]:

$$\begin{aligned} A &\underset{k_2'}{\overset{k_2}{\rightleftharpoons}} 2X \\ X &\underset{k_1'}{\overset{k_1}{\rightleftharpoons}} B \end{aligned} \tag{8.3}$$

on a one-dimensional ring. At equilibrium, the fluctuations are Poissonian and there is no spatial correlation. But if we supply A in large excess and eliminate B continuously, the reactions are only direct and one finds a linear macroscopic rate law:

$$\frac{dx}{dt} = 2k_2 a - k_1 x \tag{8.4}$$

with stationary solution

$$x = \gamma a \tag{8.5a}$$

where

$$\gamma = 2\frac{k_2}{k_1} \tag{8.5b}$$

If we partition the ring in n cells of length L, indexed by i, each of them including always A molecules of species A, and if $\mathcal{D}$ is the diffusion (jump) frequency for species X, the multivariate master equation can be written as:

$$\frac{\partial}{\partial t} P(\{X_\ell\},t) = k_2 A \sum_j \left(P(\{X_\ell - 2\delta^{Kr}_{\ell j}\},t) - P(\{X_\ell\},t) \right)$$
$$+ \; k_1 \sum_j \left((X_j + \delta^{Kr}_{\ell j}) P(\{X_\ell + \delta^{Kr}_{\ell j}\},t) - X_j P(\{X_\ell\},t) \right)$$
$$+ \; \frac{\mathcal{D}}{2} \sum_j \left((X_{j-1} + 1) P(\{X_\ell + \delta^{Kr}_{\ell,j-1} - \delta^{Kr}_{\ell j}\},t) \right.$$
$$+ \; (X_{j+1} + 1) P(\{X_\ell + \delta^{Kr}_{\ell,j+1} - \delta^{Kr}_{\ell j}\},t)$$
$$\left. - \; 2X_j P(\{X_\ell\},t) \right) . \tag{8.6}$$

Exact time-dependent solutions have been found; they evolve towards the stationary solution. This stationary solution is characterized by its expected population in each cell:

$$< X_j > \; = \; \gamma A \tag{8.7}$$

in agreement with (8.5). The second cumulant $\pi^{(2)}_{j\ell}$ is a function of n, k_1, $\mathcal{D}$, γ and A, and turns out to be non-Poissonian. In the continuum limit, it takes the simple form:

$$< \delta x(\underline{r}) \delta x(\underline{r}') > \; = \; \alpha(\underline{r} - \underline{r}') \exp\left[\frac{-|\underline{r}-\underline{r}'|}{\ell_{cor}}\right] \tag{8.8}$$

where $\alpha(\underline{r})$ decreases less fast than exponentially and

$$\ell_{cor} = \sqrt{\frac{D}{2k_1}} \tag{8.9}$$

is the macroscopic correlation length. On integrating this quantity over space we find a non-Poissonian variance for the system as a whole.

In addition to the departure from a Poisson law, this result indicates that diffusion and reaction cooperate in building spatial correlations: the length ℓ_{cor} has the same order of magnitude as the root-mean-square distance travelled by a particle X between its creation <u>in a pair</u> and its destruction by a chemical reaction (with rate k_1). Of course, this correlation length increases as the lifetime k_1^{-1} increases, but for small k_1, the concentration of X will grow in such a way that the inverse nonlinear reaction 2X → A will no longer be negligible. As a consequence, nonlinear modes come into play and the problem becomes much more involved. In this respect, the cancellation characterizing the equilibrium (and leading to very simple distributions) is very exceptional.

For a general reaction-diffusion scheme, the situation should remain very similar. Away from the critical point, the behaviour of the spatial correlation function should be typically described by an exponential decay such as (8.8). For instance, for a phenomenological evolution equation

$$\frac{\partial x}{\partial t} = F(x) + D\nabla^2 x \tag{8.10}$$

one expects near a stationary state x_s:

$$\ell_{cor} \sim \sqrt{-\frac{D}{F'(x_s)}} \tag{8.11}$$

where $F'(x_s)$ is the eigenvalue of the linear stability analysis. Near a bifurcation point ($F'(x_s) \to 0$), the correlation length diverges, indicating the possibility of a nonequilibrium transition. Since by transversality (Section V.C) $F'(x_s)$ is proportional to the distance $\lambda - \lambda_c$ from the bifurcation point, ℓ_{cor} diverges like:

$$\ell_{cor} = O(|\lambda - \lambda_c|^{-\frac{1}{2}}) \quad . \tag{8.12}$$

We call this a classical law of divergence. At this point, however, one is no more allowed to neglect the nonlinear modes. The latter can modify the law of divergence of the correlation length or even rule out the transition.

B. H-Theorem for Markov Processes

The theory of Markov processes on a discrete (countable) state space is well established and cannot be detailed here. We refer the reader to the outstanding books of Feller [1967, 1971]. The main results of the general theory of the master equations (7.1) and (7.12) are the following theorems:

- the master equation always admits a "minimal solution" $\overline{P}$ such that any solution P evolving from the same initial conditions satisfies

$$\forall \{X_i\},t: \quad P(\{X_i\},t) \geq \overline{P}(\{X_i\},t) \tag{8.13}$$

- conditions can be found, under which this minimal solution is the unique solution of the master equation satisfying given initial conditions;

- if the minimal solution is the unique solution of the master equation, and if the master equation admits a (normalized) stationary solution, then all solutions will evolve asymptotically towards this stationary state.

In the case of finite sets of states ($\{X_i\}$), the existence of a stationary solution and its global stability (attractivity) can also be demonstrated by means of a Lyapounov functional [Schlögl, 1971; Schnakenberg, 1976]:

Theorem:

If the master equation has the unique stationary solution $P_o(\{X_i\})$, we define the functional

$$K(t) \;=\; K[P(\{X_i\},t)] \;=\; -\sum_{\{X_i\}} P(\{X_i\},t)\,\ln\frac{P(\{X_i\},t)}{P_o(\{X_i\})} \tag{8.14}$$

for any solution P. Then:

$$\forall\, t \in \mathbb{R}: \qquad \frac{dK}{dt} \geq 0 \tag{8.15}$$

and

$$\frac{dK}{dt} \;=\; 0$$

if, and only if

$$P(\{X_i\},t) \;=\; P_o(\{X_i\}) \quad . \tag{8.16}$$

This result is the analogue of the well-known Boltzmann's H-theorem in the kinetic theory of gases. It also reflects the fact that the eigenvalues of the transition matrix $W(\{X_i\}|\{X_i'\})$ have a nonpositive real part and are not purely imaginary, as a simple application of Gershgorin's theorem indicates. Note, however, that the extension of these properties to infinite sets is delicate.

C. Connection with Macroscopic Descriptions: Kurtz' Theorem

The passage from the discrete cell picture to the continuous phenomenological description involves the very delicate thermodynamic limit. Several theorems (reviewed by Kurtz [1976, 1978]) partly specify this connection for extensive Markov chains and their continuous analogues; the physical variables pertaining to this connection are[6] the population X in a volume V, and the concentration x in the cell.

The evolution equation for the distribution P(X,t) is the master equation

$$\frac{\partial}{\partial t} P(X,t) = \sum_{Y} W(Y|X)P(Y,t) \tag{8.17}$$

where

$$W(X|Y) \geq 0 \quad \text{if} \quad X \neq Y \tag{8.18a}$$

$$W(X|X) = - \sum_{Y \neq X} W(X|Y) \quad . \tag{8.18b}$$

This is a linear homogeneous equation in P, with constant coefficients. With given initial conditions, this equation defines a Cauchy problem: its solution is therefore unique over some interval $[0,T_V]$, where T_V is the first time at which the distribution of X becomes singular in the volume V. The "explosion time" for the whole system is by definition

$$T = \lim_{V \to \infty} T_V \quad . \tag{8.19}$$

Extensivity requires that

$$w(\frac{X}{V}, Y-X) = V^{-1}W(X|Y) \tag{8.20}$$

remains bounded for $V \to \infty$. In the special case of reaction-diffusion processes, the transition rates $W(X|Y)$ are polynomials in X with bounded (usually small) values of the jump Y-X. We introduce the differential moments associated to w:

$$F(x) = \sum_{\nu} \nu\, w(x,\nu) \tag{8.21}$$

$$Q(x) = \sum_{\nu} \nu^2\, w(x,\nu) \quad . \tag{8.22}$$

The first expression measures the net variation of x due to chemical reactions; it is easily shown to coincide with the macroscopic rate law (3.19). The second expression is a definite positive quantity which indicates the sensitivity of the system to chemical

[6] For simplicity, we shall limit this presentation to one-variable systems.

fluctuations; it is analogous to a variance for the transition rates w. These two functions will determine the evolution of the concentration

$$x_t = \frac{X_t}{V} \tag{8.23a}$$

and its average

$$\overline{x}(t) = \frac{< X_t >}{V} \tag{8.23b}$$

resulting from the microscopic approach. This evolution will be compared to the deterministic evolution of a variable $x^*(t)$ satisfying the rate law:

$$\frac{d}{dt} x^*(t) = F(x^*) \quad . \tag{8.24}$$

In the thermodynamic limit, we shall also compare the continuous stochastic process x_t defined by (8.23) to the continuous Markov process z_t with probability density $p(z,t)$ satisfying the "non-linear" Fokker-Planck equation (the coefficients are nonlinear functions of z):

$$\frac{\partial}{\partial t} p(z,t) = - \frac{\partial}{\partial z} (F(z)p(z,t)) + \frac{1}{2V} \frac{\partial^2}{\partial z^2} (Q(z)p(z,t)) \quad . \tag{8.25}$$

Theorem P1:

With all the hypotheses stated above, and with the initial condition

$$x^*(0) = \overline{x}(0) \tag{8.26a}$$

then for any $\varepsilon > 0$, for any $t \in [0,T]$:

$$\lim_{V \to \infty} P\left(\sup|\overline{x}_V(t) - x^*(t)| > \varepsilon\right) = 0 \quad . \tag{8.26b}$$

Theorem P2:

With all the hypotheses stated above, and with the initial condition

$$\lim_{V \to \infty} VP(zV,0) = p(z,0)$$

then for any time $t \in [0,T]$, the stochastic process $x_t = X_t/V$ converges to the Markov process z_t in the thermodynamic limit $(V \to \infty)$.

The first theorem is analogous to a law of large numbers, whereas the second theorem is a central limit theorem (for a class of stochastic processes).

Estimates have been found to express the accuracy of these limits; they determine the order of corrections in the limit of large volumes [Kurtz, 1978]:

$$\sup_{0 \leq t < T} \left| \frac{X_t}{V} - z_t \right| = 0\left(\frac{\ln V}{V}\right) \text{ (almost surely (a.s.))} \tag{8.27}$$

$$\sup_{0 \leq t < T} \left| \frac{X_t}{V} - x^*(t) - \frac{u_t}{\sqrt{V}} \right| = 0\left(\frac{\ln V}{V}\right) \quad \text{(a.s.)} \tag{8.28}$$

and subsequently

$$\sup_{0 \leq t < T} \left| \frac{X_t}{V} - x^*(t) \right| = 0\left(\frac{1}{\sqrt{V}}\right) . \quad \text{(a.s.)} \tag{8.29}$$

Here the fluctuation u_t, approximating the discrepancy between the realization x and the deterministic value $x^*(t)$, is a Markov process with density $\rho(u,t)$ obeying the linearized Fokker-Planck equation:

$$\frac{\partial}{\partial t} \rho(u,t) = -F'(x^*(t)) \frac{\partial}{\partial u} (u\rho(u,t)) + \frac{1}{2} Q(x^*(t)) \frac{\partial^2}{\partial u^2} \rho(u,t) \tag{8.30}$$

where

$$F'(x) = \frac{dF}{dx}(x) .$$

From these theorems, it follows that for finite times (up to T_V), the birth-and-death process x_t can be approximated by a non-linear diffusion process z_t, or by a linear diffusion process

$$x^*(t) + V^{-\frac{1}{2}} u_t$$

equally well in the thermodynamic limit. The former approximation (z_t) corresponds to keeping the same intensive variables as in the phenomenological description and adding a random force to the deterministic evolution equation: the result is a stochastization of the state variable x itself. The latter approach (u_t) corresponds to studying first the deterministic evolution (x^*) and then the fluctuations around it by linearization. As far as almost sure convergence between the processes x_t, z_t and $x^*(t) + V^{-\frac{1}{2}} u_t$ is concerned, no expansion in V^{-1} of the master equation (8.17) appears to be possible, since the corrections should have the order lnV/V [Horsthemke, 1980].

The behaviour of the system in the long-time limit is more complicated, since there may be several (asymptotic) steady states and the system may reach an absorbing boundary. It has been proved [Kurtz, 1978] that if the deterministic equation (8.24) admits only one asymptotically stable solution (globally attractive) then all the previous considerations hold for any time. Much remains to be done in the case of multiple steady states.

In the following section, we shall see how these theorems are verified on simple models and for asymptotic solutions of the master equations.

IX. ASYMPTOTIC EXPANSIONS OF THE MASTER EQUATION

Exact solutions of the master equation are very exceptional; in most cases they are limited to stationary states of simplistic systems. One must therefore turn to approximate solutions. In this section we present some perturbative methods for extracting the relevant information contained in the master equation, which are free of uncontrollable assumptions like truncation. These methods, which apply equally well to the mean-field master equation (Sections IX. A-D) and to the multivariate master equation (Sections IX. E-H), rely upon a systematic use of the thermodynamic limit [Nicolis and Turner, 1977; Malek Mansour et al, 1980].

A. Perturbative Analysis of the Mean-Field Master Equation

In view of the peculiar role of the thermodynamic limit ($V \to \infty$) in the connection with macrophysics, one is tempted to determine an asymptotic expansion of the probability distribution $P(\{X_i\},t)$ in the small parameter

$$\varepsilon = \frac{1}{V} \tag{9.1}$$

where V is the volume of the system. To simplify the notations, we present the method on the simple example of systems with one state variable X changing by jumps of ± 1; more complicated situations were also treated by the same techniques [Turner, 1979]. The most general scheme satisfying these conditions is the following sequence of chemical reactions

$$A_j + jX \underset{\hat{\ell}_{j+1}}{\overset{\hat{k}_j}{\rightleftharpoons}} (j+1)X \quad , \quad 0 \leq j \leq n \tag{9.2}$$

where we suppose that the populations of all independent precursors $A_0, A_1, \ldots A_n$ are controlled from outside.

The phenomenological rate law for this set of coupled reactions is obviously

$$\frac{d}{dt} x = F(x) \tag{9.3a}$$

where

$$F(x) = \sum_{j=0}^{n+1} (k_j - \ell_j) x^j \tag{9.3b}$$

and

$$\begin{aligned} k_{n+1} &= 0 \\ \ell_0 &= 0 \\ k_j &= \hat{k}_j \frac{A_j}{V} \quad , \quad 0 \leq j \leq n \\ \ell_j &= \hat{\ell}_j \quad , \quad 1 \leq j \leq n+1 \end{aligned} \tag{9.4}$$

The birth-and-death master equation corresponding to (9.2) is easily constructed following the methods developed in VII.B:

$$\begin{aligned} \frac{\partial}{\partial t} P(X,t) &= \lambda(X-1)P(X-1,t) + \mu(X+1)P(X+1,t) \\ &\quad - (\lambda(X) + \mu(X))P(X,t) \end{aligned} \tag{9.5}$$

with the birth and death transition rates

$$\lambda(X) = V w\left(\frac{X}{V}, +1\right) = \sum_{j=0}^{n+1} k_j \varepsilon^{j-1} \frac{X!}{(X-j)!} \tag{9.6a}$$

$$\mu(X) = V w\left(\frac{X}{V}, -1\right) = \sum_{j=0}^{n+1} \ell_j \varepsilon^{j-1} \frac{X!}{(X-j)!} \quad . \tag{9.6b}$$

We will assume that $k_o \neq 0$ and $\ell_{n+1} \neq 0$. As it turns out, this rules out the existence of an absorbing state at zero or at infinity. It also guarantees the uniqueness of solutions of (9.2) and the ergodicity of the process for any finite V [Karlin and McGregor, 1957].

We now want to determine the behaviour of the solutions of the master equation (9.5) in the limit of a macroscopic system ($V \to \infty$ or $\varepsilon \to 0$). To this end it will be useful to switch to the generating function defined as:

$$\mathfrak{F}(s,t) = \sum_{X=0}^{\infty} s^X P(X,t) \quad , \quad |s| \leq 1 \quad . \tag{9.7}$$

From (9.5) and (9.6) we then obtain

$$\frac{\partial \mathfrak{F}(s,t)}{\partial t} = (s-1) \sum_{i=0}^{n+1} (k_i s - \ell_i) \varepsilon^{i-1} s^{i-1} \frac{\partial^i \mathfrak{F}(s,t)}{\partial s^i} \quad . \tag{9.8}$$

Note that $\mathfrak{F}$ is a convex monotonously increasing function in the interval $s \in [0,1]$. As $\varepsilon \to 0$, we expect that $P(X,t)$ will be extremely small for all small and moderate values of X. On the other hand, for large X, the factor s^X in expression (9.7) is also going to be very small, unless s remains in the immediate vicinity of $s = 1$ (see Fig. 9.1).

In order to explore this vicinity, we scale the speed at which s approaches unity when ε goes to zero, with a suitable power of the inverse volume

$$s = 1 + \varepsilon^a \xi \ ,$$

$$\xi = O(1) \quad , \quad 0 \leq a \leq 1 \tag{9.9a}$$

and study the equation obeyed by the new generating function

$$\phi_a(\xi,t) = \mathfrak{F}(1 + \varepsilon^a \xi, t) \tag{9.9b}$$

assumed to be analytic in the vicinity of $\xi = 0$. This will allow a regularization of the generating function when written in terms of the scaled variable, and an identification for the dominant terms in ε in the equation of evolution. In the original form (9.8), such an identification cannot be made in general: the higher powers of the smallness factor ε multiply higher derivatives of the unknown function $\mathfrak{F}$, and this constitutes a singular perturbation problem which must be handled with special care [Cole, 1968].

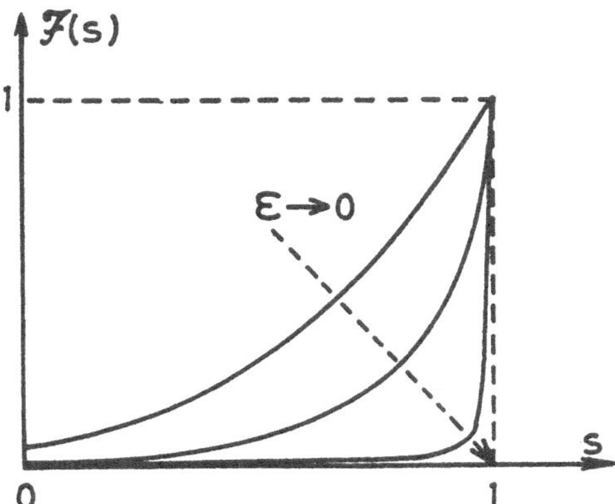

Fig. 9.1. The generating function $\mathcal{F}(s)$ becomes singular in the thermodynamic limit $\varepsilon \to 0$.

B. Macroscopic Limit

The value of the scaling exponent a in (9.9) is an intrinsic property of the system as it reflects the importance of the fluctuations relative to the results of the phenomenological description. It will have, therefore, to be determined by the system itself, as it will be shown in IX.C. Before we do this, however, we examine the structure of the master equation when in (9.9) a is set equal to unity. To the lowest order in ε, (9.8) written in terms of $\phi_a(\xi) \equiv \phi_1(\xi)$ gives then the result:

$$\frac{\partial \phi_1}{\partial t} = \xi \sum_{j=0}^{n+1} (k_j - \ell_j) \frac{\partial^j \phi_1}{\partial \xi^j} \quad . \tag{9.10}$$

The meaning of ϕ_1 is easily ascertained on the basis of (9.7) and (9.9), together with the use of the Euler-Mac Laurin formula replacing the infinite sum by an integral [Abromowitz and Stegun, 1964]:

$$\phi_1(\xi,t) = \int_0^\infty (1+\varepsilon\xi)^X P(X,t)\,dX \quad .$$

Introducing the intensive variable

$$x = \varepsilon X \tag{9.11a}$$

and its probability density

$$p(x,t) = \varepsilon^{-1} P(\varepsilon^{-1}x,t) \tag{9.11b}$$

we can write $\phi_1(\xi,t)$ as the Laplace transform

$$\phi_1(\xi,t) = \int_0^\infty e^{\xi x}\, p(x,t)\,dx \quad . \tag{9.12}$$

The scaling (9.9) for a = 1 thus amounts to the study of the intensive variable

$$x = \frac{X}{V} \quad .$$

The equation for p(x,t) to lowest order in ε is easily obtained from (9.10) and (9.12):

$$\frac{\partial}{\partial t} p(x,t) = -\frac{\partial}{\partial x}\left[F(x)p(x,t)\right] \quad . \tag{9.13}$$

A standard initial condition for the master equation is to impose the total number of particles $X = \overline{X}_o = \varepsilon^{-1}\overline{x}_o$ at $t = 0$. The solution of (9.13) subject to this condition, $P(x,o) = \delta(x-x_o)$, is the Green function

$$P(x,t) = \delta\left(x - \overline{x}(t)\right) \tag{9.14}$$

where $\overline{x}(t)$ obeys (9.3) with the initial condition $\overline{x}(t=0) = \overline{x}_o$.

It is instructive to formulate this result in terms of the generating function. The initial condition reads

$$\mathcal{F}(\xi,t=0) = s^{\overline{X}_o} = s^{\varepsilon^{-1}\overline{x}_o}$$

which to the lowest order in ε becomes

$$\phi_1(\xi,t=0) = e^{\xi\overline{x}_o} \quad . \tag{9.15a}$$

The solution of (9.10) is then

$$\phi_1(\xi,t) = e^{\xi\overline{x}(t)} \quad . \tag{9.15b}$$

We can thus conclude that in the thermodynamic limit ($\varepsilon \to 0$) the stochastic trajectory converges in distribution to the deterministic path, if this is true at the initial time. For finite t this result is contained in the more general theorem P1 of Kurtz (VIII.C). It remains valid in the limit $t \to \infty$ taken prior to the thermodynamic limit, provided that the macroscopic rate equation (9.14) has a globally stable stationary solution, i.e.

$$\lim_{t\to\infty} \overline{x}(t) = \overline{x}_s$$

independent of the initial condition. In fact, as (9.10) together with the initial condition (9.15a) constitutes a Cauchy problem, the uniqueness of the solution (9.15) follows for some finite time. This result can be extended to arbitrary long times provided the limit $t \to \infty$ of the r.h.s. of (9.15b) exists, is time-independent and coincides with the stationary solution of (9.10). This clearly covers the case of marginal stability right at a bifurcation point. In the region where the macroscopic rate equation admits multiple steady states, however, this property will break down for $t \to \infty$.

C. Fluctuations around the Deterministic Path

i) General Theory

We have seen that in the thermodynamic limit $\varepsilon \to 0$, the value a = 1 of the scaling exponent of (9.9) allows us to recover the macroscopic regime in which fluctuations are wiped out. We now want to analyze the corrections to the macroscopic behaviour, which are described by the random variable $X - \langle X \rangle$. Since this is an extensive quantity, it is more convenient to introduce the scaled variable

$$u = \frac{X - \langle X \rangle}{V^a} = \varepsilon^{a-1}(x - \overline{x}) \tag{9.16a}$$

and its probability distribution

$$\rho(u,t) = \lim_{\varepsilon\to 0} \varepsilon^{-a} P(\overline{X} + \varepsilon^{-a} u, t) \quad . \tag{9.16b}$$

The scaling exponent a will be determined later, but it is expected to satisfy two requirements:

$$a > 0 \quad \text{as} \quad \lim_{V\to\infty} |X - \overline{X}| = \infty \tag{9.16c}$$

$$a \leq 1 \quad \text{as} \quad \lim_{V\to\infty} \left|\frac{X - \overline{X}}{\overline{X}}\right| < \infty \tag{9.16d}$$

almost surely.

The Laplace transform

$$\psi(\xi,t) = \int_{-\infty}^{+\infty} e^{-\xi u} \rho(u,t)du \tag{9.17}$$

is related to the generating function $\mathfrak{F}$:

$$\psi(\xi,t) = \mathfrak{F}(s,t)s^{-\overline{X}} \tag{9.18a}$$

and

$$\xi = \varepsilon^{-a}(s-1) \tag{9.18b}$$

to dominant order in ε. Because of the elimination of $\overline{X}$ by (9.16), the scaling exponent is now $a < 1$.

An equation for ψ follows by simple insertion of these definitions into the evolution equation (9.8); if we retain only the lowest order terms in ε and perform the inverse Laplace transform of (9.17), we find the evolution equation for the fluctuations:

$$\frac{\partial}{\partial t}\rho(u,t) = \frac{\partial}{\partial u}\left(-\sum_{k=1}^{n+1} \varepsilon^{(k-1)(1-a)} F^{(k)}(\overline{x}(t))\frac{u^k}{k!} + \right.$$

$$\left. + \frac{1}{2}\varepsilon^{2a-1} Q(\overline{x}(t))\frac{\partial}{\partial u}\right)\rho(u,t) \tag{9.19}$$

where we have defined the macroscopic functions

$$F^{(k)}(x) = \frac{\partial^k F}{\partial x^k}(x) = \sum_{j=k}^{n+1} \frac{j!}{(j-k)!}(k_j - \ell_j)x^{j-k} \tag{9.20a}$$

$$Q(x) = \sum_{j=0}^{n+1}(k_j + \ell_j)x^j \quad . \tag{9.20b}$$

Note that not all terms in (9.19) will contribute, depending on the value of a. Remarkably, this relation is a Fokker-Planck equation (compare with (8.25) and (8.30) with a nonlinear drift coefficient and a constant (and positive definite) diffusion coefficient Q. Thus, $\rho(u,t)$ remains non-negative for all t and can be considered as a true probability, provided it satisfies all these requirements at the initial time [Feller, 1971].

ii) Continuous Stochastic Processes

The interest of (9.19) is that it allows us to do a systematic perturbation calculus in ε and to understand the relative role of the nonlinearity of the drift coefficient ($F'' \neq 0$) compared to the diffusion coefficient. Once this information is utilized, however, it is instructive to switch back to the intensive variable $x = X/V$. Utilizing (9.20b) as well as the Taylor expansion of the rate function F(x) around $\bar{x}$ we can then write (9.19) in the simpler form:

$$\frac{\partial p}{\partial t} = \frac{\partial}{\partial x}\left[-F(x) + \frac{1}{2}\varepsilon \frac{\partial}{\partial x} Q(\bar{x})\right]p(x,t) \quad . \tag{9.21}$$

According to the theory of stochastic differential equations [Arnold, 1973], this relation is equivalent to a Langevin equation as (7.14)

$$\frac{\partial x}{\partial t} = F(x) + \varepsilon^{\frac{1}{2}} q(t) \tag{9.22}$$

q being a Gaussian white noise defined by:

$$\langle q(t)\rangle = 0 \tag{9.23a}$$

$$\langle q(t_1)q(t_2)\rangle = Q(\bar{x})\delta(t_1 - t_2) \quad . \tag{9.23b}$$

From a physical standpoint, Q appears as a measure of the fluctuations, which are affected by the frequency of the transitions (k_j, ℓ_j) and the population of the reactive species. This result establishes the validity of the nonlinear Langevin equation with a process independent noise, so long as the perturbation expansion adopted remains valid. This means that both the initial condition chosen for the probability distribution and instantaneous values of the latter for $t > 0$, must be compatible with the requirement $\frac{1}{2} \leq a < 1$. For sufficiently large volumes V this will be so for all finite times, if valid at the initial time. Its validity for $t \to \infty$, however (i.e. up to the steady state), is guaranteed only when the deterministic equation has a unique stable stationary solution. This result is clearly related to the theorems of Kurtz (Section VIII.C) on the validity of a Markov process with continuous realizations (diffusion process) as an asymptotic representation of the master equation for finite times.

The connection between master equations and Fokker-Planck equations in the region of multiple steady states (beyond the bifurcation point) constitutes an interesting open question. Two special cases in (9.19) can be solved completely to dominant order in ε.

Let us come back to (9.21). Consider first the case where the first derivative $F'(\overline{x}(t))$ is finite. The dominant scaling for the fluctuations is then given by $a = 1/2$ and (9.19) reduces to a Fokker-Planck equation with linearized coefficients. Its solution is a propagating Gaussian distribution provided it is initially Gaussian. In particular, its stationary solution near a stable state $\overline{x}_s$ is to dominant order

$$\rho_{st}(u) = (2\pi < \delta u^2 >_{st})^{-\frac{1}{2}} \exp\left(- \frac{u^2}{2 < \delta u^2 >_{st}}\right) \tag{9.24a}$$

with the variance

$$< \delta u^2 >_{st} = - \frac{Q(\overline{x}_s)}{F'(\overline{x}_s)} \quad . \tag{9.24b}$$

This expression diverges near a critical point as an inverse power of the distance from the bifurcation point: this is reminiscent of the behaviour of systems undergoing equilibrium phase transition (see also Section VIII.A).

iii) Critical Fluctuations

Let $\overline{x}_s$ be the unique sationary solution of the deterministic equation, and suppose that for some "critical" value of the control parameters we have:

$$\begin{aligned} F(\overline{x}_s) &= F'(\overline{x}_s) = \cdots F^{(k-1)}(\overline{x}_s) = 0 \\ F^{(k)}(\overline{x}_s) &\neq 0 \end{aligned} \tag{9.25}$$

for a certain (odd) value $k \geq 1$ (the case where $k = 1$ was considered in the previous subsection, where a Gaussian distribution was shown to result). At such a point one has marginal stability of $\overline{x}$ and hence when the parameters are greater than their critical value, one has bifurcation (degenerate for $k \geq 5$).

From (9.19) we see that the drift and diffusion terms contribute equally when the scaling exponent a has the value

$$a = \frac{k}{k+1} \quad . \tag{9.26}$$

We thus obtain to dominant order in ε:

$$\frac{\partial}{\partial \tau} \rho(u,\tau) = -\frac{\partial}{\partial u}\left[F^{(k)}(\bar{x}_s)\frac{u^k}{k!} - \frac{1}{2}\frac{\partial}{\partial u} Q(\bar{x}_s)\right] \rho(u,\tau) \tag{9.27}$$

with

$$\tau = t\,\varepsilon^{\left(\frac{k-1}{k+1}\right)} \quad . \tag{9.28}$$

At the stationary state, one finds

$$\rho_{st}(u) = Z \exp\left(-\frac{F^{(k)}(\bar{x}_s)}{\frac{1}{2}Q(\bar{x}_s)}\;\frac{u^{k+1}}{(k+1)!}\right) \tag{9.29}$$

Z being determined by normalization. It follows that [Abramowitz and Stegun, 1964]:

$$\lim_{V\to\infty} \langle \delta X^2 \rangle V^{-\frac{2k}{k+1}} = \langle \delta u^2 \rangle$$
$$= \left(-\frac{1}{2} Q(\bar{x}_s)\,\frac{(k+1)!}{F^{(k)}(\bar{x}_s)}\right)^{\frac{2}{k+1}} \frac{\Gamma(\frac{3}{k+1})}{\Gamma(\frac{1}{k+1})} \tag{9.30}$$

where Γ denotes the Euler gamma function.

This expression specifies to what extent the central limit theorem breaks down near a bifurcation point. Indeed, the latter theorem is based upon the assumption that the global population X is the sum of a large number of independent cell populations showing the same distribution. Its breakdown means therefore that the system can in no way be subdivided into uncorrelated subsets (and that the probability distribution law is not infinitely divisible [Feller, 1971]. We have thus found the mechanism by which the <u>coherence</u> associated with bifurcation emerges at the stochastic level. Moreover, the scaling (9.28) also reflects the critical slowing down observed in the phenomenological description (Section IV.C).

The radical change of the stochastic properties of the system implied by the transition to critical behaviour is also expected to subsist beyond the bifurcation. Unfortunately, as it was pointed out earlier, the asymptotic analysis of the master equation outlined in this section has not yet been extended to this region.

iv) Illustration: The Schlögl Model

The stochastic version of the Schlögl model can be analyzed exactly [Nicolis and Turner, 1977]; the results agree completely with the approximate solutions described here. Indeed, the phenomenological functions are

$$F(y) = -y^3 + 3y^2 - (\delta + 3)y + (\delta' + 1) \tag{9.31a}$$

$$Q(y) = y^3 + 3y^2 + (\delta + 3)y + (\delta' + 1) \tag{9.31b}$$

for

$$y = \frac{X}{m} = \xi + 1 \quad . \tag{9.31c}$$

Along the line $\delta = \delta'$, one finds:

- if $\delta > 0$: the steady state $y = 1$ is stable and the fluctuations are Gaussian ($k = 1$, $a = \frac{1}{2}$) with a variance:

$$< \delta u^2 >_{st} = \lim_{\varepsilon \to 0} \varepsilon < \delta X^2 > = \frac{4}{\delta} + 1 \ ; \tag{9.32}$$

- if $\delta = 0$: the steady state $y = 1$ is stable and the fluctuations are distributed according to a quartic law ($k = 3$, $a = 3/4$) with a variance

$$< \delta u^2 >_{st} = \lim_{\varepsilon \to 0} \varepsilon^{3/2} < \delta X^2 > = 4 \frac{\Gamma(3/4)}{\Gamma(1/4)} = 1.35 \ . \tag{9.33}$$

This analysis indicates that as long as there is only one stationary state in the phenomenological version, the distribution $P(X,t)$ will present a single peak around this state (in the limit $\varepsilon \to 0$), and it can be approximated by a Gaussian law in agreement with Kurtz' theorems. In the region where several steady states are allowed, the stochastic description leads, in general, to a singly-peaked distribution, except along a "coexistence line": this is due to the fact that the most stable state "traps" the fluctuations; however, this coexistence line bears no trivial relation to the phenomenological rate laws [Malek Mansour, 1978].

D. Time-Dependent Behaviour

A satisfactory analysis of fluctuations must clearly go beyond the static properties, to which most of this section was limited. Indeed, the mechanisms by which a system evolves, in a spontaneous fashion, away from a reference state, are essential for the elucidation of nonequilibrium transition phenomena.

In recent years, this question attracted considerably more attention. The typical case to which most of the investigations refer is a system involving an unstable state surrounded by two stable ones, like the Schlögl model. Two complementary approaches have been followed.

i) Eigenvalue Problem of the Evolution Operator

Matsuo [1977] has developed an asymptotic evaluation of the master equation using a WKB type of method. Most of the recent results, however, refer to the eigenvalue problem of the Fokker-Planck equation in a system involving a single variable and an unstable state surrounded by two stable ones. The Fokker-Planck equation is written in the form (9.21):

$$\frac{\partial p}{\partial t} = \frac{\partial}{\partial x}\left[-F(x) + \varepsilon \frac{\partial}{\partial x}\right]P \qquad (9.34)$$

where $0 < \varepsilon << 1$ and the diffusion coefficient has been set equal to unity. In a bistable system the simplest drift coefficient is a cubic. Moreover, when only one variable is present, $F(x)$ can always be written as a force deriving from a "kinetic potential" $\Phi(x)$:

$$F(x) = -\frac{\partial \Phi}{\partial x}(x) \quad . \qquad (9.35)$$

For the cusp bifurcation of a cubic rate law, this potential can be reduced to a canonical form

$$\Phi(x) = -(\lambda - \lambda_c)x^2 + \mu x^4$$

where μ is some positive constant. Before the bifurcation point, Φ has one minimum and the solution $x = 0$ is stable; beyond the bifurcation point, Φ has two minima and the solution $x = 0$ is the unstable maximum. A typical bistable potential is pictured in Fig. 9.2.

In discussing the eigenvalue problem for (9.34), Caroli et al [1979] applied a WKB approximation, whereas Van Kampen [1977] was able to develop a model amenable to a solvable equation of the

Schrödinger type. An interesting result of these investigations is the justification of Kramers' theory, which describes the transition between the two stable states on the potential wells, over the "potential barrier" constituted by an intermediate unstable state. As it turns out, this theory corresponds to the regime of the first "excited states" above the "ground state" corresponding to the stationary solution of (9.34). A singular perturbation approach extending the Kramers problem to many coupled variables has been developed by Matkowsky and Schuss [1979].

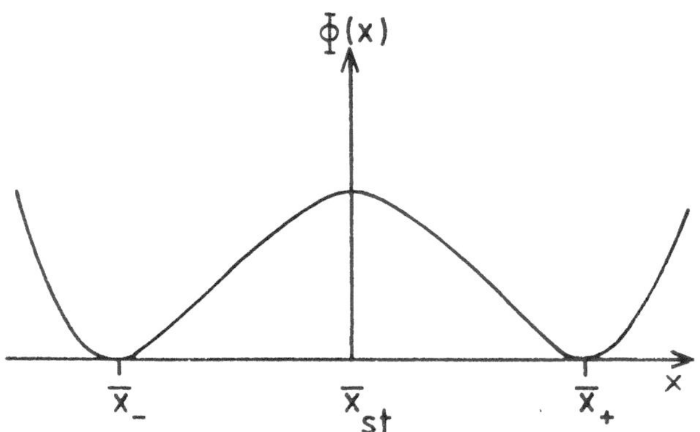

Fig. 9.2. A bistable potential $\Phi(x)$ admits three extrema, corresponding to the stationary states: two minima $\bar{x}_-$ and $\bar{x}_+$, and one maximum $\bar{x}_{st}$. The relative weights of the solutions in the asymptotic $(t \to \infty)$ distribution depends also on the relative depths of the wells.

ii) Initial Value Problem

The question asked here is how an initial probability distribution centered on an unstable state is deformed in time, until it develops maxima centered on the stable states. Suzuki [1976, 1980; see also de Pasquale and Tombesi, 1979] proposed an elegant scaling theory, which consists in splitting the evolution into an "initial regime", amenable to a linearized Fokker-Planck equation, and an "intermediate" regime amenable to a macroscopic description. A matching between the two regimes is required. During the first stage the distribution is continuously broadened, but remains centered on the unstable state. For long times the variance of this distribution ceases to be extensive, and one has to switch to the second stage. A rough estimation of this transition time leads to

$$t_o \sim \log \frac{1}{\varepsilon} \ . \tag{9.36}$$

In the thermodynamic limit, $\varepsilon \to 0$, this time becomes, therefore, exceedingly long. This indicates that the evolution of the system rests then on the dynamics of inhomogeneous fluctuations, which may occur in small volumes with appreciable probability, despite the fact that the total volume of the system, $V = \varepsilon^{-1}$, may become very large. We discuss some aspects of inhomogeneous fluctuations in the following subsections.

E. Perturbative Methods for the Inhomogeneous Fluctuations

Since the description of reacting media by the global master equation is an idealization, we now turn to the multivariate master equation (7.1). For the chemical reactions described by (9.2), it takes the form

$$\begin{aligned}\frac{\partial}{\partial t} P(\{X_{\underline{r}}\},t) = & \sum_{\underline{r}} \Big(\lambda(X_{\underline{r}}-1)P(X_{\underline{r}}-1,t) + \mu(X_{\underline{r}}+1)P(X_{\underline{r}}+1,t) \\ & - \Big(\lambda(X_{\underline{r}}) + \mu(X_{\underline{r}})\Big) P(X_{\underline{r}},t)\Big) \\ & + \frac{\mathcal{D}}{2d} \sum_{\underline{r}} \sum_{\underline{L}} \Big((X_{\underline{r}}+1)P(X_{\underline{r}}+1, X_{\underline{r}+\underline{L}}-1,t) \\ & - X_{\underline{r}} P(X_{\underline{r}}, X_{\underline{r}+\underline{L}},t)\Big)\end{aligned} \tag{9.37}$$

where we have omitted all numbers $X_{\underline{r}}$ unchanged during these processes.

All difficulties encountered with the mean-field master equation remain in the chemical part of (9.37), but the diffusive coupling between neighbouring cells complicates the description. A perturbative approach is highly desirable: a few small parameters are tempting.

In each cell, the thermodynamic limit can be defined by means of a small parameter

$$\varepsilon = \frac{\ell_{mic}^{d}}{\Delta V} = \left(\frac{\ell_{mic}}{L}\right)^{d} \qquad (9.38)$$

where: d is the dimensionality;

ΔV is the volume of a cell;

ℓ_{mic} is a microscopic characteristic length;

L is the length of a cell.

The limit $\varepsilon \to 0$ requires that the microscopic properties of the medium do not vary significantly within distances involving many particles; otherwise the description of the system would depend directly on the partition in cells.

A second (independent) small parameter can be introduced as

$$\varepsilon^{*} = \frac{\Delta V}{V} \qquad (9.39)$$

where V is the volume of the whole system. The "continuum limit" $\varepsilon^{*} \to 0$ describes the passage from the cell picture to a field picture, after the thermodynamic limit; this is conceptually important but usually straightforward.

In the following analysis, we shall use the thermodynamic limit $\varepsilon \to 0$ in each cell to generate a perturbative expansion [Malek Mansour et al, 1980; see also Van den Broeck, Houard and Malek Mansour, 1980, for an alternative approach].

F. The Multivariate Master Equation and the Thermodynamic Limit

In (9.37), the partition of space was initially conceived in such a way that spatial coherence be ensured within each cell. Typically, therefore, each cell has a linear dimension of the order of the mean free path: this could hardly be regarded as a large parameter playing a role analogous to the volume V in the global description.

However, the discussion of Section VIII.A indicates that the cooperation of reaction and diffusion ensures the coherence within a characteristic length scale

$$\ell_{cor} \simeq \sqrt{-\frac{D}{F'(\bar{x}_s)}} \qquad (\text{cf.}\,8.11)$$

around a stationary solution $\overline{x}_s$. This quantity is expected to be of macroscopic order, and it even diverges near a critical point ($F' = 0$). In this case, much of the macroscopic behaviour of the system will be independent of the details of the initial partition into cells.

We are now in position to apply the same procedure as in Section IX.B for the global master equation, using the inverse of the cell size ΔV as a perturbation parameter.

The dominant contribution to the evolution of the concentrations $x_{\underline{r}}$ comes from the deterministic relation

$$\frac{\partial}{\partial t} x_{\underline{r}} = F(x_{\underline{r}}) + \frac{\mathcal{D}}{2d} \sum_{\underline{L}} (x_{\underline{r}+\underline{L}} - x_{\underline{r}}) \tag{9.40}$$

reducing to the reaction-diffusion equation (3.19) in the continuum limit ($L \to 0$) [Arnold, 1979].

The fluctuations around this deterministic path are still described by means of an auxiliary function corresponding to the scaled variables

$$u_{\underline{r}} = (\Delta V)^{-a} (X_{\underline{r}} - \overline{X}_{\underline{r}}) \tag{9.41}$$

for which a Fokker-Planck equation is again derived from the master equation as (9.19). Here, of course, the noise term $Q(\overline{x}_{\underline{r}})$ will be complemented by a noise term $Q_{\underline{r}\underline{r}'}(\{x_{\underline{r}}\})$ proportional to $\mathcal{D}$ and linear in the concentrations $x_{\underline{r}}$: this expresses the spatial covariances generated by diffusion. As in the mean-field theory, a nonlinear Langevin equation with a process-independent noise can be associated to this Fokker-Planck equation [Malek Mansour et al, 1980], substantiating the more macroscopic approach of Section VIII.C: this result can be viewed as an extension of the well-known fluctuation-dissipation theorem to far-from-equilibrium systems.

The validity of the Fokker-Planck equation for $t \to \infty$ is guaranteed only when the deterministic equation has a unique stable stationary solution. This is true before and at the bifurcation point.

G. Critical Behaviour

i) The Distribution of Fluctuations

Around a stationary state $\overline{x}_s$, the Fokker-Planck equation admits a stationary solution $p(\{x_r\})$. Far from critical points, its linearization leads to a multi-Gaussian with spatial correlations: this corresponds to a value $a = 1/2$ of the scaling exponent, as in the mean-field theory (Section IX.C.ii)).

In the vicinity of a bifurcation point, however, this description breaks down and another value of a is required to regularize the perturbation expansion for the fluctuations (see IX.C, ii)).

$$a = \frac{k}{k+1} \quad . \tag{9.42}$$

However, with such a scaling for the fluctuations, we observe that the term in $Q_{\underline{r}\,\underline{r}'}$, responsible for the diffusive noise contribution, will become negligible. The Fokker-Planck equation takes then the following simple form:

$$\frac{\partial}{\partial t} p(\{x_{\underline{r}}\},t) = \sum_{\underline{r}} \frac{\partial}{\partial x_{\underline{r}}} \left(- F(x_{\underline{r}}) - \frac{\mathcal{D}}{2d} \sum_{\underline{L}} (x_{\underline{r}+\underline{L}} - x_{\underline{r}}) + \frac{1}{2} \varepsilon \, Q(\overline{x}_{\underline{r}}) \frac{\partial}{\partial x_{\underline{r}}} \right) p(\{x_{\underline{r}}\},t) \tag{9.43}$$

which can be solved exactly at the stationary state; one finds

$$p_{st}(\{x_{\underline{r}}\}) = Z^{-1} \exp\left(- \frac{2\,\varepsilon^{-1}}{Q(\overline{x}_s)} \sum_{\underline{r}} \left(\Phi(x_{\underline{r}}) + \frac{\mathcal{D}}{8d} \sum_{\underline{L}} (x_{\underline{r}+\underline{L}} - x_{\underline{r}})^2 \right)\right) \tag{9.44}$$

where Z is a normalization constant and

$$\Phi(x_{\underline{r}}) = - \int_{\overline{x}_s}^{x_{\underline{r}}} F(x)\,dx \tag{9.45}$$

is the "kinetic potential".

Equations (9.43) and (9.44) bear some striking similarities with the theory of equilibrium critical phenomena. In thermodynamic equilibrium, the probability distribution of fluctuations is given by the exponential of the free energy in a system at constant temperature and volume [see Nicolis and Prigogine, 1977, Section 9.4]. The modern theory of critical phenomena [Ma, 1976] establishes that, in the vicinity of the critical point, this expression reduces to the exponential of a form known as the Landau-Ginzburg functional, which has the same structure as (9.44). The powerful renormalization group techniques are nowadays used to extract the properties of the fluctuations near the bifurcation [Dewel, Walgraef and Borckmans, 1977; Borckmans et al., 1980]. Among the various results of this approach, let us mention the existence of a critical dimensionality and the concomitant appearance of non-classical exponents describing the divergence of variances, correlation functions, and so forth [Malek Mansour et al, 1980].

ii) Qualitative Theory of Nonequilibrium Phase Transitions

The critical dimensionality d_c plays a fundamental role in the study of bifurcations, for it describes the very possibility of coherent behaviour. In reaction-diffusion systems, its value can be inferred directly, without resorting to the solution (9.44) and to renormalization group analysis. Indeed, the characteristic time of diffusion is given by (7.10) as:

$$\tau_d = \mathcal{D}^{-1} = \frac{1}{2d}\frac{L^2}{D} \tag{9.46}$$

whereas the characteristic time scale of reaction results from (9.28),

$$\tau_{ch} = \frac{1}{k_o}\,\varepsilon^{-\frac{k-1}{k+1}} \tag{9.47}$$

where k_o is a typical rate constant. The ratio of these two values is

$$\varepsilon' = \frac{\tau_d}{\tau_{ch}} = \frac{1}{2d}\frac{k_o}{D}(\Delta V)^{2/d}\,\varepsilon^{\frac{k-1}{k+1}} \,. \tag{9.48}$$

Defining the numbers

$$d_c = 2\,\frac{k+1}{k-1} \tag{9.49}$$

$$\alpha^2 = \frac{1}{2d}\frac{k_o}{D}\,\ell_{mic}^{\;2} \tag{9.50}$$

we observe that

$$\varepsilon' = \alpha^2\,\varepsilon^{(2/d_c)-(2/d)} \,. \tag{9.51}$$

Therefore the competition of diffusion and reaction will yield different results according to the dimensionality d.

For dimensionalities d higher than d_c, ε' decreases when ε decreases. Diffusion is then very effective in reestablishing the spatial homogeneity that would be perturbed by localized fluctuations provoked by the chemical kinetics. Hence, we expect to recover in this limit the results of mean field theory discussed earlier.

For $d < d_c$, on the other hand, diffusion is not strong enough to correlate the cells perfectly. We expect therefore the appearance of spatial correlations showing non-classical behaviour or even the suppression of bifurcation altogether: the regime is dominated by chemical reaction.

For $d = d_c$, of course, diffusion and reaction may compete in the thermodynamic limit.

This critical dimensionality depends on the order of the critical point. Here we present a few values from (9.49) and (9.26):

k	1	3	5	7	$(+\infty)$
d_c	$(-\infty)$	4	3	8/3	(2)
a	1/2	3/4	5/6	7/8	(1)

Note that a degenerate bifurcation can be interpreted as the confluence of several successive bifurcation points in a system with high nonlinearities.

In conclusion, the stochastic analysis of nonequilibrium transition phenomena reveals some striking analogies with phase transitions. This should not mask, however, some basic differences between these two types of phenomena. An equilibrium phase transition is the result of a competition between the intermolecular forces, which tend to order the system, and the thermal noise which has the opposite effect. This competition is described by an appropriate thermodynamic potential which measures the effect of internal energy versus the entropy contributions. Equilibrium transitions are then induced by, for example, increasing the density (as it favours the effect of intermolecular interactions) or by decreasing the temperature (as it diminishes the thermal noise).

In a nonequilibrium transition, however, the essential ingredient is the applied constraint which maintains the system in nonequilibrium conditions. Thanks to the nonequilibrium, chemical kinetics becomes then capable of generating, locally, large non-Poissonian fluctuations. Subsequently, these are spread by diffusion. As discussed earlier in this section, if the rates of generation and of spreading are comparable, the system can sustain long range correlations and show critical behaviour.

H. Symmetry-Breaking Bifurcations and Conclusions

The problem of symmetry-breaking is qualitatively different because of the appearance of at least two coupled variables (see the linear analysis of Section IV.D), whereas the bifurcations just discussed can take place in the presence of only one state variable obeying a nonlinear dynamics. Still, the results accumulated so far tend to confirm the deep and unexpected relations between bifurcation phenomena and stochastic processes. Consider first the stochastic aspects of time symmetry-breaking bifurcation leading to limit cycles in the mean-field approximation [Turner, 1979]. The main result is that in the presence of a limit cycle the master equation admits, in the thermodynamic limit $\varepsilon \to 0$, time-dependent solutions $P(t)$ in the form of probability peaks rotating along the limit cycle. In addition to these solutions, the system admits a steady-state distribution centered on the limit cycle, as shown in Fig. 9.3, which can be viewed as a time average $\overline{P}$ of the time-dependent solutions.

The existence of both steady-state and time-dependent solutions reflects the symmetry-breaking associated with the formation of a limit cycle: Each individual realization breaks the gauge symmetry, as it has a definite phase. In a statistical ensemble, on the other hand (represented by $\overline{P}$), phases are averaged out and one obtains a stationary distribution over different portions of the limit cycle. The analogy with phase transitions in ferromagnetism and superfluidity is striking. The function $\overline{P}$ turns out to be the envelope of $P(t)$ as it moves in time, and has a number of remarkable properties. First, for any sufficiently smooth function $g(\{X_i\})$ one has an interesting form of an ergodic theorem: the time average $\overline{g}$ of g over a long interval is identical to the statistical average $< g >$ evaluated with the aid of the distribution $\overline{P}$. And secondly, the height of $\overline{P}$ integrated over the direction normal to the limit cycle is inversely proportional to the tangential speed along the macroscopic limit cycle. In other words, the faster a system moves along a portion of the limit cycle, the smaller the statistical weight of that portion will be. We see again how some remarkable relations between macroscopic behaviour and fluctuations appear, in the limit when the size of the system becomes very large.

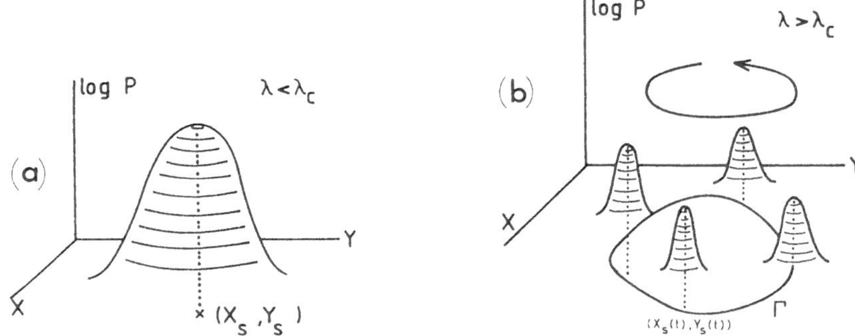

Fig. 9.3. Transition to a limit cycle in a two-variables stochastic model: (a) For $\lambda < \lambda_c$, there is only one stationary distribution of probability P, peaking on the stable steady state (x_s, y_s); (b) For $\lambda > \lambda_c$, there is a family of time-periodic distributions (differing by a phase), peaking on the stable limit cycle solution $(x_s(t), y_s(t))$.

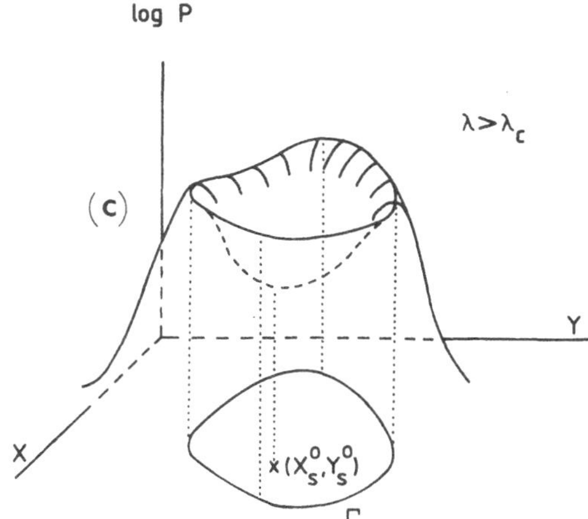

Fig. 9.3. (c) For $\lambda > \lambda_c$, a suitable average of the time-dependent distributions yields a stationary distribution forming a crater with its crest along the deterministic limit cycle trajectory Γ, and its bottom at the unstable steady state (x_s^o, y_s^o).

The effect of spatially inhomogeneous fluctuations on limit cycles has also been analyzed [Hentschel, 1978; Walgraef et al, 1981]. As it turns out, their properties are quite analogous to those of equilibrium systems such as the X - Y model of ferromagnetism.

Finally, the properties of fluctuations in the vicinity of bifurcations breaking the spatial symmetries have been analyzed [Walgraef et al, 1980]. It can be shown that fluctuations destroy all patterns predicted by the phenomenological rate equations in infinite one and two dimensional systems. On the other hand, in three dimensions stable patterns with an intrinsic wavelength subsist and are only slightly affected by the fluctuations. These patterns correspond therefore to a symmetry-breaking quite analogous to that arising in certain equilibrium phase transitions like, for instance, the freezing transition. Again, however, as pointed out in the previous subsection, despite these appealing analogies one should be fully aware of the basic differences between the two types of transition phenomena.

X. FINAL REMARKS - CONCLUSIONS AND PERSPECTIVES

In the beginning of these lectures, we insisted on the rich variety of nonequilibrium structures present in nature. Then, starting from "reasonable" descriptions, we observed that such organized states could arise from a structureless background through a number of transition phenomena induced by nonequilibrium constraints. At both the macroscopic phenomenological level and the more refined stochastic level we developed systematic perturbative techniques allowing us to calculate the new behaviours associated to the bifurcation, and to elucidate the mechanisms governing these phenomena.

As a consequence of these theoretical investigations, various predictions were made, referring to the stability of bifurcating branches, the merging of branches, but also the validity of the mean-field probabilistic description, the Langevin and Fokker-Planck modellization or the renormalization group theory. Some of them lead to experimental predictions. But the gap between experiment and theory remains still large; on many questions, new experimental results are urgently needed.

In all these new fields, several challenging problems remain open. In bifurcation theory, a fascinating possibility is to achieve a more and more complete knowledge and classification of the various transitions, including the transition to chaotic behaviour. Also the phenomenological mechanisms governing the exchange of stability deserve a deeper analysis.

In fluctuation theory, the link between stochastic master equations for nonequilibrium systems and statistical mechanics needs further elucidation. Furthermore, many of the properties of these stochastic equations, and in particular the behaviour of their solutions past the bifurcation, remain largely unexplored. Also, the extension of the whole perturbative approach to include inhomogeneous fluctuations in systems involving two or more variables is necessary to understand the mechanism of symmetry-breaking bifurcations. Finally, a most important question is the development of an approach capable of handling the initial value problem associated with the multivariate master equation or the Fokker-Planck equation.

REFERENCES

The following textbooks will provide general presentations and analyses of the various questions discussed in these notes:

A. General References

De Groot, S., and Mazur, P., 1962, "Nonequilibrium Thermodynamics," North-Holland, Amsterdam.

Glansdorff, P., and Prigogine, I., 1971, "Thermodynamic Theory of Structure, Stability and Fluctuations," Wiley-Interscience, London.

Nicolis, G., and Prigogine, I., 1977, "Self-Organization in Nonequilibrium Systems," Wiley-Interscience, New York.

For the mathematical analyses, we mainly refer to:

Feller, W., 1967 (first edition), 1971 (second edition), "An Introuction to Probability Theory and its Applications," Wiley, New York.

Minorski, N., 1962, "Nonlinear Oscillations," Van Nostrand, Princeton.

Sattinger, D., 1973, "Topics in Stability and Bifurcation Theory," Lecture Notes in Mathematics 309, Springer-Verlag, Berlin.

Sattinger, D., 1979, "Group Theoretic Methods in Bifurcation Theory," Lecture Notes in Mathematics 762, Springer-Verlag, Berlin.

B. Specialized References

Abramowitz, M., and Stegun, I., 1964, "Handbook of Mathematical Functions," Dover Publications, New York.

Ahlers, G., and Walden, R.W., 1980, Turbulence near the onset of convection, Phys. Rev. Lett., 44:445.

Andronov, A., Vitt, A., and Khaikin, C., 1966, "Theory of Oscillators," Pergamon Press, Oxford.

Arnold, L., 1973, "Stochastic Differential Equations," Wiley, New York.

Arnold, L., 1979, in: "Dynamics of Synergetic Systems," H. Haken, ed., Springer-Verlag, Berlin.

Bedeaux, D., Mazur, P., and Pasmanter, R., 1977, The ballast resistor; an electrothermal instability in a conducting wire I; the nature of the stationary states, Physica, 86 A:355.

Bell, G., and Primbley, P., 1978, "Theoretical Immunology," Dekker, New York.

Bergé, P., Dubois, M., Manneville, P., and Pomeau, Y., 1980, Intermittency in Rayleigh-Bénard convection, Jour. de Physique - Lettres, 41:341.

Berggren, K., and Huberman, B., 1978, Peierls state far from equilibrium, Phys. Rev. B, 18:3369.

Boiteux, A., Goldbeter, A., and Hess, B., 1975, Control of oscillating glycolysis of yeast by stochastic, periodic and steady source of substrate: A model and experimental study, Proc. of the Nat. Acad. of Sciences of the USA, 72:3829.

Borckmans, P., Dewel, G., and Walgraef, D., 1980, Concepts in the theory of nonequilibrium phase transitions, Jour. de Physique - Colloque C4, 41:101.

Briggs, T., and Rauscher, W., 1973, An oscillating iodine clock, Jour. of Chem. Educ., 50:496.

Caroli, B., Caroli, C., and Roulet, B., 1979, Diffusion in a bistable potential: a systematic WKB treatment, Jour. of Stat. Phys., 21:415.

Chandrasekhar, S., 1961, "Hydrodynamic and Hydromagnetic Stability," Oxford University Press.

Chorin, A., Marsden, J., and Smale, S., 1977, "Turbulence Seminar (Berkeley 1976/77)," Lecture Notes in Mathematics 615, Springer-Verlag, Berlin.

Clavin, P., and Guyon, E., 1978, La flamme, La Recherche, 94:954.

Cole, J., 1968, "Perturbation Methods in Applied Mathematics," Blaisdell Publishing Company, Waltham, Mass.

Creel, C.L., and Ross, J., 1976, Multiple stationary states and hysteresis in a chemical reaction, Jour. of Chem. Phys., 65:3779

Decker, D., and Keller, H., 1980, Multiple limit point bifurcation, Jour. of Math. Anal. and its Appl., 75:417.

Degn, H., 1972, Oscillating chemical reactions in homogeneous phase, Jour. of Chem. Educ., 49:302.

de Pasquale, F., and Tombesi, P., 1979, The decay of an unstable equilibrium state near a "critical point", Phys. Lett., 72 A:7.

Dewel, G., Walgraef, D., and Borckmans, P., 1977, Renormalization group approach to chemical instabilities, Zeits. für Phys., B 28:235.

Dubois, M., and Bergé, P., 1980, Experimental evidence for the oscillators in a convective biperiodic regime, Phys. Lett., 76 A:53.

Eigen, M., and Schuster, P., 1979, "The Hypercycle, a Principle of Natural Self-Organization," Springer-Verlag, Berlin.

Ermentrout, G., and Cowan, J., 1978, Large scale spatially organized activity in neuronal nets, SIAM Jour. on Appl. Math., 38:1.

Erneux, T., and Herschkowitz-Kaufman, M., 1977, Rotating waves as asymptotic solutions of a model chemical reaction, Jour. of Chem. Phys. 66:248.

Erneux, T., and Hiernaux, J., 1980, Transition from polar to duplicate patterns, Jour. of Math. Biol., 9:193.

Field, R., 1975, Limit cycle oscillations in the reversible oregonator, Jour. of Chem. Phys., 63:2289.

Fife, P., 1978, Asymptotic states for equations of reacting and diffusion, Bull. of the Amer. Math. Soc., 84:693.

Fraedrich, K., 1978, Structural and stochastic analysis of a o-dimensional climate system, Quart. Jour. of the Royal Meteor. Soc., 104:461.

Gardiner, C., 1976, A comment on chemical Langevin equations, Jour. of Stat. Phys., 15:451.

Gerisch, A., and Hess, B., 1974, Cyclic-AMP-controlled oscillations in suspended Dictyostelium cells: their relation to morphogenetic cell interactions, Proc. of the Nat. Acad. of Sci. of the USA, 71:2118.

Gierer, A., and Meinhardt, H., 1972, A theory of biological pattern foundation, Kybernetik, 12:30.

Goldbeter, A., 1980, Models for oscillations and excitability in biochemical systems, in: "Mathematical Models in Molecular and Cellular Biology," L.A. Segel, ed., Cambridge Univ. Press.

Golubitsky, M., and Schaeffer, D., 1979, A theory for imperfect bifurcation via singularity theory, Commun. on Pure and Appl. Math., 32:21.

Grossman, S., 1976, Langevin forces in chemically reacting multicomponent fluids, Jour. of Chem. Phys., 65:2007.

Haberman, R., 1977, On the behavior of spatially dependent nonlinear wave envelopes, SIAM Jour. on Appl. Math., 32:154.

Haberman, R., 1979, Slowly varying jump and transition phenomena ssociated with algebraic bifurcation problems, SIAM Jour. on Appl. Math., 57:69.

Haken, H., 1978, "Synergetics, an Introduction," Springer-Verlag, Berlin.

Haken, H., 1980, editor, "Dynamics of Synergetic Systems," Proceedings of the International Symposium of Synergetics (Bielefeld 1979), Springer-Verlag, Berlin.

Hentschel, H., 1978, Application of the dynamic renormalization group to non-equilibrium instabilities: The k = o hard mode transition, Zeits. für Phys., B 31:401.

Horsthemke, 1980, in: "Proceedings of the XVIIth Solvay Conference on Physics (Washington 1980)," G. Nicolis, G. Dewel and J.W. Turner, eds., Wiley, N.Y.

Jacobs, J.A., 1975, "The Earth's Core," Academic Press, London.

Joseph, D., 1973, Remarks about bifurcation and stability of quasiperiodic solutions which bifurcate from periodic solutions of the Navier-Stokes equation, in: "Nonlinear Problems in the Physical Sciences and Biology," I. Stakgold, D. Joseph, and D. Sattinger, eds., Springer-Verlag, Berlin.

Joseph, D., and Sattinger, D., 1972, Bifurcating time-periodic solutions and their stability, Archive for Rational Mechanics and Analysis, 45:79.

Karlin, S., and Mc Gregor, J., 1957, The differential equations of birth-and-death processes, and the Stieltjes moment problem, Trans. of the Amer. Math. Soc., 85:489.

Kauffman, S., and Wille, J., 1975, The mitotic oscillator in Physarum polycephalum, Jour. of Theor. Biol., 55:47.

Keener, J., 1976, Secondary bifurcation in nonlinear diffusion reaction equations, Stud. in Appl. Math., 55:187.

Kopell, N., and Howard, L., 1975, Bifurcations and trajectories joining critical points, Adv. in Math., 18:306.

Kubo, R., 1966, The fluctuation-dissipation theorem, Rep. on Prog. in Phys., 29:255.

Kuramoto, Y., and Tsuzuki, T., 1975, On the formation of dissipative structures in reaction-diffusion systems: Reductive perturbation approach, Prog. of Theor. Phys., 54:687.

Kurtz, T.G., 1976, Limit theorems and diffusion approximations for density dependent Markov chains, Math. Prog. Study, 5:67.

Kurtz, T.G., 1978, Strong approximation theorems for density dependent Markov chains, Stoch. Proc. and Their Appl., 6:223.

Landau, L., and Lifschitz, E., 1959, "Fluid Mechanics," Addison-Wesley, Reading, Mass.

Landauer, R., 1962, Fluctuations in bistable tunnel diode circuits, Jour. of Appl. Phys., 33:2209.

Lanford III, O., 1973, Bifurcation of periodic solutions into invariant tori: the work of Ruelle and Takens, in: "Nonlinear Problems in the Physical Sciences and Biology," I. Stakgold, D. Joseph and D. Sattinger, eds., Springer-Verlag, Berlin.

Langer, J.S., 1980, Eutectic solidification and marginal stability, Phys. Rev. Lett., 44:1023.

Langer, J.S., 1980, Instabilities and pattern formation in crystal growth, Rev. of Mod. Phys., 52:1.

Lefever, R., Herschkowitz-Kaufman, M. and Turner, J.W., 1977, Dissipative structures in a soluble non-linear reaction-diffusion system, Phys. Lett., 60 A:389.

Lefever, R., and Garay, R., 1978, A mathematical model of the immune surveillance against cancer, in: "Theoretical Immunology," G. Bell and P. Primblely, eds., Dekker, N.Y.

Lorenz, E., 1963, Deterministic nonperiodic flow, Jour. of Atmos. Sci., 20:130.

Lovett, R., Ortoleva, P., and Ross, J., 1978, Kinetic instabilities in first order phase transitions, Jour. of Chem. Phys., 69:947.

Ma, S.K., 1976, "Modern Theory of Critical Phenomena," Benjamin, Reading, Mass.

Malek Mansour, M., and Nicolis, G., 1975, A master equation description of local fluctuations, Jour. of Stat. Phys., 13:197.

Malek Mansour, M., 1978, Sur la description stochastique des systèmes de non-équilibre, Jour. de Phys. - Colloque C5, 39:79.

Malek Mansour, M. and Van Den Broeck, C., 1980a, Inhomogeneous fluctuations in reaction-diffusion systems, Preprint.

Malek Mansour, M., Van Den Broeck, C., Nicolis, G., and Turner, J.W., 1980, Asymptotic properties of markovian master equations, Ann. of Phys., in press.

Mandel, P., 1979, Lasers with saturable absorbers driven by a coherent field, Zeits. für Phys., B 33:205.

Manneville, P., and Pomeau, Y., 1980, Different ways to turbulence in dissipative dynamical systems, Physica, 1 D:219.

Matijevic, E., 1978, "Surface and Colloid Science," Vol. 10, Plenum Publishing Corporation.

Matsuo, K., 1977, Relaxation mode analysis of nonlinear birth and death processes, Jour. of Stat. Phys., 16:169.

May, R., 1976, Simple mathematical models with very complicated dynamics, Nature, 261:459.

Miller, C.A., 1978, Stability of interfaces, in: "Surface and Colloid Science," Vol. 10, E. Matijevic, ed., Plenum, N.Y.

Misra, B., Courbage, M., and Prigogine, I., 1979, From deterministic dynamics to probabilistic descriptions, Proc. of the Nat. Acad. of Sci. (USA), 76:3607.

Moreau, M., 1980, Note on the reaction rate in non-ideal mixtures, Physica, 102 A:389.

Mori, H., 1965, Transport, collective motion and brownian motion, Prog. of Theor. Phys., 33:423.

Mori, H., 1975, Stochastic processes of macroscopic variables, Prog. of Theor. Phys., 53:1617.

Nakamura, K., 1977, Nonlinear fluctuations associated with instabilities in dissipative systems, Prog. of Theor. Phys., 57:1874.

Nakaya, J., 1954, "Snow Crystals," Harvard University Press, Cambridge.

Nicolis, G., and Turner, J.W., 1977, Stochastic analysis of a nonequilibrium phase transition: some exact results, Physica, 89 A:326.

Nicolis, G., and Turner, J.W., 1979, Effects of fluctuations on bifurcation phenomena, Ann. of the N.Y. Acad. of Sci., 316:251.

Nicolis, G., and Malek Mansour, M., 1980, Systematic analysis of the multivariate master equation for a reaction-diffusion system, Jour. of Stat. Phys., 22:495.

Nicolis, G., Dewel, G., and Turner, J.W., 1980, eds., "Proceedings of the XVIIth Solvay Conference on Physics (Washington 1980)," Wiley, New York.

Normand, C., Pomeau, Y., and Velarde, M., 1977, Convective instabilities: a physicist's approach, Rev. of Mod. Phys., 49:581.

Prigogine, I., and Lefever, R., 1968, Symmetry breaking instabilities in dissipative systems. II, Jour. of Chem. Phys., 48:1695.

Resibois, P., and de Leener, M., 1977, "Classical Kinetic Theory of Fluids," Wiley-Interscience, New York.

Rice, S., Freed, K., and Light, J., 1972, eds., "Statistical Mechanics: New Concepts, New Problems, New Applications," Proceedings of the Sixth IUPAP Conference on Statistical Mechanics (Chicago 1971), University of Chicago Press.

Richter, F., 1973, Sea floor spreading, Rev. of Geophys. and Space Phys., pp. 233-287.

Risken, H., and Nummedal, K., 1968, Self-pulsing in lasers, Jour. of Appl. Phys., 39:4662.

Roux, J.C., Rossi, A., Bachelart, S., and Vidal, C., 1980, Representation of a strange attractor from an experimental study of chemical turbulence, Phys. Lett., 77 A:391.

Ruelle, D., and Takens, F., 1971, On the nature of turbulence, Commun. in Math. Phys., 20:167; 23:343.
Ruschin, S. and Bauer, S., 1979, Bistability, hysteresis and critical behavior of a CO_2 laser, with SF_6 intracavity as a saturable absorber, Chem. Phys. Lett., 66:100.
Scalapino, D., and Huberman, B., 1977, Onset of an inhomogeneous state in a nonequilibrium superconducting film, Phys. Rev. Lett. 39:1365.
Schloegl, F., 1971, On stability of steady states, Zeits. für Phys., 243:303.
Schloegl, F., 1972, Chemical reaction models for nonequilibrium phase transitions, Zeits. für Phys., 253:147.
Schnakenberg, J., 1976, Network theory of microscopic and macroscopic behavior of master equation systems, Rev. of Mod. Phys., 48:571.
Schuss, Z., and Matkowsky, B., 1979, The exit problem: A new approach to diffusion across potential barriers, SIAM Jour. on Appl. Math., 36:604.
Segel, LA., 1980, ed., "Mathematical Models in Molecular and Cellular Biology," Cambridge University Press.
Stakgold, I., Joseph, D., and Sattinger, D., 1973, eds., "Nonlinear Problems in the Physical Sciences and Biology," Proceedings of a Battelle Summer Institute, Seattle, July 3-28, 1972, Lecture Notes in Mathematics 322, Springer-Verlag, Berlin.
Suzuki, M., 1976, Scaling theory of nonequilibrium systems near the instability point. I. General aspects of transient phenomena, Prog. of Theor. Phys., 56:77.
Suzuki, M., 1976, Scaling theory of nonequilibrium systems near the instability point. II. Anomalous fluctuation theorems in the extensive region, Prog. of Theor. Phys., 56:477.
Suzuki, M., 1980, in: "Proc. of the XVIIth Conf. on Phys.," G. Nicolis, G. Dewel and J.W. Turner, eds., Wiley, New York.
Thom, R., 1972, "Stabilité Structurelle et Morphogénèse," Benjamin, New York.
Turner, J.S., 1980, Private communication.
Turner, J.W., 1979, Stationary and time-dependent solutions of master equations in several variables, in: "Dynamics of Synergetic Systems," H. Haken, ed., Springer-Verlag, Berlin.
Tyson, J., and Kauffman, S., 1975, Control of mitosis by a continuous biochemical oscillation: Synchronization; Spatially inhomogeneous oscillations, Jour. of Math. Biol., 1:289.
Tyson, J., 1976, "The Belousov-Zhabotinski Reaction," Lecture Notes in Biomathematics 10, Springer-Verlag, Berlin.
Van Den Broeck, C., Horsthemke, W., and Malek Mansour, M., 1977, On the diffusion operator of the multivariate master equation, Physica, 89 A:339.
Van Den Broeck, C., Houard, J., and Malek Mansour, M., 1980, Chapman-Enskog development of the multivariate master equation, Physica, 101 A:167.

Van Der Pol, B., 1930, Oscillations sinusoidales et de relaxation, L'Onde Electrique, 9:245,293.

Van Kampen, N.G., 1977, A soluble model for diffusion in a bistable potential, Jour. of Stat. Phys., 17:71.

Velarde, M., and Normand, C., 1980, Convection, Scientific American, July.

Vidal, C., and Roux, J., 1980, La turbulence chimique existe-t-elle?, La Recherche, 107:66.

Walgraef, D., Dewel, G., and Borckmans, P., 1980, Fluctuations near nonequilibrium phase transitions to nonuniform states, Phys. Rev., A 21:397.

Walgraef, D., Dewel, G., and Borckmans, P., 1981 (in press), Nonequilibrium phase transitions and chemical instabilities, Advanc. in Chem. Phys.

Zhabotinski, A., 1974, "Self-Oscillating Concentrations," Nauka, Moscow.

Zwanzig, R., 1972, Collective modes in classical liquids, in: "Statistical Mechanics: New Concepts, New Problems, New Applications," S. Rice, K. Freed and J. Light, eds., U. of Chicago Press.

Zwanzig, R., 1972, Nonlinear dynamics of collective modes, in: "Statistical Mechanics: New Concepts, New Problems, New Applications," S. Rice, K. Freed and J. Light, eds., U. of Chicago Press.

CHEMICAL OSCILLATIONS

Louis N. Howard

Department of Mathematics
Massachusetts Institute of Technology
Cambridge, Mass. 02139
U.S.A.

TABLE OF CONTENTS

CHEMICAL OSCILLATIONS

Louis N. Howard

Department of Mathematics
Massachusetts Institute of Technology
Cambridge, Mass. 02139
U.S.A.

I. CHEMICAL OSCILLATORS

A. Some Model Systems

1. The general point of view of my lectures will be that of the applied mathematician interested in problem-solving. Some general techniques, such as bifurcation methods and various aspects of singular perturbations, will be discussed, but mostly in the context of specific examples related to chemical oscillations and waves. The more general background can be found in the paper by Professor Nicolis [1973].

Consider the following imaginary system of chemical reactions:

$$* + X \to 2X + * \,, \tag{1.1a}$$

$$X + Y \to 2Y + * \,, \tag{1.1b}$$

$$Y \to * \,. \tag{1.1c}$$

The *'s represent reactants present in such large concentrations that they may be regarded as constant, or inert products. The first reaction (1.1a) is "autocatalytic" in X whereas the second reaction (1.1b) is autocatalytic in Y but is "activated" by X. The corresponding chemical kinetic equations obtained by using the macroscopic mass action model are given by

$$\frac{dx}{dt} = K_1 x - K_2 xy \ , \tag{1.2a}$$

$$\frac{dy}{dt} = K_2 xy - K_3 y \ . \tag{1.2b}$$

These equations are a thinly disguised version of the Lotka-Volterra model for two populations in interaction where X = rabbits and Y = foxes.

In dimensionless form, the equations become

$$\frac{d\xi}{d\tau} = \xi(1-\eta) \ , \tag{1.3a}$$

$$\frac{d\eta}{d\tau} = K\eta(\xi - 1) \ , \tag{1.3b}$$

with a nontrivial critical point at $\xi = 1$, $\eta = 1$. This system has a first integral obtained by multiplying (1.3a) by $(1 - 1/\xi)$ and (1.3b) by $(1 - 1/\eta)$ giving

$$(1 - \frac{1}{\xi}) \frac{d\xi}{d\tau} = (\xi - 1)(1 - \eta) \ , \tag{1.4a}$$

$$(1 - \frac{1}{\eta}) \frac{d\eta}{d\tau} = K(\eta - 1)(\xi - 1) \ . \tag{1.4b}$$

Thus,

$$\frac{d}{d\tau} \left\{ K[\xi - 1 - \ell n\xi] + \eta - 1 - \ell n\eta \right\} = 0 \ , \tag{1.5}$$

or

$$Kf(\xi) + f(\eta) = C \tag{1.6}$$

where $f(\xi) = \xi - 1 - \ell n\xi$. Since $f'(\xi) = 1 - 1/\xi$, $f''(\xi) = 1/\xi^2$, then $f(1) = f'(1) = 0$, $f''(1) = 1$, and $f(\xi) \simeq \frac{1}{2}(\xi - 1)^2$ near $\xi = 1$ (see Fig. 1).

The critical point (c.p) is a center (see Fig. 2) and small nearby trajectories are nearly ellipses. For the full system, the trajectories have the following properties:

(a) There is a family of periodic solutions, no individual one of which is an attractor.

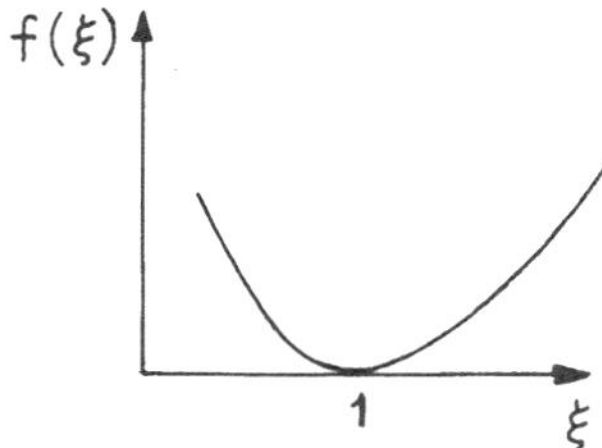

Fig. 1. Sketch of $f(\xi) = \xi - 1 - \ell n\xi$ vs ξ near $\xi = 1$.

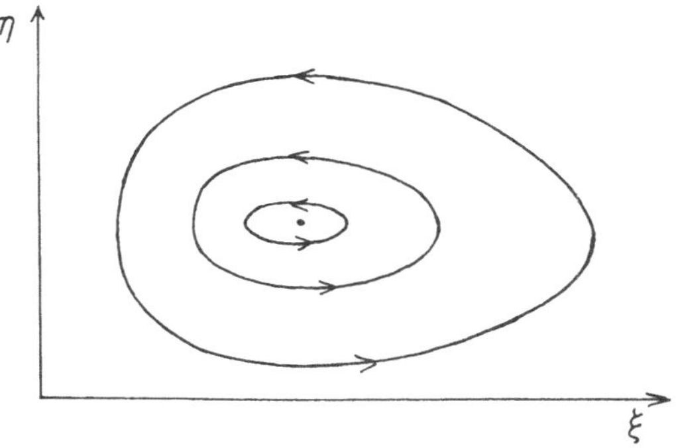

Fig. 2. Phase plane diagram in the neighborhood of the critical point of the system of equations (1.3).

(b) Any perturbation away from the stationary solution leads to sustained oscillations. (Ill-advised intervention leading to the killing of many rabbits will lead to a plague of them later.)

(c) The system is <u>structurally unstable</u>. Small perturbations in the vector field can (and usually will) qualitatively change the phase-portrait.

2. Although not satisfactory as a model for a system expected to have stable hehavior under various naturally occurring perturbations, such idealizations as (1.3) are often very useful for problem solving when other effects - which actually change the system to a structurally stable one - are small. The idealizations do not tell the whole story, but make possible simple and effective techniques for accurately accounting for these other effects.

As an example, consider the following generalization of Eqs. (1.3):

$$\dot{\xi} = \xi(1-\eta) + \varepsilon F(\xi,\eta) \ , \tag{1.7a}$$

$$\dot{\eta} = K\eta(\xi-1) + \varepsilon G(\xi,\eta) \ . \tag{1.7b}$$

Then, we obtain

$$\frac{d}{d\tau}\left\{Kf(\xi) + f(\eta)\right\} \equiv \frac{dI}{d\tau} = \varepsilon\left\{K(1-\frac{1}{\xi})F + (1-\frac{1}{\eta})G\right\} \ . \tag{1.8}$$

Evaluating F and G along a particular loop of the $\varepsilon = 0$ system, which is close to the solution of the full system <u>for a finite time</u>, we can approximately calculate the change in I which is $O(\varepsilon)$ over one turn about this loop. This gives an approximate description of the slow drift, on a time-scale $O(1/\varepsilon)$, through the family of periodic solutions (or part of it) of the idealized model. An approximate description of this is obtained by calculating the integral around the loop for I = C, giving the "long time scale differential equation"

$$\frac{dC}{d(\varepsilon t)} = \mathcal{G}(C) \ . \tag{1.9}$$

(In the present context, this is the "method of averaging".) Different special cases can be discussed:

(a) If $\mathcal{G}(C) \leq -kC$ for some negative constant $-k$, then the perturbed system has a stable critical point toward which all solutions are attracted.

(b) If $\mathcal{G}(C) < 0$ for $C > C_* > 0$ and $\mathcal{G}(C) > 0$ for $0 < C < C_*$, then the perturbed system has an attracting limit cycle close to the curve $I = C_*$, for sufficiently small ε.

(c) If $\mathcal{G}(C) \equiv 0$, then no conclusion can be drawn in general.

This general idea can be made completely rigorous by calculating the exact orbit as a regular perturbation in ε for one loop around. This can be done uniformly for a <u>finite</u> time. The long time scale differential equation is then replaced by a difference equation, but conclusions as in cases (a) and (b) can frequently be drawn.

3. Consider next the model of chemical systems studied by Prigogine and Lefever [1968],

$$* \to X \tag{1.10a}$$

$$* + X \to Y + * \tag{1.10b}$$

$$2X + Y \to 3X + * \tag{1.10c}$$

$$X \to * \tag{1.10d}$$

the steps (a) - (d) being shown schematically in Fig. 3. (The feedback loop suggests the possibility of oscillations.)

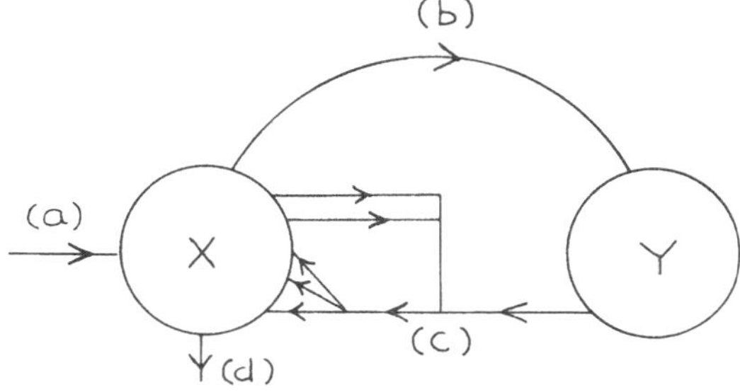

Fig. 3. Schematic representation of chemical reactions (1.10) (a) - (d).

In dimensionless form, the kinetic equations are:

$$\frac{dx}{dt} = 1 - x - Ax(1 - xy) , \tag{1.11a}$$

$$\frac{dy}{dt} = Bx(1 - xy) , \tag{1.11b}$$

where the time scale is chosen as the decay-time of X via (d); the scale of X is chosen so that the rate of production by (a) is unity and the scale of Y is chosen so that when x = 1 (i.e., when (a) and (d) just cancel on production of X) then y = 1 when (b) and (c) just cancel on production of Y.

The (unique) critical point of this system is at x = 1 = y. We examine its neighborhood by a local linearization, viz. setting $x = 1 + \xi$, $y = 1 + \eta$ and linearizing,

$$\begin{bmatrix} \dot{\xi} \\ \dot{\eta} \end{bmatrix} = \begin{bmatrix} -1 + A & A \\ -B & -B \end{bmatrix} \begin{bmatrix} \xi \\ \eta \end{bmatrix} \tag{1.12}$$

with the coefficient matrix having Trace = -1 + A - B and determinant = B.

Since B > 0, the stability of the c.p. cannot change by an eigenvalue passing through zero; but it can change by a conjugate

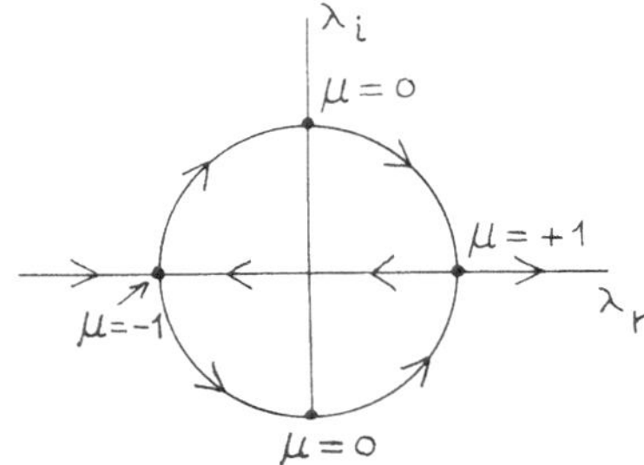

Fig. 4. Eigenvalue locus.

pair crossing the imaginary axis. This happens when the Trace = 0, for then the eigenvalues are pure imaginary (= $\pm i\sqrt{B}$).

For definiteness, consider the case B = 1, and set A = 2 + 2μ, so that the Trace = 2μ. The eigenvalues are (see Fig. 4)

$$\lambda_{\pm} = \mu \pm i\sqrt{1-\mu^2} \quad . \tag{1.13}$$

If $\mu < 0$, the c.p. is stable ($\mu \leq -1$, however, is irrelevant since $A > 0$) and it becomes unstable as μ crosses 0, remaining unstable for $\mu > 0$. (For $\mu > 1$ it becomes an unstable node rather than an unstable spiral point.)

On general grounds (Hopf Bifurcation Theorem) we may conclude that in the neighborhood of x = y = 1 and μ = 0 in (x,y,μ) space there is a family of periodic orbits of the system of differential equations, a family branching off from the point (1,1,0). If this family exists for $\mu > 0$, i.e., on the side on which the c.p. is unstable, these periodic orbits will be stable limit cycles, at least for small enough μ. (Further calculation shows this to be in fact the case.)

This bifurcation method is one way of showing the existence of limit cycles and approximately calculating them. It has its limitations, most notably in being restricted to the neighborhood of the bifurcation point. Its advantage is simplicity and fairly general applicability, and in more complicated problems it may be the only more or less analytical method available. Such analytical methods, despite their limitations, can be valuable adjuncts to numerical work, particularly when a number of other parameters are involved. Particularly useful is the clarification that bifurcation methods provide of the process by which the limit cycle "comes into existence" as the parameter is varied.

B. The Hopf Bifurcation Theorem

A more general statement of the Hopf Bifurcation Theorem is as follows:

(a) Consider an autonomous system of n differential equations (x an n-column vector)

$$\dot{x} = F_{\mu}(x) \tag{1.14}$$

depending on a real parameter μ and having a critical point for all μ (of interest) which we take to be at x = 0: $F_{\mu}(0) = 0$. (F depends analytically, or differentiably up to a suitable order, on μ and the components of x.)

(b) For small x and μ,

$$F_\mu(x) = L_o x + \mu L_1 x + Q(x,x) + C(x,x,x) + \text{higher order terms.} \quad (1.15)$$

(c) L_o (i.e., linearization at $\mu = 0$) has a conjugate pair $\pm i\omega_o$ of simple pure imaginary eigenvalues and no other eigenvalues on the imaginary axis. Let $L_o r = i\omega_o r$ and $\ell L_o = i\omega_o \ell$, where r and ℓ are right and left eigenvectors for the eigenvalue $i\omega_o$. Let $\overline{r}$ and $\overline{\ell}$ be the eigenvector for $-i\omega_o$, $\ell \cdot \overline{r} = 0$ and we may assume $\ell \cdot r = 1$.

(d) If $\lambda(\mu)$ is the continuous extension to $\mu \neq 0$ of the eigenvalue $i\omega_o$, then

$$\mathrm{Re}\left(\frac{d\lambda}{d\mu}\right)_{\mu=0} \neq 0 .$$

(This corresponds to a transversal crossing of the imaginary axis and a change of stability of the critical point.) This is equivalent to

$$\mathrm{Re}(\ell L_1 r) \neq 0, \text{ indeed } \left(\frac{d\lambda}{d\mu}\right) = \ell L_1 r .$$

Conclusion: In the neighborhood of the origin in (x,μ) space, there always exists a 1-parameter family $x(t,\varepsilon)$, $\mu(\varepsilon)$ of periodic solutions of $\dot{x} = F_\mu(x)$ (with $x(t,\varepsilon) \to 0$, $\mu \to 0$ as $\varepsilon \to 0$), having periods $T(\varepsilon)$ (with $T \to 2\pi/\omega_o$ as $\varepsilon \to 0$). When $F_\mu(x)$ is analytic, ε can be chosen so that

$$x = \varepsilon x_1(t) + \dots , \quad (1.16)$$

$$\mu = \mu_2\varepsilon^2 + \dots , \quad (1.17)$$

$$T = 2\pi/\omega_o(1 + T_2\varepsilon^2 + \dots) . \quad (1.18)$$

If $\mu_2 \neq 0$ (the generic case), than for small ε the periodic solutions exist either supercritically, when $\mu_2 \mathrm{Re}\lambda'(0) > 0$, or else subcritically when $\mu_2 \mathrm{Re}\lambda'(0) < 0$, and then the periodic solutions have a Floquet exponent $\beta(\varepsilon) = \beta_2\varepsilon^2 + \dots$ with $\beta_2 = -2\mu_2 \mathrm{Re}\lambda'(0)$. (There is an exchange of stabilities if all other eigenvalues of L_o are in the left half-plane.)

To calculate the periodic solutions for small ε, let $t = Ts/2\pi$, so that the period in s is 2π, and the equation becomes:

$$\frac{dx}{ds} = \frac{T}{2\pi} F_\mu(x) . \quad (1.19)$$

Let

$$x = \varepsilon y = \varepsilon y_o + \varepsilon^2 y_1 + \dots , \tag{1.20}$$

$$T = \frac{2\pi}{\omega_o} (1 + \varepsilon^2 T_2 + \dots) , \tag{1.21}$$

$$\mu = \varepsilon^2 \mu_2 + \dots , \tag{1.22}$$

and define ε by the normalization $\ell \cdot y(0) = 1/2$. Then Eq. (1.19) yields

$$\omega_o \frac{dy_o}{ds} = L_o y_o , \tag{1.23}$$

$$\omega_o \frac{dy_1}{ds} = L_o y_1 + Q(y_o, y_o) , \tag{1.24}$$

$$\omega_o \frac{dy_2}{ds} = L_o y_2 + T_2 L_o y_o + \mu_2 L_1 y_o + 2Q(y_o, y_o) + C(y_o, y_o, y_o) . \tag{1.25}$$

Eq. (1.19) has a unique real 2π-periodic solution satisfying the normalization condition $\ell \cdot y_o(0) = 1/2$, namely

$$y_o = \mathrm{Re}(re^{is}) . \tag{1.26}$$

With this

$$Q(y_o, y_o) = \frac{1}{2} Q(r, \bar{r}) + \frac{1}{2} \mathrm{Re}(e^{2is} Q(r, r)) \tag{1.27}$$

so the periodic solutions of Eq. (1.24) have the form

$$y_1 = a + \mathrm{Re}(ce^{2is}) + \mathrm{Re}(C_1 e^{is} r) \tag{1.28}$$

where the vectors a and c satisfy

$$-L_o a = \frac{1}{2} Q(r, \bar{r}) , \tag{1.29}$$

$$(2i\omega_o - L_o) c = \frac{1}{2} Q(r, r) . \tag{1.30}$$

These equations can be solved uniquely for a and c since $\pm i\omega_o$ are the only eigenvalues of L_o on the imaginary axis. The complex number C_1 is then found uniquely from the normalization condition $\ell \cdot y_1(0) = 0$, and in fact one finds

$$C_1 = (i\omega_o)^{-1}\ell[Q(r,\bar{r}) - \tfrac{1}{2}Q(r,r) + \tfrac{1}{6}Q(\bar{r},\bar{r})] \,. \qquad (1.31)$$

Equation (1.25) then becomes:

$$\omega_o \frac{dy_2}{ds} = L_o y_2 + \mathrm{Re}\Big\{ C_1 Q(r,\bar{r}) + e^{is}\Big[(i\omega_o T_2 + \mu_2 L_1)r + 2Q(r,a) + Q(\bar{r},c) + \tfrac{3}{4}C(r,r,\bar{r})\Big] + e^{2is} C_1 Q(r,r) + e^{3is}[Q(r,c) + \tfrac{1}{4}C(r,r,r)]\Big\} \,. \qquad (1.32)$$

This equation has no periodic solutions at all unless the forcing terms in e^{is} and e^{-is} are orthogonal to ℓ and $\bar{\ell}$, respectively. Thus μ_2 and T_2 are to be determined from the relation,

$$\mu_2(\ell L_1 r) + i\omega_o T_2 = -2\ell Q(r,a) - \ell Q(\bar{r},c) - \tfrac{3}{4}\ell C(r,r,\bar{r}) \,. \qquad (1.33)$$

The transversality condition $\mathrm{Re}(\ell L_1 r) \neq 0$ shows that this uniquely determines μ_2 and T_2.

C. Application to the Prigogine-Lefever Model

To apply this to the Prigogine-Lefever model (with $B = 1$ and $A = 2 + 2\mu$) we rewrite it in terms of $\xi = x - 1$ and $\eta = y - 1$ as,

$$\frac{d}{dt}\begin{bmatrix}\xi\\ \eta\end{bmatrix} = \begin{bmatrix}1 & 2\\ -1 & -1\end{bmatrix}\begin{bmatrix}\xi\\ \eta\end{bmatrix} + \mu\begin{bmatrix}2 & 2\\ 0 & 0\end{bmatrix}\begin{bmatrix}\xi\\ \eta\end{bmatrix} + (2\xi\eta + \xi^2)\begin{bmatrix}2\\ -1\end{bmatrix} + \xi^2\eta\begin{bmatrix}2\\ -1\end{bmatrix} + \dots \,. \qquad (1.34)$$

Therefore,

$$L_o = \begin{bmatrix}1 & 2\\ -1 & -1\end{bmatrix}$$

with $\omega_o = 1$ and

$$r = \frac{1}{2}\begin{bmatrix} 2 \\ -1+i \end{bmatrix}, \quad \ell = \frac{1}{2}[1-i,-2i] \quad . \tag{1.35}$$

Thus,

$$\ell L_1 r = \frac{1}{2}[1-i,-2i]\begin{bmatrix} 1+i \\ 0 \end{bmatrix} = 1 \neq 0 \quad , \tag{1.36}$$

verifying the transversality condition, and consistent with

$$\ell L_1 r = \left(\frac{d\lambda}{d\mu}\right)_o \quad , \text{ since } \lambda = \mu + i\sqrt{1-\mu^2} \quad .$$

Next, we have

$$\frac{1}{2}\, Q(r,\bar{r}) = \frac{1}{8}\Big(2(-1-i) + 2(-1+i) + 2\cdot 2\Big)\begin{bmatrix} 2 \\ -1 \end{bmatrix} = \begin{bmatrix} 0 \\ 0 \end{bmatrix} \tag{1.37}$$

and

$$\frac{1}{2}\, Q(r,r) = \frac{1}{8}\Big(2\cdot 2(-1+i) + 2\cdot 2\Big)\begin{bmatrix} 2 \\ -1 \end{bmatrix} = \frac{i}{2}\begin{bmatrix} 2 \\ -1 \end{bmatrix} \quad . \tag{1.38}$$

Thus, $a = 0$, and

$$\begin{bmatrix} 2i-1 & -2 \\ 1 & 2i+1 \end{bmatrix}\begin{bmatrix} c_1 \\ c_2 \end{bmatrix} = \frac{i}{2}\begin{bmatrix} 2 \\ -1 \end{bmatrix}$$

so

$$c = \frac{1}{6}\begin{bmatrix} 4 \\ -2+i \end{bmatrix} \quad . \tag{1.39}$$

Since

$$Q(\bar{r},\bar{r}) = -i\begin{bmatrix} 2 \\ -1 \end{bmatrix} \quad ,$$

we get for C_1 the value $-2/3$. Finally, to determine μ_2 and T_2, we compute:

$$Q(r,a) = 0 ,$$

$$Q(\overline{r},c) = \frac{1}{12}\,[2(-2+i) + (-1-i)4 + 2\cdot 4]\begin{bmatrix}2\\-1\end{bmatrix} = -\frac{i}{6}\begin{bmatrix}2\\-1\end{bmatrix} ,$$

$$\begin{aligned} C(r,r,\overline{r}) &= \frac{1}{8\cdot 3}\,[2\cdot 2(-1+i) + 2(-1+i)2 + 2\cdot 2(-1-i)]\begin{bmatrix}2\\-1\end{bmatrix} \\ &= \frac{1}{24}\,(-12+4i)\begin{bmatrix}2\\-1\end{bmatrix} . \end{aligned} \tag{1.40}$$

Therefore,

$$\begin{aligned} \mu_2 + iT_2 &= -\frac{1}{2}(1-i,-2i)\left\{\left[-\frac{i}{6} + \frac{3}{4}\cdot\frac{1}{6}\,(-3+i)\right]\begin{bmatrix}2\\-1\end{bmatrix}\right\} \\ &= -\frac{1}{2}(2)(-\frac{3}{8} - \frac{i}{24}) = \frac{3}{8} + \frac{i}{24} \end{aligned} \tag{1.41}$$

so that

$$\mu_2 = \frac{3}{8} \quad \text{and} \quad T_2 = \frac{1}{24} .$$

Thus the bifurcation is supercritical and we have periodic solutions for $\mu > 0$ with period

$$T = 2\pi(1 + \varepsilon^2/24 + \ldots) , \tag{1.42}$$

$$\mu = \frac{3}{8}\,\varepsilon^2 + \ldots , \tag{1.43}$$

$$x = \varepsilon\mathrm{Re}\left\{e^{is}\,\frac{1}{2}\begin{bmatrix}2\\-1+i\end{bmatrix}\right\} + \varepsilon^2\mathrm{Re}\left\{e^{2is}\cdot\frac{1}{6}\cdot\begin{bmatrix}4\\-2+i\end{bmatrix} - \frac{1}{3}\,e^{is}\begin{bmatrix}2\\-1+i\end{bmatrix}\right\} \tag{1.44}$$

Periodic solutions for the Prigogine-Lefever model, as well as other aspects of it, have been studied extensively, not least by two of our colleagues at this meeting, Nicolis and Hershkowitz-Kauffman. In the book of Tyson [1976], one will find a demonstration of the existence of at least one periodic solution whenever the critical point is unstable. This is done by a topological method based on the Poincare-Bendixson theorem and is not restricted to a neighborhood of the bifurcation point. The main limitation of this approach is that it does not give much quantitative information about the periodic solution, and does not establish that it is a stable limit cycle or that it is unique. Further references are given in Tyson's book [1976].

D. The Belousov-Zhabotinskii Reaction and the Field-Noyes Model

In this section, I would like to discuss a real oscillating reaction, the Belousov-Zhabotinskii reaction [Belousov, 1958; Zhabotinskii, 1964]. It involves some organic acid. I think that Belousov [1958] originally used citric acid, but malonic acid, $CH_2(COOH)_2$, is more popular in terms of recent studies. The malonic acid is mixed with cerium ammonium nitrate and sulphuric acid. (The correct proportions in this convenient recipe given by Field [1972] may be found on p. 30 of Tyson's book [1976].) The resulting solution will be first yellow, then after a few minutes, clear. On then adding sodium bromate, the solution will oscillate between a yellow colored state and a clear one[1], with a period of about a minute depending on the rate of stirring of the mixture. The concentrations of certain substances, which you didn't put in to start with but which result from the various reactions, oscillate in a periodic way.

A simple kinetic model invented by Field and Noyes [1974] which embodies the most important features of the actual chemistry taking place in the Belousov-Zhabotinskii reaction is the following:

$$BrO_3^- + Br^- + 2H^+ \rightarrow HBrO_2 + HOBr \tag{1.45a}$$

$$HBrO_2 + Br^- + H^+ \rightarrow 2HOBr \tag{1.45b}$$

$$2HBrO_2 \rightarrow BrO_3^- + HOBr + H^+ \tag{1.45c}$$

$$2Ce^{3+} + BrO_3^- + HBrO_2 + 3H^+ \rightarrow 2Ce^{4+} + 2HBrO_2 + H_2O \tag{1.45d}$$

$$4Ce^{4+} + BrCH(COOH)_2 + 2H_2O \rightarrow 4Ce^{3+} + Br^- + HCOOH + 2CO_2 + 5H^+ \tag{1.45e}$$

[1] A more dramatic color change, between red and blue, can be achieved by adding Ferroin.

or, letting $X = HBrO_2$, $Y = Br^-$ and $Z = Ce^{4+}$ label the important substances in the oscillating reaction, the five equations may be written as

$$* + Y \rightarrow X + * \quad (1.46a)$$

$$X + Y \rightarrow * \quad (1.46b)$$

$$2X \rightarrow * \quad (1.46c)$$

$$* + X \rightarrow 2X + 2Z + * \quad (1.46d)$$

$$Z \rightarrow hY \quad (1.46e)$$

this last equation being a schematic representation of (1.45e) <u>and</u> certain other things which are going on. The parameter h is a "fudge factor" which is thought to be about ½ (which clearly doesn't follow from (1.45e) alone). Fig. 5 is a schematic representation of (1.46a-e).

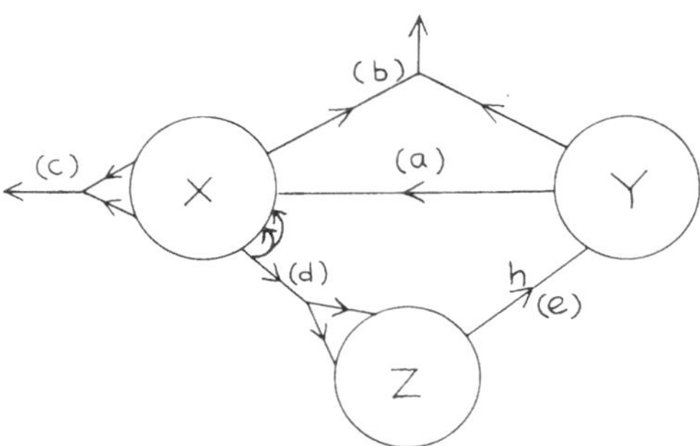

Fig. 5. Schematic representation of the reactions in Field-Noyes model, equations (1.46) (a) - (e).

Field and Noyes constructed from this sort of picture a kinetic model and made estimates of the reaction rates from the available information. They ended up with a set of differential equations in dimensionless form which, writing then as they appear in Tyson's book [p. 43], are as follows:

$$\varepsilon \frac{d\xi}{d\tau} = \xi + \eta - q\xi^2 - \xi\eta \qquad (1.47a)$$

$$\frac{d\eta}{d\tau} = -\eta + 2h\zeta - \xi\eta \qquad (1.47b)$$

$$p \frac{d\zeta}{d\tau} = \xi - \zeta \qquad (1.47c)$$

with ξ, η and ζ representing the concentrations of $HBrO_2$, Br^- and Ce^{4+}, and

$$\varepsilon \sim 2\times 10^{-4}, \; q \sim 8\times 10^{-6}, \text{ and } p \sim 300 \quad .$$

Although there is a critical point at $\xi = \eta = \zeta = 0$, and one with $\xi < 0$, there is only one critical point of chemical interest. This is determined from

$$\begin{aligned} \zeta &= \xi \;, \quad \eta = 2h\xi/(1+\xi) \\ q\xi^2 - \xi + 2h\xi \frac{\xi-1}{\xi+1} &= 0 \end{aligned} \qquad (1.48)$$

or

$$1 - q\xi = 2h \frac{\xi-1}{\xi+1} \quad . \qquad (1.49)$$

There is only one interesting c.p. as seen in Fig. 6. The c.p. depends sensitively on the parameters ε, p and q when h is near ½; but in any case, $1 < \xi < 1/q$. For h between about ¼ and 1.2, the c.p. is unstable unless p is quite small ($< \sim 10^{-3} - 10^{-4}$).

Field and Noyes [1974], found by direct numerical integration that this system has a very stable limit cycle for reasonable values of the parameters. There is trouble from "stiffness", which they circumvented by using good sense, and later using a suitable numerical method.

Fig. 7 shows the Field-Noyes results for the 3-dimensional limit cycle projected onto the $HBrO_2$-Br^- and Ce^{4+}-Br^- concentration planes. (Note that the plots are log-log plots.)

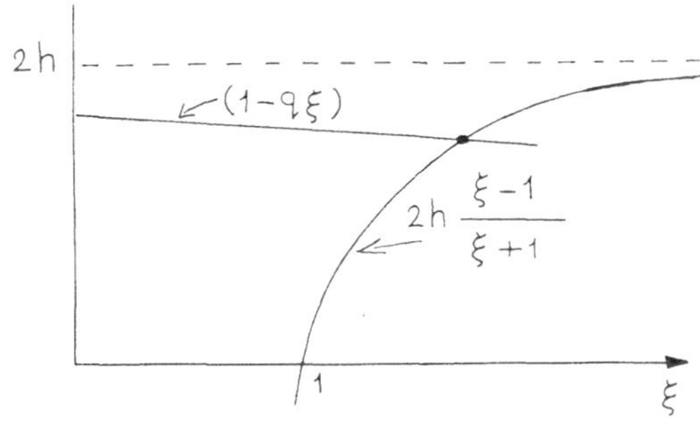

Fig. 6. Qualitative solution of equation (1.49).

Stanshine [1976] calculated the limit cycle by a singular perturbation technique, essentially using a large parameter s with $\varepsilon = 1/s^2$, $q = Qs^{-3}$, $p = s/w$, and Q, $w = O(1)$. This required 12 regions of different scaling to go all around the limit cycle. The results were in very good agreement with numerical calculation for $s = 77.27$ and with another calculation at $s = 20$. He found the Floquet exponents to be 0, -159, -8.6×10^6.

For h near ¼ there is a subcritical Hopf bifurcation at the c.p. In a certain range of h, the c.p. is stable; near it there is an unstable periodic solution and far away there is a limit cycle similar to that for h near ½. Thus, in this case bifurcation theory is of little help in understanding the physically interesting limit cycle. There is also a range where the c.p. is globally asymptotically stable, but a very small finite displacement sends the phase point off on a very long excursion before it returns to the c.p. The system is "excitable but not self-excited".

At about the same time similar results were obtained by Tyson [1975] who used the approximation obtained by setting $\varepsilon = 0$, eliminated ξ using the first equation, and studied the other two equations using a singular perturbation method based on large p. This also works well, though it seems to be quantitatively less accurate. This is discussed in Tyson's book [1976].

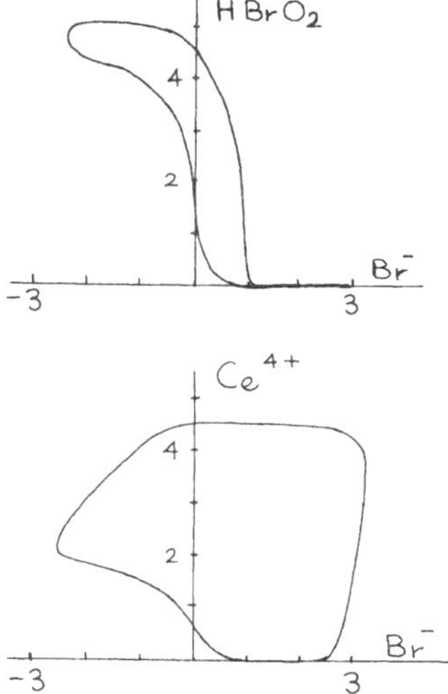

Fig. 7. Field-Noyes calculation of variation in concentrations of $HBrO_2$ and Ce^{4+} vs. Br^- concentration ($\log_{10}$ - $\log_{10}$ plots).

A proof of the existence of a periodic solution was given by Hastings and Murray [1975] using an interesting topological method. They showed that the critical point could be surrounded by a large box into which all trajectories eventually entered and from which none left. Furthermore, this box could be divided into q sub-boxes, two of which were of little interest. Assuming the critical point to be unstable, it is found to have a one-dimensional stable manifold and a two-dimensional unstable manifold. The two trajectories which approach the critical point pass through the two uninteresting boxes, but all others in them eventually leave and never return. Trajectories in any of the other 6 boxes eventually leave each of them and enter another, ultimately always cycling through the 6 in a definite order. This leads to a mapping of one face of a sub-box into itself, from which the existence of a fixed point, and hence a periodic solution is deduced using the Brouwer fixed-point theorem.

II. WAVE PATTERNS

A. Introductory Remarks

I would now like to discuss the patterns of waves that occur also in this same real reaction and models of which can be made with the aid of the Field-Noyes model or other models of oscillating reactions. The "target" patterns of waves were first reported by Zaiken and Zhabotinskii [1970] who used the above set of substances in a thin layer in a dish. [A convenient recipe due to Winfree [1972] for creating target patterns as well as spirals may be found on p. 71 of Tyson's book.] They noticed that patterns of concentric rings appeared, the rings from a given centre moving outwards until they encountered those from another centre. For a given centre, the propagation speed and spacing of the rings was found to be quite constant, but could differ from one centre to another. When the rings from two different centres met, a fairly sharp boundary or transition region occurred. Unlike the situation for water waves, the patterns do not overlap but simply end at the boundary. In time, most of the boundaries are observed to move, moving from the higher frequency target to the lower one. Such features were, of course, the subject of further theoretical and experimental investigation. The interested reader is referred to Ch. IV of Tyson [1976].

B. Formulation of the Plane Wave-Train Problem

Let me now formulate the problem of describing waves in this kind of system. One expects, I think, that what's happening is partly the chemical reaction which is described by a system of differential equations (reaction-diffusion equations) of the stucture

$$c_t = F(c) + K\nabla^2 c \tag{2.1}$$

where c is a vector of concentrations and F(c), eg., could be described by the right-hand sides of the three Field-Noyes model equations. In the diffusion term, the diffusion constant K might be a non-diagonal matrix but it's fairly likely that K is not too far from being a scalar matrix because most of the substances of interest are relatively small molecules.

We assume that

$$\frac{dc}{dt} = F(c)$$

has a linearly stable limit cycle solution $c = y_o(\sigma_o t)$ with period $2\pi/\sigma_o$ (so $y_o(\theta)$ is 2π-periodic). Then for all sufficiently small α, we conjecture that there is a 2π-periodic function $y_{\alpha^2}(\theta)$ and an angular frequency $\sigma = H(\alpha^2)$, such that $c = y_{\alpha^2}(\sigma t - \alpha x)$ solves the reaction-diffusion equations (2.1), i.e.,

$$y = y_{\alpha^2}$$

solves

$$\sigma y' = F(y) + \alpha^2 K y'' \tag{2.2}$$

where K is a positive definite matrix. This formulation has advantages in the proof and in the construction of the solution, but the problem can be stated also a little more simply by not insisting on period 2π from the start. We hypothesize simply that $c_t = F(c)$ has a stable limit cycle solution $c = Y_o(t)$ and then assert that for all small enough β there is a solution of the reaction-diffusion equations of the form $c = Y_\beta(t - \beta x)$, Y_β being periodic in its argument, with a period that tends to the period of Y_o as $\beta \to 0$. In this form Y_β satisfies the equation

$$Y' = F(Y) + \beta^2 K Y'' \quad . \tag{2.3}$$

This second-order system for the n-components of Y can be written explicitly as a 2n-order system by setting $KY'' = u$ so that we have

$$Y' = F(Y) + \beta^2 u \ , \tag{2.4}$$

and

$$\beta^2 u' = K^{-1}u - N(Y)(F(Y) + \beta^2 u) \ , \tag{2.5}$$

where N is the matrix F_Y of partial derivatives of the vector F with respect to the components of its argument. This puts the problem into a form to which we can apply an important singular perturbation result called the "Flatto-Levinson Theorem" [see Flatto and Levinson, (1955)]. This theorem is about systems of n + m equations of the form (x in R^n, y in R^m):

$$\frac{dx}{dt} = f(x,y,\varepsilon) \quad , \tag{2.6}$$

$$\varepsilon \frac{dy}{dt} = g(x,y,\varepsilon) \; , \tag{2.7}$$

where f and g are analytic in ε, or possess uniform asymptotic expansions in ε. Here it is assumed that the m equations $g(x,y,0) = 0$ define an n-dimensional manifold, the "slow manifold", and in fact that these equations can be solved for y:

$$y = h_o(x) \; , \tag{2.8}$$

at least in the range of interest (to be specified more explicitly below). To assure this we assume a little more, namely that all eigenvalues of $g_y(x,y)$ have real parts exceeding some positive number δ in absolute value for all x and y in the range of interest. It is also assumed that the equation

$$\frac{dx}{dt} = f(x,h_o(x),0) \tag{2.9}$$

has a periodic solution $x_o(t)$ which is a stable limit cycle, or at any rate which is such that all but one of the Floquet multipliers of the linearization about it are off the unit circle. The "range of interest" is some neighborhood of the orbit of this limit cycle in (x,y) space, i.e., of the curve $(x_o(t),\, h_o(x_o(t)))$.

Note that these hypotheses are fulfilled for the plane wave problem: here $g_y = K^{-1}$, and $h_o = K\,N\,F$.

Flatto and Levinson [1955] proved that for all small ε there is a locally unique (up to a time translation) solution $x(t,\varepsilon)$, $y(t,\varepsilon)$, periodic in t, whose orbit is close to $(x_o(t),\, h_o(x_o(t)))$ and whose period is close to that of $x_o(t)$.

It is worth remarking that although this is a singular perturbation problem, the periodic solution does not have any "boundary layers". It is true that most of the solutions of the equations for $\varepsilon \neq 0$ do have boundary layers - representing a rapid approach to, or flight from, the slow manifold, but a few do not, and among these is the periodic solution which is here of greatest interest.

The proof of the Flatto-Levinson theorem cannot be given here but the following way of looking at it will perhaps make it plausible. Introduce the 'fast' time scale $\tau = t/\varepsilon$, so that the equations become

$$\frac{dx}{d\tau} = \varepsilon f(x,y,\varepsilon) \quad , \tag{2.10}$$

$$\frac{dy}{d\tau} = g(x,y,\varepsilon) \quad . \tag{2.11}$$

At $\varepsilon = 0$ the phase portrait of this system is given schematically in Fig. 8. The vector field is 'vertical', i.e., in x = constant surfaces, and there is an invariant slow manifold consisting entirely of critical points, $y = h_o(x)$. This manifold is "hyperbolic" in the sense that any center manifold of any of these critical points is n dimensional and tangent to $y = h_o(x)$ - this is because g_y is non-singular. It is then geometrically plausible that if the vector field is perturbed a little by taking $\varepsilon \neq 0$ and small, there will still be an invariant manifold of dimension n, $y = h(x,\varepsilon)$ close to $y = h_o(x)$. Restricting the vector field to this manifold we obtain

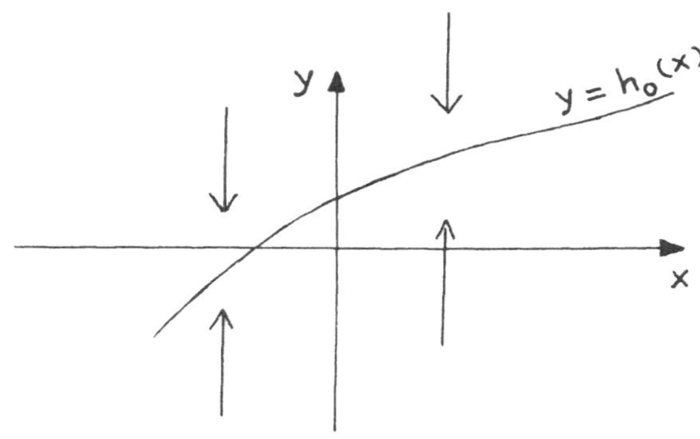

Fig. 8. Phase diagram of equations (2.10), (2.11) at $\varepsilon = 0$.

$$\frac{dx}{dt} = f(x,h(x,\varepsilon),\varepsilon) \tag{2.12}$$

which is now a <u>regular</u> perturbation of

$$\frac{dx}{dt} = f(x,h_o(x),0) \ , \tag{2.13}$$

and so by well-known results has a nearby periodic solution.

The Flatto-Levinson proof does not make use of such invariant-manifold ideas, but goes directly after the periodic solution by an iterative process, each step of which requires only the solution of relatively straightforward linear problems. This technique can be made the basis of a numerical algorithm for effectively computing the periodic solution, which is quite difficult to compute directly from the 2n-order system. (This is partly because it is a stiff system, but more importantly because, as a solution of these ordinary differential equations, it is generally unstable both forward and backward.) A sketch of the proof and some discussion of certain issues involved in making it into a numerical algorithm can be found in Howard [1979].

Some examples of plane waves for the chemical kinetics of the Prigogine-Lefever model are given in Kopell and Howard [1973]. Plane waves for the Field-Noyes model have been obtained by Stanshine [1976] using an extension of the singular perturbation method mentioned above, which exploits the relaxation-oscillation character of this model. Further application of the relaxation-oscillation feature, to target patterns, is made by Tyson and Fife [1980].

Suppose now that the family of plane wave solutions $y_{\alpha^2}(\theta)$, and the associated nonlinear dispersion relation $\sigma = H(\alpha^2)$, is known for some system of reaction-diffusion equations

$$c_t = F(c) + K\nabla^2 c \ . \tag{2.1}$$

To have a simple specific example, one may consider the system

$$\begin{bmatrix} c_1 \\ c_2 \end{bmatrix}_t = \begin{bmatrix} \lambda & -\omega \\ \omega & \lambda \end{bmatrix} \begin{bmatrix} c_1 \\ c_2 \end{bmatrix} + \nabla^2 \begin{bmatrix} c_1 \\ c_2 \end{bmatrix} \tag{2.14}$$

where $\lambda = 1 - c_1{}^2 - c_2{}^2$,

$$\omega = \omega_o + (\omega_1 - \omega_0)(c_1{}^2 + c_2{}^2) \ ,$$

or some qualitatively similar functions λ, ω of $c_1{}^2 + c_2{}^2$. (One may think of c_1 and c_2 as deviations of the concentrations from some stationary but unstable state, but this model is not really satisfactory as a model of any reasonable chemical system.) Here the limit cycle of the 'kinetics' is given by $c_1 = \cos(\omega_1 t)$, $c_2 = \sin(\omega_1 t)$, and it is readily verified that the plane waves are (given parametrically in terms of an "amplitude" parameter $A = c_1{}^2 + c_2{}^2$),

$$y_A = \sqrt{A}\begin{bmatrix}\cos\theta \\ \sin\theta\end{bmatrix}, \quad \sigma = \omega(A), \quad \alpha^2 = \lambda(A), \tag{2.15}$$

ie,

$$H(\alpha^2) = \omega(\lambda^{-1}(\alpha^2)) \qquad (= \omega_1 + (\omega_0 - \omega_1)\alpha^2) \quad . \tag{2.16}$$

This family of plane wave solutions can be used as a centre of perturbations to find more general solutions of the reaction-diffusion equations in the form of "slowly-varying waves". This has been discussed at some length in Howard and Kopell [1977]; only a brief sketch can be given here. Let $X = \varepsilon x$ and $T = \varepsilon t$ be "slow" space and time variables. Here ε is the measure of slowness. A slowly-varying wave is a family of functions, parameterized by ε, expressible in the form

$$c(x,t,\varepsilon) = Y(\theta(x,t,\varepsilon),X,T,\varepsilon) \tag{2.17}$$

where

(1) $Y(\theta,X,T,\varepsilon)$ is a C^2 periodic function of θ with least period 2π,

(2) $Y = Y^0(\theta,X,T) + o(1)$ (or else has an asymptotic expansion in ε up to some order) uniformly for X and T in some compact region, and

(3) $\Theta(X,T,\varepsilon) \equiv \varepsilon\theta(x,t,\varepsilon) = \Theta^0(X,T) + \varepsilon\,\Theta'(X,T) + o(\varepsilon)$ (or else has a corresponding asymptotic expansion to one order higher than Y), also uniformly.

It is helpful to normalize the phase of the periodic functions by some condition such as "the last component of Y having the value k at $\theta = 0$". With such a normalization it can be shown that the representation as a slowly varying wave, if it exists, is unique. When is a slowly-varying wave a solution of the reaction-diffusion equation (2.1)? Substituting in, we get:

$$Y_\theta(\Theta/\varepsilon)\Theta_T - F(Y) - KY_{\theta\theta}(\Theta/\varepsilon)\Theta_X^2 + o(1) = 0 \qquad (2.18)$$

From this equation and the uniqueness theorem one can show that if we have a solution, then

$$Y_\theta^0(\theta,X,T)\Theta_T^0 - F(Y^0(\theta,X,T)) - KY_{\theta\theta}^0(\theta,X,T)(\Theta_X^0)^2 = 0 , \qquad (2.19)$$

where now θ,X,T are <u>independent</u> variables. Assuming that our family of plane waves has been phase-normalized in the same way as for the slowly-varying waves, it follows that the slowly-varying waves can be a solution only if

$$Y^0(\theta) = Y_{\left(\Theta_X^0\right)^2}(\theta) , \qquad (2.20)$$

and

$$\Theta_T^0 = H(\Theta_X^{0\,2}) . \qquad (2.21)$$

This Hamilton-Jacobi equation for the phase function Θ^0 makes possible an investigation of the (slow) evolution of an initial concentration distribution which locally is close to a plane wave but varies slowly in wave number.

The characteristic equations in the "strip space" for this first-order partial differential equation are

$$\frac{dX}{ds} = \frac{\partial}{\partial p}H(p^2) = 2pH'(p^2) , \qquad (2.22)$$

$$\frac{dT}{ds} = 1 , \qquad (2.23)$$

$$\frac{d\Theta}{ds} = q - 2p\,H'(p) , \qquad (2.24)$$

$$\frac{dp}{ds} = 0 , \qquad (2.25)$$

$$\frac{dq}{ds} = 0 . \qquad (2.26)$$

These are easily solved; the characteristics are straight lines and their projections into the X,T space move with the <u>group velocity</u>

$$c_g = \frac{\partial H}{\partial p} , \qquad (2.27)$$

which is constant along a given characteristic since p is.

To solve an initial-value problem for the Hamilton-Jacobi equation, say

$$\Theta^0(X,0) = \Theta_0(X) , \qquad (2.28)$$

the recipe of the method of characteristics is to solve the characteristic ordinary differential equations with initial data $T = 0$, $X = X_0$, $\Theta^0 = \Theta_0(X_0)$, $p = -\Theta_{0X}(X_0)$, $q = H(\Theta_0(X_0))$, thus finding, in this case:

$$X = X_0 + c_g(\Theta_0(X_0))s , \qquad (2.29)$$

$$T = s , \qquad (2.30)$$

$$\Theta^0 = \Theta_0(X_0) + [H(\Theta_0(X_0)) + \Theta_{0X}c_g(\Theta_{0X})]s , \qquad (2.31)$$

etc.

To find the actual solution $\Theta^0(X,T)$, one has to solve the first two of these equations for the parameters X_0 and s in terms of X and T, and then substitute them into the third. Geometrically, for a given X,T, one projects the characteristics in the strip space down to the X,T space, follows back the one going through the desired X,T to the initial line to find the appropriate value of X_0, and then finds Θ^0, etc. In fact, the local wave number $p = -\Theta^0_X$ is constant along the characteristic so its value at (X,T) is the same as at the appropriate point $(X_0,0)$ on the same characteristic.

Now this procedure provides a description of the evolution of the solution of the initial-value problem for the Hamilton-Jacobi equation, but generally, this can be continued forward in T for only a finite time - it is, in general, the case that the characteristics, though they exist for all s and never intersect in the strip-space, have projections into X,T space which do intersect. Thus, in general, discontinuities in the wave number $-\Theta^0_X$ will develop, and the solution to the Hamilton-Jacobi equation cannot be continued. The essential point here is that the phase-gradient becomes very large, and the solution can no longer be described as "slowly-varying".

This situation is mathematically analogous to the development of shocks in gas dynamics. As in that field, it is possible to extend the usefulness of the slowly-varying wave theory beyond the point at which a shock appears by accepting these discontinuities and finding a rule for describing how they propagate - the Hamilton-Jacobi equation is then used between the shocks. In the present case, a plausible argument can be made that a shock moves with velocity $\Delta\sigma/\Delta\alpha$, where $\Delta\sigma$ and $\Delta\alpha$ are the jumps in angular frequency and wave number across it. (In addition to this condition, which is the analogue of the Rankine-Hugoniot relations in gas dynamics, there is also an analogue of the entropy condition, namely: characteristics should be converging toward a shock as T increases.)

These shocks appear to be an appropriate description of the 'transition regions' (toward which waves are converging) which separate the different 'targets' seen in the Belousov-Zhabotinskii patterns.

To verify this picture one would like to show from the full reaction-diffusion equations that at least in idealized cases there are real solutions of them of this kind, that is, for instance, a solution which tends to each of two different wave trains as $x \to \pm\infty$, but has a transition region in-between, propagating with the velocity suggested above. To formulate such an "ideal shock" problem one may ask for a solution to the reaction-diffusion equations, asymptotic to two different wave train solutions as $x \to \pm\infty$, and which, when observed in a suitably moving coordinate system, appears to be everywhere periodic, with the same period, in time. In terms of a formula this says:

$$c(x,t) = C(x - c_s t, t) \tag{2.32}$$

where the functions $C(\xi,t)$ are periodic in t (for fixed ξ) and for large positive x,

$$C(x - c_s t, t) \sim y_{\alpha_+}(\sigma_+ t - \alpha_+ x + \phi_+) \tag{2.33}$$

and for large negative x

$$C(x - c_s t, t) \sim y_{\alpha_-}(\sigma_- t - \alpha_- x + \phi_-) \ . \tag{2.34}$$

If one writes $\sigma_\pm t - \alpha_\pm x + \phi_\pm$ in terms of the moving system, one obtains

$$(\sigma_\pm - \alpha_\pm c_s)t - \alpha_\pm(x - c_s t) + \phi_\pm \ ,$$

so in order to have the same Doppler-shifted frequency everywhere we must have $\sigma_+ - \alpha_+ c_s = \sigma_- - \alpha_- c_s$ which gives the shock speed $\Delta\sigma/\Delta\alpha$ alluded to above. The question then is, do the reaction diffusion equations really have such solutions? With certain reasonable additional assumptions - notably stability of the asymptotic plane waves - this was shown to be true at least for sufficiently 'weak' shocks (sufficiently small $\Delta\alpha$) in the paper referred to above; rigorously for a special class of reaction diffusion equations, and formally in general. The methods used were bifurcation methods, exploiting the 'weak shock' idea.

One feature of interest, especially in the context of this conference, is that the weaker the shock is, the more extended is the transition region. This suggests that an extended form of the basic idea of slowly-varying waves might be developed which would also include the structure of weak shocks, not only the tendency toward their formation exhibited by the above Hamilton-Jacobi equation. This is the case, and the resulting equation, written in terms of the original unscaled x,t and phase function θ has the form

$$\theta_t - H(\theta_x^2) = K_1(\theta_x^2)\theta_{xx} . \qquad (2.35)$$

Like H, $K_1(\theta_x^2)$ is determined by the properties of the plane wave solutions. In the simplest cases it turns out to be just the diffusion constant, but in the general case it may depend also on the local wave number. This equation is a generalization of the integrated form of the Burgers equation, to which it reduces if K_1 is a constant and H is a linear function of its argument.

These shock-structures provide an example, for reaction-diffusion equations with oscillatory kinetics, of a persistent, localized structure, having some features in common with a solitary wave. It is not 'persistent' in the sense of being time-independent in a suitable moving system, but is rather time-periodic in such a system.

C. Conclusion

To conclude, I can merely mention some different kinds of persistent structures which have been found in some of these systems. These appear to have similarities to the centres of the targets and spirals of the Belousov-Zhabotinskii patterns though at present this identification is only qualitative and tentative. An analogue of a 'centre' in one space dimension would be a solution of the reaction-diffusion equations asymptotic to two outgoing plane waves of equal but opposite wave numbers. Here "outgoing" is in the sense of group velocity, (which for the Zhabotinskii waves appears to have the same direction as phase velocity). For any (sufficiently small) pair of equal and opposite incoming (group velocity) wave numbers, there is a shock structure solution, but this is not true in the outgoing case. This is analogous to the non-existence of expansion-shock structures in viscous heat conducting gas dynamics. However, while expansion-shocks never exist, for a certain class of reaction-diffusion equations centre structures do exist, but only for special wave numbers. These have an entirely different structure from the shocks, differing greatly from the plane waves in the centre, and do not appear to be accessible by bifurcation methods. They have been found by a singular perturbation method. These methods have also enabled us to demonstrate the existence of analogues of these centre structures in finite domains with Neumann boundary conditions, provided the domain is large enough. (If it is too small, it has been shown by Othmer [1977] and by Conway, Hoff, and Smoller [1978], that such spatially heterogeneous and temporally periodic solutions cannot exist.) In addition, these methods have uncovered a large class of temporally periodic but spatially irregular solutions of this special class of reaction-diffusion equations on the infinite line. These results are presented in a forthcoming paper by Kopell and Howard [1981].

REFERENCES

Belousov, B.P., 1958, in: "Collection of Abstracts on Radiation Medicine," Medgiz, Moscow.

Conway, E., Hoff, D., and Smoller, J., 1978, Large time behavior of solutions of nonlinear reaction-diffusion equations, SIAM J. Appl. Math., 35:1.

Field, R.J., 1972, A reaction periodic in time and space. A lecture demonstration, J. Chem. Educ., 49:308.

Field, R.J., and Noyes, R.M., 1974, Oscillations in chemical systems, IV. Limit cycle behavior in a model of a real chemical reaction, J. Chem. Phys., 60:1877.

Flatto, L., and Levinson, N., 1955, Periodic solutions of singularly perturbed systems, J. Rat. Mech. Annal., 4:943.

Hastings, S.P., and Murray, J.D., 1975, The existence of oscillating solutions in the Field Noyes model for the Belousov-Zhabotinskii reaction, SIAM J. Appl. Math., 28:678.

Howard, L.N. and Kopell, N., 1977, Slowly varying waves and shock structures in reaction-diffusion equations, Stud. in Appl. Math., 56:95.

Howard, L.N., 1979, Nonlinear oscillations, in: "Lectures in Applied Mathematics, Vol. 17," American Math. Society.

Kopell, N., and Howard, L.N., 1973, Plane wave solutions to reaction-diffusion equations, Stud. in Appl. Math., 52:291.

Kopell, N., and Howard, L.N., 1981, Target patterns and horseshoes from a perturbed central-force problem: Some temporally periodic solutions to reaction-diffusion equations, Stud. in Appl. Math., 64:1.

Nicolis, G., and Portnow, J., 1973, Chemical oscillations, Chem. Rev., 73:365.

Othmer, H.G., 1977, Current theories of pattern formation, in: "Lectures on Mathematics in the Life Sciences, Vol. 9," Amer. Math. Soc., Providence, R.I.

Prigogine, I., and Lefever, R., 1968, Symmetry breaking instabilities in dissipative systems. II., J. Chem. Phys., 48:1695.

Stanshine, J.A., and Howard, L.N., 1976, Asymptotic solutions of the

Field-Noyes model for the Belousov reaction - I. Homogeneous oscillations, Stud. in Appl. Math., 55:129.
Stanshine, J.A., 1976, Asymptotic solutions of the Field-Noyes model for the Belousov reaction - II. Plane waves, Stud. in Appl. Math., 55:327.
Tyson, J.J., 1975, Classification of instabilities in chemical reaction systems, J. Chem. Phys., 62:1010.
Tyson, J.J., 1976, "The Belousov-Zhabotinskii Reaction," Lecture Notes in Biomathematics 10, Springer-Verlag, N.Y.
Tyson, J.J., and Fife, P.C., 1980, Target patterns in a realistic model of the Belousov-Zhabotinskii reaction, J. Chem. Phys., 73:2224.
Winfree, A.T., 1972, Spiral waves of chemical activity, Science, 175:634.
Zaikin, A.N., and Zhabotinskii, A.M., 1970, Concentration wave propagation in two-dimensional liquid-phase self-oscillating system, Nature, 225:535.
Zhabotinskii, A.M., 1964, Periodic course of oxidation of malonic acid in solution (investigation of the kinetics of the reaction of Belousov), Biophysics, 9:329.

MODELS IN NEUROBIOLOGY*

John Rinzel

Mathematical Research Branch
National Institutes of Health
Bethesda, Maryland 20205

* This paper is a slightly revised version of the author's paper in Mathematical Aspects of Physiology (ed., F.C. Hoppensteadt), Amer. Math. Soc., to appear.

TABLE OF CONTENTS

MODELS IN NEUROBIOLOGY

John Rinzel

Mathematical Research Branch
National Institutes of Health
Bethesda, Maryland 20205

I. INTRODUCTION

The most familiar mode of neural communication is electrical and synaptic signaling by individual nerve cells. Here we shall consider a few stereotypical phenomena of such signaling. A primary observable is the potential V (deviation from rest) across the cell membrane which responds to applied stimulating current I_{app} and to changes in membrane permeability to the various ion species. These permeability changes also usually depend on V and, for excitable membrane, result in the generation of the nerve impulse. We will describe and present results for models (some quantitative but others more qualitative) of excitable membrane behavior.

In the first half of this paper we consider space-clamped dynamics, i.e., dynamics of an isolated membrane patch or, alternatively, of a spatially lumped neuron model. We introduce the four-variable Hodgkin-Huxley (HH) model for squid giant axon membrane and reduced, two-variable, versions of it which include the FitzHugh-Nagumo (FHN) model. We outline threshold behavior for a brief stimulus and for a slowly increasing ramp of I_{app}. The latter leads to the appearance of periodic solutions of the dynamic equations through a Hopf bifurcation; some of these solutions correspond to repetitive firing of the nerve. This behavior in the full HH model is characterized as a hard oscillation. That is, with I_{app} as a parameter, there is a range of values for I_{app} over which the membrane may be either in a stable repetitive firing mode or in a stable state of steady depolarization ($V = \text{const} > 0$). These theoretical results have motivated corresponding experimental observations for squid giant axon. In contrast, however, to regular beating with equally-spaced periodic pulses, as described above, some neurons are observed to fire in repetitive bursting patterns. Among the proposed explanations for bursting are: 1) the oscillator hypothesis which suggests that bursting is intrinsic to the membrane dynamics of an individual cell; 2) the network hypothesis which maintains that mutual coupling of different cells in a network leads to alternating firing patterns from one cell to the other. Simplified models will be described to illustrate these two hypotheses.

We go beyond the membrane patch description in the second half of this paper and consider the effect of a spatio-temporal distribution of V and ionic permeabilities. The single space dimension corresponds to distance x along the nerve axon or along a dendritic branch. One obtains a nonlinear parabolic partial differential equation analogous to some types of electrical transmission line or cable equations. An ubiquitous feature of excitable membrane cable responses is impulse propagation. Some phenomenology along with mathematical results on propagation will be described for the case of an axon with uniform properties. Although threshold behavior in the cable case is difficult to analyze rigorously, we will formally describe the onset of repetitive activity for a slowly rising stimulus current applied at one location. This result, a bifurcation calculation, is analogous to the space-clamped case. Finally, we consider a simple model problem to study impulse propagation through a region of changing geometry. A pulse upstroke may fail to propagate into a region where the axon diameter suddenly increases too much.

As suggested earlier there are several modes of potential communication in the nervous system. Barker and Smith [1979] discuss the effect of chemical substances, so-called neuromodulators and neurohormones, which directly alter the synaptic and excitability characteristics of a neuron. If the release of such substances is under neural control then one imagines additional avenues of communication. Also, since electrical and synaptic transmission involve ion flows they result in intra- and extracellular changes in ionic composition. Such changes, if significant, can alter, for example, the driving forces for subsequent electrical signaling and thereby perhaps serve a functional role in communication; as a general reference see Nicholson [1980]. Yet another possible mode of communication is axoplasmic transport (e.g., see Schwartz [1980]) which may carry information on trophic function requirements. Readable and interesting accounts of various aspects of nervous system function and communication appear in Scientific American, September, 1979, and at the textbook level, e.g., Kuffler and Nicholls [1977]; references to mathematical modeling aspects may be found in Rinzel [1978a].

II. BEHAVIOR OF THE MEMBRANE PATCH OR LUMPED NEURON

A. Physical and Mathematical Model

Here we neglect spatial differences in V and membrane properties so $V \equiv V(t)$. In some cases this can be achieved experimentally by the space-clamp procedure [Hille, 1976]. For some invertebrate neurons, relative electrical isolation of the cell body may not be an unreasonable assumption a priori. In any case, the properties we explore are attributed to a patch of excitable membrane and are thought to be shared qualitatively by a variety of neurons.

Membrane current $I_m(t)$ (positive direction is outward) has a capacitive and ionic component, the latter carrying a contribution from the individual species to which the membrane is permeable:

$$I_m(t) = C_m\dot{V} + I_{ion} \ . \tag{2.1}$$

In many applications the ionic currents are represented as ohmic with V and t dependent conductances

$$I_m(t) = C_m\dot{V} + \sum_j g_j(V,t)(V-\bar{V}_j) \tag{2.2}$$

where $\bar{V}_j$ is (deviation from rest of) the Nernst-Planck equilibrium potential for ion species j as determined by its interior and exterior concentrations [Hille, 1976]. In a few cases, clever experimental techniques (the voltage-clamp along with ion substitution or use of pharmacological agents to selectively block ion channels) have led to the separation of the individual components and to quantitative descriptions, via curve-fitting, for the $g_j(V,t)$. The most well-studied case is the Hodgkin-Huxley (HH) model [Hodgkin and Huxley, 1952] for squid giant axon in which the principal ionic currents are for sodium (inward) and potassium (outward). It takes the form

$$\begin{aligned} C_m\dot{V} &= -\bar{g}_{Na}m^3h(V-\bar{V}_{Na}) - \bar{g}_K n^4(V-\bar{V}_K) - \bar{g}_L(V-\bar{V}_L) + I_{app}(t) \\ \dot{m} &= [m_\infty(V) - m]/\tau_m(V) \\ \dot{h} &= [h_\infty(V) - h]/\tau_h(V) \\ \dot{n} &= [n_\infty(V) - n]/\tau_n(V) \ . \end{aligned} \tag{2.3}$$

Here I_{app} represents current applied by the experimenter. The sodium conductance, $\bar{g}_{Na}m^3h$, is expressed in terms of m, Na-activation, and h, Na-inactivation, while potassium conductance, $\bar{g}_K n^4$, involves the single auxiliary variable n, K-activation. Each of the dynamic variables, m, h, n lies between 0 and 1 and satisfies a first order rate law with V-dependent "time constant" and "steady-state" functions. For the activation variables, m_∞ and n_∞ are monotonic increasing with V and saturating, with shapes qualitatively like $(1 + \tanh V)/2$; i.e., m and n are "turn-on" variables. The function h_∞ is monotonic decreasing and is qualitatively like $(1 - \tanh V)/2$; h describes a "turn-off" process. The time scale of $\dot{m}$ is roughly ten times faster than $\dot{n}$ and $\dot{h}$. The leakage current $\bar{g}_L(V-\bar{V}_L)$ is a relatively small contribution and $\bar{g}_L$ is constant. Specific expressions for the empirical functions and constants in equations (2.3) may be found in FitzHugh [1969]. Hille [1976] offers a good biophysical review of membrane ionic currents and also discusses models for other excitable membranes which may involve additional ionic components.

At rest, the membrane is relatively more permeable to K^+ than to Na^+ so the rest potential V = 0 is closer to $\bar{V}_K$ ($\approx$ -12 mV) than to $\bar{V}_{Na}$ ($\approx$ 115 mV). The rest state of the model, given by V = 0, $m_o = m_\infty(0)$, $h_o = h_\infty(0)$, $n_o = n_\infty(0)$, is linearly stable; the dynamics, however, are excitable. For a brief, but adequate-sized, $I_{app}(t) > 0$, there is a large excursion of V, called the action potential, followed by an eventual return to rest; time scale of the pulse is O(msec) and

$$\max_t V - \min_t V \text{ is } O(100\,\text{mV}).$$

This pulse is effected by large swings in g_{Na} and g_K. The initial positive displacement of V leads to a rapid increase in m and inward sodium current which regeneratively further increases V towards $\bar{V}_{Na}$. The rapid V-upstroke is halted and the downstroke is initiated because of two slower factors; first, g_{Na} decreases due to h decreasing towards $h_\infty(V)$, and second g_K increases as n rises towards $n_\infty(V)$. During this downstroke phase g_K begins to dominate and V falls below rest towards $\bar{V}_K$. As n and h try to follow this lowered V, the rest state is slowly approached.

Threshold behavior for pulse initiation can be illuminated by considering approximate two-variable HH reduced models, first studied by FitzHugh [1960, 1961]. For example, the upstroke of the excitation process is uncovered by first neglecting the slower recovery processes, i.e., assume $h \equiv h_o$, $n \equiv n_o$. This V-m system has two stable states (the rest state with V = 0 and the "excited state" with V slightly below $\bar{V}_{Na}$) separated by a saddle type singular point ($\approx$ 3 mV). The two incoming trajectories of the saddle form a separatrix curve which distinguishes subthreshold from superthreshold initial conditions.

In order to describe a pulse with upstroke and downstroke, and to describe threshold behavior for longer duration I_{app}, we need at least one recovery variable. A useful approximation is obtained by retaining $h = h_o$, constant, and by setting $m = m_\infty(V)$, i.e., m is assumed to be so fast that it relaxes instantaneously to m_∞. As a function of V, I_{ion} is no longer linear as in (2.3), but rather is N-shaped with a region of negative resistance and a region of inward current ($I_{ion} < 0$). This V-n system for $I_{app} = 0$ behaves qualitatively as the two-variable FitzHugh-Nagumo (FHN) model [FitzHugh, 1961; Nagumo et al, 1962]

$$\begin{aligned} \dot{v} &= -f(v) - w + I_{app} \\ \dot{w} &= \varepsilon(v - \gamma w) \end{aligned} \tag{2.4}$$

Here v is like membrane potential and w, the single recovery variable, plays the role of h and n in the HH model; f(v) is an N-shaped instantaneous current-voltage law, sometimes taken to be a cubic polynomial [FitzHugh, 1961]. The parameters ε, γ are non-negative where γ is such that (2.4) has a unique singular point for any value of I_{app}. The phase plane for (2.4) with $I_{app} = 0$ is shown in Fig. 1; the nullclines $\dot{w}=0$, $w=-v/\gamma$, and $\dot{v} = 0$, $w = -f(v)$ are shown dashed. Trajectories for sub and superthreshold initial conditions are shown for two values of ε.

B. Repetitive Activity-Beating

In response to a step of current, $I_{app} = 0$ for $t < 0$, I_{app} = constant > 0 for $t > 0$, some nerves will fire an infinite train of pulses; for references, see [Rinzel, 1978b]. After an early transient phase, the interpulse intervals may approach uniformity (regular beating) and the steady frequency depends on I_{app}. Correspondingly, the HH and FHN models have periodic solutions for I_{app} in a certain range of values. This range contains the interval (I_1, I_2) over which the steady state (which depends on I_{app}) is unstable. At the critical value I_1 (I_2), the steady state loss (gains) stability; I_1 is the threshold for repetitive activity in response to a very slowly rising I_{app}; the restabilization at I_2 is thought to correspond to "nerve block" for large I_{app}. For $I_{app} = I_1, I_2$, the model equations when linearized about the steady state have a coefficient matrix with a pure imaginary pair of eigenvalues. According to the Hopf bifurcation theorem, a branch of small amplitude periodic solutions bifurcates from the steady state at I_1 and I_2 [Rinzel, 1978b; Troy, 1978; Hassard, 1978]. Whether the emergent branch is sub or supercritical (i.e., leads to hard or soft oscillation) depends on the model parameter values; this

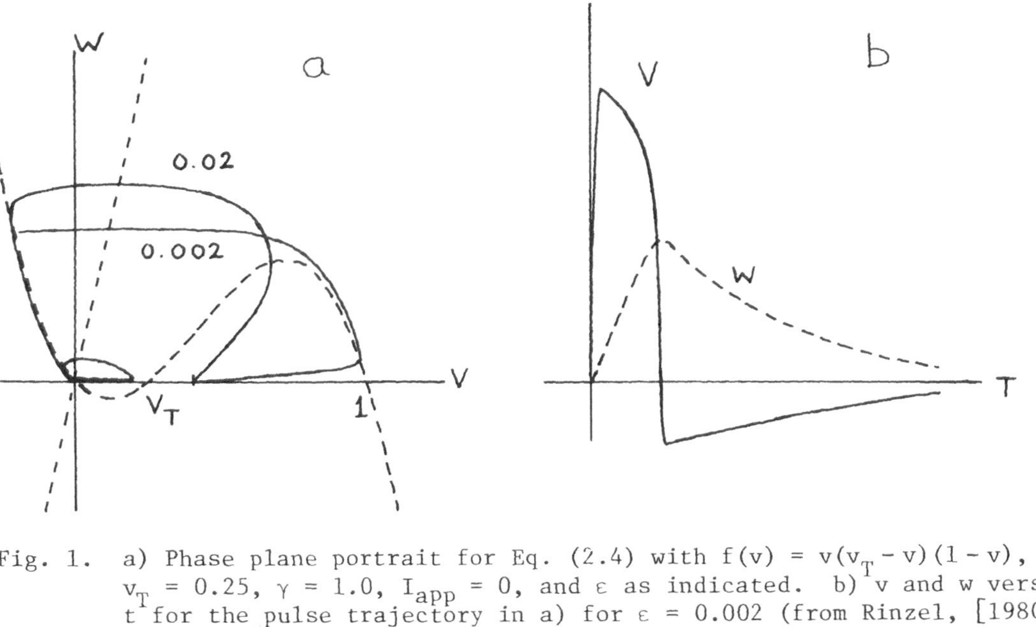

Fig. 1. a) Phase plane portrait for Eq. (2.4) with $f(v) = v(v_T - v)(1 - v)$, $v_T = 0.25$, $\gamma = 1.0$, $I_{app} = 0$, and ε as indicated. b) v and w versus t for the pulse trajectory in a) for $\varepsilon = 0.002$ (from Rinzel, [1980]).

bifurcation direction has been computed [Troy, 1978; Hassard, 1978].

Consider the FHN model and let $(\bar{v},\bar{w})$, functions of the parameter I_{app}, denote the steady state. For $\varepsilon\gamma^2 < 1$, it is unstable if

$$f'(\bar{v}) + \varepsilon\gamma < 0 \tag{2.5}$$

and therefore I_1, I_2 correspond to values $\bar{v}$ (there will be two) such that

$$f'(\bar{v}) = -\varepsilon\gamma \ . \tag{2.6}$$

Thus, loss of stability occurs when the time scale of the destabilizing, negative resistance, action of f(v) matches that of the stabilizing recovery process. One may also understand the development of repetitive activity from the phase plane (Fig. 1) for Eqs. 2.4. First observe that $I_{app} > 0$ simply translates the v-nullcline upward. If I_{app} is small (large) then $(\bar{v},\bar{w})$ is on the left (right) branch and is stable. For an intermediate range of I_{app} values, $(\bar{v},\bar{w})$ is on the middle branch and is unstable if condition (2.5) is satisfied. In the case $0 < \varepsilon << 1$, the periodic solution is a relaxation oscillation with, 1) sharp upstroke and downstroke which are seen as rapid transitions from the left to the right, or right to left, branches of the v-nullcline (at its knees), and 2) slow phases, plateau and recovery, during which the phase point hugs these branches.

The local stability analysis of the singular point for the HH model parallels that above except in this fourth-order case, most evaluations are carried out numerically; early work was done by Cooley, Dodge, and Cohen [1965] (see Rinzel [1978b] for additional references). Hopf bifurcation of periodic solutions occurs for $I_{app} = I_1, I_2$ and the direction is subcritical at I_1 (hard oscillation) but supercritical at I_2 (soft oscillation). Corresponding to the hard oscillation which arises at I_1, a stable periodic response to a step is observed numerically for $I_\nu < I_{app} < I_1$. Thus the system is bistable for the I_{app}-interval (I_ν, I_1); there is a large stable periodic solution and a stable steady state. There is also an unstable (in some cases more than one) periodic solution which may be calculated numerically [Rinzel, 1978b; Hassard, 1978; Rinzel and Miller, 1980]. The bifurcation diagram for these solutions at three temperatures is shown in Fig. 2. These observations suggest that it should be possible to annihilate repetitive firing in response to a just-superthreshold step, $I_{app} \in (I_\nu, I_1)$, by a superposed shock $I_{app} + I_s\,\delta(t - t_s)$ of appropriate timing t_s and proper amplitude I_s. Theoretically this is seen by numerical calculations (Best [1979], Guttman, Lewis, and Rinzel [1980]). Annihilation has also been demonstrated experimentally for the squid giant axon, as in Fig. 3 from Guttman et al [1980] (also Chapman, personal communication) as well as for other preparations (Jalife [1979]). These features suggest an efficient mechanism by which to control

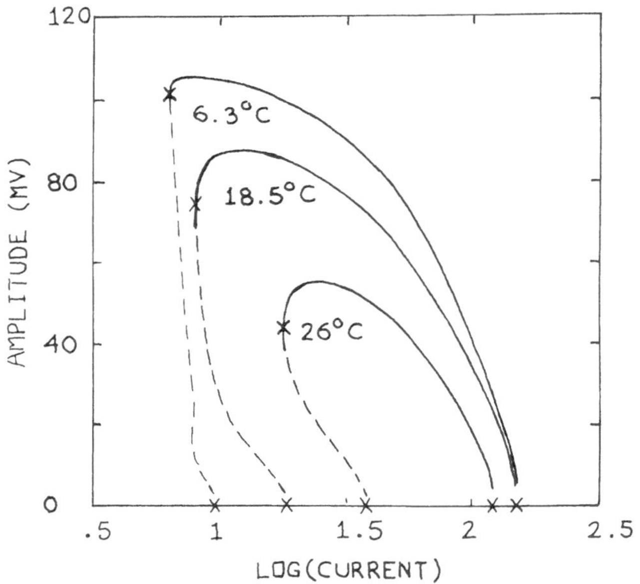

Fig. 2. Amplitude ($\max_t V - \min_t V$) of periodic solutions to HH equations as functions of applied current for three temperatures. Dashed portions correspond to unstable limit cycles. Points of Hopf bifurcation and knees on amplitude curves for neutrally stable limit cycles are indicated by X (from Rinzel and Miller [1980]).

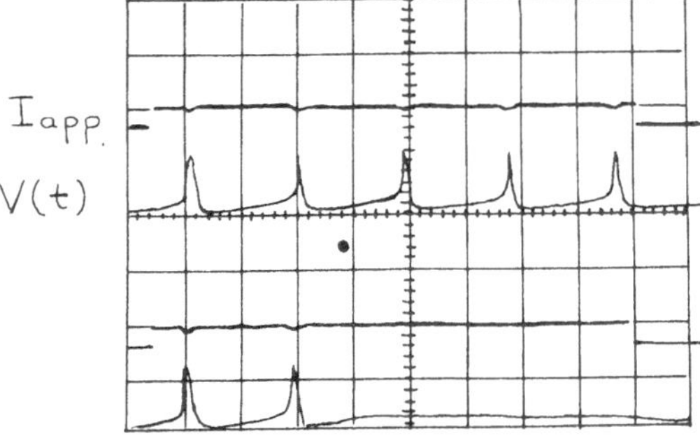

Fig. 3. Annihilation of repetitive firing of space-clamped squid axon membrane by a depolarizing shock. Upper two traces are control run without the shock (from Guttman et al [1980]).

a neuronal pacemaker; namely, brief, rather than maintained, inputs (e.g., synaptic events) would be used to turn on or turn off a pacemaker. Finally we remark that a reasonably accurate quantitative prediction for I_1 and ω_1, the frequency of the emergent periodic solution, is obtained from the linearized two variable V-n system [Rinzel, 1978b].

C. Repetitive Activity - Bursting

In contrast to the regular beating mode of repetitive activity as described above, some neurons fire with periodic bursts of spikes. Examples are the R15 abdominal ganglion cell of Aplysia (see Plant [1978, 1981] for references), and neurons in the Lobster stomatogastric ganglion (Selverston [1976]). In some cases, the bursting behavior is attributed to the endogenous characteristics of an individual cell, the oscillator hypothesis, while in other situations bursting is thought to arise from mutual coupling by several cells, each of which if isolated would not be a burster, the network hypothesis [Selverston, 1976; Perkel and Mulloney, 1974]. We will sketch examples corresponding to these two different hypotheses.

As mentioned above, repetitive firing of the standard HH model does not provide a mechanism for the oscillator hypothesis. By including additional ionic components, Plant has proposed and subsequently has refined HH-type membrane models which exhibit bursting behavior. Experimental data suggest that bursts are generated by an underlying slow (period = O(sec)), smaller amplitude, oscillation which during its depolarizing (V above rest) phase activates the spiking components. In particular, if the sodium current is eliminated (pharmacologically, by adding the Na^+ channel blocker TTX) then spikes are eliminated and the underlying slow oscillation is revealed. A recent version of Plant's model [Plant, in press] is based on some evidence which suggests that the biophysical basis for this oscillation is a calcium activated potassium current. His membrane description (a simplified version) takes the form

$$C_m\dot{V} = -(g_{Na}^{HH} + g_{Ca})(V - \overline{V}_I) - (g_K^{HH} + g_p)(V - \overline{V}_K) - g_L(V - \overline{V}_L) \tag{2.7}$$

where g_{Na}^{HH}, g_K^{HH} are more-or-less the HH sodium and potassium conductances governed by m, h, n dynamics, g_{Ca} is the calcium current, g_p is the "pacemaker", Ca^{++}-mediated, potassium conductance, and g_L is the constant leakage conductance. The equilibrium batteries (both large and positive) for sodium and calcium are lumped into $\overline{V}_I$. The dynamics for g_{Ca} are given by an activation variable x (analogous to the HH variables m and n): $g_{Ca} = \overline{g}_{Ca}x$ where

$$\dot{x} = (x_\infty(V) - x)/\tau_x . \tag{2.8}$$

The conductance g_p is modelled to depend instantaneously on c, the free calcium concentration just interior to the membrane: $g_p = \overline{g}_p\, c/(K_p + c)$ where K_p is the dissociation constant for the rapid (assumed to be in equilibrium) first order binding of calcium to the "p" channel; a "bound channel" is in the conducting state. Finally to complete the model, free Ca^{++} flows in through the membrane channel and is removed either by internal uptake or by active pumping back outward through the membrane so that

$$\dot{c} = k_{in} x(\overline{V}_{Ca} - V) - k_{out} c \tag{2.9}$$

(see Plant [1978] for parameter values and more detailed motivation of the model). This latter step occurs on a much slower time scale than the other dynamic processes. Thus, to expose the slow oscillation, Plant first sets $g_{Na}^{HH} \equiv 0$, to mimic the application of TTX, and then, allows x and n to relax to their equilibrium values (i.e., $x \equiv x_\infty(V)$, $n \equiv n_\infty(V)$) because of their relatively rapid time scales. The reduced system, with two dynamic variables, V and c (a slow variable), may then be studied in the phase plane. By adjustment of $\overline{g}_{Ca}$ and other parameters, sufficient conditions are found to yield a stable limit cycle solution. When the more realistic time scales are reconsidered and "TTX is removed" from the model, the full V, m, h, n, x, c system exhibits the desired feature, a burst of pulses (with absolute peaks ≈ 30 mV) riding on the depolarizing phase of the underlying oscillation (absolute peak ≈ -40 mV). The biophysical and mathematical understanding of how the spike generation and underlying oscillation are dynamically coupled and interact is still not complete.

To consider the network hypothesis for bursting, imagine two pacemaker cells which are mutually coupled by inhibitory synaptic input. One supposes that while cell (1) is firing it inhibits the activity of cell (2) and then this situation reverses. Such switching, if it continues, then leads to the two cells firing in alternating patterns of bursts. Let us attempt a simple model. Let synaptic current I_s, delivered to (1) by a depolarization in (2) ($V_2 > 0$), be represented as $I_s = g_s(V_2)(V_1 - \overline{V}_s)$ where g_s is the post-synaptic conductance (e.g., a monotone increasing and saturating function of V_2) and $\overline{V}_s$ is the lumped equilibrium battery for inhibitory synaptic currents ($\overline{V}_s < 0$). Next, for simplicity, suppose a cell's firing rate is proportional to "mean" (say in the temporal sense over the duration of a few spikes) membrane potential to be denoted by V here. Then we might consider the model

$$\begin{aligned} \dot{V}_1 &= -V_1 - g_s(V_2)(V_1 - \overline{V}_s) + E_1 \\ \dot{V}_2 &= -V_2 - g_s(V_1)(V_2 - \overline{V}_s) + E_2 \end{aligned} \tag{2.10}$$

where $-V_i$ represents membrane leakage and recovery, and $E_i > 0$ provides constant input for a self-sustained base firing rate. A qualitative phase plane analysis of this simple model reveals, over a range of parameter values, two stable steady states separated by a saddle. This means a steady state is approached with either, (1) firing and (2) relatively quiet, or vice versa, but not a continued oscillatory exchange of activity between (1) and (2). For this model, once a cell starts firing fast enough and gains control there is no way for it to quit firing and relinquish control to its partner. To allow for this, Perkel and Mulloney [1974] include an additional variable to describe how a cell might fatique during repetitive activity and so stop firing. To further enhance the possibility of exchange of firing between (1) and (2), they include a mechanism to mimic postinhibitory rebound (PIR). That is, a cell when released from inhibition may be more excitable (e.g., have a lower threshold) than at rest. This phenomenon bears some resemblance to anodal break reexcitation [FitzHugh, 1961]. Perkel and Mulloney describe numerical simulations with their model and obtain alternating burst patterns; they offer examples even when the individual cells are not self-sustained pacemakers. For purposes of illustration, we might model a single cell, which exhibits PIR, by starting with a simple "integrate-and-fire" description:

$$\dot{V} = -V + I_{app}, \quad V(0) < 1 , \tag{2.11}$$

where $V(t^+) = 0$ if $V(t) = 1$, i.e., the membrane "fires" and its state is reset to rest ($V = 0$) when V reaches the threshold value one. Next include a variable Q to represent an inward (depolarizing) current which slowly increases if the cell experiences sufficient inhibition (say $V \leq V_Q < 0$). For example,

$$\begin{aligned} \dot{V} &= -V + aQ + I_{app} \\ \dot{Q} &= \varepsilon(H(V - V_Q) - Q) \end{aligned} \tag{2.12}$$

where $0 < \varepsilon << 1$, $a > 1$ and $H(\cdot)$ is the Heaviside step function. If such a cell is released from adequate inhibition at $t = 0$ ($I_{app} < 0$ for $t < 0$), then Q will be large enough so that $\dot{V} > 0$, for some time, as Q slowly decays back to zero; this can be seen in the V-Q phase plane. During this stage several "firings" will occur before V falls below threshold. Mutual synaptic inhibition between two such cells could be included and might be expected to lead to bursting although no explicit calculations are presented here for this crude model.

III. RESPONSES OF AN EXCITABLE CABLE

A. Phenomenology and Cable Model

Spatio-temporal differences in membrane potential generate current flows through the (ohmic) intracellular medium. To model this, one usually treats an axon or dendritic branch as a membrane cylinder with x denoting distance along the cylinder, e.g., see [Rinzel, 1978a; Rall, 1976]. If R_i is the intracellular resistivity (ohm-cm) and d is the diameter, then axial current is

$$i_i = -(\pi d^2/4R_i)V_x \quad . \tag{3.1}$$

Conservation of current yields the balance statement

$$C_m V_t = \frac{d}{4R_i} V_{xx} - I_{ion} + I_{app} \quad . \tag{3.2}$$

When this cable equation is non-dimensionalized and the auxilliary variables are included, one obtains a nonlinear parabolic partial differential equation which, in the FHN case, may be written as:

$$\begin{aligned} v_t &= v_{xx} - f(v) - w + I_{app}(x,t) \\ w_t &= \varepsilon(v - \gamma w) \quad . \end{aligned} \tag{3.3}$$

The corresponding HH cable model is analogous.

The most straightforward mathematical problem might be for a uniform cable which has passive membrane with constant ionic conductance in which recovery processes are neglected. Solutions for this model equation

$$v_t = v_{xx} - v \tag{3.4}$$

are well known for either the infinite or finite-length cable (under a variety of boundary conditions). The results have been applied to describe axonal and dendritic subthreshold signals. In the case of some dendrites, the assumption of passive membrane may be reasonable. The mathematical problems are made interesting in such applications by the extensively branched geometry. A good reference on this, as well as cable theory in general, is Rall [1976]. Here we shall be more concerned with the case of excitable membrane.

A meaningful stimulus for Eq. (3.3) is $I_{app}(x,t) = I(t)\delta(x)$ which corresponds to the application of the current I(t) by an intracellular electrode at location x = 0. For simplicity, assume even symmetry about x = 0 and suppose the axon has infinite length. The rest state (for (2.4): v = w = 0 when $I_{app} \equiv 0$) is linearly stable (e.g., see Rinzel [1978a]). For a finite duration square pulse of current I(t), a minimum intensity (threshold) is necessary to generate a single impulse. This strength-duration relation is described in [Rinzel, 1978a, Hille, 1976]. After a transient phase, the impulse travels with constant velocity c_o along the axon; the asymptotic speed and shape are independent of the stimulus. If a second adequate stimulus follows the first, a second impulse is generated. Because it advances into the recovery wake of the first impulse, rather than into a region of resting membrane, its velocity will not in general equal c_o. One might expect, if the nerve is less excitable than at rest, that the trailing pulse travels slower. Indeed, this occurs in many cases. But for some axons, it appears that, for certain ranges of interpulse intervals, the trailing pulse may actually travel faster than the leading pulse. If instead of a finite duration stimulus, one applies a current step, then repetitive firing may result for an adequately-sized step; the firing frequency depends on the value of the constant I. Also analogous to the behavior of the membrane patch, when a slowly increasing I(t) is the stimulus, repetitive firing may occur when I reaches some critical value.

B. Speed of Impulse Propagation

A steadily propagating impulse or impulse train corresponds to a traveling wave solution of Eqs. (3.3), i.e. v and w are functions of the single variable z = x + ct where c is the speed. For $I_{app} = 0$, such a solution satisfies

$$\begin{aligned} cv' &= v'' - f(v) - w \\ cw' &= \varepsilon(v - \gamma w) \; . \end{aligned} \qquad (3.5)$$

In this formulation, c is an unknown which must be determined along with the wave solution. For example, a stable solitary pulse satisfies (3.5) and v, v', and w tend to zero as $|z| \to \infty$ for a unique value of c corresponding to c_o. There may be several types of solutions to (3.5) for different values of c. A useful set of solutions is the family of periodic pulse trains (equally spaced pulses) which may be parametrized by, for example, the spatial period P. For these solutions, the speed c depends on P and this defines the so-called dispersion relation c = c(P). Typically, this relation is multi-valued with at least two different solutions (and speeds) for each P greater than some minimum value of P.

Certain of the waves with slower speeds are evidently unstable solutions to Eqs. (3.3) (with $I_{app} \equiv 0$). As $P \to \infty$, $c(P)$ tends to the speed of a solitary pulse solution. The above statements on existence and stability are verifiable for a simple FHN model in which $f(v)$ is a piecewise linear function [Rinzel and Keller, 1973]. In some parameter ranges, solutions of Eqs. (3.5) (if $0 < \varepsilon << 1$) and corresponding, time-scaled, versions of HH-like equations, may be constructed or their existence demonstrated by singular perturbation methods; for example, see Carpenter [1979] and the references in Rinzel [1980]. In other cases, numerical calculation can provide the solutions. Fig. 4 (from Miller and Rinzel [1981]) shows computed dispersion relations for (regularly spaced) periodic wave train solutions of the HH equations; the dashed lines represent speeds of the fast and slow solitary pulse solutions (the slow one being unstable [Evans and Feroe, 1977]). These numerical dispersion curves, if interpreted to describe approximately the dependence of speed on spacing of an individual pulse in a train, illustrate the degree of slowing for short interpulse intervals. They also reveal that, for certain interpulse spacings, supernormal speed ($c > c_o$) may be expected. The dispersion relation has been employed quantitatively to obtain an approximate kinematic description for changes in impulse speed and spacing during propagation [Rinzel, 1980; Miller and Rinzel, 1981]; singular perturbation techniques on Eqs. (3.5) also lead to a kinematic description [Keener, 1980]. More details on the above, and related propagation results, with additional references to the relevant experimental and mathematical work may be found in [Rinzel, 1978a; Rinzel, 1980; Miller and Rinzel, 1981].

C. Onset of Repetitive Activity

Here we outline an analysis for the onset of repetitive activity in response to the maintained stimulus, $I_{app}(x,t) = I\delta(x)$, where I is constant. This analysis, for the cable case, parallels that for the spaced-clamped case where I_{app} is independent of x and t. As in that previous analysis, we seek a steady state solution to Eqs. (3.3) which depends on I; in this case, the steady state $(\bar{v}(x;I), \bar{w}(x;I))$ depends on x. From the second of (3.3), $\bar{w} = \gamma^{-1}\bar{v}$, and substitution into the first equation yields

$$0 = \bar{v}_{xx} - f(\bar{v}) - \gamma^{-1}\bar{v} + I\delta(x) \qquad (3.6)$$

where $-\infty \leq x \leq \infty$. This equation may be integrated twice and one finds that $\bar{v}(x;I)$ is monotone decreasing with x (tending to zero as $|x| \to \infty$) and $\bar{v}(x;I)$ increases with I; also $\bar{v}(x;I) = \bar{v}(-x;I)$. Next, we question the stability of this steady state solution. We linearize Eq. (3.3) about $(\bar{v},\bar{w})$ and seek a solution to this linear variational equation of the form $(e^{\lambda t} V(x), e^{\lambda t} W(x))$. By eliminating W from the second equation (which is again merely

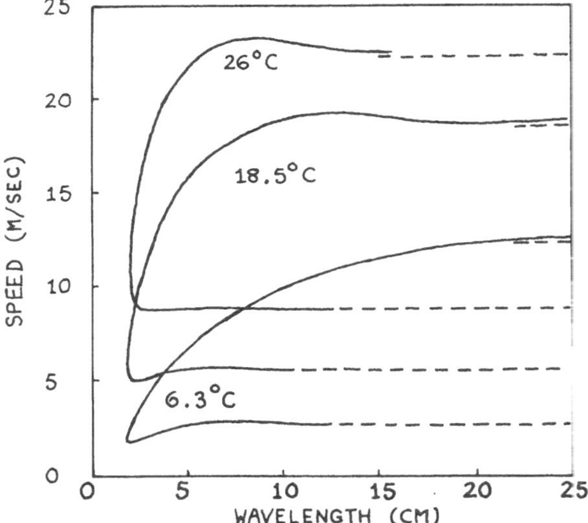

Fig. 4. The relation between propagation speed c and spatial period P (wavelength) for periodic traveling wave solutions to the HH equations (from Miller and Rinzel [1981]).

algebraic) we obtain the following problem for V(x):

$$V_{xx} - [\psi(x) - \mu]V = 0 \;, \; -\infty \leq x \leq \infty \tag{3.7a}$$

where

$$\psi(x) \;=\; f'(\overline{v}(x;I)) \;, \tag{3.7b}$$

and

$$\mu \;=\; \mu(\lambda) \;=\; -\lambda - \varepsilon(\lambda + \varepsilon\gamma)^{-1} \;. \tag{3.7c}$$

Thus the analysis of stability of $(\overline{v}(x;I), \overline{w}(x;I)$ reduces to a Schroedinger eigenvalue problem and we must examine the discrete spectrum μ_j and the corresponding values of λ from (3.7c). If (3.7) has a bounded solution for some λ with $\text{Re}\lambda > 0$, then $(\overline{v},\overline{w})$ is an unstable steady state solution. On the other hand, if bounded solutions to Eqs. (3.7) occur only for values of λ with $\text{Re}\lambda < 0$, then $(\overline{v},\overline{w})$ is linearly stable. Because of the quadratic relation (3.7c) between μ and λ we find that stability ($\text{Re}\lambda < 0$) corresponds to $\mu > -\varepsilon\gamma$. Moreover, at destabilization (i.e., the value of I such that $\mu_o = \min \mu_j = -\varepsilon\gamma$) we have $\text{Re}\lambda = 0$ but $\text{Im}\lambda \neq 0$. Hence we conclude that a branch of periodic solutions emerges through Hopf bifurcation and this corresponds to the onset of repetitive activity at a critical value of I.

To see how destabilization occurs, let us consider the qualitative shape of the "well" $\psi(x)$ in (3.7). For small I, $\psi(x) = f'(\overline{v}(x;I)) > 0$ and so each $\mu_j > 0$ and therefore $(\overline{v},\overline{w})$ is stable. If I is of moderate size, then for some region near $x = 0$, $f'(\overline{v}(x;I)) < 0$ so $\psi(x) < 0$ in that region. If this "well" is wide enough and deep enough then μ_o may drop below $-\varepsilon\gamma$ and $(\overline{v},\overline{w})$ would be unstable. For large values of I, the negative troughs of $\psi(x)$ (one for $x < 0$ and one for $x > 0$) migrate, as I increases, away from $x = 0$. In contrast to the space-clamped case, $(\overline{v},\overline{w})$ may or may not, depending on parameter values, regain stability for large I. As in the spaced-clamped case, the mechanism for instability is the negative resistance of f(v). However, in this cable case, stability depends on how large an x interval has $f' < 0$ and on how close this region is to the electrode. For a specific piecewise linear f(v), which leads to a square-well $\psi(x)$, these stability questions can be answered analytically to a large extent. For details on this and physiological interpretations see Rinzel [Rinzel, 1978c]. The case of a more general N-shaped f(v) has been worked out with J. Keener (in preparation).

D. Propagation Failure at Branch Points

An impulse which approaches a region where there is a sudden, and too large, increase in diameter may fail to propagate beyond the region (as in Fig. 5). In a variety of cases, this is equivalent to what may happen in a region where an axon branches into several daughters. For additional discussion, examples, and references, see Rinzel [1977], and Goldstein and Rall [1974].

Here, we consider such phenomena by studying a simple model problem. We use the FHN model and neglect recovery so that we are studying approximately the propagation of a pulse upstroke. Suppose the axon has diameter d_o for $x < 0$ and $d_1 \geq d_o$ for $x > 0$. Then, with respect to the scaled distance, $z = x\sqrt{d_o}$ for $x < 0$ and $z = x/\sqrt{d_1}$ for $x > 0$, the model equation and matching conditions may be written as

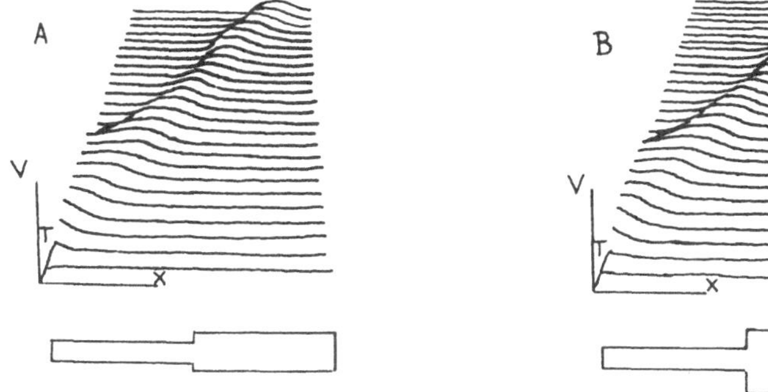

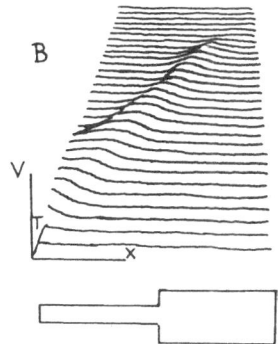

Fig. 5. Propagation of single impulse along FHN axon model with jump in diameter by factor D. Successful propagation beyond jump for D = 1.4 (A) and failure for D = 1.45 (B). (See Rinzel [1977] for model parameter values.)

$$v_t = v_{zz} - f(v) , \tag{3.8a}$$

$$v(0^-,t) = v(0^+,t) , \tag{3.8b}$$

$$v_z(0^-,t) = D^{3/2}\, v_z(0^+,t) \tag{3.8c}$$

where $D = d_1/d_o \geq 1$. Equation (3.8c) is the matching condition for axial current at $x = 0$. Our concern is the fate of a wave which approaches from the far left where now the wave is a transition front satisfying $v \to 1$ as $z \to -\infty$, $v \to 0$ as $z \to \infty$. If D is too large, we expect the front will fail to propagate into $z > 0$. If the upstroke is blocked in this way we expect v to tend toward a (t-independent) steady state as $t \to \infty$ with $v \approx 1$ for $x << 0$ and $v \approx 0$ for $x >> 0$. Below, we show that such a steady state exists if D is large enough.

To further simplify the problem, for this exposition, we take $f(v) = -v + H(v-a)$, where $0 < a < 1/2$. With this choice, a steady state solution to (3.8) may be explicitly constructed, by matching exponentials, provided $D \geq \overline{D} = [(1-a)/a]^{2/3}$. Furthermore, there are two solutions $u(z), U(z)$. They are shown in Fig. 6 and given by

$$\begin{aligned} u(z) &= 1 + (a-1)e^{z+\zeta} && \text{for } z \leq -\zeta \\ &= [(2a-1)e^{z+\zeta} + e^{-z-\zeta}]/2 && -\zeta < z \leq 0 \\ &= u(0^-)e^{-z} && 0 < z , \end{aligned} \tag{3.9a}$$

$$\begin{aligned} U(z) &= ae^{-z+\zeta} && \text{for } \zeta \leq z \\ &= 1 + [(2a-1)e^{-z+\zeta} - e^{z-\zeta}]/2 && 0 \leq z < \zeta \\ &= 1 - [1 - U(0^+)]e^{z} && z < 0 \end{aligned} \tag{3.9b}$$

where

$$e^{2\zeta} = (1-2a)^{-1}\, (D^{3/2} - 1)/(D^{3/2} + 1) . \tag{3.9c}$$

The inequality $D \geq \overline{D}$ follows from the requirement $\zeta \geq 0$ in (3.9c). One can additionally show that $u(z)$ is linearly stable and $U(z)$ is unstable.

The significance of $u(z)$ follows from an application of the maximum principle (e.g., as discussed by Aronson [1976] for problems with discontinuous coefficients; also, see Pauwelussen [1980]) from which we conclude: if $v(z,0) \leq u(z)$ then $v(z,t) \leq u(z)$ for $t > 0$. Thus $u(z)$ blocks propagation into $z > 0$ of any initial data which lies below $u(z)$; this class of data includes $v(z,0)$ which may be close to a front far to the left of $z = 0$. Moreover, because

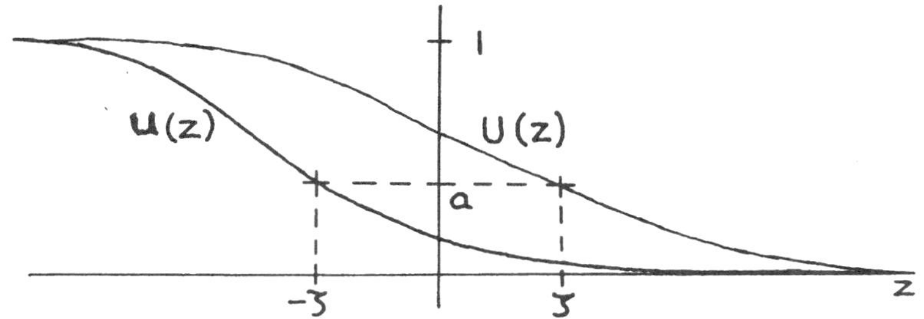

Fig. 6. Steady state solutions to Eq. (3.8) with $f(v) = v - H(v-a)$ when $D > \bar{D}$.

$U(z)$ is unstable, one expects that if $u(z) < v(z,0) < U(z)$ then $v(z,t) \to u(z)$ as $t \to \infty$. Also, if $v(z,0) > U(z)$, then v should eventually approach a front propagating to the far right. Finally, one expects intuitively, since the inhibitory effect of w has been neglected, that the critical value of D for $\varepsilon > 0$ will be less than $\bar{D}$; i.e., less of a diameter jump could be tolerated for $\varepsilon > 0$.

The above analysis was motivated by similar techniques used in population genetics for cline problems [Aronson, 1976]. Rigorous results (existence and stability) on the above nerve conduction problem (3.8) for a general $f(v)$ have also been obtained by J.P. Pauwelussen [1980].

REFERENCES

Aronson, D.G., 1976, "Lectures on Nonlinear Diffusion," Univ. of Houston, unpublished notes.

Barker, J.L., and Smith, T.G., 1979, Three modes of communication in the nervous system, in: "Modulators, Mediators, and Specifiers in Neuronal Function," Y.H. Ehrlich, ed., Plenum Press, N.Y.

Best, E.N., 1979, Null space in the Hodgkin-Huxley equations: a critical test, Biophys. J., 27:87.

Carpenter, G.A., 1979, Bursting phenomena in excitable membranes, SIAM J. Appl. Math., 36:334.

Cooley, J., Dodge, F., and Cohen, H., 1965, Digital computer solutions for excitable membrane models, J. Cell. Comp. Physiol., 66, Suppl. 2:99.

Evans, J.W., and Feroe, J., 1977, Local stability theory of the nerve impulse, Math. Biosci., 37:23.

FitzHugh, R., 1960, Thresholds and plateaus in the Hodgkin-Huxley nerve equations, J. Gen. Physiol., 43:867.
FitzHugh, R., 1961, Impulses and physiological states in models of nerve membrane, Biophys. J., 1:445.
FitzHugh, R., 1969, Mathematical models of excitation and propagation in nerve, in: "Biological Engineering," H.P. Schwan, ed., McGraw-Hill, N.Y.
Goldstein, S.S. and Rall, W., 1974, Changes of action potential shape and velocity for changing core conductor geometry, Biophys. J., 14:731.
Guttman, R., Lewis, S., and Rinzel, J., 1980, Control of repetitive firing in squid axon membrane as a model for a neuron oscillator, J. Physiol. (Lond.), 305:377.
Hassard, B., 1978, Bifurcation of periodic solutions of the Hodgkin-Huxley model for the squid giant axon, J. Theoret. Biol., 71:401.
Hille, B., 1976, Ionic basis of resting and action potentials, in: "Handbook of Physiology, The Nervous System, Vol. I," E.R. Kandel, ed., Am. Physiol. Soc.
Hodgkin, A.L., and Huxley, A.F., 1952, A quantitative description of membrane current and its application to conduction and excitation in nerve, J. Physiol. (Lond.), 117:500.
Jalife, J., and Antzelevitch, C., 1979, Phase-resetting and annihilation of pacemaker activity in cardiac tissue, Science, 206:695.
Keener, J., 1980, Waves in excitable media, SIAM J. Appl. Math., 39:528.
Kuffler, S.W., and Nicholls, J.G., 1977, "From Neuron to Brain," Sinauer, Sunderland, Mass.
Miller, R.N., and Rinzel, J., 1981, The dependence of impulse propagation speed on firing frequency, dispersion, for the Hodgkin-Huxley model, Biophys. J.
Nagumo, J.S., Arimoto, S., and Yoshizawa, S., 1962, An active pulse transmission line simulating nerve axon, Proc. IRE., 50:2061.
Nicholson, C., 1980, Dynamics of the brain cell microenvironment, Neurosciences Research Program Bulletin, 18.
Pauwelussen, J.P., 1980, Nerve impulse propagation in a branching nerve system: a simple model, Technical Report, Mathematics Center, Amsterdam.
Perkel, D.H., and Mulloney, B., 1974, Motor pattern production in reciprocally inhibitory neurons exhibiting postinhibitory rebound, Science, 185:181.
Plant, R.E., 1978, The effects of calcium^{++} on bursting neurons, Biophys. J., 21:217.
Plant, R.E., 1981, Bifurcation and resonance in a model for bursting nerve cells, J. Math. Biology, 11:15.
Rall, W., 1976, Core conductor theory and cable properties of nerve cells, in: "Handbook of Physiology, The Nervous System, Vol. I," E.R. Kandel, ed., Am. Physiol. Soc.
Rinzel, J., and Keller, J.B., 1973, Traveling wave solutions of a nerve conduction equation, Biophys. J., 13:1313.

Rinzel, J., 1977, Repetitive nerve impulse propagation: Numerical results and methods, in: "Nonlinear Diffusion," W.E. Fitzgibbon (III) and H.F. Walker, eds., Research Notes in Mathematics, Pitman.

Rinzel, J., 1978a, Integration and propagation of neuroelectric signals, in: "Studies in Mathematical Biology," S.A. Levin, ed., Math. Assoc. America.

Rinzel, J., 1978b, On repetitive activity in nerve, Federation Proc., 37:2793.

Rinzel, J., 1978c, Repetitive activity and Hopf bifurcation under point-stimulation for a simple FitzHugh-Nagumo nerve conduction model, J. Math. Biology, 5:363.

Rinzel, J., and Miller, R.N., 1980, Numerical calculation of stable and unstable periodic solutions to the Hodgkin-Huxley equations, Math. Biosci., 49:27.

Rinzel, J., 1980, Impulse propagation in excitable systems, in: "Proceedings of Math. Res. Center Advanced Seminar, Dynamics and Modelling of Reactive Systems, Vol. 44," W.E. Stewart, W.H. Ray, and C.C. Conley, eds., Academic Press, N.Y.

Schwartz, J.H., 1980, The transport of substances in nerve cells, Sci. Amer., 242:152.

Selverston, A., 1976, A model system for the study of rhythmic behavior, in: "Simpler Networks and Behavior," J.C. Fentress, ed., Sinauer, Sunderland, Mass.

Troy, W.C., 1978, The bifurcation of periodic solutions in the Hodgkin-Huxley equations, Quart. Appl. Math., 36:73.

NONLINEAR WAVES IN NEURONAL CORTICAL STRUCTURES*

Robert M. Miura

Department of Mathematics
University of British Columbia
Vancouver, B.C., Canada
V6T 1Y4

* This work was supported in part by the National Sciences and Engineering Research Council Canada under Grant No. A4559.

TABLE OF CONTENTS

NONLINEAR WAVES IN NEURONAL CORTICAL STRUCTURES

Robert M. Miura

Department of Mathematics
University of British Columbia
Vancouver, B.C., Canada V6T 1W5

I. PRELIMINARY REMARKS ON MATHEMATICAL MODELS AND MODELLING IN NEUROPHYSIOLOGY

> "the term 'mathematical model' ... will be used for any complete and consistent set of mathematical equations which is thought to correspond to some other entity, its prototype. The prototype may be a physical, biological, social, psychological or conceptual entity, perhaps even another mathematical model ..."

This quote, from Aris' [1978] interesting book on mathematical modelling, gives a basic description of a mathematical model. As Aris points out, mathematical modelling largely remains an art and it is difficult to communicate modelling skills to the uninitiated.

Here we will be concerned with the development of mathematical models (metaphor[1]) for macroscopic phenomena in neuronal cortical structures (a more complete discussion will be given later). In order to give the reader a better idea of the type of mathematical models to be developed, we give a brief perspective view of several different types of mathematical models. In particular, we consider a classification of mathematical models into four basic types - mechanistic models, models of mechanistic models, analog models, and general models. The type of model to be derived or studied in any specific situation will depend on what is already known, on what information (qualitative or quantitative) is desired, and, of course, on the modelling and mathematical abilities of the researcher.

[1] Attributed to the late Professor Henry D. Block, Cornell University.

Mechanistic models are the ultimate goal for understanding a given phenomenon. They incorporate and can be used to test various mechanisms based on physical and chemical laws. Of course, the justification for the validity of a mechanistic model must come from a detailed comparison with experimental data. In practice, except for relatively simple or well understood phenomena, completely mechanistic models are seldom achieved. Some empiricism, ad hoc mechanisms or suitable approximations may be needed to obtain a usable and useful model until such time when there is a better understanding of the phenomena. Furthermore, completely mechanistic models may be undesirable when the number of components in the system become large as we will see with the nerve cells in the brain. For example, a basic understanding of most fluid dynamical phenomena does not require statistical mechanics.

Models of mechanistic models are mathematically simpler systems which are either pared down versions of the mechanistic models or are in some sense a caricature of them. Such models are developed for at least two different reasons. One is to obtain a mathematically simpler system to analyze but which retains some qualitative and perhaps quantitative aspects of the phenomena. A second is to develop mathematical and/or numerical techniques which can be generalized or extended to the mechanistic models.

Analog models are models which are not based on mechanisms but rather obtain from models of similar phenomena or from models that exhibit similar types of solutions. Such models are generally useful in organizing and unifying data but have several distinct disadvantages. Most importantly is that they may not be reliable for predictive purposes. They do not generally lead to an explanation of the phenomenon but can serve as the first step in a realistic modelling program.

Finally, a general model represents a class of mathematical models and is useful for determining some general properties of that class of problems. In this way, one need only test to see if a particular model falls into this class to determine if it has all these general properties.

II. SOME NEUROANATOMICAL FEATURES AND TERMINOLOGY

Before proceeding with a discussion of the modelling of a neurophysiological phenomenon called spreading depression, we will acquaint the reader with some elements of neuroanatomy and terminology which are necessary to understand the presentation here. The reader can consult standard textooks such as Stevens [1966], Miles [1969], Junge [1976], and Kuffler and Nicholls [1976] for more details. An excellent collection of papers on the brain appeared recently in Scientific American [1979].

We will be concerned with phenomena which occur principally in the cortex or outermost layer of a given brain structure. This layer is also referred to as gray matter (see Fig. 1) and consists principally of the soma (cell bodies), of neurons (nerve cells) with their processes (branches), and glia (glial cells). A typical neuronal soma is about 100μ (microns) in diameter [1μ = 10^{-6} meter]. The many branches in the gray matter are called the dendrites and a single branch called the axon leaves the soma to enter the white matter, so called because of the myelin sheath, a fatty substance, covering each axon. Information is sent out from the cell body along the axon.

It is important to point out that the cells sketched in Fig. 1 are not to scale. In man, the typical nerve cell density is 100 cells per 0.001 mm^3, leading to about 10^{11} nerve cells in the brain. Glia typically outnumber neurons by 10 to 1, comprising about half the volume in the brain. Our interest here is focussed mainly on the neurons, partly because glia are thought to play a passive role in spreading depression and partly because not much is known about them and their function.

Important for the discussion here is the synapse, shown circled in Fig. 1, which is one of the principal ways in which nerve cells communicate with each other. Synapses at a given cell are formed with nearby and more distant cells, usually with an axon terminating at a dendrite, e.g., see the collateral axon shown in Fig. 1. There may be as many as 10,000 synapses on a single neuron coming from 1,000 different neurons. The synapse is particularly important for the model developed here and details are shown in Fig. 2. The axonal terminus is called the presynaptic terminal and the dendrite forming the remainder of the synapse is part of the postsynaptic neuron. Note that the presynaptic terminal and postsynaptic neuron are not in physical contact but are separated by a synaptic cleft which is typically 150 Å (angstrom) wide [1 Å = 10^{-8}centimeter].

Cells are bounded by a cell membrane consisting of a lipid bilayer, typically 5 Å thick. The region inside the cell membrane is called the intracellular space and the region between cells is called the extracellular space. Both the intracellular and extracellular spaces are filled with various charged ions, the most abundant being sodium (Na^+), potassium (K^+), and chloride (Cl^-). This nerve cell membrane is capable of separating ionic charges, and differences in the ion concentrations across the cell membrane lead to a membrane potential. For most neurons, the capacitance is about 1 μfarad. Typically the resting membrane potential is about -70 mV (millivolts), negative inside relative to the outside. When the membrane potential is increased, i.e., moved towards zero membrane potential, a depolarization of the membrane is said to occur, whereas if the membrane potential is further decreased, a hyperpolarization occurs.

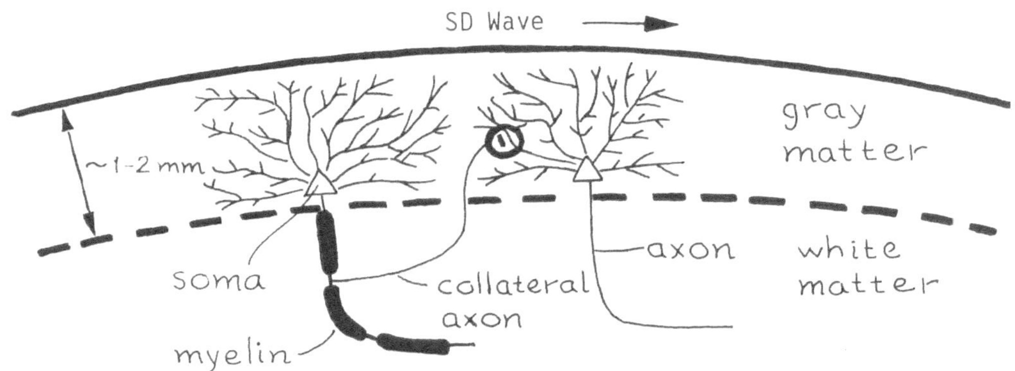

Fig. 1. Schematic sketch of the cortex.

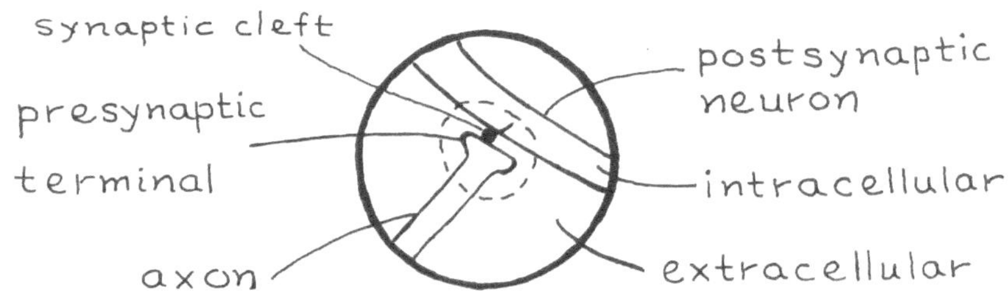

Fig. 2. The synapse.

The negative resting membrane potential occurs in spite of the fact that the membrane contains many open channels allowing passive transport of ions through it. By passive transport we mean that the charged ions are driven through the membrane as a consequence of having two forces acting on them, namely, the electric potential difference and the ion concentration difference across the membrane. The net force on a given ion species is modelled to be proportional to $V - V_N^i$ where V is the membrane potential and V_N^i is the Nernst potential for the given ith ion species. The Nernst potential of an ion species is the electric potential difference across the membrane necessary to maintain a given ion concentration difference. Formulas for these quantities will be given later. The membrane potential depends collectively on the concentrations of all ions on the two sides of the membrane while the Nernst potential for each ion species depends only on the concentrations of that ion species. As is clear on physical grounds and which we shall see later, one solution is that the concentrations of each ion species are equal on the two sides. Why is this not the normal resting state of the membrane? It is generally accepted that active transport mechanisms (pumps) which consume energy are contained within the membrane to transport ions from one side to the other, possibly against the electro-diffusion gradient.

Some of the passive channels are sensitive to membrane potential in a nonlinear way and thus the membrane conductivity to a given ion species can vary. These conductivity changes lead to the well studied action potential and, as we shall see here, can account for spreading depression. The action potential is the information carrying wave[2] which propagates down the axon. It corresponds to a propagating rapid depolarization of the axonal membrane, followed by a slower and lesser hyperpolarization, and then an even slower recovery to the resting membrane potential. All of this occurs in about 1 millisecond. The speeds of the wave vary from .1 to 100 meters/second.

Signalling across the synapse is initiated by the arrival of an action potential at the presynaptic terminal causing a local depolarization. The depolarization opens up calcium channels and the resulting influx of Ca^{2+} (calcium) ions moves vesicles (small packets) of neurotransmitter substance from the interior of the presynaptic terminal to the membrane wall adjacent to the synaptic cleft. These packets merge with the membrane and spill their contents of chemicals into the synaptic cleft. This neurotransmitter

[2] Although we are familiar with water waves, the general concept of a wave is difficult to define. The best description I have heard of to date is from J. Scott Russell [1844] who described a wave as the motion of the motion!

substance is then free to diffuse across and along the synaptic cleft and attaches to receptor sites on the postsynaptic membrane. Activation of these sites result in increases or decreases in the membrane conductivity to certain ion species.

One other concept which may be useful for the reader is that of an excitable medium. Aidley [1978] describes an "excitable cell" as a "cell which readily and rapidly responds to suitable stimuli, and in which the response includes a fairly rapid electrical change at the cell membrane." From a more mathematical point of view, an excitable medium is one in which the behaviour of the system is markedly different depending on the strength (and duration) of the stimulus. For a sufficiently weak (and short) stimulus, the system returns directly to the rest state when the stimulus is removed. If the stimulus is strong (and long), then the system deviates considerably from the rest state before returning to it. This introduces the concept of a threshold, although, in general, the threshold is not sharply defined. From the point of view of propagation, an excitable medium will either have a finite amplitude propagating wave far from the stimulus or it will not depending on the strength and duration of the stimulus. therefore, the word "excitable" is used in both physiological and mathematical senses but they do not correspond to the same ideas. The physiological usage is more a psychological description whereas the mathematical usage is more a qualitative description.

III. A DESCRIPTION OF SPREADING DEPRESSION

A. Introduction

Here we give a description of the main attributes of the neurophysiological wave phenomenon of concern in this paper, namely, spreading depression (SD for short). The reader is referred to Bureš et al [1974] for a comprehensive discussion of SD up to 1974 and to Nicholson and Kraig [1981] for a more recent summary. The remainder of this paper is mainly based on the paper by Tuckwell and Miura [1978] and the reader is referred to that paper for further references.

SD was discovered in experimental studies of epilepsy by the Brazilian physiologist A.A.P. Leão [1944] during his Ph.D. thesis work at Harvard. His studies were carried out on the rabbit. SD is a slowly travelling wave phenomenon which has been observed in a variety of brain structures, e.g., neocortex, hippocampus, cerebellum, retina, and olfactory bulb, in a variety of animals, e.g., rat, cat, catfish, and alligator.

The name "spreading depression" comes from Leão's observation that a wave of depression of neuronal activity propagates outward from the region where the wave is instigated. A direct manifestation of this is a depressed electroencephalogram (EEG), see Fig. 3. Note that the time scale of the phenomenon is on the order of a minute. SD travels in a speed range which varies by an order of magnitude, from .5 mm/min. in the catfish cerebellum (Kraig and Nicholson [1978]) to 15 mm/min. in the rat cerebellum (Nicholson, private communication). However, the speed range in a given brain structure for a particular animal, e.g., neocortex of the rat, is much more restricted. In fact, this makes this phenomenon particularly interesting because the speed of an SD wave appears to be independent of the method of instigation of the wave. (This should be contrasted with certain other dynamical systems in which the wave speed is dependent on the manner of starting the wave.) Since the time scale is on the order of a minute, the space scale is on the order of millimeters. Therefore, SD occurs in populations of neurons and any study of this phenomenon must take this into account.

B. Phenomenological Description of SD

The set-up for a typical SD experiment is illustrated in Fig. 4, which shows the top view of half of the cerebral cortex of a rat brain. An initial stimulus at A produces a wave or waves of SD spreading outward from A. Various types of electrodes placed at B, some distance from A, then record details of the phenomenon as it passes by. (We have already mentioned one of these, the EEG, which requires two neighboring electrodes measuring the synchrony of neuronal activity in neighboring regions.)

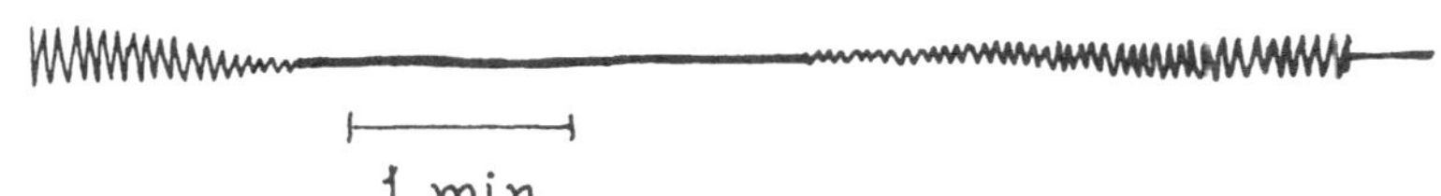

Fig. 3. Depressed electroencephalogram (EEG) associated with spreading depression (SD).

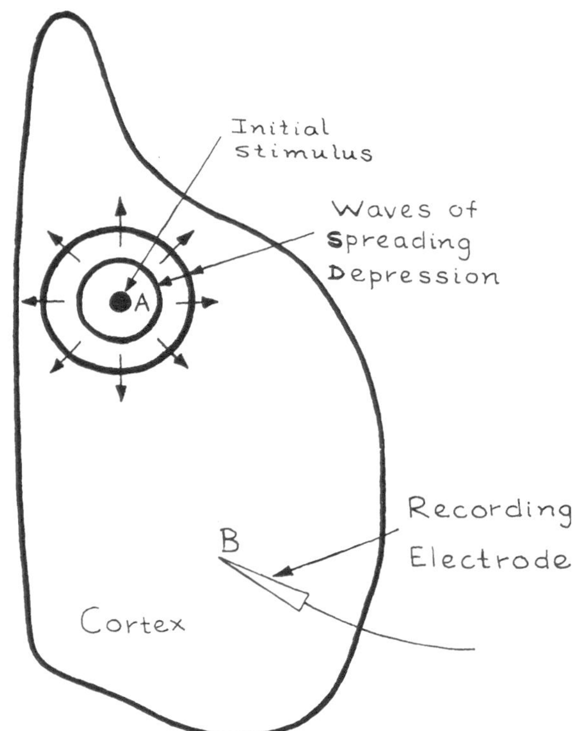

Fig. 4. Set-up for a typical SD experiment.

The stimuli which instigate waves of SD are also varied; among these are: a) application of various chemicals, e.g., KCl, glutamate (a suspected neurotransmitter), and ACTH (a hormone); b) mechanical means, e.g. by dropping a solid object on the cortex surface or by creating a wound in the cortex; c) application of an electrical stimulus, d.c. or a.c. currents at various amplitudes, intervals, and frequencies; and d) evoked responses, i.e., stimulation of the brain in one location causing a stimulation of the brain at another location. One of the major effects common to all these methods of stimulation is a large local change in the ion concentrations. It is presumably this common feature resulting from the various stimuli that leads to SD waves which propagate independently of the method of instigation.

In the depressed region a variety of changes occur. Two of these changes are: 1) the neuronal membrane undergoes a depolarization to zero membrane potential (see Fig. 5), and 2) there is a large increase in the extracellular potassium (K^+) ion concentration accompanied by large decreases in the sodium (Na^+), chloride (Cl^-), and calcium (Ca^{2+}) ion concentrations. The large changes in ion concentrations are particularly important for the model developed here; specifically, we present a simplified model in which K^+ and Ca^{2+} are the essential ions, see Fig. 6 taken from Nicholson et al [1977].

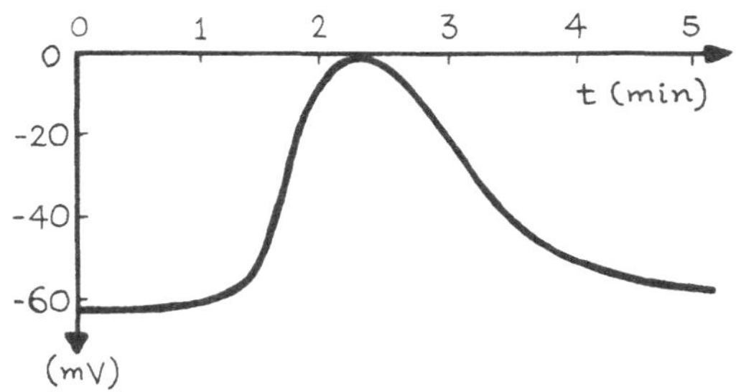

Fig. 5. Neuronal membrane potential associated with SD.

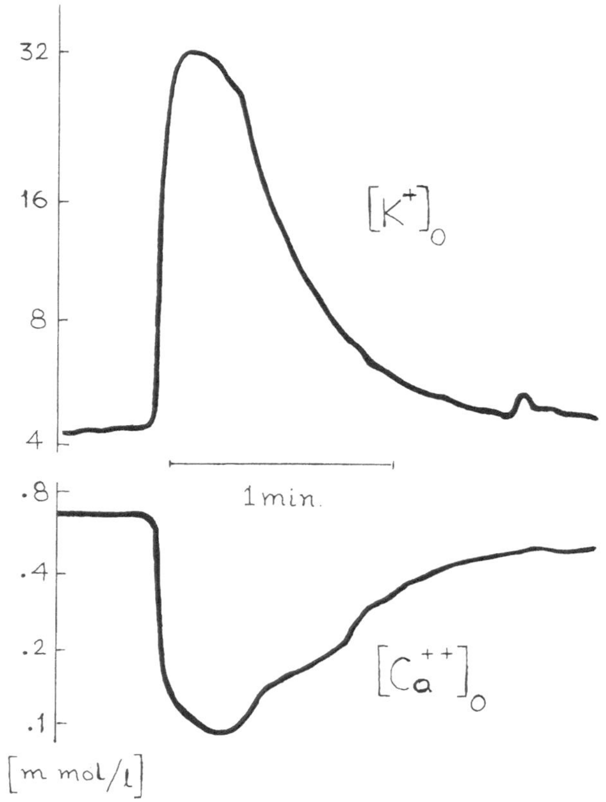

Fig. 6. Extracellular ion concentrations associated with SD.

C. Comparison with Action Potentials

As mentioned earlier, SD occurs in populations of neurons and yet it has a number of characteristics which are similar to propagated action potentials in neuronal axons, as well as waves in other excitable media. We point out some of these similarities and, of course, there are differences which must be accounted for in any mathematical model of SD. Both action potentials and SD waves involve ion fluxes through neuronal membrane.

We first give a brief description of the action potential so the comparisons given later are more easily understood. The discussion will focus on an action potential propagating on an unmyelinated axon since this represents the kind of wave most analogous to a one-dimensional SD wave. The propagated action potential corresponds to a moving pulse in the membrane potential, V. At a fixed position, one sees the following chain of events as the pulse travels by. Initially, the membrane potential is at its resting level, it rapidly depolarizes to nearly the sodium Nernst potential, then moves not quite as rapidly to nearly the potassium Nernst potential, and finally undergoes a slower recovery back to the resting membrane potential. We will not go into the detailed mechanisms causing these specific membrane potential changes; the reader is referred to Miles [1969]. At a fixed location on the axon, the duration of the main depolarization and movement towards the potassium Nernst potential is on the order of one millisecond. As mentioned earlier, the wave speed has a wide range depending on the axon diameters and membrane properties but typically is on the order of tens of meters per second. This leads to a length scale on the order of centimeters.

We now list a number of similarities between action potentials and SD waves: 1) Both waves involve ion fluxes through neuronal membrane; action potential waves have ions passing through axonal membrane and as we claim below, SD waves have most of the ions passing through postsynaptic membrane. 2) In terms of the wave properties, both phenomena are approximately all or none phenomena, i.e., either a wave is generated from an initial stimulus or it is not generated. 3) There are either solitary waves or multiple waves, depending on the strength and duration of the stimulus. 4) There is a depolarization of neuronal membrane in both waves although the action potential depolarization is due mainly to a capacitive current whereas SD depolarization is caused by large changes in ion concentrations. 5) There is both absolute and relative refractory periods in both waves. Refractoriness means that after a pulse has been generated, some time must pass before a second pulse can be created. The absolute refractory period means no matter how large the second stimulus is, no second wave is generated during this time. A large enough stimulus in the relative refractory period can generate a second pulse. 6) If two waves are fired at each other, they annihilate one another upon collision. (This is in sharp contrast with solitons which preserve their shapes and speed following a collision.) 7) The wave shapes, amplitude, and speeds are independent of the method of generating them.

These similarities might lead one to attempt to model SD by adapting models already derived for action potentials. Such an approach is completely inappropriate because the following essential differences between action potentials and SD waves must be taken into account: 1) The time scale for SD is much longer than for action potentials whereas the length scales for SD are much shorter than for action potentials. 2) The ion fluxes through the membrane are through postsynaptic membrane in SD (as explained later) and through nonsynaptic membrane in action potentials. 3) There are substantial changes in extracellular ion concentrations in SD whereas the ion concentration changes during an action potential are negligible. 4) Active transport mechanisms are essential for recovery of the SD waves but not for action potential waves. 5) SD waves have interesting phenomena in one, two, or three space dimensions but action potentials are one dimensional phenomena.

Of course, the fundamental difference between these phenomena is that action potentials occur in isolated axons whereas SD is a collective phenomenon involving populations of neurons.

IV. PREVIOUSLY PROPOSED THEORIES

Although SD has been studied experimentally since 1944, no completely satisfactory theory has yet been proposed. Here we mention some previously proposed theories which are all basically analog-type models. They all suffer from not being based on physiological mechanisms.

The first theory is due to Grafstein [1963] who proposed that potassium ions were responsible for SD. She claimed that SD was instigated by potassium ions released during neuronal firings which are produced by an external stimulus. This is undoubtedly true to some extent, but to date, no detailed proposals for the instigation mechanisms have been given. The main part of Grafstein's theory concerned the propagation of SD where she proposed that the diffusion of K^+ away from a region of high concentration caused neuronal firing in neighboring cells which in turn released more K^+ into the extracellular space. This process then continues and causes the spread of SD. We believe that diffusion of ions does play an important role in SD but that neuronal firings play a minor role at best. The model she used was proposed by Hodgkin and solved by Huxley in private communications. It consisted of the single nonlinear diffusion equation

$$\frac{\partial K}{\partial t} = \frac{\partial^2 K}{\partial x^2} + f(K) \qquad (4.1)$$

where K represents the local concentration of K^+, x and t are the spatial and temporal variables, D is the constant diffusion coefficient, and f(K) is a nonlinear function of K which represents the nonlinear response of the nerve cell aggregate to the locally increased K^+ concentration. The function f has a cubic-like shape and therefore the only stable steady solutions K are wavefront solutions which propagate at a constant velocity but asymptotically approach a value as $x \to \infty$ which is different from the value as $x \to -\infty$ (Fife [1977]). The model can be modified to

$$\frac{\partial K}{\partial t} = D\frac{\partial^2 K}{\partial x^2} + f(K) - \varepsilon\int_0^t K(t')dt', \quad \varepsilon > 0 \qquad (4.2)$$

which is the integro-differential form of the FitzHugh-Nagumo equation. The integral term represents a recovery mechanism which returns K back to its resting level. As noted earlier, these models suffer from the lack of incorporating physiological mechanisms.

Shibata and Bureš [1972] proposed that the theory by Wiener and Rosenbluth [1946] for impulses in cardiac muscle might describe SD. This was a discrete model and is analogous to a later model solved in computer simulations by Reshodko and Bureš [1975].

The model developed here is due to Tuckwell and the author [1978] and does attempt to incorporate proposed neurophysiological mechanisms.

V. CONJECTURED MECHANISMS FOR SD

Before deriving a mathematical model for SD, we need to propose a plausible physiological model taking into account mechanisms believed to be important in the propagation of SD. Such mechanisms should account for a variety of the observed experimental phenomena. Ultimate justification of these mechanisms must come from quantitative agreement between the experimental results and solutions to the mathematical model. As we have already indicated, the diffusion of K^+ ions probably plays an important role in the propagation of SD. However, experiments by Sugaya et al [1975] showed that the neuronal firing in neighboring cells proposed by Grafstein does not play an essential role in SD. In particular, they propagated an

SD wave into and through a region of cortex treated with tetrodotoxin (TTX), a poison from the balloon fish which is known to prevent neurons from firing.

Without neuronal firings, a new mechanism for release of potassium is needed. Tuckwell and the author conjectured that synaptic mechanisms causing the release of potassium were triggered by means other than action potentials. Some evidence for this conjecture are: a) SD is blocked by the divalent ions Mg^{2+} which are known to block synaptic release of neurotransmitter substance by preventing the inward Ca^{2+} currents needed for this release. b) As we shall see later, Ca^{2+} plays an important role in synaptic mechanisms, however, in high concentrations it too can block SD. Possible explanations are that when introduced as $CaCl_2$, there may be a "chloride block", or that high Ca^{2+} concentration provides enough Ca^{2+} ions to immediately replenish Ca^{2+} ions removed from postsynaptic receptor sites by neurotransmitters, rendering it ineffective. c) TTX does not alter the postsynaptic conductance changes effected by the suspected neurotransmitter L-glutamate and it does not interfere with nonsynaptic K^+ conductance. d) It has been observed that the amplitude of the SD wave varies with depth in the cortex. The maximum amplitude occurs near the top of the cortex, occupied mainly by apical dendrites, those dendrites extending upwards towards the surface of the cortex. These dendrites have a high density of synapses, thus providing strong evidence for the importance of synaptic mechanisms.

As already mentioned, normal synaptic mechanisms operate by depolarization of the presynaptic terminal upon the arrival of an action potential. Without neuronal firings, the synaptic mechanisms could operate by another method of presynaptic depolarization. We conjecture that an initially high extracellular concentration of K^+ in the instigation region will diffuse outward and provide this depolarization. Neurotransmitter is then released, more K^+ is released from postsynaptic neurons, and these local ion concentration changes propagate through the cortex in the form of an SD wave. Of course, this is an overly simplified view of the SD wave and in reality many other competing events are occurring simultaneously.

Alternatively, one could conjecture that the synaptic mechanisms are bypassed completely by a large diffusing population of neurotransmitter substance, i.e., the primary chemical involved in SD is not K^+. Such a proposal has been put forth by Van Harreveld [1959], who proposed glutamate, a suspected neurotransmitter, as the primary chemical. Obviously, there would be concomitant ion concentration changes and experimentally it would be difficult to distinguish which chemical was the primary one. We do not pursue this controversial alternative but rather concentrate on the changes and movement of K^+ ions.

VI. DERIVATION OF THE MATHEMATICAL MODEL

It is clear that our conjectured mechanism for the propagation of SD is incomplete. In this section, we will clarify and quantify the various mechanisms which are believed to be essential to a basic understanding of SD. As will become evident, numercus simplifications and approximations will be used to arrive at a "reasonable" mathematical model.

A. Continuum Approximation

The main difficulty in dealing with the brain is the large number of cells. To account for each individual cell and its interactions with neighboring cells is more difficult than the basic problem of statistical mechanics which treats the interactions of many idealized particles. However, we can introduce a major simplification by treating the cortex as a continuum of overlapping extracellular and intracellular spaces. This means that at each point in the mathematical cortex, we can identify the local concentrations of the extracellular and intracellular ions separately. Such an approximation is valid since we have already indicated that spatial variations occur on the millimeter scale and involve large populations of neurons. Thus, neighboring neurons "see" essentially the same extracellular and intracellular ion concentrations.

On the scale of the neurons, the ion concentrations are "slowly varying". To interpret the overlapping of extracellular and intracellular spaces, we mean that at a point in the real cortex, the extracellular ion concentrations are the values at the nearest extracellular points and the intracellular ion concentrations are the values at the nearest intracellular points. Neither the extracellular nor the intracellular spaces occupy the entire volume of the cortex so we must also take into account the relative ratio of these volumes.

Although we are in some sense smearing out the neurons by making the continuum approximation, we do take into account mechanisms which occur at the neuronal level. Note further that this type of continuum modelling of the cortex does not involve starting with networks of neurons. It does have the limitation of being restricted to phenomenon which are on a space scale which is large compared with the scale of individual neurons.

B. Diffusion of Ions

One of the important mechanisms to be modelled is the diffusion of ions in the extracellular and intracellular spaces. We set up a three-dimensional cartesian coordinate system with two dimensions parallel to the cortex surface and one dimension perpendicular to it.

Properties of the cortex in the two directions parallel to the surface can be treated as homogeneous, but vary in the direction perpendicular to the surface. In particular, there is anisotropic diffusion of the ions and a simple way to reflect this in the model is to have different diffusion coefficients in the parallel and perpendicular directions to the cortex surface. Furthermore, the extracellular and intracellular media are different and this must be reflected in the diffusion coefficients as well.

Although this anisotropy in cortex properties and the three dimensions are important in the explanation of some of the phenomena associated with SD, our objective here will be to derive a simple model whose solutions exhibit the two basic properties of pulse propagation and annihilation upon collision. Some remarks will be made later about models for other phenomena.

C. One-Dimensional Approximation

Since SD waves propagate mainly in the two directions parallel to the surface with only structure in the third direction, we simplify the model by treating the waves as one dimensional, parallel to the cortex surface. Effectively we are approximating the SD wave by a wave which does not have any structure through the cortex. Furthermore, in this one-dimensional approximation, because the intracellular space of each neuron is not connected with the intracellular spaces of neighboring neurons, there is no diffusion of ions in the intracellular space. An ion which moves from one point in the intracellular space to a distant point must first enter the extracellular space, and then re-enter the intracellular space. This may have to be repeated several times before reaching the distant point.

D. Importance of Potassium and Calcium

In the simplified model to be developed here, we omit the effects of neuronal firings and mainly account for the movements of ions in the extracellular spaces in the dendritic tree. This corresponds to the case of a TTX-treated cortex. These ion movements are not caused by the arrival of action potentials at the presynaptic terminals but rather by a local increase in extracellular potassium from diffusion.

In Fig. 7, we have sketched the subsystem which is conjectured to be important for the propagation of SD waves. The cross-hatched region represents the intracellular space of the presynaptic terminals and the dendrites of the postsynaptic neuron. The sequence of events leading to the spread of a high extracellular potassium concentration is roughly as follows. Suppose we are studying a wave moving from left to right; in other words, there is a high

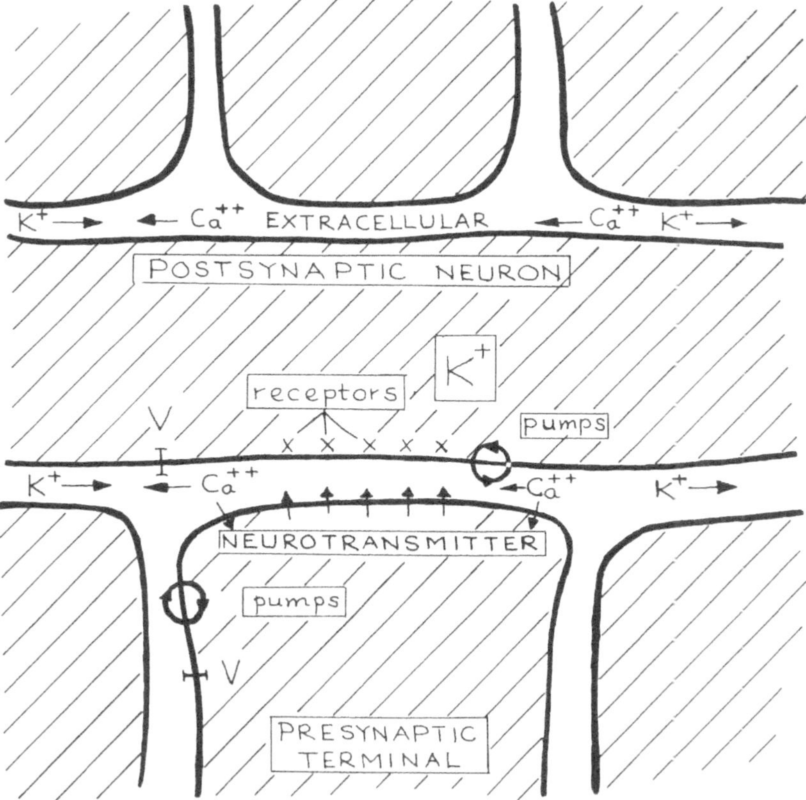

Fig. 7. The subsystem conjectured to be important for the propagation of SD waves.

concentration of potassium on the left and the level is essentially at its resting value in the middle and to the right. Potassium, therefore, diffuses through the extracellular space to the middle of Fig. 7 where the increased extracellular concentration of potassium has the effect of depolarizing the presynaptic and postsynaptic membranes.

To see how this occurs, we assume both membranes have membrane potentials given by the Goldman-Hodgkin-Katz formula

$$V = \frac{RT}{F} \ell n \left[\frac{p_K K^o + p_{Na} Na^o + p_{Cl} Cl^i}{p_K K^i + p_{Na} Na^i + p_{Cl} Cl^o} \right] \qquad (6.1)$$

where R is the universal gas constant, T is the absolute temperature, F is the Faraday, p_K, p_{Na}, p_{Cl} are the membrane permeabilities to K^+, Na^+, Cl^-, respectively, and K^o, Na^o, Cl^o, and K^i, Na^i, Cl^i are the extracellular and intracellular concentrations of K^+, Na^+, and Cl^+, respectively. This formula is valid for a membrane which has a slowly-varying membrane potential relative to the time scale of action potentials, i.e., capacitive currents are not important for SD. In the resting state V is approximately -70 mV so the argument of the logarithm in (6.1) is less than 1. The effect of increasing K^o is to increase this ratio and thereby depolarize V.

Depolarization of the presynaptic terminal membrane causes a local influx of calcium ions which cause neurotransmitter substance to be released into the synaptic cleft. The neurotransmitter substance diffuses across the cleft and attaches to receptor sites on the postsynaptic membrane, causing local increases in the conductance to potassium. Since potassium is in higher concentration inside the neuron, potassium ions pour out into the extracellular space and cause a further increase in the local extracellular potassium concentration. This sequence of events is then repeated further to the right and the SD wave propagates. In this scenario, we have ignored the changes in the other ion concentrations which must accompany the changes in the potassium concentration. These other ions are not expected to greatly modify the picture we have just given, however.

Therefore, a greatly simplified model in which only potassium and calcium are included will be developed here. We include calcium because it is important for the release of neurotransmitter substance and hence the changes in the potassium conductance on the post-synaptic membrane. With a model accounting only for changes in potassium and calcium concentrations, we cannot expect quantitative agreement with experimental data. On the other hand, it would be partially successful if some of the qualitative features can be reproduced, as we shall do below.

The above discussion has concentrated on the leading part of the SD wave and we must also account for the return of the ion concentration levels back to their resting levels. Unlike the action potential, one needs active transport mechanisms for this return. The circles in Fig. 7 which cross the presynaptic and postsynaptic membranes represent energy consuming pumps which transport potassium back into the intracellular space and calcium out to the extracellular space.

In Fig. 8, we have sketched an idealized version of Fig. 7. The vertical bars in the cross-hatched region denote the fact that no free diffusion occurs through the intracellular space. The vertical scale indicates the proportion of the entire space occupied by the extracellular and intracellular spaces. The one space dimension is denoted by x.

E. Nernst Potential

One other concept we will need before writing down the evolution equations for the ion concentrations is the Nernst potential. Consider a single ion species denoted by subscript j with valence z_j and with extracellular and intracellular concentrations denoted by C_j^o and C_j^i, respectively. If $C_j^o > C_j^i$, then for a neuronal membrane, there would be a flow of j ions from the extracellular space into the intracellular space. The Nernst potential given by

$$V_j = \frac{RT}{z_j F} \ln \left[\frac{C_j^o}{C_j^i} \right] \tag{6.2}$$

is the membrane potential needed to prevent this flow of j ions across the membrane. Therefore, the electric potential driving the j ions across the membrane is $V - V_j$. When $V = V_j$, no j ions flow across the membrane due to electrodiffusional forces, although flow could occur from active transport mechanisms.

F. Equations for Spreading Depression

We are now in a position to write down the governing equations. Initially we need not specialize the equations to potassium and calcium. The equations are

$$\frac{\partial C_j^o}{\partial t} = D_j \frac{\partial^2 C_j^o}{\partial x^2} + I_j + P_j \quad , \tag{6.3}$$

$$\frac{\partial C_j^i}{\partial t} = - \frac{\alpha}{1-\alpha} [I_j + P_j] \quad , \tag{6.4}$$

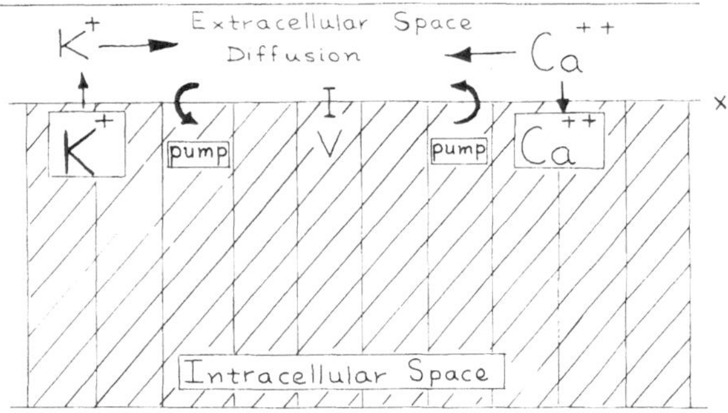

Fig. 8. Idealized version of Fig. 7.

where C_j^o, C_j^i are the extracellular and intracellular concentrations of the j ions, respectively, D_j is the diffusion coefficient, and α is the fraction of the space which is extracellular. The index j = K, Ca, Na, Cl, etc. The time rate of change of the extracellular concentration of ion j is due to three effects: diffusion of ions in the extracellular space represented by the first term on the right side, passive ion transport through the membrane represented by I_j, and active ion transport through the membrane represented by P_j. Since we have assumed no diffusion in the intracellular space, the intracellular concentration of ion j changes only from ion transport through the membrane. Therefore the right side of Eq. (6.4) is identical to the membrane ion transport terms in (6.3), the coefficient $\alpha/(1-\alpha)$ accounts for the difference between the intracellular and extracellular volumes, and the negative sign indicates a gain by the extracellular space is a loss from the intracellular space and vice versa.

Obviously, these equations are incomplete since we have not specified I_j and P_j. The passive ion transport term I_j is modelled by assuming the electrodiffusional forces lead to a current flow across the membrane through an equivalent electrical resistance, i.e.,

$$I_j = \rho_j g_j (V - V_j) \tag{6.5}$$

where ρ_j is a constant and g_j is the conductance of the membrane which will vary because of changes in the membrane potential as well as the presence of neurotransmitter substance.

The rate of release of neurotransmitter substance from the presynaptic terminals is proportional to the calcium current into the terminals. Thus

$$\frac{\partial Tr}{\partial t} \propto g_c(V)(V - V_c) \tag{6.6}$$

where Tr is the concentration of neurotransmitter substance, $g_c(V)$ is the membrane conductance for calcium ions which is assumed to depend only on the membrane potential, and V_c is the calcium Nernst potential. The amount of neurotransmitter which is acting on the receptors on the postsynaptic membrane at a given time is assumed to depend on the amount released in some previous time interval τ, neglecting the diffusion time across the synaptic cleft. After the time interval τ the neurotransmitter substance is rendered ineffective and the receptors return to their rest state. Thus at some spatial point x,

$$Tr^* \propto \int_{t-\tau}^{t} g_c(V)(V - V_c)d\bar{t} \tag{6.7}$$

where Tr* is the amount of effective transmitter at time t and V and V_c depend on $\bar{t}$. For simplicity we assume the postsynaptic membrane conductance is linearly proportional to the amount of effective neurotransmitter

$$g_j \propto Tr^* \quad . \tag{6.8}$$

The above picture has been simplified further by not accounting for the fact that depending on the synapse, neurotransmitter can have either an excitatory or an inhibitory effect on the postsynaptic membrane, i.e., causing the membrane potential to move in a depolarizing or hyperpolarizing direction, respectively. We lump these opposite effects into a single term with a net excitatory effect which is needed to produce a propagating SD wave. A net inhibitory effect from the neurotransmitter would not produce the high levels of potassium needed for an SD wave.

We now specialize the model to include only potassium and calcium as the important ions, keeping the concentrations of sodium and chloride fixed at their resting levels. The extracellular concentrations of sodium and chloride do change considerably during SD but, as mentioned earlier, our aim here is to produce a mathematical model with solutions having the basic qualitative features of SD waves. Therefore, we have four dependent variables K^o, K^i, C^o, C^i corresponding to extracellular and intracellular concentrations of potassium and calcium, respectively.

If we use (6.8) as stated, our system of equations would be integro-differential equations. To avoid this, we take advantage of the fact that τ, the time interval over which neurotransmitter is effective, is very small compared with the characteristic rise time of the phenomenon which is on the order of seconds. Therefore, the integral in (6.7) is approximated by

$$\int_{t-\tau}^{t} g_c(V)(V - V_c)d\bar{t} \simeq \tau g_c(V)(V - V_c) \tag{6.9}$$

where V and V_c on the right side are evaluated at time t to avoid differential-difference equations.

The lack of detailed knowledge of the flow of ions through the membrane requires some empiricism in the model. Specifically we assume the calcium conductance is given in terms of the membrane potential by

$$g_c(V) = g_o\{1 + \tanh[p_g(V + V_g)]\} \tag{6.10}$$

where g_o, p_g, V_g are constants. This corresponds to a sigmoidal function of V which is near zero for $V << -V_g$ and near $2g_o$ for $V >> -V_g$. As V increases through the number $-V_g < 0$, the conductance begins to increase rapidly at a rate measured by $1/P_g$. The saturating behaviour of the conductance is consistent with a picture of the membrane consisting of a large but finite number of pores opening up to allow ions to pass through.

For the pump terms, P_K and P_C for potassium and calcium, respectively, we assume the expressions

$$P_K = P_K(K^o) = f_K\{1 - \exp[-r_K(K^o - K_R^o)]\} + f_K^* \quad , \tag{6.11}$$

$$P_C = P_C(C^i) = f_C\{1 - \exp[-r_C(C^i - C_R^i)]\} + f_C^* \quad , \tag{6.12}$$

where f_K, r_K, f_C, r_C are constants and K_R^o and C_R^i are the resting levels of extracellular potassium and intracellular calcium. Since at the resting membrane potential there is a leakage of both potassium and calcium ions, f_K^* and f_C^* are the pumping rates needed to maintain the resting levels of these ions. One generally associates the active transport of potassium with a coupled active transport of sodium. Here we are holding the sodium concentration fixed, therefore our assumption is that the potassium pump works to maintain the usual extracellular resting level of potassium. Similarly, the calcium pump works to maintain the usual intracellular concentration of calcium. In both cases, we again have saturating behaviour for $K^o >> K_R^o$ and $C^i >> C_R^i$.

Finally, the potassium conductance is taken to be on the postsynaptic membrane and we assume it is proportional to the amount of effective transmitter which in turn is proportional to the calcium current.

Putting all of these ideas together we arrive at the following simplified mathematical model for spreading depression

$$V = A \ln\left[\frac{K^o + \gamma}{K^i + \delta}\right] \quad , \quad A \equiv \frac{RT}{F} \quad , \tag{6.13}$$

$$V_K = A \ln\left[\frac{K^o}{K^i}\right] \quad , \quad V_C = \frac{1}{2} A \ln\left[\frac{C^o}{C^i}\right] \quad , \tag{6.14}$$

$$\frac{\partial K^o}{\partial t} = D_K \frac{\partial^2 K^o}{\partial x^2} + \rho_K g_C(V)(V - V_C)(V - V_K) + P_K \quad , \tag{6.15}$$

$$\frac{\partial K^i}{\partial t} = -\frac{\alpha}{1-\alpha}\left[\rho_K g_C(V)(V-V_C)(V-V_K) + P_K\right] , \qquad (6.16)$$

$$\frac{\partial C^o}{\partial t} = D_C \frac{\partial^2 C^o}{\partial x^2} + \rho_C g_C(V)(V-V_C) + P_C , \qquad (6.17)$$

$$\frac{\partial C^i}{\partial t} = -\frac{\alpha}{1-\alpha}\left[\rho_C g_C(V)(V-V_C) + P_C\right] , \qquad (6.18)$$

where $g_C(V)$, P_K, and P_C are given by (6.10)-(6.12), γ and δ involve the permeabilities and concentrations of sodium and chloride, and ρ_K and ρ_C are constants. The model consists of two nonlinear diffusion equations (6.15), (6.17) coupled with two nonlinear ordinary differential equations (6.16), (6.18). In this regard, it is basically different from the Hodgkin-Huxley equations which consist of one nonlinear diffusion equation coupled with three ordinary differential equations. One other difference is the nonuniqueness of the uniform rest state of the system which is obtained by setting all derivatives identically to zero leaving two algebraic equations to determine four unknowns

$$K_R^o , K_R^i , C_R^o , C_R^i .$$

The nonuniqueness can be avoided in the travelling wave case by specifying the rest state at infinity.

VII. NUMERICAL RESULTS AND CONCLUSION

No serious attempt has yet been made in the mathematical analysis of this system. However, numerical solutions of the equations have shown qualitative agreement with two basic features of SD, namely, solitary wave propagation and annihilation upon collision of two pulses. The results of these computations for extracellular potassium and calcium are shown in Figs. 9 and 10.

In Fig. 9, the potassium level is at a resting value of 2 mM/ℓ except near x = 0 where a large quantity of potassium has been placed. The calcium level is at 1 mM/ℓ. A pulse is soon generated which propagates off to the right at a speed of about 1.5 mm/min for the parameter values chosen in these computations, see Tuckwell and Miura [1978]. Note that the calcium drop and return to the resting level lags behind the rise and fall of potassium. The computation was performed assuming a zero derivative boundary condition at x = 0 so in fact, by symmetry, there is also a wave generated which moves off to the left.

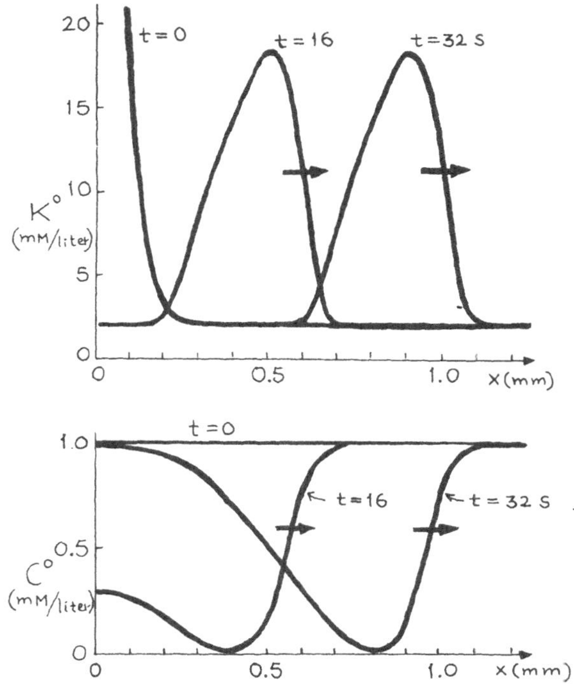

Fig. 9. Time evolution of K^o and C^o ion concentrations illustrating solitary wave propagation.

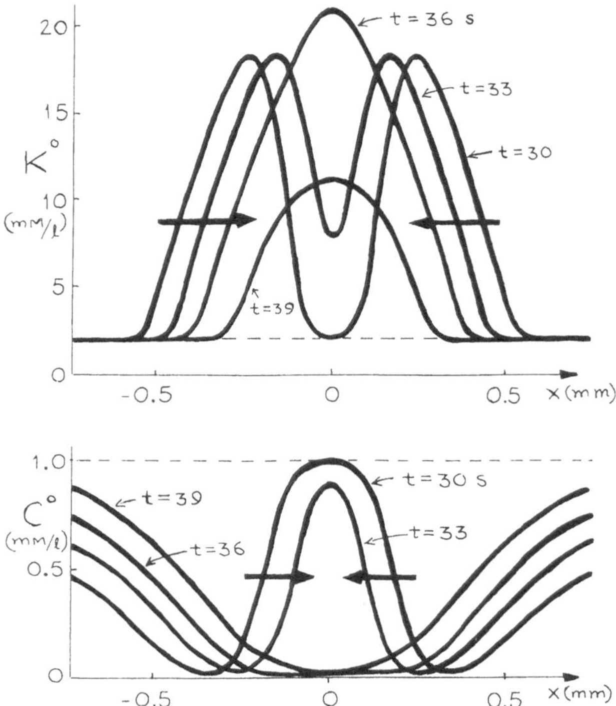

Fig. 10. Time evolution of K^o and C^o ion concentrations illustrating annihilation upon collision of two pulses.

In Fig. 10, we have generated a right moving pulse at $x = -1$ and a left moving pulse at $x = 1$. These waves begin to collide at $t = 30$ and are annihilated upon collision as in the experiments. These pulses do not act like solitons for conservative systems.

In conclusion, two of the basic features of SD waves are qualitatively reproduced by a mathematical model designed on the basis of synaptic depolarizations from increased extracellular potassium and not from action potentials. Such a model represents the starting point for further modelling studies which incorporate more variables, more mechanisms, and new knowledge gained about mechanisms already contained in the model. Also, quantitative comparisons with data must await these more complete models.

Mathematical analyses of these equations do not appear to be simple, and even accurate, speedy computations are not standard for such equations with small diffusion coefficients. Some of the mathematical and computational questions one would hope to answer are:

1. Is the initial-boundary value problem well posed? Does there exist a unique solution which depends continuously on the initial and boundary data?

2. What can be determined about special solutions such as the space clamped solutions (Have any experiments been performed?), steady progressing waves such as the solitary wave and periodic waves, multiple waves generated from a large stimulus, and spiral-like solutions in higher space dimensions?

3. What can be said about the stability of any of the special solutions to spatial and temporal perturbations?

4. How can the parameters be estimated to obtain a good fit with data? Will these tell us something about the neuroanatomy and neurophysiology?

VIII. HIGHER SPACE DIMENSIONS

We mentioned earlier that SD is a wave phenomenon with structure in three space dimensions. In order to account for some of the observed phenomena, one must take into account more than one space dimension. For example, to account for the measured surface d.c. potential, one needs to take into account the variations of ionic currents with depth in the cortex. One could model this for ion j by

$$\frac{\partial c_j^o}{\partial t} = D_j \frac{\partial^2 c_j^o}{\partial x^2} + \tilde{D}_j \frac{\partial^2 c_j^o}{\partial z^2} + I_j + P_j \quad , \tag{8.1}$$

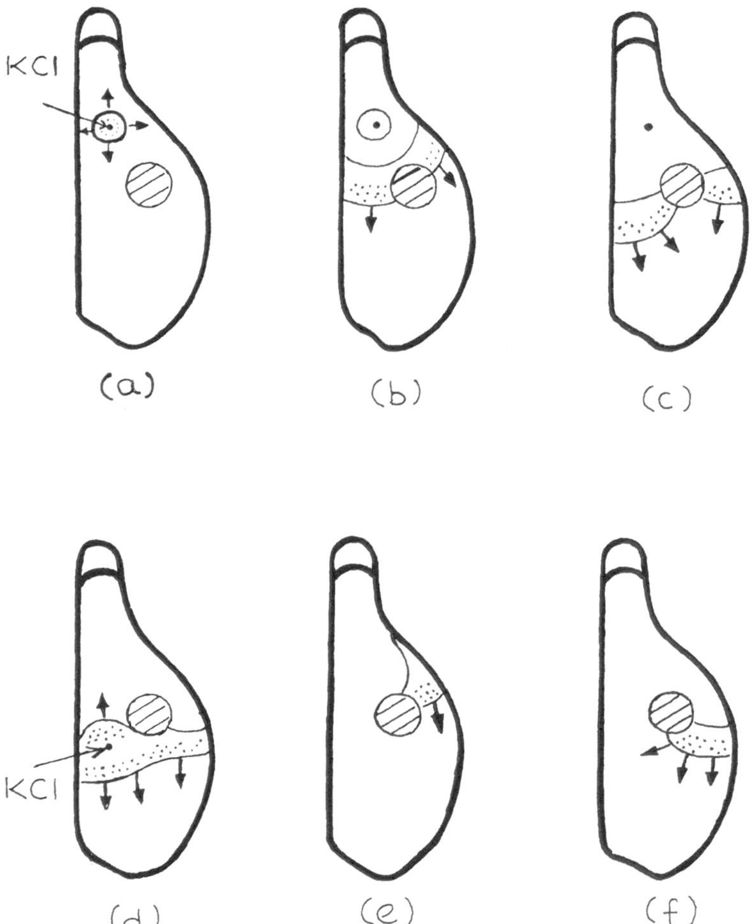

Fig. 11. Rotation of spiral waves around a lesion in the cortex.

$$\frac{\partial C_j^i}{\partial t} = \hat{D}_j \frac{\partial^2 C_j^i}{\partial z^2} - \frac{\alpha}{1-\alpha} [I_j + P_j] \quad , \tag{8.2}$$

where $\tilde{D}_j$ and $\hat{D}_j$ are the diffusion coefficients for ion j in the z direction extracellularly and intracellularly, respectively.

In the experiments, the stimulus is usually applied to a localized region, almost a point stimulus, and therefore the waves propagate outward in a radial direction. To account for such waves a polar coordinate system with axially symmetric solutions could be assumed. This reduces to a one-dimensional problem in space but modified to account for the polar coordinates.

A more interesting two-dimensional phenomenon is the spiral waves which rotate around a lesion in the cortex, a region of dead cells. Such an experiment was carried out by Shibata and Bureš [1974], see Fig. 11. They used an idea proposed by Wiener and Rosenblueth [1946] for generating such spiral waves. First they stimulated an SD wave (a) which propagated outwardly in a radial direction encountering the lesion, the cross-hatched region (b). The SD wave goes around the lesion (c) and upon meeting on the other side they annihilate one another leaving only a wavefront propagating away from the lesion (d). However, immediately behind the SD wavefront, the wave is refractory whereas a little further back it has recovered sufficiently so that it is no longer refractory. A stimulus in the overlap region between these two will generate a half circle wave moving into the nonrefractory region (e). Now no opposing wave exists to annihilate this single wave so it simply continues to rotate about the lesion (f), (g). It has been reported that such waves can survive 55 revolutions before dying out. It is not known why the wave died - whether it is because of metabolic exhaustion or a slow modification of the background state of the cortex which eventually cannot support an SD wave. A model which could be used in the study of such a phenomenon would only have to include two-dimensional diffusion instead of one-dimensional diffusion.

REFERENCES

Aidley, D.J., 1978, "The Physiology of Excitable Cells," 2nd Edition, Cambridge University Press, Cambridge.

Aris, R., 1978, "Mathematical Modelling Techniques," Pitman Publishing Ltd., London.

Bureš, J., Burešova, O., and Křivánek, J., 1974, "The Mechanism and Application of Leão's Spreading Depression of Electroencephalographic Activity," Academic Press, Inc., New York.

Fife, P.C., 1977, Stationary patterns for reaction-diffusion equations, in: "Nonlinear Diffusion," W.E. Fitzgibbon and H.F. Walker, eds., Res. Notes in Math. 14, Pitman, London.

Grafstein, B., 1963, Neuronal release of potassium during spreading depression, in: "Brain Function," Vol. I, M.A.B. Brazier, ed., University of California Press, Berkeley.

Junge, D., 1976, "Nerve and Muscle Excitation," Sinauer Assoc., Inc., Sunderland, Mass.

Kraig, R.P., and Nicholson, C., 1978, Extracellular ionic variations during spreading depression, Neuroscience, 3:1045.

Kuffler, S.W., and Nicholls, J.G., 1976, "From Neuron to Brain," Sinauer Assoc., Inc., Sunderland, Mass.

Leão, A.A.P., 1944, Spreading depression of activity in the cerebral cortex, J. Neurophysiol., 7:359.

Miles, F.A., 1969, "Excitable Cells," William Heinemann Medical Books Ltd., London.

Nicholson, C., and Kraig, R.P., 1981, The behaviour of extracellular ions during spreading depression, in: "The Application of Ion-Selective Electrodes," T. Zeuthen, ed. (in press).

Nicholson, C., Tenbruggencate, G., Steinberg, R., and Stöckle, H., 1977, Calcium modulation in brain extracellular microenvironment demonstrated with ion-selective micropipette, Proc. Natl. Acad. Sci. U.S.A., 74:1287.

Reshodko, L.V., and Bureš, J., 1975, Computer simulation of reverberating spreading depression in a network of cell automata, Biol. Cybern., 18:181.

Russell, J.S., 1844, Report on waves, in: "Rep. 14th Meeting of the British Association for the Advancement of Science," John Murray, London.

Scientific American, 1979, The Brain, Vol. 241 (September).

Shibata, M., and Bureš, J., 1972, Reverberation of cortical spreading depression along closed-loop pathways in rat cerebral cortex, J. Neurophysiol., 35:381.

Shibata, M., and Bureš, J., 1974, Optimal topographic conditions for reverberating cortical spreading depression in cats, J. Neurobiol., 5:107.

Stevens, C.F., 1966, "Neurophysiology: A Primer," John Wiley & Sons, Inc., New York.

Sugaya, E., Takato, M., and Noda, Y., 1975, Neuronal and glial activity during spreading depression in cerebral cortex of cat, J. Neurophysiol., 38:822.

Tuckwell, H.C., and Miura, R.M., 1978, A mathematical model for spreading cortical depression, Biophysical J., 23:257.

Van Harreveld, A., 1959, Compounds in brain extracts causing spreading depression of cerebral cortical activity and contraction of crustacean muscle, J. Neurochem., 3:300.

Wiener, N., and Rosenblueth, A., 1946, The mathematical formulation of the problem of conduction of impulses in a network of connected excitable elements, specifically in cardiac muscle, Arch. Inst. Cardiol. Mex., 16:205.

BIFURCATIONS IN INSECT MORPHOGENESIS I*

Stuart A. Kauffman

Department of Biochemistry and Biophysics
University of Pennsylvania
School of Medicine
Philadelphia, Pennsylvania 19104

* It is a pleasure to thank Drs. Russell, Winfree, Hunt, Kernevez, Herschkowitz-Kaufman and Auchmuty, and M. Weir for many stimulating discussions of these problems. This work was partially supported by grants ACS CD-30 and NIH GM22341.

TABLE OF CONTENTS

BIFURCATIONS IN INSECT MORPHOGENESIS I

Stuart A. Kauffman

Department of Biochemistry and Biophysics
University of Pennsylvania
School of Medicine
Philadelphia, Pennsylvania 19104

I. INTRODUCTION: POSITIONAL INFORMATION

The past decade has witnessed a renewal of deep interest in the problem of pattern formation in developmental biology. In large measure, the resurgence of enthusiasm coincides with Wolpert's reformulation of this fundamental problem in terms of the concept of positional information [Wolpert, 1969, 1971]. Novel features of Wolpert's theory have led both to proposals of alternative "coordinate systems" supplying positional information, and to a rich variety of experiments designed to test the alternatives. The present article provides a brief review of the major alternatives which leads to the formulation of a new class of models based on the unrecognized, but common capacity of biochemical reaction-diffusion systems to generate transverse (cross) gradients in growing asymmetrical tissue domains.

Among the important experimental techniques used to study pattern formation are those which rely on surgical intervention, often coupled with grafting, and analysis of subsequent pattern regulation. Two broad classes of pattern regeneration have long been distinguished; morpholactic, in which a truncated tissue regenerates within its bounds the full range of pattern elements normally located in the entire tissue; and epimorphic, in which a truncated tissue forms a regeneration blastema which undergoes cell proliferation enlarging the initial truncated fragment, and reforms missing pattern elements among the new cells of the regeneration blastema [Morgan, 1901; reviewed by Slack, 1980]. This article focuses on epimorphic regulation.

Prior to Wolpert's introduction of the positional information concept, the dominant theory available postulated the existence of developmental fields [Weiss, 1939] possessing "prepatterns" [Stern, 1968; Tokunaga, 1978], non-uniform spatial distributions of biochemical substances in a tissue, whose local peaks would induce the formation of pattern elements such as specific digits, sensillae, or bristles.

In contrast to Stern [1968], Wolpert [1969, 1971] proposed the more abstract idea that cells within a developmental field possess positional information about their locations with respect to the boundaries of the field, through access to an underlying positional coordinate system. The behavior of each cell in the field was assumed to be due to two independent processes. The cell assesses its position, then interprets its positional information according to the type of cell it is, and forms a specific structural element in the overall pattern.

The chief difference between Wolpert's positional information and Stern's prepattern, is that positional information is free of assumptions about the existence of specific morphogen peaks underlying the subsequent differentiation of specific pattern elements. This freedom in one sense makes the theory of positional information less predictive, yet allows for two important possibilities; first, that the positional information system in all the developmental fields of one organism are identical; and more radically, that the positional information system in all organisms, or all epimorphic fields, is identical.

Strong evidence supports the more restricted hypothesis that a single system of positional information may underlie all the developmental fields of a single organism. A number of well studied homeotic mutants [Ouweneel, 1976] have been found in the fruit fly Drosophila melanogaster. Homeotic mutants replace part or all of one adult ectodermal structure with another. For example, Aristapedia [Roberts, 1964] and Nasobemia [Gehring, 1966] convert part or all of the antenna to a mesothoracic leg, tumorous head converts part or all of the head to posterior abdomen and genitalia, eyeless opthalmoptera [Postlethwait, 1974] converts eye to wing. A particularly striking observation in such homeotic tissue is that even if a very small patch of distal antenna is converted to leg, it is converted to distal leg. Conversely, if proximal antenna is converted to leg, it forms proximal leg [Postlethwait and Schneiderman, 1974; Ouweneel, 1976]. Analysis of such partially transformed appendages leads to serious difficulties for a prepattern model. If even a small adult patch derived from less than a score of cells forms proximal leg bristles in the proximal antenna, then either the small group of preleg cells forms its own small prepattern with appropriate morphogen peaks or the prepattern in the antenna is not specific to the antenna, but a more general appendage prepattern whose peaks no longer correspond to specific positions of antenna, leg, genital or wing pattern elements. The former interpretation has seemed to most workers improbable, while the latter shades smoothly into the concept that the prepattern is some system supplying positional information subsequently interpreted by cells. Hence, if the positional information in an antenna and leg are identical, proximal antennal cells in the antennal tissue will interpret that information and form appropriate antenna pattern elements, while neighboring homeotically transformed leg cells will interpret the same information and form appropriate proximal leg structures [Wolpert, 1971].

The general success of the positional information concept led to a search for the coordinate system which supplies positional information. At present, polar [French et al, 1976; Bryant et al, 1977, 1980], spherical [Russell, 1978], Cartesian [Kauffman, 1978; Slack, 1980; Kauffman and Ling, 1981; Winfree, 1980], and general orthogonal [Cummins and Prothero, 1978] models have been proposed. Differences among these alternatives are far from trivial. Although it is always possible mathematically to transform from one to another coordinate system, the "forces" or tissue properties which must be postulated to explain the observed features of the pattern regulation differ sharply in the different models. While the morphogens have yet to be found, one task in this area of biology consists in efforts to discover the coordinate system and "forces" or requisite cell properties which most simply account for the observations, with the hope that the proper formulation will provide both macroscopic laws at the tissue level, and aid in discovery of the underlying molecular variables.

II. THE PHENOMENA

To assess the relative success of the alternatives which have been proposed, it is necessary to review briefly at least some of the major phenomena of pattern formation and regeneration.

A. Intercalary Regeneration of Intervening Structures

If an amphibian limb capable of regeneration is transected proximal and distal to the elbow, and the distal wrist fragment grafted to the proximal shoulder stump, cell proliferation forms in the wound area, followed eventually by regeneration of the missing elbow region. This basic phenomenon is fairly ubiquitous [Sengel, 1953; Needham, 1965; Stocum, 1975; Slack, 1980] and fundamental. Juxtaposition of normally non-adjacent tissues from a single developmental field is generally followed by regeneration, in proper spatial order, of the structures normally lying between the juxtaposed tissue edges. This notion of "betweenness" is necessarily central to any theory of pattern formation.

The simplest physical model to account for "betweenness" in intercalary regeneration posits the existence of one or more chemical concentration gradients spanning the tissue, whose concentration levels specify the positional information of cells at each point in the domain. As shown in Figure 1, in which a proximal distal gradient along an amphibian limb is envisioned, surgical removal of the elbow and grafting of wrist to shoulder creates a discontinuity in the gradient, [S], at the graft junction. If one imagines that gradient concentrations remain fixed in the "old" tissue fragments, while diffusion occurs in the new cells of the wound blastema, then simple diffusive averaging of the concentration discontinuity at the graft junction smooths over the discontinuity, recreating all the intervening gradient values in proper spatial order.

For the remainder of this paper, I shall adopt the postulate that position is specified by graded scalar properties in tissues, although it is important to stress that alternative discrete models [Apter, 1966], have been formulated. Given the postulate of positional gradients, a fundamental question is the extent to which the simple property of diffusive-like averaging of gradient discontinuities can account for the phenomena of pattern formation and regeneration. This simple property turns out to be sufficiently powerful to account for the bulk, but not all, of the available data.

B. Sequential Formation of Positional Axes in Development

In several systems, positional axes appear to be established sequentially during development. In classical experiments, Harrison [1918, 1921] removed the right forelimb bud of *Amblystoma* and grafted in its stead the left forelimb bud. Such grafts must invert either the anterior-posterior limb axis while keeping donor and host dorsal and ventral axes aligned; or invert dorsal-ventral donor and host axes, while keeping anterior and posterior axes aligned. Harrison found that if very early left limb buds were grafted to the right, they developed into normal right limbs. If late left limb buds were grafted, they formed normal left limbs with that axis inverted with respect to the host that was inverted at surgery. But if left to right grafts were made at an intermediate stage, the outcome depended upon which axis was inverted at the graft junction. If the anterior-posterior axis remained normally aligned and the dorsal-ventral axis was inverted, the donor left limb bud formed a right limb, while if the dorsal-ventral axis remained aligned and the anterior-posterior axis was inverted at the graft junction, the donor left limb bud formed a left limb which remained inverted at the host donor junction. Harrison interpreted his results to imply that the donor anterior-posterior axis becomes autonomously self-sustaining prior to the dorsal-ventral axis. Similar data suggest that amphibian eye [Hunt, 1975; Jacobson, 1968] and ear [Harrison, 1969] axes, as well as limb axes are established sequentially, although the status of the data on the eye is in dispute [Chung and Cooke, 1975].

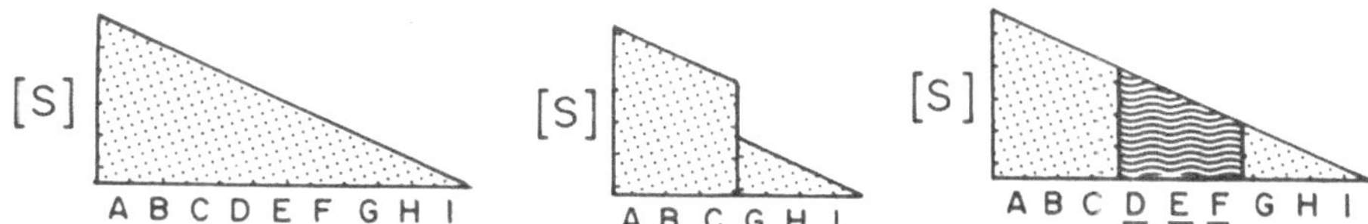

Fig. 1. Gradient in concentration of [S] provides proximal-distal positional information in amphibian limbs. Serial threshold levels specify pattern elements A, B, ...I. Removal of mid region of limb, D, E, F, and grafting, creates discontinuity, which stimulates cell proliferation. Diffusive smoothing of gradient discontinuity regenerates missing gradient levels (wavy lines), and structures DEF.

In a variety of insects, the available evidence suggests that the longitudinal body plan of the entire embryo becomes established prior to fixation of dorsal ventral commitments [reviewed by Sander, 1977].

C. Distal Transformation

If an amphibian limb or tail is transected at the elbow, the proximal stump can form a regeneration blastema and regenerate the distal limb [Harrison, 1918]. If the digits of the transected distal fragment are implanted into a host flank to establish an adequate blood supply to the distal fragment, the cut surface at the elbow, which initially faced proximally, forms a regeneration blastema and regenerates a second distal wrist and hand structures which are mirror symmetrical to the initial implanted distal limb [Dent, 1954; Butler, 1955; Carlson et al, 1974; Harrison, 1918]. In these experiments, both the proximal stump and the implanted distal limb fragment regenerate the *same* set of distal limb structures from the cut surfaces at the elbow, identified as a regenerate in the proximal stump and duplicate on the implanted distal limb fragment. The fact that both fragments form distal limb has been called the rule of distal transformation [Rose, 1962]. Similar results have been found in many insect legs [Bulliere, 1970, 1971; French, 1976; Bohn, 1965] and in the imaginal discs of *Drosophila* [French et al, 1976; Haynie and Schubiger, 1979].

D. Supernumerary Limbs

Among the most striking observations in pattern regulation is the induction of extra or supernumerary limbs following grafting operations. After both the anterior-posterior and dorsal-ventral amphibian limb axes are fixed, transplantation of a left distal limb to a right proximal stump which reverses the anterior-posterior axis of donor relative to host but leaves the dorsal-ventral axis aligned, typically results in formation of two supernumerary limbs at the anterior and posterior margins of the donor host junction. If instead the anterior-posterior axes of host and graft are aligned, but dorsal-ventral axes are inverted, the two supernumeraries emerge from the dorsal and ventral margins of the host donor junction. These supernumerary limbs generally have the handedness of the proximal stump [Harrison, 1918, 1921; reviewed in Bryant and Iten, 1976]. Similar results have been obtained in transplantation of cockroach limbs [Bohn 1965, 1972; French, 1976; Bulliere, 1970].

Rotation of a left distal limb by 180° and regrafting to its own proximal stump results in a more variable range of results. After such rotations, the donor may partially rotate back toward its normal alignment; sometimes no, one, two, or more supernumerary limbs are formed at the graft site, with the same or opposite handedness [Bulliere, 1970; Bohn, 1965, 1972; Bryant and Iten, 1976; French et al, 1976; Milojevic, 1924; Swett, 1924].

E. Duplication and Regeneration by Complementary Fragments of a Tissue

Distal transformation by both proximal and distal limb fragments of amphibians is one example of duplication and regeneration by complementary fragments of a developmental field. The phenomenon is common, however, and has been studied in greatest detail in the imaginal discs of *Drosophila melanogaster*. The bulk of the remaining discussion of data pertaining to pattern regulation will focus on this system.

F. Drosophila Imaginal Discs

Drosophila melanogaster is a holometabolous insect with egg, larva, pupa and adult stages. During metamorphosis, the larval ectoderm lyses, and the ectoderm of the adult is formed by the terminal differentiation of special larval organs called imaginal discs [Gehring and Nöthiger, 1973]. In the late third instar larva, each imaginal disc is a two dimensional sheet of cells forming the surface of a hollow sphere. The columnar cells on one hemisphere form the imaginal disc proper, while the thin squamous cells on the remaining hemisphere form the peripodal membrane which is lost during metamorphosis (ibid.). Imaginal discs are found as bilaterally symmetric pairs, each destined to form specific left and right regions of the adult ectoderm: the left and right first leg discs form the two prothoracic legs; the two wing-thorax discs form the left and right mesothoraces and wings, etc.

By injecting specific fragments of each disc into host larvae for metamorphosis, it has been possible to construct a fate map of each type of imaginal disc [Schubiger 1971; Bryant 1975; Van der Meer and Ouweneel, 1974; Bryant and Hsei, 1977]. For example, the fate map of the wing-thorax (hereafter wing) disc, Figure 2, shows that the upper and lower margins of the disc along its longitudinal axis form ventral and dorsal thoracic structures, while the mid region of the disc forms wing structures. During metamorphosis, the wing disc folds along a bent arc running from the anterior to posterior disc margin, and apposes ventral and dorsal thorax areas, ventral and dorsal wing hinge areas, and ventral and dorsal wing blade areas, creating a "bag" which everts through the peripodal membrane. The center of the disc maps to the distal wing tip, while an arcing line from anterior to posterior disc edge through the disc center corresponds to the wing margin.

Grafting experiments like those in amphibian or cockroach limbs are not yet feasible in *Drosophila*. However, closely analogous experiments can be performed by cutting the wing disc into known

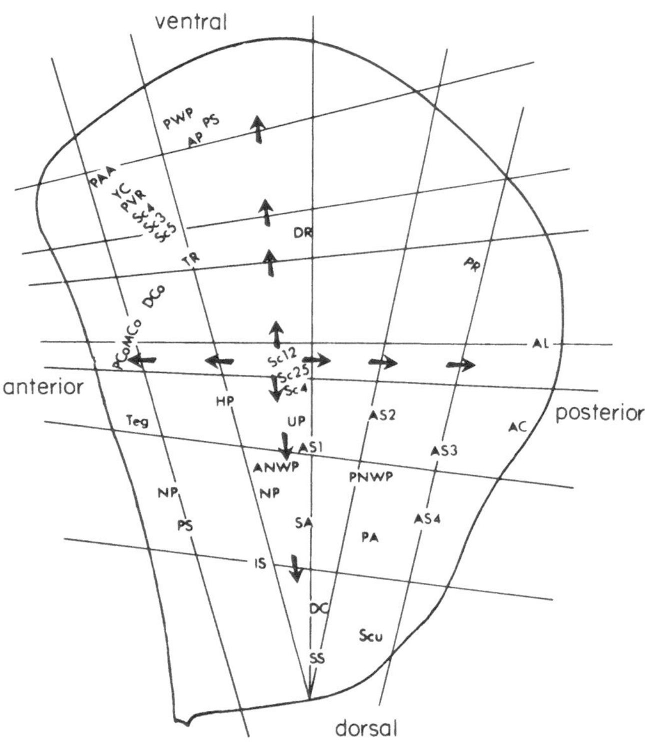

Fig. 2. Fate map of Drosophila wing disc. Lines show positions of single cuts. Arrow across each line points from the fragment created by that cut which regenerates, into the fragment which duplicates. From Bryant [1975]. Abbreviations for wing disc pattern elements: PST, presutural bristle; NP, notopleural bristles; SA, supraalar bristles; PA, postalar bristles; DC, dorsocentral bristles; Scu, scutellar bristles; ANWP, PNWP, anterior and posterior notal wing processes; Teg, tegula; HP, humeral plate; UP, unnamed plate; AS1-4, first to fourth axillary sclerites, Pco, Mco, Dco, proximal, medial and distal costa; TR, triple bristle row (anterior wing margin); DR, double bristle row (distal wing margin); PR, posterior row of hairs; Sc4, Sc25, Sc12, groups of sensilla campaniformia on the proximal dorsal radius; AL, alar lobe; AC, axillary cord; PAA, prealar apophysis; YC, yellow club; PVR, proximal ventral radius; PWP, pleural wing process; PS, pleural sclerite, AP, axillary pouch; Sc4, Sc3, Sc5, groups of sensilla campaniformia on the proximal ventral radius; Anterior, Posterior, Ventral and Dorsal disc margins.

fragments, and injecting each fragment into the abdomen of a fertilized adult female. In that environment, the disc fragment heals its cut edge, apposing tissue regions which are normally non-adjacent [Reinhardt et al, 1977] and new cells form in the wound area. Over a week in culture, the mass of a disc fragment typically doubles. After culture in adult abdomens, fragments may be recovered and injected into host larvae for metamorphosis, then recovered from the emerged host adult. By comparison of the patterns of hairs, sensillae and bristles formed when a known disc subfragment is injected directly into larvae for immediate metamorphosis, it is possible to characterize the pattern regulation which occurs in the cultured fragment.

The following are the dominant results:

a. If the wing disc is cut in two fragments by a straight cut, the roughly smaller fragment duplicates some or all of its structures, in favorable cases yielding a mirror symmetrical duplicate whose symmetry line lies near the cut on the wing disc; the larger complementary fragment regenerates structures normally formed by its smaller complement [Bryant, 1975]. Thus complementary fragments often exhibit complementary behavior, one duplicating, one regenerating. Consequently, it is possible to draw an arrow across each single straight cut on the disc, pointing from the fragment which regenerates into that which duplicates. As shown in Figure 2, such arrows point radially outward from a small region in the interior of the disc. The polarity of regeneration and duplication reverses about this interior point, termed the "high point".

b. If the disc is cut into arbitrary 3/4 and 1/4 "pie" sectors, the 3/4 fragment regenerates, the 1/4 fragment duplicates [Bryant et al, 1977].

c. If an interior "distal" circular region containing the "high point" is cut and cultured, it duplicates, the duplicate forming a mirror symmetric hemisphere to the original high point region. If the outer "proximal" annulus remaining after the high point region is removed is cultured, the outer annulus regenerates the central "distal" high point region [Bryant, 1975; Haynie and Schubiger, 1979].

d. If two narrow normally duplicating crescent fragments from opposite edges of the disc are mixed, they regenerate the intervening pattern elements spanning across the disc [Haynie and Bryant, 1976].

e. Complementary fragments do not strictly duplicate and regenerate. Under different experimental conditions, and with varying frequencies normally duplicating fragments can regenerate most of the pattern elements in the complementary fragment [Duranceau, 1977; Haynie and Schubiger, 1979; Bryant, 1975; Kauffman and Ling, 1981; Krivi and Schneiderman, 1980].

f. Anterior-posterior and ventral-dorsal asymmetries exist in the capacity of narrow duplicating fragments to regenerate: narrow anterior margin and ventral thorax margin structures have been found by a number of workers to regenerate extensively, while narrow posterior and dorsal margin fragments do not [Duranceau, 1977; Haynie and Schubiger, 1979; Kauffman and Ling, 1981]. The results in other imaginal discs are fundamentally similar [Van der Meer and Ouweneel, 1974; Schubiger, 1971; Bryant and Hsei, 1977], although an interior "high point" has not been demonstrated.

g. Mutants causing pattern duplication in different imaginal discs have been analyzed by a number of workers [Jürgens and Gateff, 1979; Bryant and Schubiger, 1971] and will be discussed further below.

III. MODELS OF POSITIONAL COORDINATE SYSTEMS

A. Polar Coordinate Model

The first major advance in predictive use of the Positional Information hypothesis lay in the formulation of the polar coordinate, or "clockface" model for pattern regulation in epimorphic fields by French et al, [1976]. The model was based on some of the results described in amphibian limbs, cockroach limbs, and imaginal discs, and is well illustrated by application to the wing disc of *Drosophila*.

The existence of a "high point" in the central (distal) region of the disc, Fig. 2, about which the direction of regeneration reverses suggested that cells might measure their distance in the tissue from this special point. This raised the possibility that the position of cells is specified in a polar coordinate system with the high point as its origin. Since the wing disc is a two dimensional surface, azimuthal angle must be measured. Were angle specified by a single scalar variable, that variable would necessarily be discontinuous along some radial line in the tissue; for example along the radial line $\phi = 2\pi = 0$. But Bryant found that any 1/4 pie wedge fragment of the wing disc duplicates, while its 3/4 complement regenerates. If an aximuthal discontinuity were present, the 1/4 wedge fragment containing it should behave differently from the rest, and regenerate. Failure to find evidence of an angular discontinuity implies that, if cells measure angle, they do so seamlessly. Therefore the model proposes that cells measure radial distance from a distal "high point" origin, and angle seamlessly modulo 2π.

In order to account for the bulk of the data on epimorhpic regeneration and duplication, the polar coordinate model initially proposed two rules of intercalary regeneration, and a third special rule for distal transformation.

Rule 1. If cells having different radial values are apposed, cell proliferation will be stimulated, and the missing intervening radial values will be restored back to a resting radial gradient, then proliferation will cease.

Rule 2a. If cells having different angular values are apposed, cell proliferation will be stimulated and the missing angular values will be intercalated back to a resting angular gradient.

Rule 2b. Since two angular arcs around a 2π circle of values join any two juxtaposed angular values, a choice rule is needed. The simplest postulates that angular intercalation occurs along the shorter arc.

Special Rule 3 - the complete circle rule. If a complete circle of angular values at a proximal radial level is exposed, distal regeneration occurs. The special nature of Rule 3 will be discussed below.

The two intercalation rules 1 and 2a, b, suffice to explain major features of epimorphic pattern regulation. For example, the classical example of grafting an amphibian hand to shoulder with regeneration of the missing elbow region is explained by rule 1. Proximal radial values in the shoulder apposed to distal radial values in the hand lead to intercalation of the intervening radial values specifying elbow.

Duplication and regeneration by complementary fragments of the wing disc is explained by rules 2a, b. Fig. 3 shows a single straight cut on the wing disc, yielding a narrow anterior fragment and a large posterior fragment. During culture, the narrow anterior fragment folds over, apposing the cut edge such that the ventral and dorsal thoracic regions heal together, ventral and dorsal wing blade regions also heal together [Reinhardt et al, 1977]. This healing juxtaposes cells with similar radial values, but discordant angular values. This discontinuity stimulates cell proliferation, and smoothing of the angular discontinuity along the shorter angular arc. Since this shorter angular arc is the arc present in the original narrow anterior fragment, the positional values in the new cells form a mirror symmetrical duplicate of those in the original anterior fragment, and the fragment duplicates.

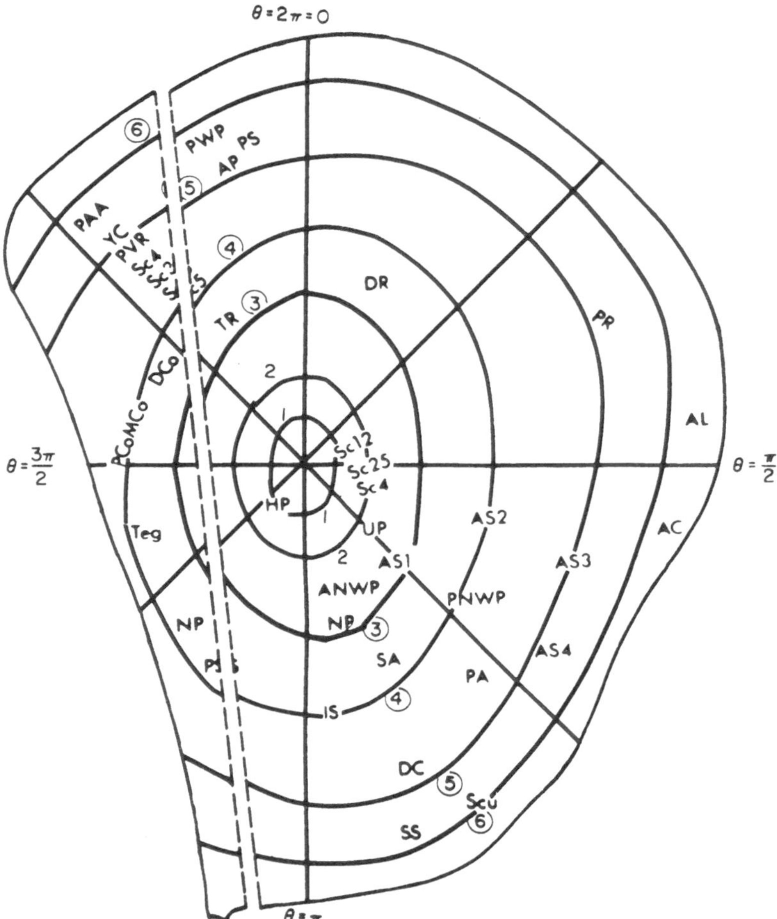

Fig. 3. The polar coordinate model of the wing disc. Radial values (1-6) are measured from the "high point". Angle, ϕ, is measured without discontinuity modulo 2π. A straight cut anterior to the high point creates a narrow anterior fragment and wide posterior fragment. The latter heals by folding along its wound margin, apposing positions with equal radial but discordant angular values. Cell proliferation and smoothing of angular discontinuities along the short arc between apposed values leads to regeneration by the posterior fragment. The narrow anterior fragment folds and heals along its wound margin, juxtaposing the same positional values as did the posterior fragment. Short arc intercalation leads to duplication by the anterior fragment.

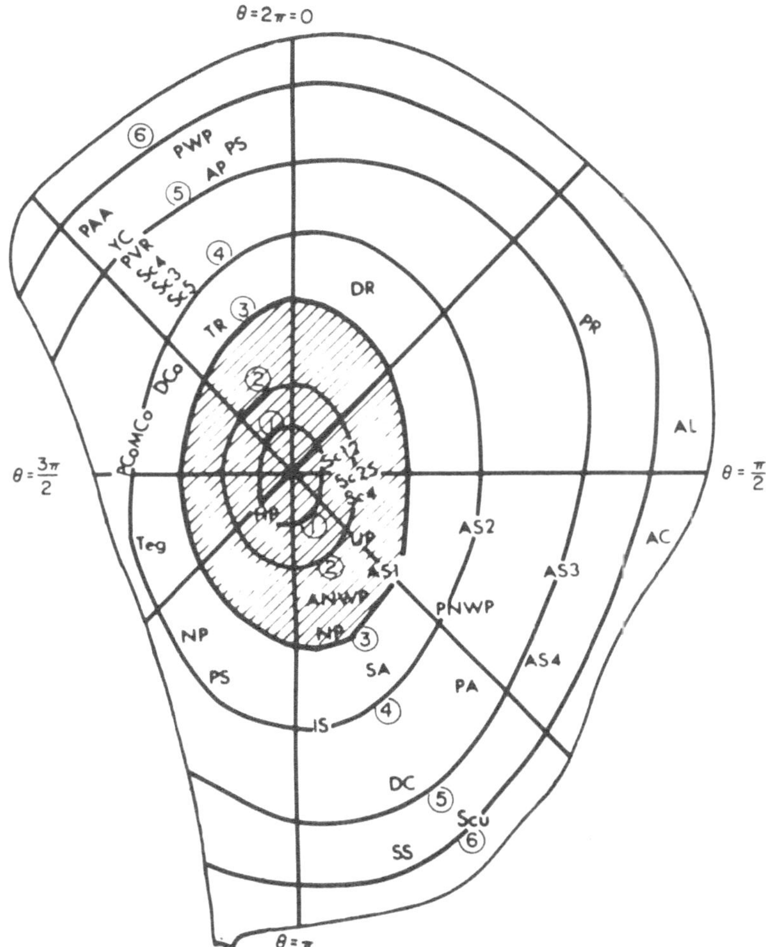

Fig. 4. Polar coordinate system on wing disc, with distal wing "high point" region removed. No parts of proximal annulus have missing distal radial values. Therefore, no tissue contact and positional averaging can intercalate the missing distal radial values. See text for discussion.

The positional values present along the cut margin of the large posterior fragment are identical to those along the cut margin of the narrow anterior fragment. Therefore, if the large fragment wound heals in a similar way, the same pairs of positional values must be apposed in the posterior fragment as in the anterior fragment. Therefore, the posterior fragment must intercalate, along the shorter angular arcs, the same intervening positional values as did the anterior fragment. Therefore the posterior fragment regenerates.

The polar model demonstrates a more general result. Whatever the coordinate system specifying position in a developmental field may be, the positional values along the two margins of a single cut are identical. If the two fragments heal in similar ways, both will appose essentially identical pairs of positional values. Therefore, if subsequent pattern regulation is governed by diffusive-like averaging of positional disparities, both complementary fragments must reform the _same_ set of structures. If one fragment duplicates, the second must regenerate. The prediction of complementary behavior in complementary fragments is a coordinate free property which follows from the postulates of graded positional values, and averaging of disparities.

Since the polar model is symmetric about the "high point", a narrow posterior fragment will duplicate, its complement will regenerate. Therefore the direction of regeneration will reverse about the high point. Further, any 1/4 wedge section will heal its two cut margins and duplicate, while its 3/4 complement will regenerate.

Intercalary Regeneration, Betweenness, and Convex Sets. A central feature of the postulates of intercalary smoothing is that only those positional values lying _between_ the apposed values can be reformed. This leads to a critical restriction in the predictive consequences of any given coordinate system, since it implies that diffusive-like smoothing can _only_ recreate positional values lying in the _convex set_ bounded by the positional values in the apposed tissue edges. This restriction, in turn, implies that different coordinate systems may demand different special cellular behaviors beyond simple diffusive-like smoothing to account for the data.

The concept of a convex set, and the limitations it imposes can be brought out in the polar coordinate model. Radial position can be visualized without loss of generality as a radially symmetric gradient whose peak is at the distal "high point". In Fig. 4, I show a wing disc from which the distal high point region has been removed, thus removing the radial gradient "peak". In the remaining outer proximal annulus, only lower values of the radial gradient are present. Therefore, no juxtaposition of tissue edges in the proximal annulus can lead to "diffusive-like" filling in of the missing high point radial peak. In a polar model, if the region containing the origin is removed, that region does not lie "between" the positional values in the proximal annulus. That is, the region around the origin is not in the convex set of all those positional values derivable by averaging any pairs of positional values present along the cut margin of the proximal annulus.

A more formal definition of a convex set may be given. Consider a two dimensional sheet of cells in which transverse gradients of two scalar properties (concentrations) X and Y exist and supply a unique positional value ($X_i;Y_i$) to each point in the tissue. Each position in the tissue can be mapped to a unique "image" point in an X,Y "Tissue Specificity Space" [Winfree, 1980], yielding an "image" of the tissue in the XY space, Fig. 5. If the tissue is cut into a fragment which heals and apposes normally nonadjacent tissue regions, diffusive-like averaging will smooth the discontinuities in X and Y in the new cells which form in the wound area. If the original X and Y values are imagined to remain fixed at the two apposed margins, diffusion will yield smooth gradients of X and Y across the new tissue, spanning between the bounding values of X and Y in the apposed tissue edges. The image of these intercalated new X and Y values lie on smooth lines in the Tissue Specificity Space, connecting the images of the two apposed old positional values. The collection of such lines, shaded area, Fig. 5, is the convex set of XY positional values which can be obtained by diffusive smoothing of the juxtaposed wound edges.

The polar coordinate model can be deformed to a cylinder by opening the tissue sheet at the high point origin and mapping it to the top edge of the cylinder, as in a mercator projection. Radial positions become latitude circles along the cylinder, angular values map seamlessly modulo 2 around the cylinder. Use of the independent radial and short arc intercalation rules of the polar model allows precise definition of the convex set obtained by apposing any two tissue regions. Removal of the high point region leaves the lower portion of the cylinder corresponding to the proximal annulus. The distal region does not lie in the convex set derivable by intercalation from the proximal annulus.

The implication of this feature of any polar coordinate model is that distal regeneration by a proximal fragment cannot be obtained by diffusive-like averaging of positional discontinuities, and some special rule is needed. In the initial formulation of the polar coordinate model, special rule 3, the complete circle rule was proposed. According to this rule, exposure of a complete circle of angular values at a proximal level leads to regeneration of distal radial values. With the assumption of this rule, the model accounts for the capacity of a truncated amphibian limb to undergo distal transformation, and regenerate distal structures. The same postulate

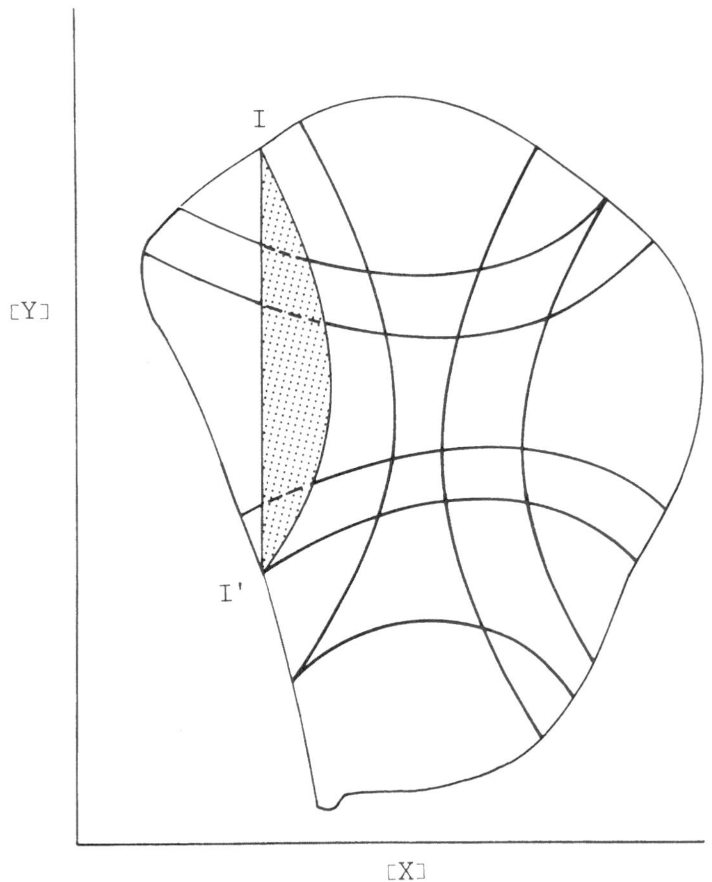

Fig. 5. Tissue Specificity Space assigning each point on wing disc a unique value of X and Y. Bowed concave lines are images of straight cuts on the wing disc like those in Fig. 2. A single cut from I to I' lies along the concave arc from I to I'. Wound healing apposes the wound margin in the posterior fragment. Juxtaposition creates positional discontinuities which are smoothed by diffusion to fill the convex set, shaded area, bounded by the cut margin.

accounts for distal regeneration by a proximal wing disc annulus. Equally, the third rule accounts for duplication of the cultured wing disc high point region from the exposed circle of angular values at a proximal radial level, and similarly explains duplication from the proximal cut surface by a newt hand whose digits are implanted into a host flank. Finally, the third rule accounts for the striking observation that grafting a left hand to a right stump yields two supernumerary limbs at the position of maximal discord of angular values. Such a graft creates two complete circles of angular values at the radial level of the graft. These undergo distal transformation and yield two supernumeraries, with the handedness of the host [French et al, 1976].

The successes of the polar model are considerable, and have led to testing several features of the model. I turn now to an evaluation of weaknesses in this model.

First, it should be stressed that special rule 3, or any modified form of rule 3, is formally equivalent to postulating a mechanism beyond diffusive-like averaging of positional values, to regenerate a missing radial gradient "peak". For example, in the current version of the polar coordinate model [Bryant, 1978; Bryant et al, 1980] its authors propose that short arc intercalation itself generates distal regeneration. While the postulate of a special mechanism to regenerate distal radial values is not a flaw, its status should be made explicit. As shown below, distal regeneration does not require special rules in other coordinate systems. If special mechanisms exist to recreate missing gradient values, then those mechanisms are of central importance. They are likely to play a role in the initial establishment of positional gradients as well as subsequent pattern regulation. The forms of special mechanisms which are suggested depend on the choice of coordinate system. Thus, different coordinate systems suggest that different cellular properties beyond diffusive averaging are required to account for the data on pattern formation.

Second, as noted by Winfree [1980], in the vicinity of the origin, the gradient of the angular variable in the tissue becomes infinitely steep, since the origin is a singularity in the angular map on the plane. If cell proliferation is proposed to occur proportional to the gradient in radial or angular values, then proliferation near the origin should be unbounded. It is not. As Winfree points out, this implies a lack of independence of radial and angular gradients near the origin.

Third, anterior-posterior and dorsal-ventral developmental axes appear to be established sequentially. This is not deducible, hence not explained, by the proposal that position is measured in polar coordinates. For example, if a radial axis is established first, the amphibian limb bud should be angularly symmetric in transplantation between the time when the first axis and second axis are formed, in contravention to Harrison's classical observation [Harrison, 1921]. If angular values are established first, then reversal of anterior-posterior or dorsal-ventral areas should yield similar, not dissimilar results, at each stage, but do not (ibid).

Fourth, a considerable body of work testing the complete circle rule has shown it to be false in imaginal tissue, cockroach limbs, and amphibian limbs [Schubiger and Schubiger, 1978; Haynie and Schubiger, 1979; Kauffman and Ling, 1981; Tank, 1978; Holder et al, 1980; Stocum, 1978; Slack and Savage, 1978]. It now appears that distal regeneration is roughly proportional to the proximal angular arc present. Confirmation of the complete circle rule would have been striking evidence in favor of a polar model. Its disconfirmation does not rule out such a model, but alternative coordinate systems have the deductive consequence that distal regeneration should be proportional to the proximal arc; hence the data disconfirming the complete circle rule leave a polar model with a postulate as rule 3, while alternatives deduce, hence explain, this phenomenon.

Fifth, accumulating evidence shows that normally duplicating imaginal disc fragments can with variable frequency regenerate some, most, or perhaps all of the complementary domain. Such results are now known in the wing, haltere, leg, and genital discs [Duranceau, 1977; Haynie and Schubiger, 1979; Bryant and Hsei, 1977; Van der Meer and Ouweneel, 1974; Schubiger, 1971; Kauffman and Ling, 1981; Krivi and Schneiderman, 1980]. No convincing comparable data are yet available on amphibian or cockroach limbs or other epimorphic systems, although divergent regeneration from symmetrical half limbs in amphibians [Holder, Tank and Bryant, 1980; Tank and Holder, 1978] might be due to regeneration by normally duplicating fragments.

The polar coordinate model is constructed such that intercalation (rules 1 and 2a,b) causes complementary fragments strictly to duplicate and regenerate, Fig. 3. The model is unable to account for regeneration by normally duplicating disc margin fragments by simple positional averaging. In fact, no proposed coordinate system, coupled with the assumptions of simple diffusive averaging of apposed positonal values, can account for extensive regeneration by both the normally duplicating and normally regenerating fragments of a tissue. All coordinate systems share the property that both complementary fragments have identical positional values along their cut margins, hence if both fragments heal in similar ways, apposing similar pairs of positional values, both fragments must reform the same set of structures lying in the convex set bounded by the apposed positional values. Evidence that each complementary disc fragment can regenerate large domains of the other fragment is evidence that mechanisms beyond simple diffusive smoothing must operate during imaginal disc pattern regulation. The polar coordinate model can account for this phenomenon by weakening the short arc intercalation rule to allow long arc intercalation. Formally, this is equivalent to an active mechanism to regenerate angular values not in the convex set reached by diffusive smoothing.

Sixth, mutants in Drosophila are available which can produce completely duplicated legs [Jürgens and Gateff, 1979; Bryant and Schubiger, 1971]. In order for a disc fragment to produce a complete duplicate appendage by apposing cut surfaces of the fragment and positional smoothing, the apposed surfaces would have to form a convex set containing the entire tissue image. Examination of the polar coordinate model, Fig. 3, shows no single fragment has this property. Within the frame of the polar model, a partial disc fragment can produce a mirror symmetric duplicate, but no fragment of a normal disc can produce both a normal leg, and a complete duplicate appendage.

Finally, the polar model has not yet received a specific molecular or mechanistic interpretation proposing what kinds of variables measure angle or radius, nor how they are established during development.

B. Cartesian Coordinate Model

Recently, several workers have independently suggested some form of a Cartesian coordinate model to account for the data on epimorphic pattern regulation [Cummins and Prothero, 1978; Winfree, 1980; Kauffman, 1978; Kauffman and Ling, 1981; Slack, 1980]. Suggestions by Winfree and myself are nearly identical. Fig. 6A shows the Drosophila wing disc with roughly orthogonal anterior-posterior and ventral-dorsal gradients of two chemicals, X and Y. Lines of constant concentration are bowed outward on the disc, symmetrically about the "high point". The model assumes that a cut wing disc fragment heals its cut margin, apposing non-adjacent positional values, and that simple diffusion smooths discontinuities in X and Y, and fills in the convex set bounded by the apposed XY pairs along the cut margin. Fig. 5 shows, for the same Cartesian model, the image in Tissue Specificity Space XY, of a straight cut on the disc. As shown in Figs. 5 and 6a the bowing of lines of constant X and Y concentration on the disc implies that a large posterior fragment of a single straight cut will fill the shaded convex set and regenerate anteriorly to the anteriormost value present along its cut margin. The complementary anterior fragment apposes the same pairs of discordant values, and duplicates to the same anteriormost value.

Symmetrical convex bowing of X and Y concentrations about the high point insures that the direction of regeneration reverses about the high point. A large anterior fragment from a straight cut will regenerate posteriorly, its posterior complement will duplicate.

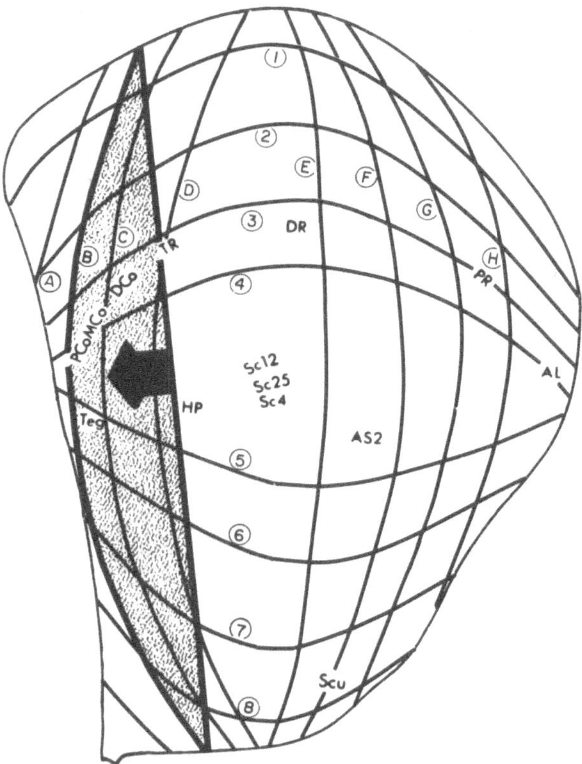

Fig. 6a. A Cartesian coordinate system with two monotonic gradients. The "Y" variable has ascending levels 0-9. The "X" variable has ascending levels A-H. A straight cut creates a large posterior fragment which wound heals by folding in half and juxtaposing opposite ends of the cut margin. For this particular cut, this leads to a discontinuity across the wound in Y values, but not in X values. The convexity in lines of constant concentration of each variable, isoconcs, assures that simple diffusive smoothing of X and Y variables yields regeneration out to the anteriormost X isoconc, B, contained along the cut margin of the posterior fragment, stippled area. Similarly, the smaller anterior fragment wound heals the same way and duplicates to the B isoconc. The symmetry of convexity in the X and Y isoconcs on the disc about a central region in the wing disc implies that the polarity of regeneration (heavy arrow) will alter about an apparent "high point" in this region, although cells do not measure position with reference to the high point. Regeneration by a normally duplicating fragment requires an active mechanism beyond simple diffusion to recreate the gradient extremum at the disc margin of the normally regenerating fragment.

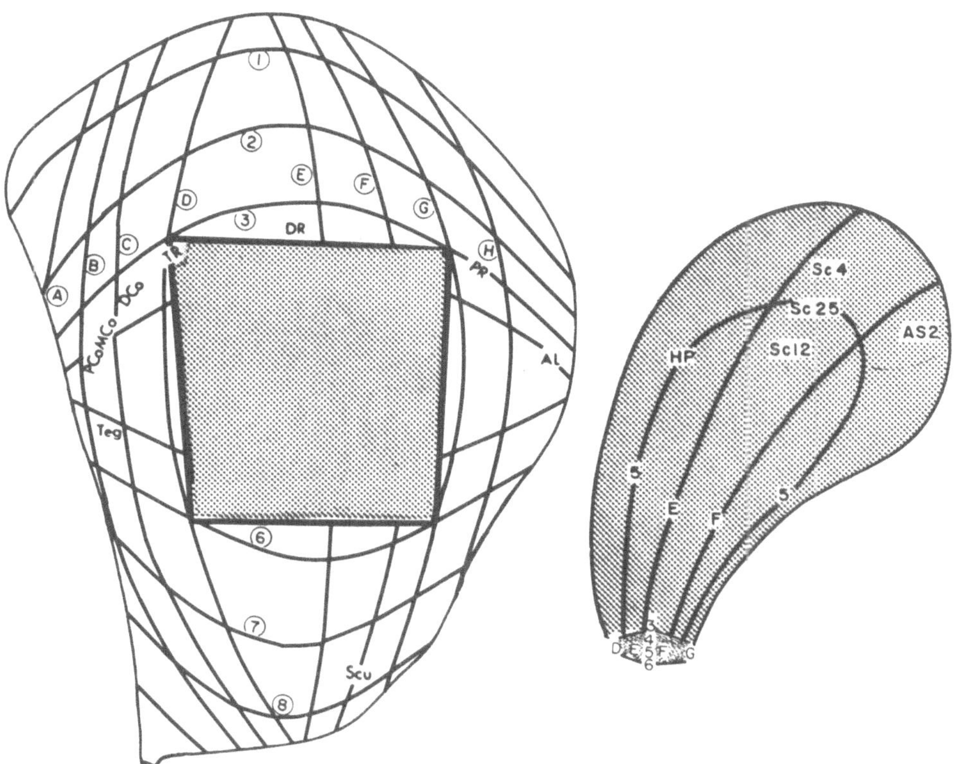

Fig. 6b. Removal of a central (distal) region leaves a proximal outer annulus. Wound healing along the cut in the annulus juxtaposes tissues, creating discontinuities in the monotonic X and Y gradients, which are smoothed over by simple diffusive-like processes, leading to distal regeneration by the proximal annulus. The central square containing the high point itself purse string closes juxtaposing tissues that again create X and Y gradient discontinuities across the wound healed zone. Smoothing of these discontinuities causes the positional values in the central square to be duplicated in the new cells of the wound zone, shown in the granular area.

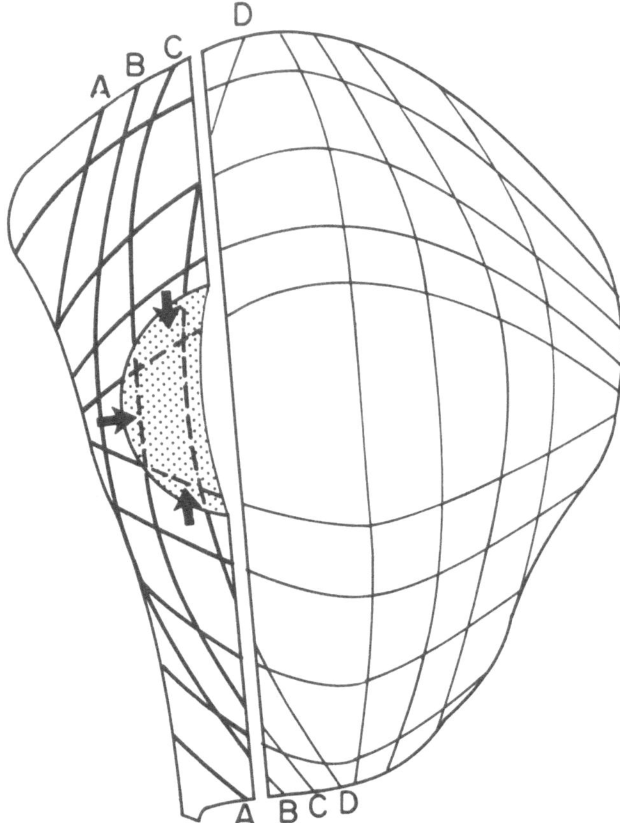

Fig. 6c. Anterior disc fragment with a distal crescent removed, fills convex set, shaded area, and undergoes partial distal regeneration.

Similarly, any 1/4 pie wedge fragment which apposes its two cut margins will duplicate, its 3/4 complement will regenerate. Therefore, a Cartesian model yields reversal of the direction of regeneration about the high point, without the assumption that the high point is a locus from which cells measure position.

Distal regeneration is a consequence of simple diffusive smoothing in a Cartesian model. As shown in Fig. 6B, deletion of the high point region leaves a proximal annulus. Wound healing apposes tissue around the circular cut margin, thereby creating discontinuities in the X and Y gradients. Since the deleted high point region lies in the convex set reached by diffusive smoothing from the proximal annulus, the distal high point region is regenerated. Similarly, the high point region itself purse string closes its circular wound margin, apposing the same pairs of positional values as did the proximal annulus, and therefore duplicates the X and Y high point values in the new cells of the wound area, Fig. 6b. The same principle predicts that any proximal arc of tissue will undergo limited distal regeneration of structure lying in the convex set of the positional values along the cut margin, Fig. 6c. Therefore, limited distal regeneration proportional to the proximal arc present, as recently demonstrated in imaginal discs, and amputated symmetric half limbs of amphibians and cockroaches, is a deductive consequence of a Cartesian model.

A Cartesian model has no singularities with infinitely steep gradients, as does the polar model for angular values near the origin. Therefore, the proposal that cell proliferation is related to gradient can be well believed throughout the domain.

As shown in Fig. 7, the Cartesian model explains the striking observations that grafts of distal left limbs to proximal right stumps can generate two supernumerary limbs with the handedness of the host. This property now appears to be a coordinate free consequence of local coordinates in the plane, and the postulate of intercalary smoothing.

With minor distortion from symmetry, the Cartesian model also has the consequence that 180° graft rotations of the left hand onto the left stump can yield variable numbers of supernumeraries, including two supernumerary limbs of opposite handedness.

Finally, unlike the polar model, a Cartesian model accords directly with the widespread data suggesting that anterior-posterior axes are established prior to dorsal-ventral axes.

Any Cartesian model suffers certain defects analogous to those of the polar model. The "high point" region in the polar model lies outside the convex set of the proximal annulus. Its regeneration requires special mechanisms. In a Cartesian model, the extreme

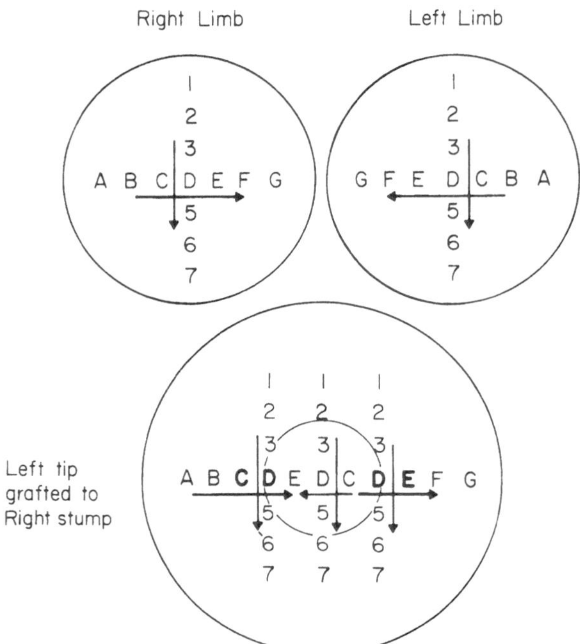

Fig. 7. Schematic picture of anterior-posterior and dorsal-ventral Cartesian X, Y coordinates for left and right limbs, projected onto the plane. The position at the mid values of the two variables, 4, D, corresponds to the distal limb tip. Grafting a left distal limb to a right proximal stump leads to smoothing of discontinuities in X and Y (bold letters, CD and DE) and the formation of two supernumerary distal limb tips with the handedness of the host proximal stump, at the positions of maximal disparity of host and graft axes.

disc margins lie outside the convex sets of their complements, Fig. 5. Thus, as shown in Fig. 6a, the large posterior fragment only regenerates partially. It does not reach the anterior disc margin. Similarly, the narrow anterior fragment duplicates partially. Either extreme bowing of lines of constant concentration or special mechanisms to recreate X and Y gradient maxima and minima are required to account for the observed capacity of disc fragments to regenerate or duplicate completely.

Like the polar model, no Cartesian model can explain the capacity of normally duplicating fragments to regenerate large domains of the complementary disc fragment by diffusive smoothing. Nor can a Cartesian model account by diffusive smoothing alone, for mutants which form complete limbs and complete duplicate limbs, since no subregion of a Cartesian coordinated tissue can have the entire tissue image in its convex set. The assumption of a Cartesian model leads to the suggestion that cells in a tissue have special mechanisms capable of reforming X and Y gradient maxima and minima lying outside the convex sets of cultured fragments. Explicit mechanisms are described below.

C. Higher Dimensional Models: Compactness

I describe only briefly the general problem associated with the assumption that 3 or more chemical gradients in a two dimensional tissue sheet supply positional information. Consider a three variable model, with a Tissue Specificity Space XYZ. The image of the tissue is a two dimensional surface embedded in the 3 chemical Tissue Specificity Space. Suppose that the image surface is curved, not planar. Then, due to the curvature, apposition of non-adjacent tissue regions can result in intercalation of positional values lying <u>outside</u> of the two dimensional image of the tissue in the XYZ space. Thus a higher dimensional model typically confronts the problem of "compactness". Are all positional values lying within the convex set of any two tissue fragments themselves "meaningful"-- i.e. part of the tissue image. The 3 variable model can meet the problem of curvature of the tissue image in two ways: 1) by assuming processes beyond diffusive smoothing which constrain the intercalated values back into the curved image; 2) by assuming lines of redundant "equivalent" positional information in X,Y,Z space, each transversal to the image and piercing it at a single spot. Transversal lines to the tissue image can fill 3 space in such a way that intercalation between normally non-adjacent wound edges leads to new positional values lying in the convex set bounded by the apposed values, which are "meaningful". The Spherical Coordinate Model discussed next uses one form of the latter solution.

D. Spherical Coordinate Model

Russell [1978] proposed a spherical coordinate model using three orthogonal X,Y and Z gradients to form a Tissue Specificity Space. Position in a tissue is specified by a solid angle, θ, ϕ corresponding to latitude and longitude angles. Each positional value is a unique ray at a constant X:Y:Z ratio emanating from the origin. The longitudinal angle ϕ, is defined by the ratios of X and Y in the equatorial plane, and the latitude angle θ, is defined by the X:Z, or Y:Z ratio. The image in XYZ space of a two dimensional tissue is a two dimensional surface pierced by a set of solid angle rays emanating from the origin, which itself does not normally lie in the physical tissue.

Russell's model is elegant in a number of respects. The rays are each lines of equivalent positional value in XYZ space, thus, the image is compact. Any two non-adjacent fragments which are apposed will fill their convex set by a simple diffusion, and the resulting values will be a surface in XYZ space pierced by rays from the origin, hence "meaningful". Consequently, the spherical model can account for almost all the available data using only the concept of diffusive smoothing. This is most easily visualized in the wing disc by remembering that the disc is topologically a spherical surface, the disc proper, backed by the peripodal membrane. Let this tissue spherical surface be embedded in XYZ space such that each solid angle pierces the closed two dimensional surface of the image once. Therefore the image surrounds the origin. If a narrow anterior fragment of the disc is cut, the margin purse string wound heals, and smoothing of X, Y, and Z discontinuities in the new cells forms a second surface in XYZ space whose edges join those of the original anterior fragment image, Fig. 8. Rays from the origin pierce both the new surface and the original anterior image surface, Fig. 8, hence the narrow fragment duplicates completely. The large posterior fragment wound heals similarly, and forms the same new image surface in XYZ space in the new cells of the wound area. These are pierced by the same rays that pierce the anterior fragment, hence the posterior fragment regenerates completely.

The model is spherically symmetric. Therefore not only do narrow anterior, posterior, dorsal or ventral fragments duplicate while their large complements regenerate, but a distal fragment containing the high point region will purse string close, creating a second image surface in the new cells in the wound area. This second surface will be pierced by the same rays that pierce the high point region, therefore it will duplicate the high point. Equally, the proximal annulus will regenerate distally. Finally, this model directly explains distal regeneration proportional to the proximal arc cultured, and the incidence of supernumerary limbs.

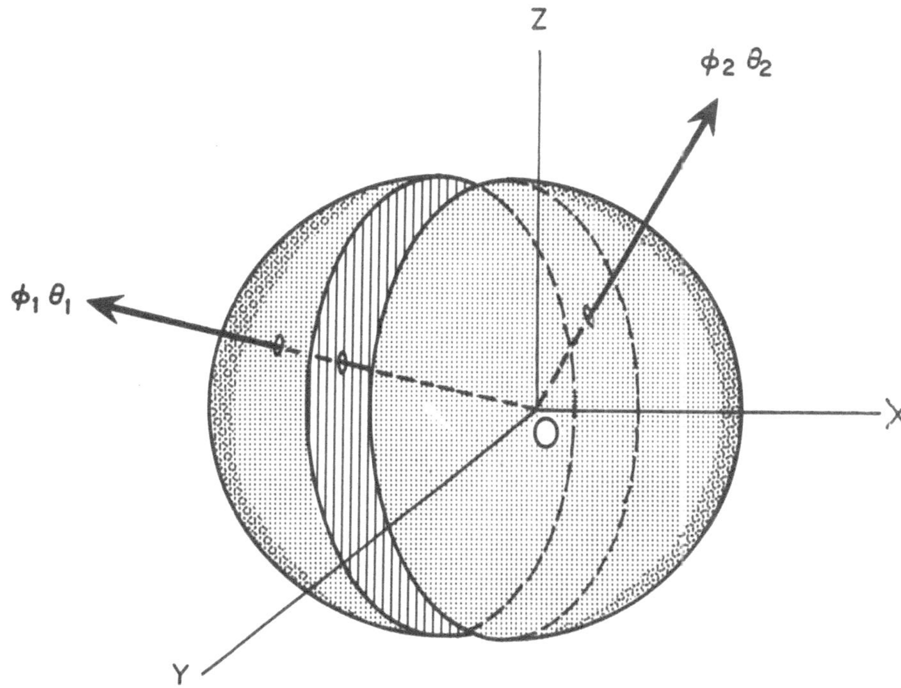

Fig. 8. Russell's [1978] spherical model. X, Y, and Z are orthogonal gradients. Position is specified by ratio of X:Y:Z: with respect to origin, 0; each ratio is a ray at a unique solid angle, ϕ, θ. Figure shows disc as a spherical surface (dotted region), with an anterior fragment cut off. Wound healing in anterior fragment creates an additional new XYZ surface (parallel lines). Rays ϕ_1, θ_1 pierce both surfaces, hence anterior fragment duplicates. Posterior fragment forms same new XYZ surface and regenerates.

Despite its strengths, the spherical model suffers some weaknesses. First, it requires cells to assess position by measuring the ratio of two or three scalar variables over a continuous range of their concentrations with respect to a reference concentration of each at the origin. Plausible mechanisms to do so need to be brought forward. Second, the spherical model, like the polar model, has a singularity at the north pole high point, where all longitudinal angular values meet. The hypothesis that positional gradients regulate proliferation confronts infinitely steep longitudinal angular gradients at the 'high point'.

Finally, while the spherical model is very appealing, it is unable to account for all the data by simple diffusive averaging of gradient discontinuities. Like the polar and Cartesian models, diffusion in the spherical model can yield only duplication and regeneration by complementary fragments. It cannot account for regeneration by normally duplicating fragments, nor the existence of mutants which form normal limbs, and complete duplicate limbs. To account for these phenomena, it is necessary to imagine cellular processes capable of recreating missing X, Y, or Z maxima and minima not in the convex set of the cultured fragment.

IV. SEQUENTIAL GENERATION OF POSITIONAL AXES

Review of the polar, Cartesian, and spherical coordinate models demonstrates that the single propery of intercalation of intervening positional values is very powerful, yet not adequate to account for regeneration by normally duplicating fragments, nor for the capacity of some mutants to produce complete normal plus duplicate appendages. In the Cartesian and spherical models, mechanisms are needed which can recreate gradient maxima and minima. One interpretation of the polar model[1] has the same requirements. In the present section, I

---

[1] One molecular interpretation of the polar model would propose three orthogonal gradients X,Y,Z. The ratio of X and Y with respect to reference X and Y values on the Z axis would define the angular measure, modulo 2π, as in the spherical model. The concentration of Z, whose maximum would be the "high point", would define radial distance. Thus <u>planes</u> of constant [Z] in XYZ space define radial distance from the high point, and rays of constant X:Y ratio in each plane, define angular measure at that radial level. Short arc angular intercalation emerges as diffusive smoothing of X and Y discontinues, radial intercalation as smoothing of Z discontinues. Active distal regeneration is the need to recreate a Z maximum, and active long arc regeneration is the need to reform missing X and Y maxima or minima not in the convex set of the cultured fragments. This interpretation makes the polar model similar to the Cartesian and spherical models, and suggests that reaction-diffusion mechanisms might be used to establish any of the three coordinate systems.

discuss the role reaction-diffusion instabilities may have in the establishment and maintenance of positional gradients during development.

It is now well known that model biochemical systems in which synthesis and degradation of chemical species are coupled, and their diffusive transport occurs, can lead to the formation of spatially inhomogeneous gradients of the constituents in the domain, [Turing, 1952; Gmitro and Sciven, 1966; Nicolis and Prigogine, 1977; Babloyantz and Hiernaux, 1975; Herschkowitz-Kaufman, 1975; Gierer and Meinhardt, 1972; Meinhardt, 1977; Kauffman et al, 1978]. As a particular model, we postulate a single biochemical system of two components, with concentrations X(r,t) and Y(r,t) at position r, time t, which are being synthesized and destroyed at rates F(X,Y) and G(X,Y) and are diffusing throughout a tissue. The equations for this system are:

$$\frac{\partial X}{\partial t} = F(X,Y) + D_1 \nabla^2 X$$
$$\frac{\partial Y}{\partial t} = G(X,Y) + D_2 \nabla^2 Y \qquad (4.1)$$

which are chosen to have a spatially homogeneous steady state X_o, Y_o at which $F(X_o,Y_o) = G(X_o,Y_o) = 0$.

Analysis of system (4.1) begins by linearizing about the spatially homogeneous steady state by substituting $X(r,t) = X_o + x(r,t)$, $Y(r,t) = Y_o + y(r,t)$ and retaining only terms up to first orders in x and y in a Taylor expansion of F(X,Y) and G(X,Y). The resulting linear equations in deviations x and y from the homogeneous state are:

$$\frac{dx}{dt} = K_{11}x + K_{12}y + D_1 \nabla^2 x$$
$$\frac{dy}{dt} = K_{21}x + K_{22}y + D_2 \nabla^2 y \qquad (4.2)$$

The stability of the spatially homogeneous steady state of the linearized equations to spatially inhomogeneous perturbations in the concentration of x or y may be analyzed by evaluating the determinant of the matrix

$$\begin{bmatrix} K_{11} - k_i^2 D_1 - \lambda & K_{12} \\ K_{21} & K_{22} - k_i^2 D_2 - \lambda \end{bmatrix} = 0 \qquad (4.3)$$

yielding the dispersion relation between the two temporal eigenvalues λ_{i+}, λ_{i-}, and the spatial eigenvalue $k^2{}_i$,

$$\lambda_{i\pm} = \frac{1}{2}[K_{11} + K_{22} - k^2(D_1 + D_2)] \pm$$

$$\pm \frac{1}{2}\sqrt{[K_{11}+K_{22}-k^2(D_1+D_2)]^2-4[k^4 D_1 D_2-k^2(D_2K_{11}+D_1K_{22})+K_{11}K_{22}-K_{21}K_{12}]} \tag{4.4}$$

Here

$$k_i = \sqrt{k_i^2}$$

is inversely proportional to the wavelength of the perturbation ℓ_i. For a given wavelength perturbation ℓ_i, corresponding to a specific value of k_i, if both λ_{i+} and λ_{i-} have negative real parts, that perturbation decays. In an unbounded domain, those perturbations with wavelengths corresponding to values of k_i for which the dispersion relation has at least one temporal eigenvalue with positive real part, will grow in amplitude in time and create a spatial pattern.

With appropriate constraints on the linearized reaction and diffusion constants [Gmitro and Sciven, 1966; Kauffman et al, 1978], the dispersion relation $\lambda = f(k)$ is positive for a restricted range of k, $k_1>k_i>k_2$, hence for a restricted range of wavelength ℓ, $\ell_1<\ell_i<\ell_2$. In this case, system (4.1) acts as an amplifier, selecting and amplifying from thermal noise those wavelength components ℓ in the range between ℓ_1 and ℓ_2.

If system (4.1) exists in a bounded spatial domain, and no-flux boundary conditions are imposed, then a chemical pattern will be established only if it simultaneously has a wavelength in the range $\ell_1 - \ell_2$, and also satisfies the boundary condition that the gradient have zero component normal to the boundaries. Assuming no-flux boundary conditions, the shapes of chemical patterns which can "fit" on to the spatial domain must satisfy the Laplacian diffusion operator and are an infinite series of specific eigenfunction patterns determined by the geometry of the domain. The first several eigenfunctions on a wing disc shape are shown in Fig. 9a-d [Bunow et al, 1980].

A. Establishment of a First Positional Axis

As the spatial domain in which system (4.1) acts grows gradually larger, a succession of the discrete eigenfunctions or their superposition will sequentially become established in the domain [Herschkowitz-Kaufman, 1975; Kauffman et al, 1978]. For any

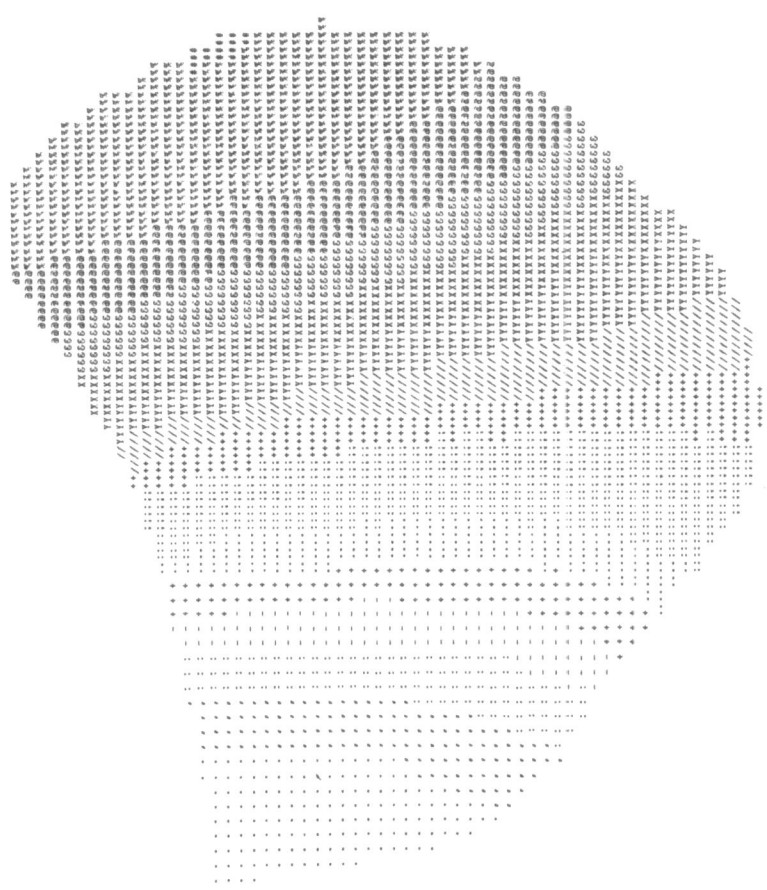

Fig. 9a. First eigenfunction pattern on wing disc as it enlarges. Concentration coded by visual density [from Bunow et al, 1980].

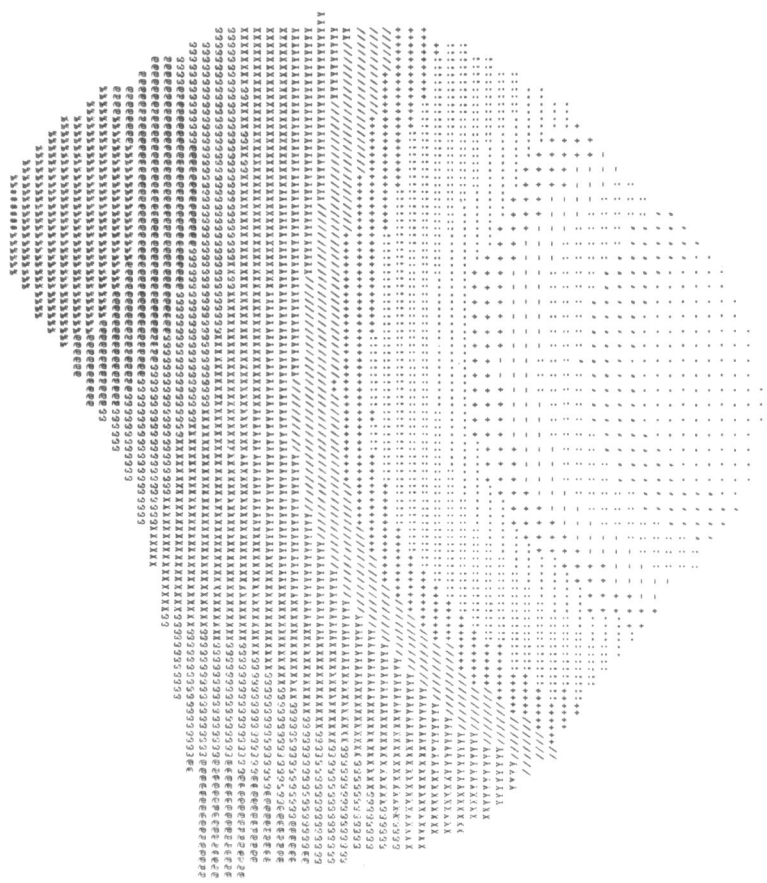

Fig. 9b. Second eigenfunction pattern on wing disc as it enlarges. Concentration coded by visual density [from Bunow et al, 1980].

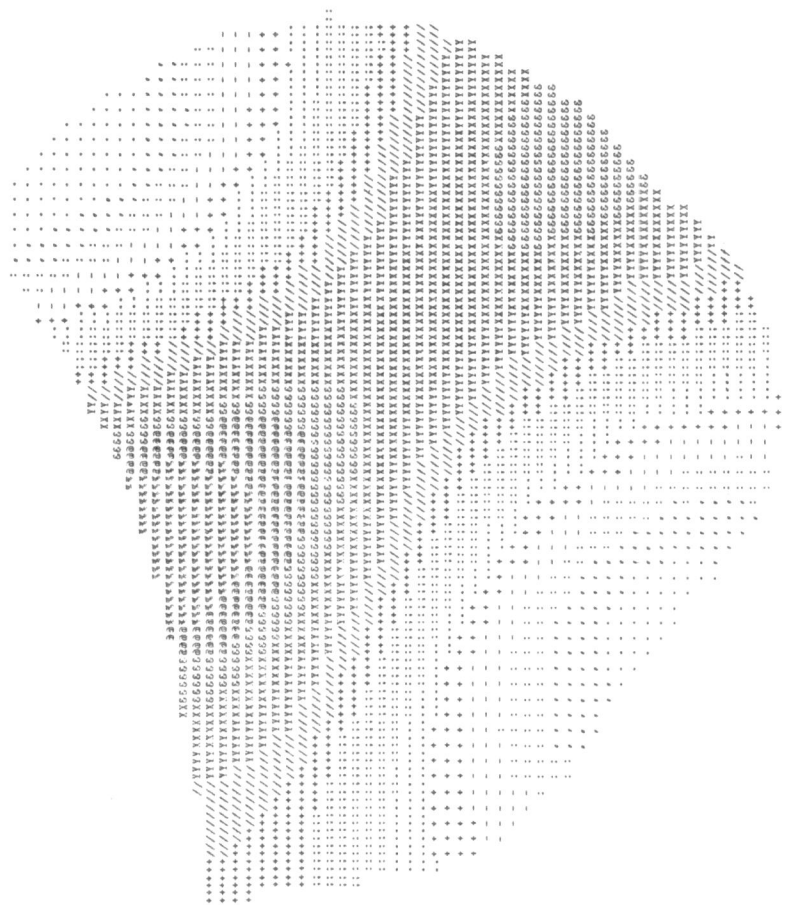

Fig. 9c. Third eigenfunction pattern on wing disc as it enlarges. Concentration coded by visual density [from Bunow et al, 1980].

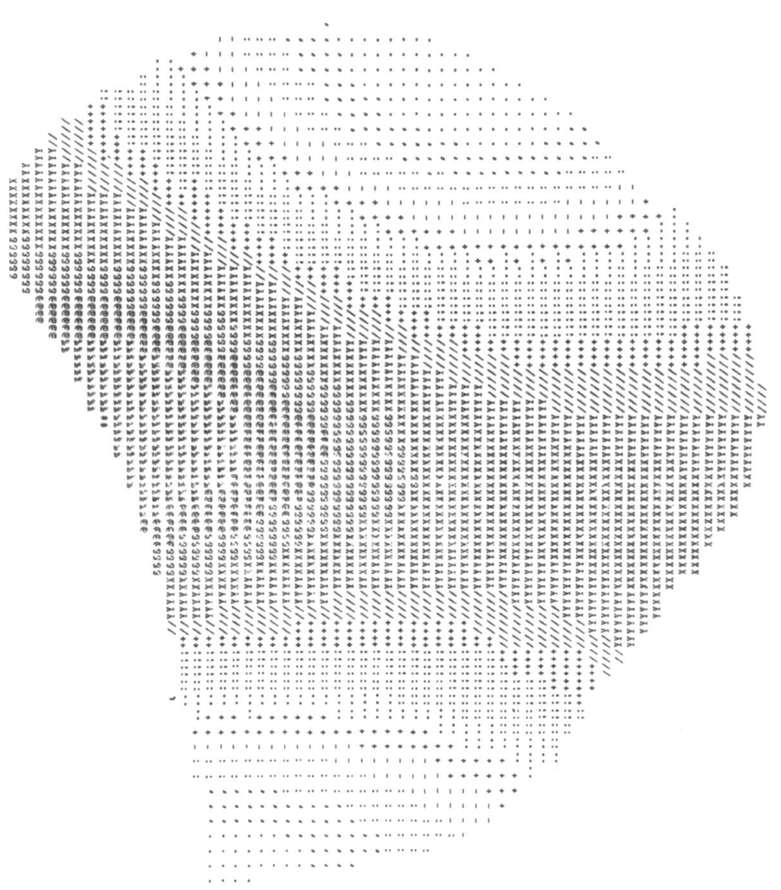

Fig. 9d. Fourth eigenfunction pattern on wing disc as it enlarges. Concentration coded by visual density [from Bunow et al, 1980].

reaction-diffusion system which acts as an amplifier, perturbations below some minimal wavelength, ℓ_1, will die away. Therefore, if the maximum physical length of the tissue is less than ℓ_1, the spatially homogeneous state is stable. As the tissue grows in length past ℓ_1, the first eigenfunction of the tissue domain will bifurcate from the homogeneous solution. A fundamental theorem of bifurcation theory guarantees that the first pattern will be a stable solution of the full non-linear equations [Sattinger, 1972]. Therefore, as a wing disc shape gradually enlarges, no chemical pattern forms for lengths below ℓ_1, then eigenfunction 1, Fig. 9a, grows out of the homogeneous state, and forms a monotonic gradient along the long axis of the tissue.

It has not been recognized for some time that this primary bifurcation provides a means to establish a first dimensional axis of positional information in a growing domain. The new question I wish to address is "How can a reaction-diffusion system like (4.1) establish two transverse axes in a growing tissue?"

In the linearized equations of (4.1) or any similar system, each eigenfunction pattern which forms in the tissue is a spatially inhomogeneous distribution of a fixed ratio of the underlying chemical variables x and y, measured as deviations from the homogeneous steady state X_o,Y_o. This property follows from the fact that the unstable wavelength ℓ_i which is amplifying on the tissue, corresponds to a unique scaling factor k_i. Each unique choice of k_i specifies a particular set of values in matrix (4.3) having at most one positive temporal eigenvalue ℓ_i; and a unique corresponding eigenvector $x_i:y_i$, which is a fixed ratio of the linearized deviations of x and y from X_o,Y_o. This constraint implies that lines of constant concentration of X on the tissue parallel those of Y. Therefore although two chemical gradients are present in a two dimensional domain, they supply at most one dimension of positional information. More generally, if N chemical species are in system (4.1) $(N \geq 2)$, all N occur as a fixed ratio when a single temporal eigenvalue λ_i is positive, hence lines of constant concentration of all N are parallel to one another and supply only one positional axis. Further, if any single eigenfunction pattern, eg. Fig. 9c, is amplified alone in the tissue through a single positive eigenvalue, the ratio of all N species is again fixed, and at most one positional gradient is established.

B. Establishment of a Second Positional Axis

It has not been recognized that the familiar class of 2 variable reaction-diffusion systems capable of establishing a first axis can sequentially establish transverse axes in growing two dimensional domains. This can be exemplified on a wing disc shape. If the range of wavelengths ℓ_1 and ℓ_2 which are amplified by the linearized system of (4.1) is made broad enough, then as the tissue domain enlarges past ℓ_1, the first eigenfunction mode, Fig. 9a, forms. As the two dimensional domain enlarges further, both the first ventral and dorsal monotonic pattern and the second anterior-posterior monotonic eigenfunction pattern, Fig. 9b, satisfy the boundary conditions and may be amplifed together. If the length and width of the tissue are unequal, each mode corresponds to a different specific wavelenth ℓ_i, ℓ_j ($\ell_i > \ell_j$), and the corresponding scaling factors k^2 are unequal $k^2_i \neq k^2_j$. Therefore, the eigenvector ratio of the chemical variables x and y, derived from matrix (4.3), differs for the two modes. The superposition of the two distinct modes, each amplifying a different eigenvector, yields non-parallel (transverse) gradients of X and Y in the tissue. As a second example, on a growing rectangle whose sides are of unequal length, $L_1 > L_2$, the linearized equations sequentially amplify the first eigenfunction creating parallel X and Y monotonic gradients in the L_1 direction, then amplify the superposition of the first two monotonic eigenfunctions in the L_1 and L_2 directions as the rectangle enlarges to an appropriate size creating transverse gradients of X and Y:

$$c_1 \begin{bmatrix} x_1 \\ y_1 \end{bmatrix} e^{\lambda_1 T} \cdot \cos\left(\frac{\pi r_1}{L_1}\right) \begin{bmatrix} x_1 \\ y_1 \end{bmatrix} + c_2 \begin{bmatrix} x_2 \\ y_2 \end{bmatrix} e^{\lambda_2 T} \cdot \cos\left(\frac{\pi r_2}{L_2}\right) \begin{bmatrix} x_2 \\ y_2 \end{bmatrix} \tag{4.5}$$

The first major property of this class of models, therefore, is the prediction that transverse positional axes should form <u>sequentially</u> as the tissue grows in size.

The linearized analysis of (4.1) is of limited value for several reasons. The amplitude of each of the two superimposed modes grow exponentially without bound. Whichever mode grows faster (λ_1 or λ_2) eventually dominates the superposition, therefore a single eigenvector $[x_1 : y_1]$ or $[x_2 : y_2]$ dominates the pattern and the underlying X and Y gradients become parallel. Second, the contributions of the first and second patterns to the superposition are given by arbitrary initial conditions, c_1, c_2. Finally, each eigenfunction pattern can occur in two opposite forms or branches, with maxima replacing minima. Which is established depends entirely on initial conditions. The conjunction of these defects implies that the relative orientation of X and Y gradients on the tissue is not uniquely fixed by tissue geometry in a linear theory.

These defects of a linear model are fully met by a fairly broad class of nonlinear reaction-diffusion systems in an asymmetric two dimensional tissue such as the wing disc. A number of workers have proposed nonlinear models which are known to bound the amplitudes of chemical patterns in one spatial dimension [Babloyantz and Hierneux,

1975; Herschkowitz-Kaufman, 1975; Meinhardt, 1977; Lacalli and Harrison, 1978] and two dimensions [Kauffman et al, 1978; Bunow et al, 1980]. Generally, the establishment of the first pattern parallels the linear analysis, except that for large amplitudes, the pattern may deviate from the eigenfunction pattern. As the spatial domain enlarges, the second pattern commonly is gradually established through a secondary bifurcation from the primary spatially inhomogeneous monotonic pattern, yielding at each tissue size a fixed mixture or amplitude of distorted first and second modes [Auchmuty, 1977; Hiernaux and Erneux, 1979]. Finally, in a spatially asymmetric two dimensional domain, the cube of the eigenfunction is almost always non-zero. From bifurcation theory [Sattiger, 1972], this implies that, as the tissue enlarges, only one of the two branches (or forms) of the primary monotonic pattern emerges smoothly from the homogeneous state, while the other branch is stable but requires a macroscopic perturbation to be attained. For at least some nonlinear systems, only one branch of the subsequent secondary bifurcation emerges smoothly from that first mode [Auchmuty, 1977; Auchmuty and Nicolis, 1975; Hiernaux and Erneux, 1979; Kernevez, 1980]. The novel implication is that the entire sequential establishment of transverse gradients in two dimensional domains, their shapes and orientations, can be uniquely determined by the changing tissue size and geometry, and the kinetic parameters of the nonlinear model.

In Figs. 10a-d, I show the results of simulations on a growing rectangle of one nonlinear model given by

$$\frac{\partial X}{\partial t} = -AX + \frac{BY^6}{1+Y^6} + D_1\nabla^2 X$$

$$\frac{\partial Y}{\partial t} = -CX + \frac{D(Y^6+b)}{1+Y^6} + D_2\nabla^2 Y \tag{4.6}$$

with the parameter values specified in the legend. As the 7 X 9 unit rectangle grows from unit area .45 to .69, a first mode is established in the longer direction, yielding parallel contour lines of X and Y concentrations, 10a. Thereafter the second mode makes increasing contributions and the X and Y contours distort, rotate, and become generally transversal. By unit size .69, Fig. 10d, the X gradient is no longer monotonic, presumably due to contributions from the first symmetrical mode.

In Figs. 11a,b, the same nonlinear equations are studied on a wing disc shape. As a wing disc shape enlarges, a first (Fig. 9a) ventral-dorsal monotonic gradient of X and parallel gradient of Y is established. As the disc enlarges further and the underlying

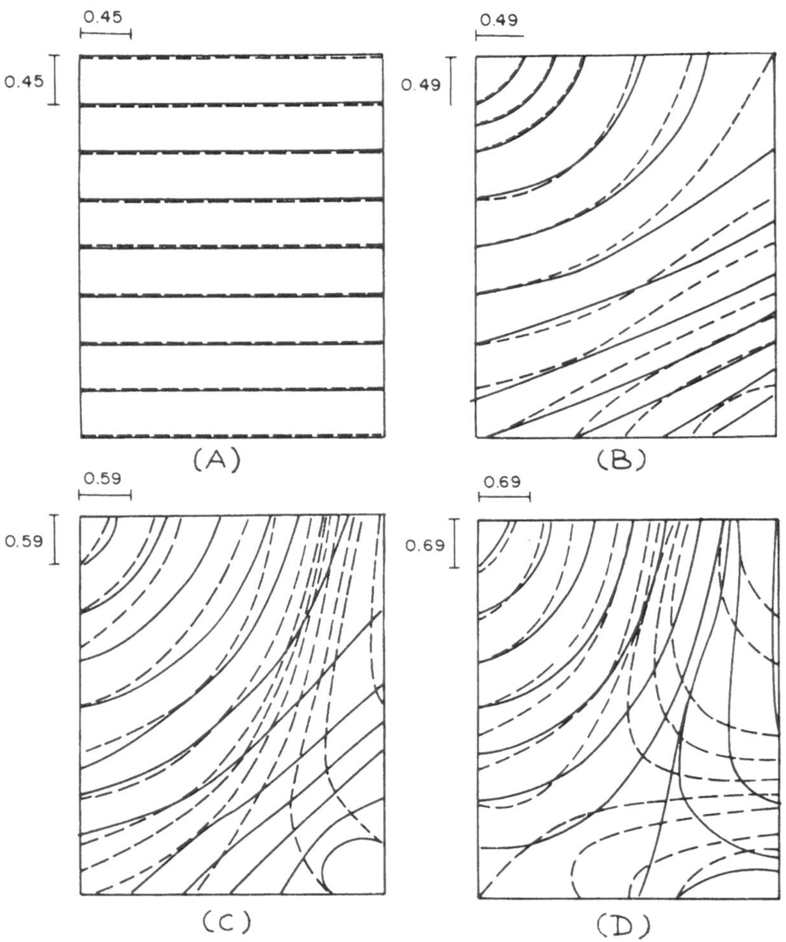

Fig. 10a-d. Establishment of transverse gradients on a growing rectangle with length ratios 7:9 using nonlinear system (4.6). As the rectangle grows in size from .45 to .69 per unit length, a first monotonic mode with parallel contours of X concentrations (solid lines) and Y concentrations (dashed lines) is established in the longer axis, 10a. By unit length .49, contributions from the second monotonic mode rotate X and Y gradients to create generally transverse contours of constant X and Y concentrations. By unit length .69, 10d, X has become slightly nonmonotonic along the lower edge. Distances between contour lines do not reflect equal concentration steps.

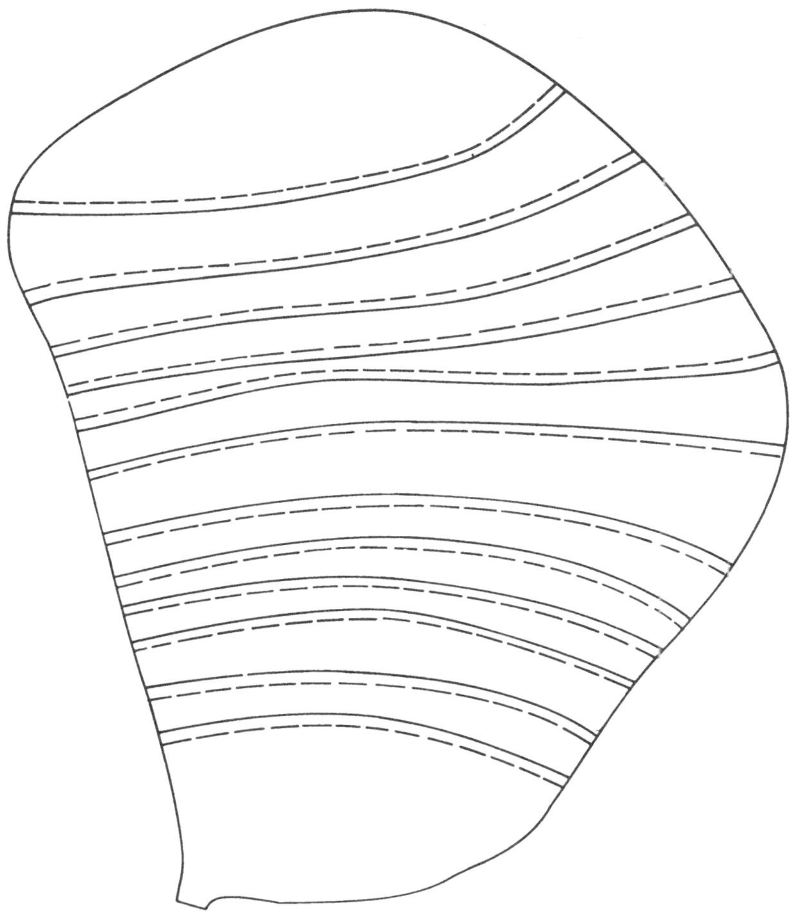

Fig. 11a. Establishment of a First Axis. First longitudinal monotonic spatially inhomogeneous steady state of nonlinear system (4.6), corresponding to first eigenfunction, Fig. 9a. Lines of constant X concentrations (solid lines) parallel lines of constant Y concentrations (dashed lines). Spacing between lines of constant X concentrations, or of Y concentrations are not strictly proportional to gradient slope, but are used to indicate shapes of X and Y gradients.

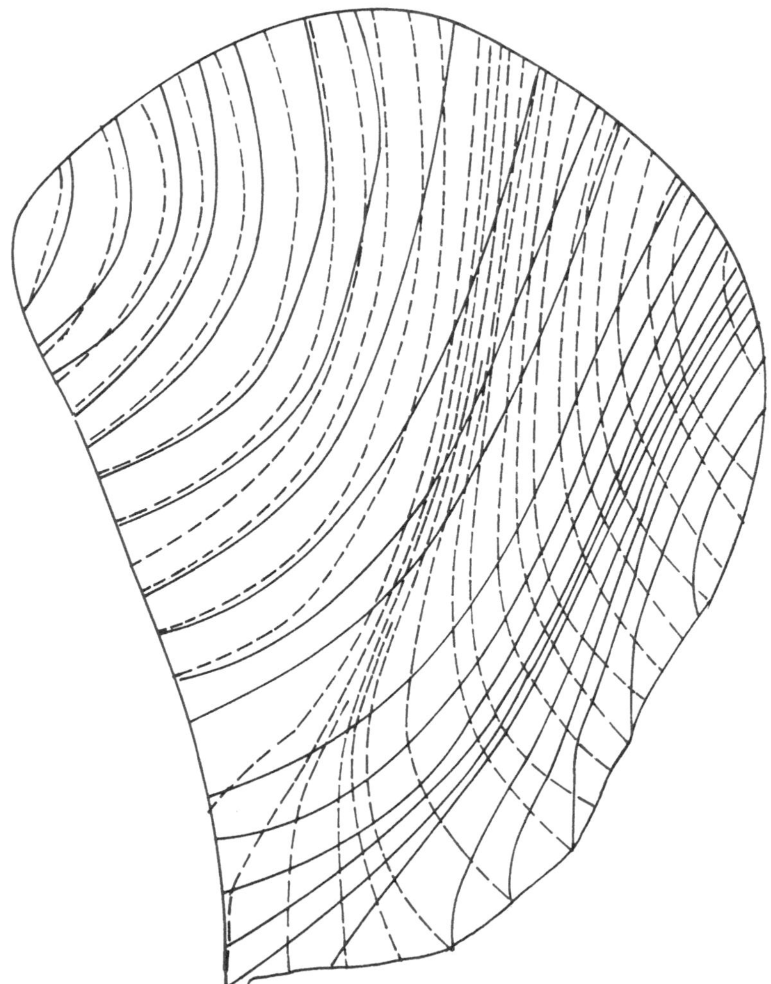

Fig. 11b. Establishment of Second Positional Axes. As wing disc grows, second eigenfunction, 9b, makes a contribution to mixed nonlinear steady state solution. X and Y gradients rotate away from positions in 10a, and lines of constant X concentrations (solid lines) separate from lines of constant Y concentration (dashed lines) creating transverse gradients and a second positional axis. See text. Parameter values used for the nonlinear model (4.6): $A = 7.8$; $B = 15.6$; $C = 1$; $D = 1.87$; $b = .2$; $D_1 = 16$; $D_2 = 1$.

anterior-posterior 2nd mode (Fig. 9b) begins to contribute to the pattern, the X and Y gradients both rotate slowly on the disc, Y faster than X, gradually establishing distinct non-parallel transverse, but not orthogonal, monotonic X and Y gradients spanning the tissue, 10b. The model sequentially establishes transverse axes and two dimensions of positional values through two gradients whose forms and orientations are initially identical, but gradually come to differ as the tissue enlarges. The positional field is uniquely determined by the shape trajectory of the tissue, and the specific nonlinear model. Several mathematical features of this class of models are now being explored.

C. Novel Predictions

The inherent simplicity and minimality of this reaction-diffusion mechanism commends it as a serious model for the establishment of transverse positional axes and positional information in developmental fields. A more complex three variable model can yield three transverse gradients which might underly Russell's spherical coordinate model. Yet more complex models with two independent two variable subsystems can generate transverse gradients. The minimal model has a number of predictive consequences.

1. Positional axes are established sequentially, as seen in several systems.

2. The longer axis in a field is established before the shorter. This expectation appears to be correct at least in the case of primary axes in insect embryos [Sander, 1977].

3. During establishment of the second axis, the positional gradients may rotate on the tissue. Evidence consistent with this prediction has been described by Duranceau [1977] for the developing wing disc. However, uncertainty about "equivalent cuts" on discs at different stages precludes clear interpretation of her results.

4. The axes are transverse, but oblique, not orthogonal.

5. Since the monotonic gradients have values above and below the homogeneous steady state values X_o, Y_o, that pair of values will typically occur at least once at an interior point in the tissue. This model predicts that very small tissue fragments cannot maintain gradients and will decay toward the steady state, therefore small fragments should drift toward the map position in the tissue corresponding to the steady state. Strub [1977] and Krivi and Schneiderman [1980] have found evidence that dissociated and reaggregated proximal leg or wing imaginal fragments from small regions transform distally. Other interpretations of these results are possible but "drift" toward a distal "steady state" is, at present, consistent with the results.

6. Predictions of the two variable nonlinear reaction-diffusion model with regard to duplication and regeneration are a combination of properties arising from simple passive diffusive averaging of positional discontinuities, and properties arising from the active dynamics and boundary conditions.

7. The X and Y transverse gradients are monotonic across the tissue. If the distal "high point" region, or any interior circle, is removed, the proximal annulus should regenerate the distal region by simple diffusive intercalation, while the distal fragment should purse string heal and duplicate. Further, any 1/4 pie wedge piece should duplicate, its 3/4 complement should regenerate by simple intercalation.

8. Lines of constant X and Y concentration are not bowed symmetrically about an interior "high point" but are determined by the eigenfunctions of the tissue shape. Therefore a narrow wing disc margin fragment and its large complement from straight cuts may not duplicate and regenerate in the observed directions by simple diffusive smoothing after apposing and healing the cut surface.

The two variable reaction-diffusion model explains the occurence of duplication and regeneration by such small and large fragments of the tissue on a novel basis. The first two modes are each monotonic eigenfunctions with wave number 1, Fig. 9a, 9b. The two next eigenfunctions are symmetric patterns of wave number 2, having one maximum and two minima, Figs. 9c and 9d. Recently Hiernaux and Erneux [1979] showed in a one dimensional domain that two variable nonlinear reaction-diffusion systems could admit the coexistence of both a monotonic gradient, and the first symmetric gradient as stable alternative solutions of the equations. Cutting of the monotonic gradient into a large and small fragment lead to the reformation of the entire monotonic gradient in the large fragment, and the assumption of the symmetrical mode in the small fragment as it grows in length, yielding its duplication.

It is not yet established, but likely that a two variable reaction-diffusion system in an asymmetric two dimensional spatial domain can admit both a mixture of the first two monotonic gradients and a coexisting mixture of the first two symmetric modes as stable alternative solutions of the equations. This model will yield duplication of a fragment, not only by intercalation of the convex set obtained by diffusive averaging, but by a macroscopic transition to a symmetric chemical mode. Thus, this class of models predicts that wound healing of apposing edges and averaging of discordant positional values may not be necessary for duplication, since other perturbations can induce transition from the monotonic to the coexisting symmetric solution. Parallel suggestions to account for homeotic mutants forming mirror symmetric double abdomen in *Drosophila* such as *bicaudal* have been made by Meinhardt [1977] and Deak [1980].

9. The coexistence of monotonic and symmetric solutions of the equations has the further consequence that the smaller tissue fragment, during growth, can regenerate the entire monotonic gradient rather than switch to the symmetric mode. Thus both large and small fragments can regenerate the entire pattern. This enables the model to account for regeneration by normally duplicating imaginal fragments [Duranceau, 1977; Haynie and Schubiger, 1979; Kauffman and Ling, 1981; Bryant and Hsei, 1977], and divergent distal regeneration in amphibian symmetrical half limbs [Holder, Tank and Bryant, 1980].

10. If monotonic and symmetric solutions coexist as stable solutions in cultured tissue fragments, the fragment may either duplicate or regenerate, or do mixtures of both, as recently observed in cultured wing disc fragments [Kauffman and Ling, 1981].

11. Finally, as the spatial domain grows, establishment of the symmetric modes can yield mirror duplicate patterns having the full range of X and Y values. Thus, the model can explain the occurrence of mutants such as those in *Drosophila* which cause formation of both a normal leg and a complete duplicate leg [Jürgens and Gateff, 1979; Bryant and Schubiger, 1971], which is inexplicable by positional smoothing alone.

These predictions have not been systematically tested. One aim of this review is to stimulate experiments further testing the polar, Cartesian and spherical models, and the postulated reaction-diffusion systems which might underlie those alternative positional coordinate systems. This discussion has focused on epimorphic pattern regulation, but it is clear that the assumption of reaction-diffusion systems which recreate gradients assimilates epimorphic and morpholactic pattern regulation into a unified spectrum of models governed by the extent to which tissue enlargement is required for reestablishment of gradient maxima and minima. I have not discussed a number of important problems such as the apparent coexistence of segmentally repetitive and whole limb gradients [Slack, 1980].

V. SUMMARY

This article has reviewed and compared the polar, Cartesian and spherical coordinate models for epimorphic pattern regulation, focusing largely on imaginal tissues of *Drosophila*. Geometric features of the three models lead to variable strengths and weaknesses and differing properties by which they can be experimentally distinguished. Two or three variable reaction-diffusion systems which establish mixtures of two eigenfunctions on the tissue are an important class of models capable of accounting for major features of the establishment and reestablishment of positional information in developmental fields.

REFERENCES

Apter, M.J., 1966, "Cybernetics and Development," International Series of Monographs in Pure and Applied Biology/Zoology Division, Vol. 29, Pergamon Press, New York, London.

Auchmuty, J.F.G., 1977, Qualitative effects of diffusion in chemical systems, Lectures on Mathematics in the Life Sciences, 10:49.

Auchmuty, J.F.G., and Nicolis, G., 1975, Bifurcation analysis of nonlinear reaction-diffusion equations. I. Evolution and the steady state solutions, Bull. Math. Biol., 37:323.

Babloyantz, A., and Hiernaux, J., 1975, Models for cell differentiation and generation of polarity in diffusion-governed morphogenetic fields, Bull. Math. Biol., 37:637.

Bohn, H., 1965, Analyse der Regenerationsfahigkeit der Insektenextremität durch Amputations und Transplantationsversuche an Larven der Afrikanischen Schabe Leucophaea maderae Fabr. (Blattaria). I. Regenerationspotenzen, Wilhelm Roux's Archives, 156:49.

Bohn, H., 1970, Interkalare Regeneration und segmentale Gradienten bei den Extremitäten von Leucophaea-Larven (Blattaria). I. Femur und Tibia, Wilhelm Roux's Archives, 165:303.

Bohn, H., 1971, Interkalare Regeneration und segmentale Gradienten bei den Extremitäten von Leucophaea-Larven (Blattaria). III. Die Herkunft des interkalaren Regenerates, Wilhelm Roux's Archives, 167:209.

Bohn, H., 1972, The Origin of the epidermis in the supernumerary regenerates of triple legs in cockroaches (Blattaria), J. Embryol. Exp. Morphol., 28:185.

Bryant, P.J., 1975, Pattern formation in the imaginal wing disc of Drosophila melanogaster: fate map, regeneration and duplication J. Exp. Biol., 193:49.

Bryant, P.J., Bryant, S.V., and French, V., 1977, Biological regeneration and pattern formation, Sci. Amer., 237:66.

Bryant, P.J., and Hsei, B.W., 1977, Pattern formation in asymmetrical imaginal discs of Drosophila melanogaster, Amer. Zool., 17:595.

Bryant, P.J. and Schubiger, G., 1971, Giant and duplicated imaginal discs in a new lethal mutant of Drosophila melanogaster, Dev. Biol., 24:233.

Bryant, S.V., 1978, Pattern regulation and cell commitment in amphibian limbs, in: "The Clonal Basis of Development." 36th Symp. Soc. Develop. Biol.

Bryant, S.V., French, V., and Bryant, P., 1980, Distal regeneration and symmetry, submitted for publication.

Bryant, S.V., and Iten, L.B., 1976. Supernumerary limbs in amphibians: Experimental production in Notophthalmus viridescens and a new interpretation of their formation, Dev. Biol., 50:212.

Bulliere, D., 1970, Interprétation des régenerats multiples chez les insectes, J. Embryol. Exp. Morphol., 23:337.

Bulliere, D., 1971, Utilisation de la régeneration intercalaire pour l'étude de la détermination cellulaire au cours de la morphogenèse chez Blabera cranifer (Insecte Dictyoptere), Dev. Biol., 25:672.

Bunow, B., Kernevez, J. P., Joly, G., and Thomas, D., 1980, Pattern formation by reaction-diffusion instabilities: Applications to morphogenesis in Drosophila, J. Theor. Biol., 84:629.

Butler, E.G., 1955, Regeneration of the urodele forelimb after reversal of its proximo-distal axis, J. Morphol., 96:265.

Carlson, B.M., Civiletto, S.E., and Goshgarian, H.G., 1974, Nerve interactions and regenerative processes occurring in newt limbs fused end-to-end, Dev. Biol., 37:248.

Chung, S.H., and Cooke, J., 1975. Polarity of structure and of ordered nerve connections in the developing amphibian brain, Nature, 258:126.

Cohen, N., 1971, Amphibian transplantation reactions: A review, Amer. Zool., 11:193.

Cooke, J., 1975, The emergence and regulation of spatial organization in early animal development, Annu. Rev. Biophys. Bioeng., 4:185.

Cummins, F.W., and Prothero, J.W., 1978, A model of pattern formation in multicellular organisms, Collective Phenomena, 3:41.

Deak, I.I., 1980, A model linking segmentation, compartmentalization and regeneration in Drosophila development, J. Theor. Biol., 84:477.

Dent, J.N., 1954, A study of regenerates emanating from limb transplants with reversed proximodistal polarity in the adult newt, Anat. Rec., 118:841.

Duranceau, C., 1977, Control of growth and pattern formation in the imaginal wing disc of Drosophila melanogaster, Ph.D. Thesis, Univ. Calif., Irvine.

Erneux, T., and Hiernaux, J., 1980, Transition from polar to duplicate patterns, J. Math. Biol., 9:193.

French, V., 1976, Leg regeneration in the cockroach, Blattella germanica. I. Regeneration from a congruent tibial graft-host junction, Wilhelm Roux's Archives, 179:57.

French, V., 1976, Leg regeneration in the cockroach, Blattella germanica. II. Regeneration from a noncongruent tibial graft-host junction, J. Embryol. Exp. Morphol. 35:267.

French, V., Bryant, P.J., and Bryant, S.V., 1976, Pattern regulation in epimorphic fields, Science, 193:969.

Gehring, W., 1966, Bildung eines vollstandigen Mittelbeins mit Sternopleura in der Antennenregion bei der Mutante Nasobemia (Ns) von Drosophila melanogaster, Arch. Julius Klaus-Stift. Vereb. Forsch, 41:44.

Gehring, W., and Nöthiger, R., 1973, The imaginal discs of Drosophila, in: "Developmental Systems: Insects," S.J. Counce and C. Waddington, eds., Academic Press, New York.

Gierer, A., and Meinhardt, H., 1972, A theory of biological pattern formation, Kybernetik, 12:30.

Gmitro, J.I., and Sciven, L.E., 1966, in: "Intracellular Transport," K.B. Warren, ed., Academic Press, New York.

Harrison, R.G., 1918, Experiments on the development of the forelimb of Amblystoma, a self-differentiating equipotential system, J. Exp. Zool., 25:413.

Harrison, R.G., 1921, On relations of symmetry in transplanted limbs, J. Exp. Zool. 32:1.
Harrison, R.G., 1969, "Organization and Development of the Embryo," S. Wilens, ed., Yale University Press, New Haven.
Haynie, J.L., and Bryant, P.J., 1976, Intercalary regeneration in imaginal discs of Drosophila melanogaster, Nature, 259:659.
Haynie, J.L., and Schubiger, G., 1979, Absence of distal to proximal intercalary regeneration in imaginal wing discs of Drosophila melanogaster, Dev. Biol., 68:151.
Herschkowitz-Kaufman, M., 1975, Bifurcation analysis of nonlinear reaction-diffusion equations. II. Steady state solutions and comparison with numerical simulations, Bull. Math. Biol., 37:589.
Hiernaux, J., and Erneux, T., 1979, Chemical patterns in circular morphogenetic fields, Bull. Math. Biol. 41:461.
Holder, N., Tank, P., and Bryant, S.V., 1980, Regeneration of symmetrical forelimbs in the axolotl, Ambystoma mexicanam, Dev. Biol., 74:302.
Hunt, R.K., 1975, Developmental programming for retinotectal patterns, in: "Cell Patterning," Ciba Foundation Symposium 29, Associated Scientific Publishers, Amsterdam.
Jacobson, M., 1968, Development of neuronal specificity in retinal ganglion cells of Xenopus, Dev. Biol., 17:202.
Jürgens, G., and Gateff, E., 1979, Pattern specification in imaginal discs of Drosophila melanogaster: Developmental analysis of a temperature-sensitive mutant producing duplicated legs, Wilhelm Roux's Archives, 186:1.
Kauffman, S.A., 1978, A Cartesian coordinate model of positional information in imaginal discs of Drosophila, 20th Annual Drosophila Conference.
Kauffman, S.A., and Ling, E., 1981, Regeneration by complementary wing disc fragments of Drosophila melanogaster, Dev. Biol., 82:238.
Kauffman, S.A., Shymko, R., and Trabert, K., 1978, Control of sequential compartment formation in Drosophila, Science, 199:259.
Kernevez, J. P., 1980, "Enzyme Mathematics," North Holland, Amsterdam, in press.
Krivi, G.G., and Schneiderman, H.A., 1980, Pattern regulation in Drosophila wing disc reaggregates, submitted for publication.
Lacalli, T.C., and Harrison, L.G., 1978, The regulatory capacity of Turing's model for morphogenesis, with application to slime moulds., J. Theor. Biol., 70:273.
Meinhardt, H.A., 1977, A model of pattern formation in insect embryogenesis, J. Cell. Sci., 23:117.
Milojević, B.D., 1924, Beiträge zur Frage über die Determination der Regenerate, Wilhelm Roux's Archives, 103:80.
Morgan, T.H., 1901, "Regeneration," MacMillan, New York.
Needham, A.E., 1965, in: "Regeneration in Animals and Related Problems," V. Kortsis and H. Trampusch, eds., North Holland, Amsterdam.

Nicolis, G., Erneux, T., and Herschkowitz-Kaufman, M., 1978, Pattern formation in reacting and diffusing systems, Adv. in Chem. Phys. 38:263.

Nicolis, G., and Prigogine, I., 1977, "Self-organization in Non-equilibrium Systems," Interscience, New York.

Ouweneel, W.J., 1976, Developmental genetics and homeosis, Adv. Gen., 18:179.

Ouweneel, W.J., 1969, Morphology and development of loboidophthalmoptera, a homeotic strain in Drosophila melanogaster, Wilhelm Roux's Archives, 164:1.

Postlethwait, J.H., 1974, Development of the temperature sensitive homeotic mutant eyeless ophthalmoptera of Drosophila melanogaster, Dev. Biol., 36:212.

Postlethwait, J.H., and Schneiderman, H.A., 1974, Developmental genetics of Drosophila imaginal discs, Ann. Rev. Genetics, 7:381.

Reinhardt, C., Hodgkin, N.M., and Bryant, P.J., 1977, Wound healing in imaginal discs of Drosophila. I. Scanning electron microscopy of normal and healing wing discs, Dev. Biol., 60:238.

Roberts, P., 1964, Mosaics involving aristapedia, a homeotic mutant of Drosophila melanogaster, Genetics, 49:593.

Rose, S.M., 1962, Tissue-arc control of regeneration in the amphibian limb, Symp. Soc. Study Develop. Growth, 20:153.

Russell, M., 1978, A spherical coordinate model of positional information, 20th Annual Drosophila Conference.

Sander, K., 1977, Current understanding of cytoplasmic control centers, in: "Insect Embryology," S.W. Visscher, ed., Montana State University Press.

Sattinger, D.H., 1972, Topics in stability and bifurcation theory, in: "Lecture Notes in Mathematics. Vol. 309," Springer-Verlag, Heidelberg.

Schubiger, G., 1971, Regeneration, duplication and transdetermination in fragments of the leg disc of Drosophila melanogaster, Dev. Biol., 67:286.

Schubiger, G., and Schubiger, M., 1978, Distal transformation in Drosophila leg imaginal disc fragments, Dev. Biol., 67:286.

Sengel, P., 1953, Sur l'induction d'une zone pharyngienne chez la planaire d'eau douce Dugesia lugubris O. Schm., Arch. d'Anat. Micr., 42:57.

Slack, J., 1980, A serial threshold theory of regeneration, J. Theor. Biol. 82:105.

Slack, J., and Savage, S., 1978a, Regeneration of reduplicated limbs in contravention of the complete circle rule, Nature, 271:760.

Slack, J., and Savage, S., 1978b, Regeneration of mirror symmetrical limbs in the axolotl, Cell, 14:1.

Stern, C., 1968, Developmental genetics of pattern, in: "Genetic Mosaics and Other Essays," Harvard University Press, Cambridge, Mass.

Stocum, D.L., 1978, Regeneration of symmetrical hindlimbs in larval salamanders, Science, 200:790.

Stocum, D.L., 1975, Regulation after proximal or distal transposition of limb regeneration blastemas and determination of the proximal boundary of the regenerate, Dev. Biol., 45:112.
Strub, S., 1977, Development potential of the cells of the male foreleg disc of Drosophila, Wilhelm Roux's Archives, 181:309.
Swett, F.H., 1977, Determination of limb axes, Quart. Rev. Biol., 12:322.
Swett, F.H., 1924, Exceptions to Bateson's rules of mirror symmetry, Anat. Rec., 28:63.
Tank, P.W., and Holder, N., 1978, The effect of healing time on the proximodistal organization of double-half forelimb regenerates in the axolotl, Abystoma mexicanum, Dev. Biol., 66:72.
Tokunaga, C., 1978, Genetic mosaic studies of pattern formation in Drosophila melanogaster, with special reference to the prepattern hypothesis, in: "Results and Problems of Cell Differentiation, Genetic Mosaics and Cell Differentiation, Vol. 9," A. Gehring, ed. Springer-Verlag, Heidelberg, New York.
Turing, A.M., 1952, The chemical basis of morphogenesis, Philos. Trans. R. Soc. London, Ser. B., 237:37.
Van der Meer, J.M., and Ouweneel, W.J., 1974, Differentiation capacities of the dorsal mesothoracic (haltere) disc of Drosophila melanogaster. II. Regeneration and duplication, Wilhelm Roux's Archives, 174:361.
Weiss, P., 1939, "Principles of Development," Holt, Reinhart and Winston, New York.
Winfree, A.T., 1980, "The Geometry of Biological Time. Biomathematics Vol. 8," Springer-Verlag, Heidelberg, New York.
Wolpert, L., 1971, Positional information and pattern formation, in: "Current Topics in Developmental Biology, Vol. 6," A.A. Moscana and A. Monroy, eds., Academic Press, New York.
Wolpert, L., 1969, Positional information and the spatial pattern of cellular differentiation, J. Theor. Biol., 25:1.

BIFURCATIONS IN INSECT MORPHOGENESIS II*

Stuart A. Kauffman
Department of Biochemistry and Biophysics
University of Pennsylvania
School of Medicine
Philadelphia, Pennsylvania 19104

* This research has been supported in part by grants GM 22341-04 from the National Institutes of Health, PCM 78-15337 from the National Science Foundation, and CD-30 from the American Cancer Society.

TABLE OF CONTENTS

BIFURCATIONS IN INSECT MORPHOGENESIS II

Stuart A. Kauffman

Department of Biochemistry and Biophysics
University of Pennsylvania
School of Medicine
Philadelphia, Pennsylvania 19104

Two of the most fundamental problems in developmental biology are the manner in which cells in different regions of an embryo come to adopt different developmental programs, and the relation between such different programs. The purpose of this lecture is to indicate how bifurcations in reaction-diffusion systems may offer new insights into these problems. The discussion will center on the fruit fly, *Drosophila melanogaster*.

Drosophila is a holmetabolous higher Dipteran, possessing egg, larval, pupal and adult stages [Gehring and Nöthiger, 1973]. The egg is an ellipsoid about 500 microns long. Following fertilization and egg deposition, the initial zygotic nucleus undergoes 13 nuclear mitoses without division of the egg itself, thereby creating a syncytium. By the ninth division, most nuclei migrate from the yolky core of the egg and arrive nearly synchronously at the cortex. After four further divisions, cell membranes extend down then beneath each nucleus, creating the first true cells which comprise the ellipsoidal monolayered cellular blastoderm [Turner and Mahowald, 1976]. Shortly thereafter, gastrulation movements begin, with the formation of bilaterally symmetric cephalic and hind gut furrows which roughly divide the egg longitudinally into thirds, and the ventral midline furrow through which prospective mesodermal tissue invaginates. The tissues which shall form the embryo proper, the germ band, deform the embryo by extending from both polar areas back along the dorsal midline, then the germ band contracts, revealing the initially segmented embryo with 12 longitudinal segments [Turner and Mahowald, 1977]. At 24 hours the embryo hatches as a first instar larva. After three instars, lasting a total of 5 days at 25°C, the third instar larva forms a pupa, the complex processes of metamorphosis ensue, and the adult emerges about 5 days later.

Part of the fascination with Drosophila is due to the fact that the entire ectoderm of the adult derives from the terminal metamorphosis of larval organs called imaginal discs [reviewed by Gehring and Nöthiger, 1973; Nöthiger, 1972]. Each disc carries a tissue and apparently cell heritable determination to form a specific part of the adult ectoderm; eye, antenna, wing-thorax, first, second, or third leg, genital structures, etc., Fig. 1. The operational definition of determination lies in experiments in which a given disc, for example, the wing-thorax disc, is dissected from a third instar larva and cultured in an adult abdomen, where the disc cells proliferate, but do not undergo terminal differentiation to produce adult cuticle. After a 7-14 day period of culture, the disc implant may be recovered from the host abdomen, cut in two and each fragment

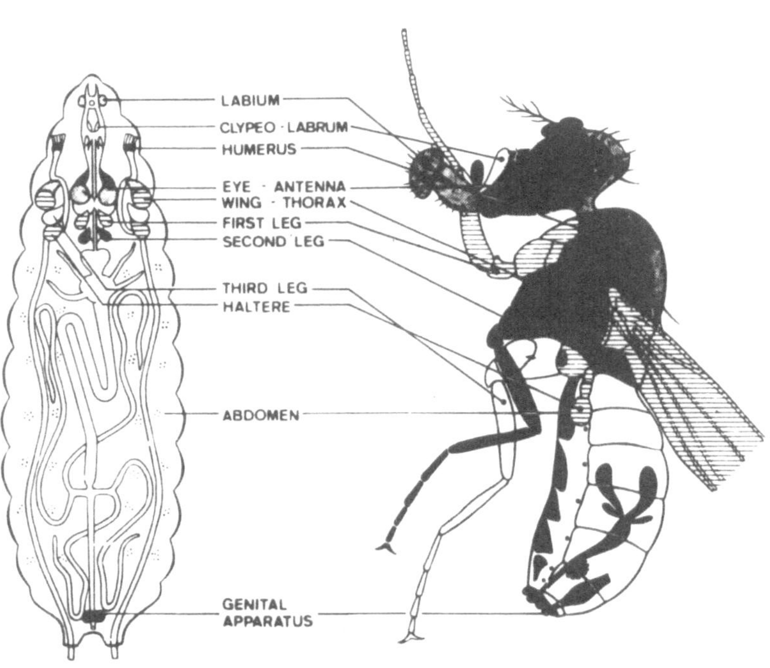

Fig. 1. Schematic representation of the larval organization and the location of the different discs. Discs and their corresponding adult derivatives are connected by lines and given the same hatching or shading.

transplanted to a second host adult, thereby establishing a tissue culture. At any stage, the differentiative capacity of a subfragment may be assayed by injection into a third instar larval host. The injected disc fragment undergoes metamorphosis with the host, and can be recovered as a mass of adult cuticle from the abdomen of the emerged adult. From the characteristic patterns of hairs, sensillae and bristles, the implant can be diagnosed as wing, first leg, genital, etc. The major result found by Hadorn and his co-workers [Hadorn, 1966] in such experiments is that a wing disc, or leg disc, subcultured for years, can still metamorphose into adult wing, or leg tissue. Important conclusions can be drawn from these series of experiments. First, disc subfragments have a heritable determination to form a specific adult structure. Second, the heritability is stable over hundreds of cell divisions. Therefore, the molecular mechanisms mediating that heritability could not be merely the initial partitioning to imaginal disc cells of some substance, since it would be diluted out over cell generations. Whatever carries the determined state must be regenerated over cell cycles. While the mechanisms remain unknown, plausible postulates include integration of transposable genetic elements at diverse loci; chromatin ligands which bind tightly to genetic control loci, are replicated or replenished once per cell cycle, and are partitioned in orderly ways to daughter cells; and systems of genes and diffusible products which activate and repress one another and possess alternative stable patterns of activity.

The progenitor cells of the different imaginal discs are now known to lie at well defined positions on the cellular blastoderm [Hotta and Benzer, 1972; Garcia-Bellido and Merriam, 1969; Janning, 1978]. Fig. 2 shows the blastoderm fate map of *Drosophila*. The features to stress at this stage are that the map is a well ordered two dimensional array of pattern elements, bilaterally symmetrical on the closed curved surface of the egg, and that the pattern elements on the fate map are essentially homotopic to the adult. The head maps at the anterior end, followed by the three thoracic fragments, the 7 abdominal segments, and the genital segment. Ventral (leg and abdominal sternites) structures map ventrally on the egg to dorsal (thoracic and abdominal tergite) structures. This ordered array of progenitor zones raises the fundamental question of *positional information* [Wolpert, 1971]. What kinds of processes underlie the establishment of ordered developmental commitments to these heritably different fates? Is there a micromosaic of molecular determinants laid down at specific positions in the oocyte by the mother? Is position in the egg specified by more general "gradients" which she preestablishes? Finally, is position specified "epigenetically" during the early stages of embryogenesis by processes within the early embryo? The answers are not clear [Sander, 1975, 1977; Kalthoff, 1979; Kauffman, 1979], but I will sketch in a later section the role dissipative structures may play.

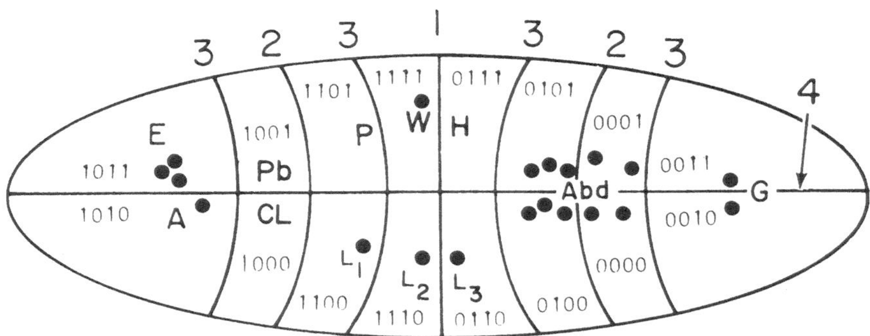

Fig. 2. Fate map of the Drosophila egg [Garcia-Bellido and Merriam, 1969; Hotta and Benzer, 1976]. A, antenna; E, eye; Pb, proboscis; W, wing and mesothorax; H, Haltere; L1, L2, L3, first, second and third legs; Cl, clypeolabrum; Abd, abdominal segments; including 2ns to 6th abdominal tergites and 2nd to 6th abdominal sternites; G, genital. Solid lines 1, 2, 3, 4, are predicted successive compartmental lines based on successive compartmental lines based on successive eigenfunction model. Unique 4 digit binary code word in each compartment reflects each different segment.

If the existence of an ordered fate map poses the problem of positional information, the striking phenomena of transdetermination and homeosis unavoidably raise the question of the relation of developmental programs in diverse imaginal discs. Transdetermination was discovered when imaginal disc fragments were cultured. Although each disc normally differentiates into structures it was initially determined to form, occasionally it forms normal structures from a different disc [Hadorn, 1966]. A broad body of evidence shows that this alteration in adult derivatives is not due to somatic mutations, and is heritable in the cultured imaginal tissue once established [Hadorn, 1966; Gehring and Nöthiger, 1973]. The term transdetermination reflects the belief that the alteration is an epigenetic change to a new heritable developmental program. Fig. 3

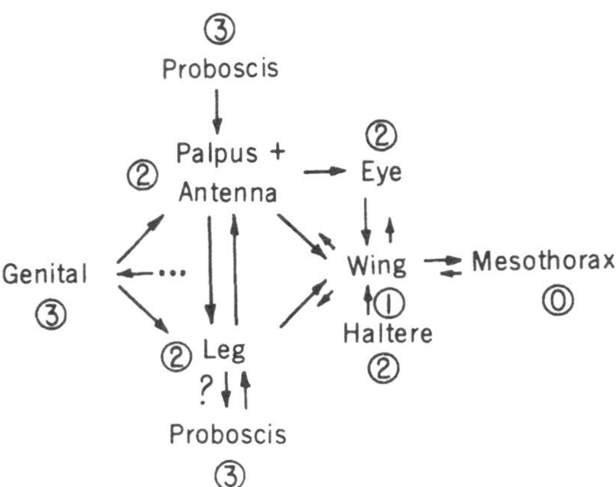

Fig. 3. Patterns of transdetermination from each disc. Length of arrows reflects transdetermination probabilities. Circled numbers, minimum number of transdetermination steps to mesothorax.

shows the known patterns of transdetermination from each disc. Lengths of arrows reflect probabilities of transdetermination. The following generalizations hold [Kauffman, 1973]: 1) Most transdetermination steps are reversible, but with asymmetric transition probabilities. 2) Each disc can transdetermine into one or a few, but not all other discs in a single step, thus there are allowed and forbidden one step transitions. 3) Pathways of sequential transdetermination exist, eg. genital to leg to wing to thorax. 4) There is a global ordering towards mesothorax; all transdeterminations which move one step closer to mesothorax are more probable than their inverse.

Transdetermination is an example of metaplasia, the transformation of tissue normally destined for one fate, to a distinct fate. The second major metaplasia in Drosophila is due to a class of mutants called homeotic mutants [reviewed by Ouweneel, 1976]. These dominant, or recessive, point, deletion, inversion, or translocation mutants convert one structure to another. Exemplars include the dominant Nasobemia, which converts antenna to the mesothoracic leg and less frequently, the eye to wing [Stepshin and Ginter, 1972; Haynie, 1980]; tumorous head, which converts head and antenna tissues to leg, abdominal and genital structures [Postlethwait et al, 1972]; and the bithorax complex which converts parts or all of some thoracic and abdominal segments into one another [Lewis, 1964, 1978]. On the basis of their phenotypes, it is useful to classify homeotic mutants as parallel, divergent, or convergent [Kauffman et al, 1978; Kauffman, 1979]. Parallel homeotics convert a single tissue in a single direction, but may act coordinately on more than one tissue. For example, Nasobemia converts antenna to mesothoracic (second) leg and eye to wing (mesothorax). Divergent homeotics, such as tumorous head, appear to cause a single tissue to diverge to two distinct fates; antenna to abdomen and leg. Convergent homeotic mutants can transform two different tissues to a common final fate, as seen in Extrasexcomb, which converts 2nd and 3rd legs to first legs [Hannah-Alava, 1958]. As discussed below, these different classes appear to require different hypotheses about the underlying mutant action. A second classification of homeotic mutants asks whether the mutant transforms a tissue to a neighboring domain on the blastoderm fate-map, or to a more distant location. Nasobemia, for example, largely transforms head and antenna to 2nd thoracic structures, skipping the intervening gnathal and 1st thoracic structures; similarly, tumorous head transforms head to genital and abdominal fates, hence from one end of the fate map to the other. In contrast, bithorax alleles largely transform between neighboring areas on the fate map, for example mesothorax to metathorax.

The importance of transdetermination and homeosis is that these metaplasias presumably reflect "neighboring" relations between the developmental programs in the discs which transform into one another.

In some sense then, antenna and leg are program neighbors. Genital, though close to both antenna and leg, is farther from eye, wing, or haltere. It is clear that these "neighboring" relations bear no simple relation to the distance separating different progenitor zones on the fate map, since genital and antenna appear to be developmental neighbors, but derive from opposite poles of the blastoderm.

The observations discussed so far lay out some of the desiderata of any good theory; to try simultaneously to understand how position may be specified in the early embryo, and also to account for the ways in which apparently neighboring developmental programs are not spatial neighbors on the blastoderm.

An important set of clues have come from analysis of the development of the wing-thorax disc. Cells in neighboring domains which have come to adopt different developmental commitments, say to wing or leg, should not thereafter jointly form either a wing or a leg. Further, since determination is cell heritable, it is to be expected that the sets of daughter cells (clones) derived from such domains with different commitments, will also not form joint structures. Such domains of cells might be expected to give rise to lines of "clonal restriction", where clones with one commitment on one side would be precluded from crossing to the other side. Such lines, called compartmental boundaries [Garcia-Bellido et al, 1973] have recently been found in Drosophila development, utilizing the genetic technique of somatic recombination which allows all the daughter cells (clone) derived from a randomly affected initial cell to be identified in the adult. The most detailed data for such lines has been amassed in the wing-thorax (hereafter, wing) disc, and I shall focus on this tissue in the present section.

At the blastoderm stage, the progenitor zone for the wing disc has about 40 cells [Merriam, 1978]. During embryogenesis, this buds inward to form a hollow ellipsoid sac of cells which remains attached to the larval trachea by a narrow stalk. During the three larval instars, the wing disc grows in size and cell number to a final 60,000 cells [Bryant, 1975]. In one "hemisphere" of the ellipsoid sac, cells fail to proliferate and stretch to form the thin peripodal membrane which is lost during metamorphosis. The remaining hemisphere of columnar cells forms the disc proper and metamorphoses into the adult wing and hemithorax.

If a cell is genetically marked at about the blastoderm stage, its clonal derivatives are found not to cross a specific line on the adult wing thorax which divides the tissue into anterior and posterior compartments [Garcia-Bellido, 1975], Fig. 4a. If one

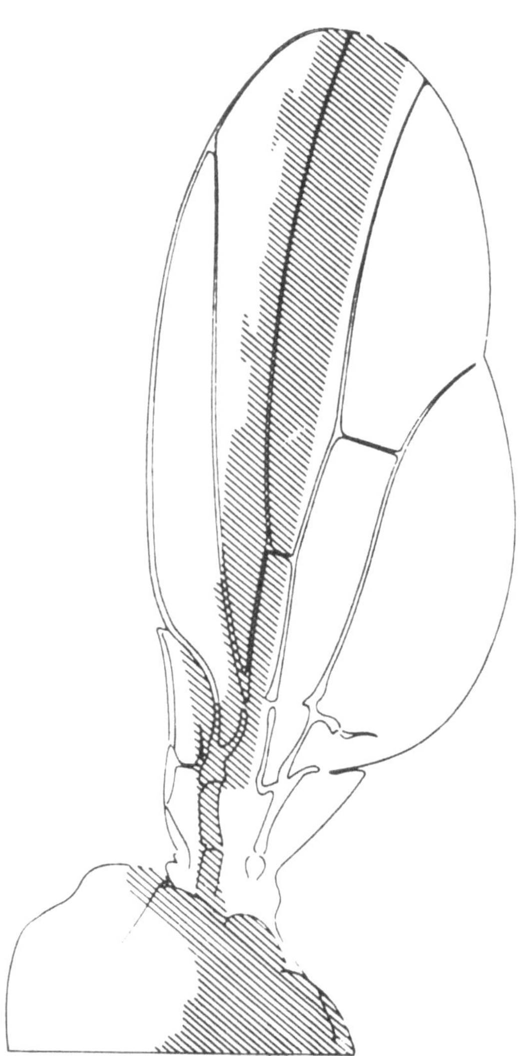

Fig. 4a. A large clone whose posterior margin runs along the anterior-posterior border of the wing and thorax. From Garcia-Bellido et al [1973].

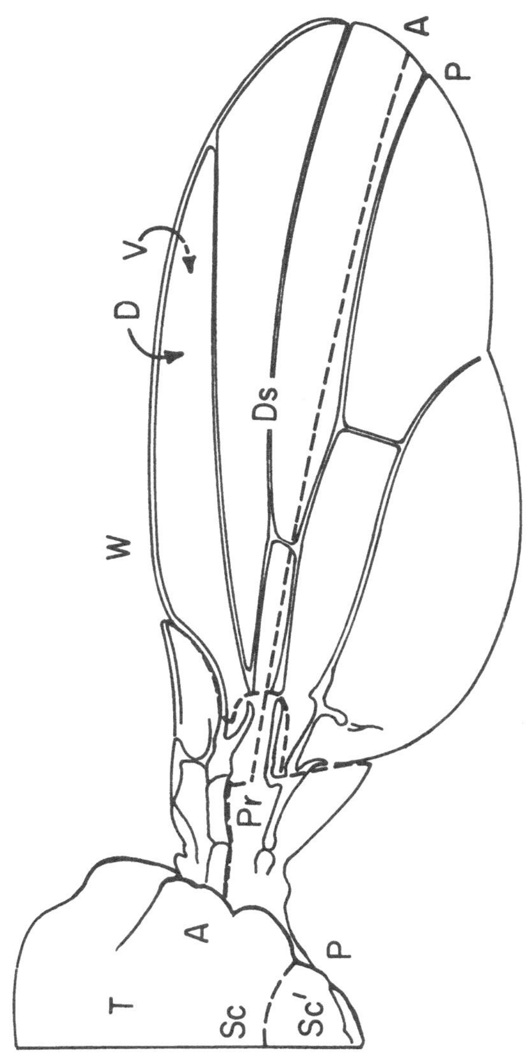

Fig. 4b. The five compartmental lines on the thorax and wing. A, anterior; P, posterior; V, ventral; D, dorsal; T, thorax, W, wing; Pr, proximal wing; Ds, distal wing; Sc, scutum; Sc', scutellum. From Garcia-Bellido [1975].

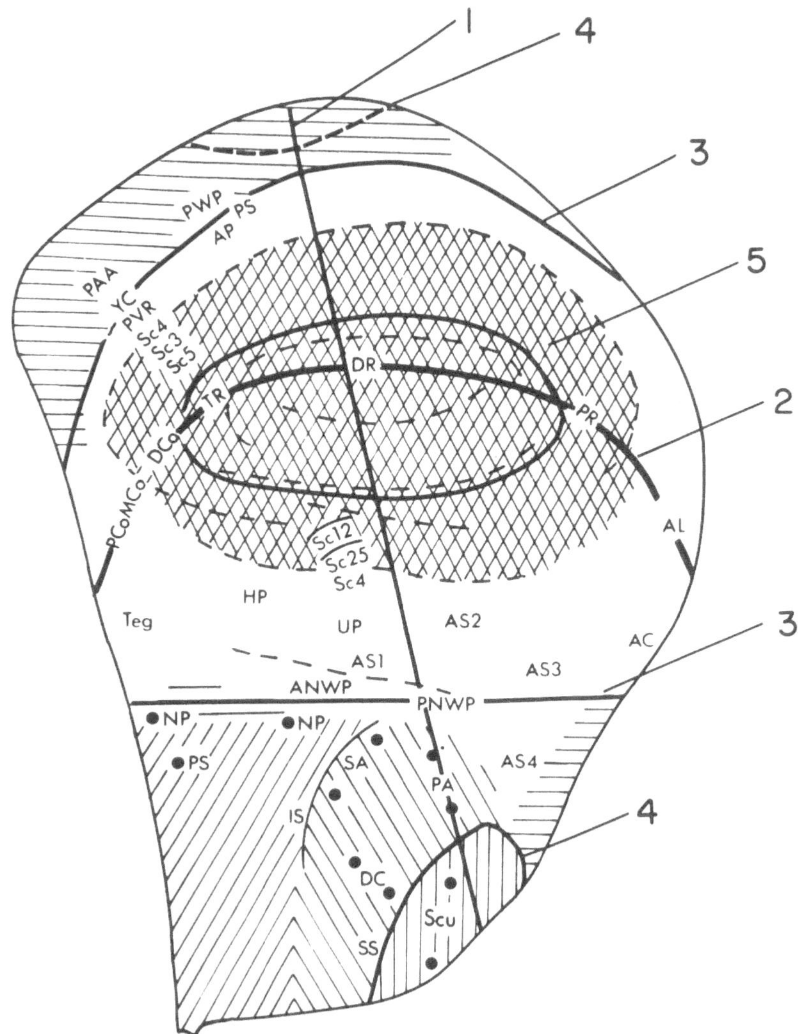

Fig. 4c. Projection of the five compartmental boundaries onto the fate map [Bryant, 1975] of the third instar wing disc. The dotted line 4 is the postulated compartmental line in the ventral thorax needed to complete the anterior-posterior, dorsal-ventral (twofold) symmetry.

examines a number of flies, however, one finds that clones initiated at that stage may cross anywhere else inside either compartment. About mid first instar, a second clonal restriction occurs and isolates clones to the dorsal or ventral halves of the wing-thorax. At about this time, each clone is further restricted to the wing or the thorax by a third boundary. Somewhat later, a fourth line subdivides the dorsal, and perhaps ventral thorax. In late third instar, a final line isolates the distal from the proximal wing [Garcia-Bellido, 1975], Fig. 4b. The basic phenomenon, then, is sequential formation of clonal restriction lines which successively subdivide the wing disc into finer subdomains as the disc grows larger in size. The locations of these lines on the adult, however, reveal little of their geometry at the times of their formation. A better approximation is seen by projecting them onto the known fate map of the third instar wing disc, Fig. 4c. The wing disc, Fig. 4c, forms the wing, by folding over the dorsal-ventral line which becomes the adult wing margin, thereby apposing dorsal and ventral thorax areas, and dorsal and ventral wing areas, creating a "bag" which everts back through its peripodal membrane to form the adult wing and thorax.

The striking feature of the compartmental lines on the wing disc is their clear two-fold symmetry. Anterior, posterior, ventral and dorsal quadrants look fundamentally similar. Good evidence suggests that these compartmental lines arise in situ in the growing disc, and do not reflect movement of cells relative to one another. Therefore, the question arises of how such symmetrical lines might arise.

The symmetries and patterns of the lines shown in Fig. 4c are closely analogous to the nodal lines in the eigenfunctions of the Laplacian operator on an ellipse [McLachlan, 1947; King and Wiltse, 1958]. The pleasant question, "Can one hear the shape of a drum?" can be transformed to ask whether one can conceive of an underlying dynamical system having the property that as the size and shape of the wing disc grows, a sequence of different eigenfunctions, with different nodal lines can arise. A wide variety of different underlying mechanisms, having similar spatial operators, hence similar eigenfunctions, would probably suffice. Figs. 5a-f show the nodal lines which would arise on a freely suspended elliptical plate excited to resonant vibration by a gradually higher pitched external sound source [King and Wiltse, 1958]. Essentially the same sequence will arise using a reaction-diffusion dissipative structure model which I now discuss briefly.

Perhaps the most general reason to consider reaction-diffusion systems and solutions to the Laplacian is that positional information in Drosophila and many organisms seems to be locally averagable.

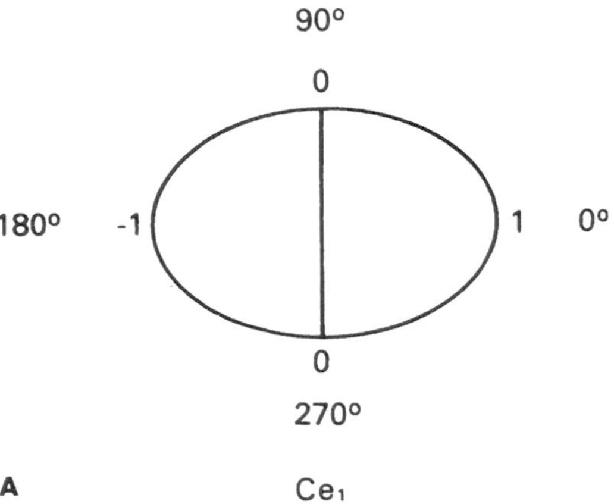

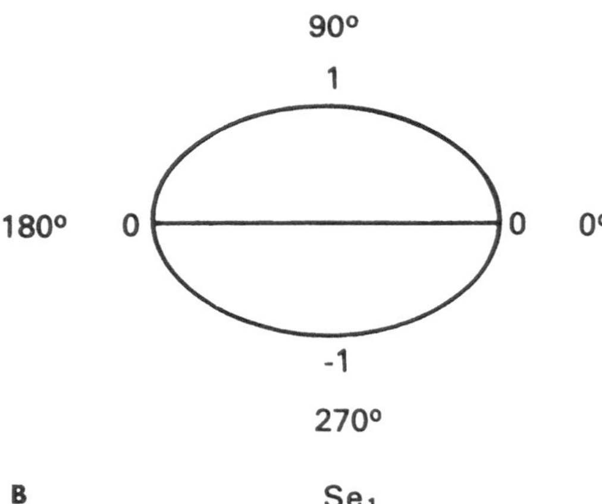

Fig. 5. Nodal lines of successive wave patterns which fit onto an ellipse as it enlarges. The patterns are similar to those on a circle. a) $Ce_1(\xi,s_{11})(n,s_{11})$ (abbreviated Ce_1) is a slight distortion of $J_1(kr)\cos\phi$ (see Fig. 7a). b) $Se_1(\xi,s_{11})se_1(n,s_{11})$ (that is Se_1) is Ce_1 rotated 90°.

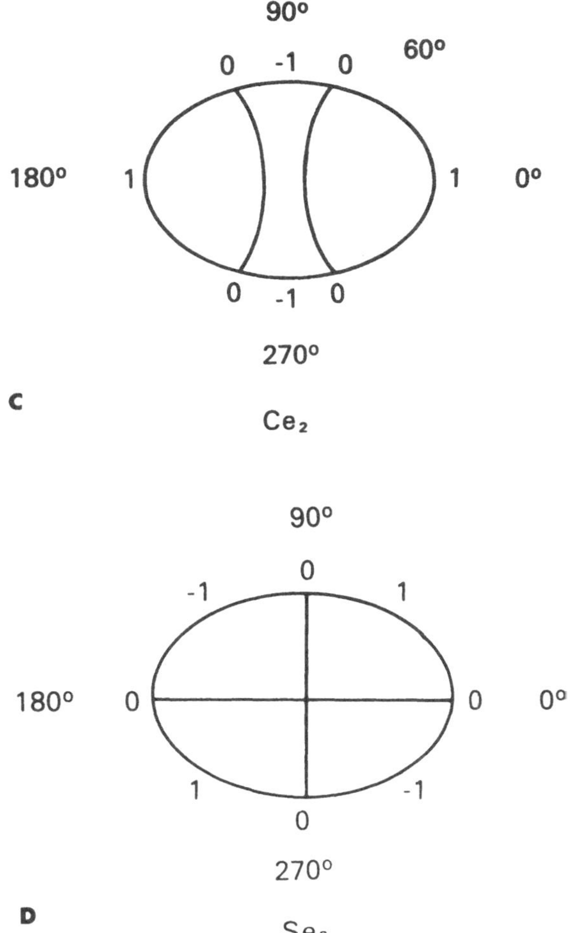

Fig. 5. c) Ce_2 is analogous to $J_2(kr)\cos 2\phi$ (see Fig. 7b) but on an ellipse the radii split to form pairs of ccnfocal hyperbolae. d) Se_2 is analogous to $J_2(kr)\sin 2\phi$.

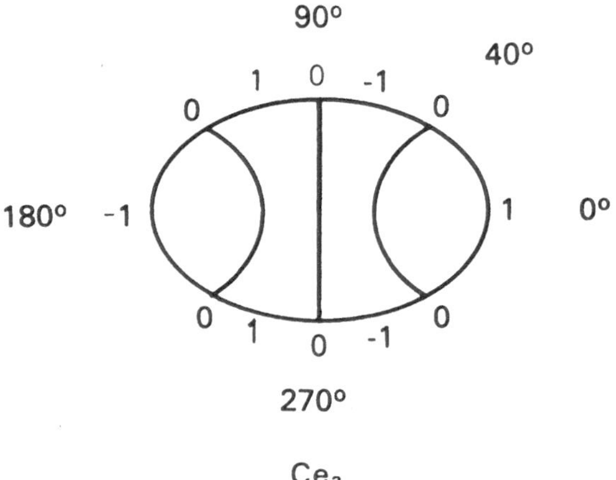

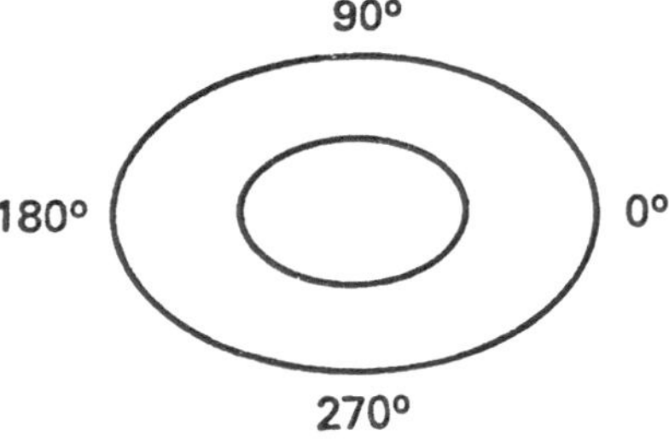

Fig. 5. e) Ce_3 is analogous to $J_3(kr) \cos 3\phi$. f) $Ce_0(\xi, s_{01}) ce_0(n, s_{01})$ or Ce_0 is analogous to $J_0(kr) \cos 0\phi$. This mode is a hill shaped pattern with an interior nodal ellipse similar to Fig. 7c. See the text for definitions of the symbols.

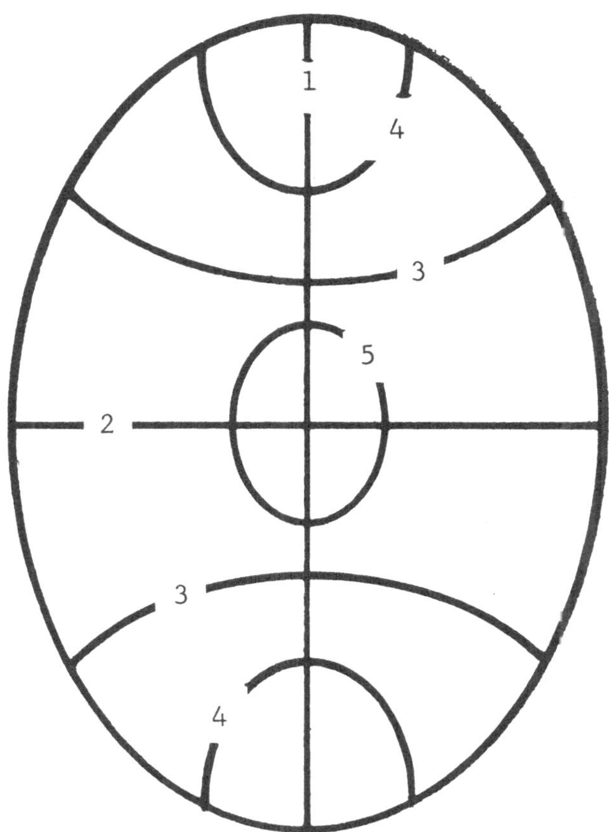

Fig. 5g. Taking account of wing disc growth, we project all five predicted compartmental boundaries onto one ellipse. The observed boundaries are shown in Fig. 4c.

Juxtaposition of tissue edges which are usually not adjacent leads to the intercalary regeneration of pattern elements normally lying between the abutted surfaces [Harrison, 1921; Lawrence, 1973; French, Bryant and Bryant, 1976]. The oldest, and still simplest, hypothesis to account for this is to postulate "gradients" of "morphogens" whose concentration specifies position. Juxtaposition of normally non-adjacent edges creates gradient discontinuities followed by a "diffusive-like" smoothing to recreate intervening gradient levels. Therefore, solutions to the Laplacian afford an intelligent first guess to the spatial shapes which such gradients might take [Trainor and Goodwin, 1980]. Gradients might be generated by a variety of means. Older hypotheses included special source or sink regions, polarized transport, or other mechanisms. However, recent evidence in Drosophila and other systems, described in the previous lecture [Kauffman, Part I] and elsewhere, [Meinhardt, 1977; Kauffman, 1980] suggests reaction-diffusion systems as a particularly valuable class of hypotheses, to establish gradients controlling positional information.

Appropriate coupling of reactions and diffusive transport in a planar sheet of cells, considered as a continuum, can readily generate a succession of differently shaped gradients of the same underlying morphogens as the tissue changes in size and shape, or other parameters alter [Herschkowitz-Kaufman, 1975; Kauffman, et al, 1978]. In Fig. 6, as in the previous lecture, we assume specific nonlinear reaction laws, and a simple diagonal diffusion matrix, such that each species' diffusion is independent of the remainder. The dynamical system possesses a spatially homogeneous, temporal steady state. The resulting equations were linearized about this steady state and analyzed in the familiar manner. As it is now well understood, Fig. 6, and discussed in more detail in the previous lecture [Kauffman, Part I], with appropriate coupling in the linearized reaction terms, and dissimilarity in the diffusion constants of at least two hypothetical morphogens, a dynamical system can be constructed which acts as an amplifier, selecting from the thermal fluctuations around the steady state, a limited range of spatial wavelengths (L_1-L_2) which will grow in amplitude in time [Turing, 1952; Gmitro and Scivin, 1966; Nicolis and Prigogine, 1977; Herschkowitz-Kaufman, 1975; Kauffman et al, 1978; Babloyantz and Hiernaux, 1975]. If such a chemical system is embedded in a specific geometric domain, and no-flux boundary conditions are imposed, Fig. 6 and its legend, then only when the size and shape of the domain allow a particular chemical pattern of the amplifiable wave-length to satisfy the boundary conditions, will that pattern grow. If the chemical system specified in Fig. 6 were placed in a circular petri dish which gradually enlarged from an initial small radius, the following general results would occur in the linearized system of equations: 1) at sufficiently small radii, diffusion overwhelms reaction induced instabilities, and the system remains at the homogeneous state. As the radius of the

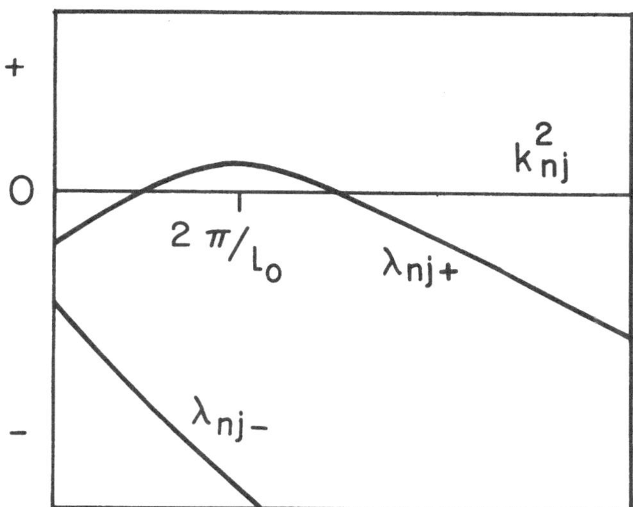

Fig. 6. Dispersion relationship for successive eigenfunction model, described in more detail in previous chapter. The dispersion relation shows the stability of the underlying two chemical reaction-diffusion system to spatially distributed perturbations away from the spatially homogeneous steady state. For those ranges of k^2_{nj} where the largest eigenvalue λ_{nj+} is positive, chemical patterns of wavelengths proportional to $1/k_{nj}$ arise. This system therefore amplifies chemical patterns with a restrictive range of wavelengths. Specific eigenfunctions arise sequentially when such a system is embedded in a growing domain having no-flux boundary conditions.

circular petri dish enlarges beyond a first bifurcation value, given by the first 0 derivative of the first Bessel function for wavelength L_1, a first chemical pattern arises, Fig. 7a. If the range of amplifiable wavelengths, L_1-L_2, Fig. 6, is sufficiently small, then when the petri dish increases past a second critical radius, such that the longer wavelength, L_2, no longer satisfies the boundary conditions, the first chemical pattern will decay toward the homogeneous steady state. As the size of the dish increases beyond a third critical bifurcation value, given by the first 0 in the

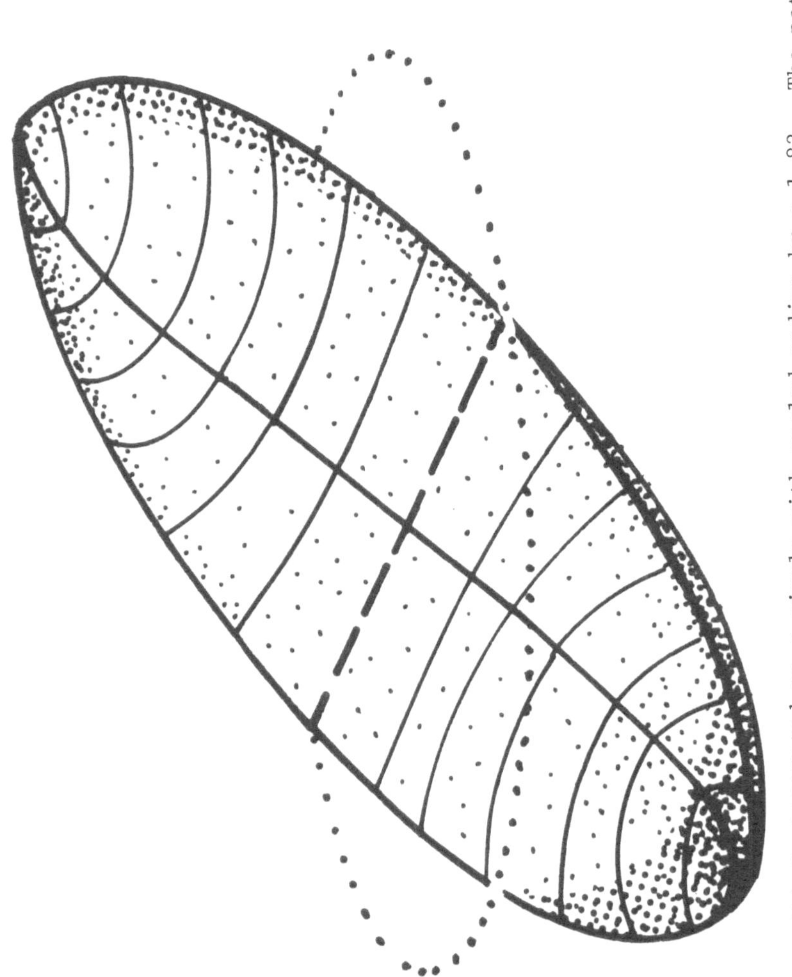

Fig. 7a. Wave pattern generated on a circle with scaled radius kr = 1.82. The pattern is the product of a radial part, $J_1(kr)$ (the first order Bessel function) and an angular part, cos ϕ. The dashed nodal line of zero (that is, steady state) concentration runs along diameter of the circle from $\phi = 90°$ to $\phi = 270°$. The dotted circle outlines the circular radius.

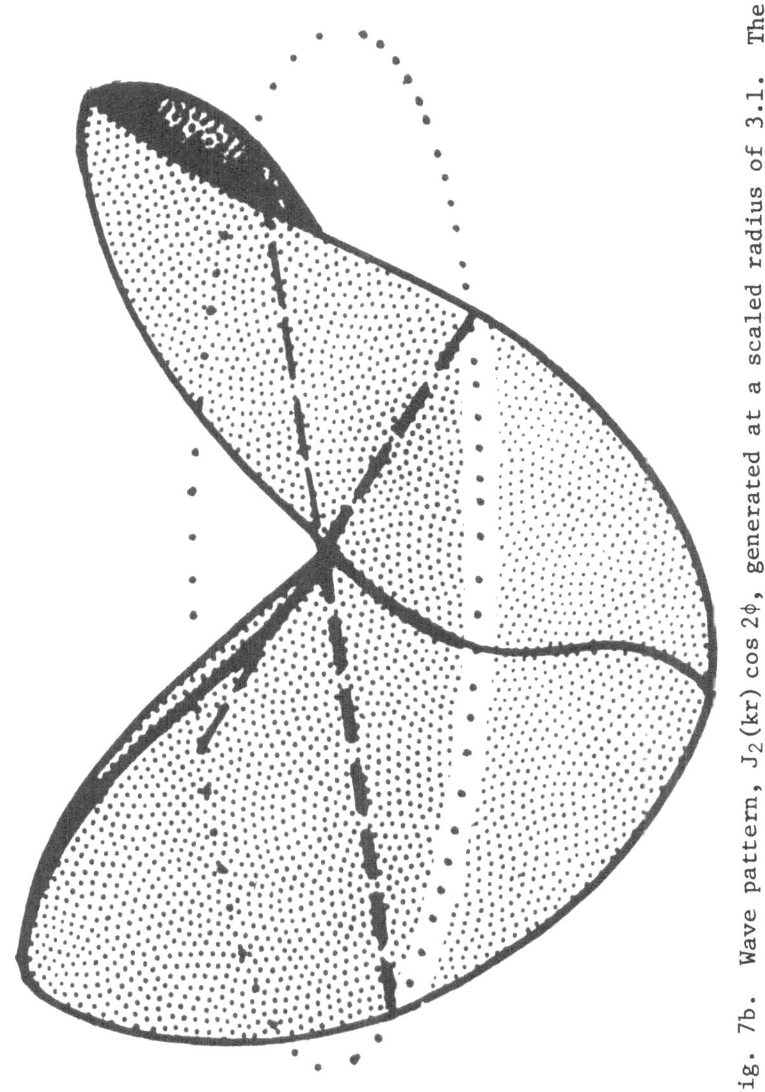

Fig. 7b. Wave pattern, $J_2(kr) \cos 2\phi$, generated at a scaled radius of 3.1. The dashed lines are crossed nodal lines on two perpendicular diamters.

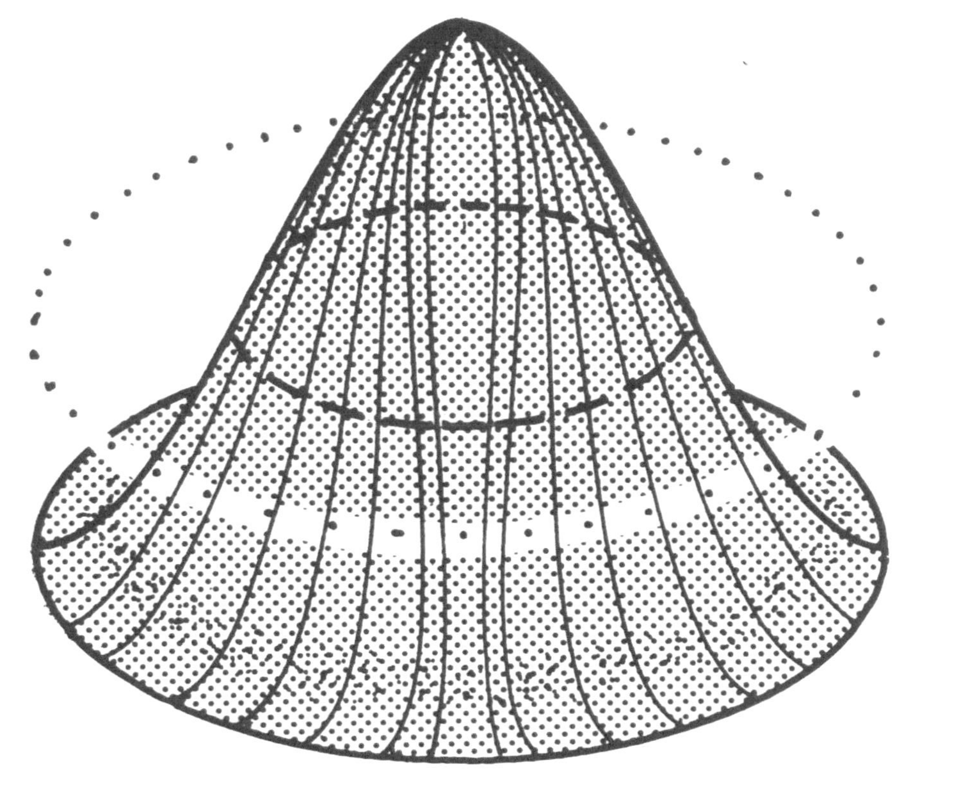

Fig. 7c. Pattern generated at a scaled radius of 3.8, where the zero in the derivative of $J_0(kr)$ matches the radial boundary condition. The pattern is $J_0(kr)\cos 0\phi$, which has no angular variation. The nodal line is concentric with the outer radius.

derivative of the second Bessel function for the smaller wavelength, L_1, a second chemical pattern, Fig. 7b will arise, then decay again as the dish continues to enlarge such that the longer wavelength, L_2 in the second pattern, does not satisfy the boundary conditions. The third pattern to arise, Fig. 7c, corresponds to a radius at which L_1 satisfies the second 0 derivative of the 0th Bessel function. In these linearized equations, when the radius becomes sufficiently large, two successive chemical patterns will simultaneously satisfy the boundary conditions and both will be amplified by the dynamical system. Thus their superposition will grow, and, as shown in Part I, establish transverse gradients of substances X and Y. For smaller circles, the existence of superposition of modes depends upon the velocity with which the circle increases in radius relative to the rate of establishment and decay of successive chemical patterns. If the range of allowed wavelengths L_1-L_2, is sufficiently small, the growing petri dish harbors very nearly a sequence of pure eigenfunction patterns. Extensions to full nonlinear analysis depend, of course, on the detailed kinetic models assumed, and can lead to persistence of an established mode beyond the size at which the linear equations predict its decay, and to bifurcation of successive modes from the homogeneous state.

Regardless of the details of the nonlinear models one might construct, the most general feature to abstract from this class of models at this stage, is that if such a dynamical system is embedded in a tissue which undergoes size and shape changes, or some other parameters such as diffusion constants change, the inevitable consequence is that a sequence of differently shaped patterns of the same morphogens arise and decay in succession at a critical sequence of bifurcation values of the size, shape and other parameters. In particular, for an appropriate sequence of elliptical shapes, the succession of chemical patterns have nodal lines as given in Figs. 5a-f.

The set of nodal lines in Figs. 5a-f provide a model of the sequence of compartmental lines, Fig. 4b-d. The first observed line, the anterior-posterior line, arises on the blastoderm prior to the existence of the wing disc as a separate entity, and must be explained in the context of the ellipsoidal egg. Therefore, the second chemical pattern, Fig. 5b, repeats a previously formed compartmental boundary as does the fourth pattern, Fig. 5d. These correspond to the Mathieu functions Se_1 and Se_2 and follow the formation respectively of Ce_1 and Ce_2 at slightly larger sizes for "wide" ellipses [McLachlan, 1947; King and Wiltse, 1958]. Se_1 and Se_2 can therefore be suppressed by nonlinearities in the kinetics by the prior establishment of Ce_1 (Fig. 4a) and Ce_2 (Fig. 4c). With such suppression, the total sequence and symmetry, Fig. 5g, closely fits the observed, Fig. 4c. The first four patterns to arise on the true wing disc shape were given in Figs. 9a-9d of Part I. A later pattern yields the proximal-distal wing boundary.

One means of testing the successive eigenfunction model is to assess its predictions on other imaginal discs with well-defined, but distinct shape histories. The haltere disc is similar in shape to the wing disc, but about .25 its size. The sequential eigenfunction model therefore predicts that only the first few patterns and compartmental lines arise. In fact, the haltere disc displays only the first three of the five lines seen on the wing disc, Fig. 4a 4b [Garcia-Bellido, 1975]. Figures 8a-c and its legend show the predicted compartmental lines on the haltere, leg, and genital discs. The observed compartmental lines on the haltere, leg [Steiner, 1976], and genital [Dubendorfer, 1977] discs are fit moderately well by the predictions.

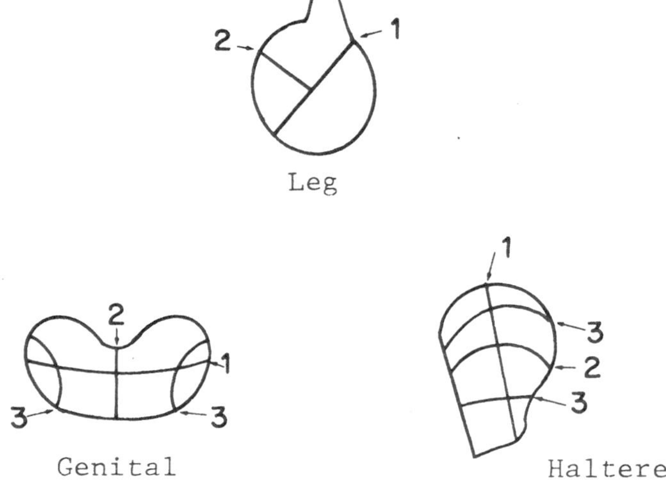

Fig. 8. Schematic compartmental lines on the a) leg, b) genital, and c) haltere.

The most intriguing predictions of the model come from its application to the developing egg. The egg does not increase in size, therefore size cannot be used as a bifurcation parameter to select different chemical patterns. However, as the nuclei migrate to the cortex of the egg, the cortical layer becomes viscous cytoplasm, rather than a relatively less viscous yolky material. It seems reasonable that diffusion constants become progressively smaller in this layer. Tuning diffusion constants smaller while holding their ratios unchanged is equivalent to tuning the amplifiable wavelengths shorter in a constant size physical domain. A succession of chemical instabilities will occur on the ellipsoidal egg whose patterns are the eigenfunctions of the Laplacian on that geometry.

Fig. 2 shows the expectations as the amplifiable chemical wavelength gradually shortens. A first mode has a maximum at one pole and a minimum at the opposite, with a first nodal line at mid-egg dividing the egg into anterior and posterior halves. A second mode, analogous to Ce_2, with minima at the poles and a maximum in the mid-egg region, has two nodal lines, creating middle versus end egg distinctions analogous to the pair of compartmental lines which isolate the mid-disc wing region from the flanking thoracic ends, Fig. 4c. A third chemical pattern creates further longitudinal subdivisions. Finally, the amplifiable range of wavelengths shortens sufficiently to fit onto the egg from dorsal to ventral, creating a fourth compartmental line along the egg's equator.

The sequential eigenfunction model creates a sequence of different chemical patterns on a defined geometry, as parameters change. At each bifurcation value allowing a new pattern to emerge, two patterns are allowed, which are inverses, peaks replacing troughs. To select a specific pattern at each choice point requires some asymmetries in the initial conditions. Therefore, a minimum added requirement of the model are such slight chemical asymmetries in the egg and the discs.

When initially formulated, this predicted sequence of compartmental boundaries on the egg was largely hypothetical. The data then available showed that at the blastoderm stage, compartmental boundaries appeared to isolate adjacent segments, but not the dorsal from ventral disc derivatives from the three thoracic segments, since genetically marked clones at that stage can straddle from mesothorax to mesothoracic leg [Steiner, 1976; Wieschaus and Gehring, 1976; Lawrence and Morata, 1977]. A short time later [Lawrence and Morata, 1977], clonal restriction lines isolate the dorsal from ventral thoracic discs, thus longitudinal lines of clonal restriction do arise on the egg before convincing dorsal ventral lines. Since the formulation of this model, we have tested the prediction that a first compartmental boundary line should arise at mid-egg, and confirmed the prediction [Kauffman and von Allmen, 1980]. The sequence in which the remaining longitudinal lines are established cannot be resolved by current techniques.

In the first part of this article I described both the fate map of the blastoderm, and the metaplasias of transdetermination and homeotic mutants. There the dominant puzzle is to understand the sense in which discs which transform into one another are program neighbors, and simultaneously to understand why that sense of "neighboring" does not map simply onto the blastoderm fate map. The transdetermination flow diagram, Fig. 3, suggests that one disc, such as genital, is simultaneously close to antenna and leg, but farther from wing. The existence of allowed and forbidden single step transitions, sequences of transdeterminations, and the global orientation of frequencies toward mesothorax as the "sink", suggest that the determined state in each disc is comprised of a combination of states of independent subentities. For example, were there two genetic systems, each with two alternative dynamical steady states, 0 and 1, then genital might be encoded (00), leg (01), antenna (10) and wing (11). This encoding predicts allowed one step transitions from genital to antenna and leg, and a sequence from genital toward wing. If the 0-1 transition is easier than the reverse, a global ordering toward wing occurs [Kauffman, 1973].

In Fig. 2, this general concept is applied to the predictions of the eigenfunction model on the egg. The first mid-egg compartmental boundary creates anterior and posterior domains, encoded with a 1 or 0 by the action of a first master gene system. The second pair of compartmental boundaries create an ends versus middle commitment, encoded by a second master gene system in state 0 in the ends, and 1 in the middle region. The third chemical pattern creates alternating 0 and 1 assignments, subdividing existing compartments. The final dorsal-ventral equatorial line assigns a fourth master gene system a 1 state dorsally and 0 ventrally. Two dominant features of this hypothesis should be noted. First, each terminal compartment is assigned a binary combinatorial code word which reflects the sequence of alternative decisions taken during its specification. These combinations supply an epigenetic code in which each terminal compartment, corresponding to a distinctly determined progenitor zone, has a specific binary combinatorial code word. Second, because the successive chemical patterns are not monotonic on the egg, the resulting code is not monotonic on the egg; that is, in passing from anterior to posterior, the number of 0s in successive code words do not increase or decrease monotonically. This feature has the consequence that domains at opposite ends of the egg can have neighboring epigenetic code words. For example, genital (0010) and antenna (1010) differ only in the state of the first master gene system reflecting the anterior versus posterior commitment. It is this feature which shall allow this combinatorial code to account for the intriguing patterns of metaplasia seen in transdetermination and homeotic mutants.

Predictions with respect to transdetermination are made on the following basis: the conversion of haltere to wing, Fig. 2, requires conversion of the first "switch" from 0 to 1. Conversion of haltere to antenna requires the same conversion of this first "switch", but also conversions of the second and fourth master gene systems from 1 to 0; hence transdetermination of haltere to wing should occur more frequently than haltere to antenna. Fig. 3 confirms this prediction. Table 1 lists 37 independent predictions of this coding model. Almost all are true, two or perhaps three are false. The _a priori_ probabilities of such a success rate are very small, and suggest that this combinatorial scheme accounts well for the sense in which developmental programs in different discs are neighbors with respect to transdetermination.

The same combinatorial code accounts well for a broad spectrum of homeotic conversions, including those which cause conversion of tissues from one end of the fate map to the other, such as _tumorous head_ which converts head to genital and abdominal tissue. If a specific homeotic mutant alters the state of a single master gene, then the following three general properties should occur: 1) Some homeotic transformations should occur between non-neighboring domains of the fate map. 2) If the same mutant alters the same master gene in two different tissues, the two should be transformed to two distinct new tissues. That is, coordinated _parallel_ homeotic transformations should occur. This is exemplified by _Nasobemia_ which converts antenna to mesothoracic leg, and eye to wing [Stepshin and Ginter, 1972]. 3) If most homeotic mutants act on single master genes, but transdetermination allows "wobble" of any decision, then the set of tissues to which any disc can transdetermine should be broader than, but inclusive of those to which any specific homeotic mutant transforms it. All these general ordering relations obtain [see review by Ouweneel, 1976]. If the specific combinatorial code in Fig. 2 is a good model, it should account for the actual neighboring relations between discs seen in homeotic mutants. That is, most such transformations should be accounted for by alterations in a single binary digit in the code. Table 2 shows that it is fair to say that this combinatorial scheme accounts for the neighboring relations of developmental programs intimated by both homeosis and transdetermination, and relates it with moderate success to the geometry of the blastoderm fate map.

It would be astonishing were this particular scheme to be correct in detail. The existence of convergent and divergent homeotic mutants shows that it probably is not. If it is assumed that a single homeotic mutant can affect at most a single master

Table 1. Predicted relative transdetermination frequencies derived from the chenical wave model applied to the blastoderm. $L_{1,2} \rightarrow A > L_{1,2} \rightarrow G$ means the model predicts transdetermination from the first or second leg to antenna is greater than to genital. Abbreviations are explained in the legend of Fig. 2.

Prediction	Status	Prediction	Status	Prediction	Status
$H \rightarrow W > H \rightarrow A$	T	$A \rightarrow W > A \rightarrow H$	T	$L \rightarrow W > L \rightarrow E$	T
$H \rightarrow W > H \rightarrow L_{1,2}$	T	$A \rightarrow L > A \rightarrow W$	F	$L_{1,2} \rightarrow W > L_{1,2} \rightarrow H$	T
$H \rightarrow W > H \rightarrow E$	T	$A \rightarrow Pb > G \rightarrow Pb$	?	$L > A > L \rightarrow E$	T
$H \rightarrow W > H \rightarrow Pb$	T	$A \rightarrow E > A \rightarrow W$	F	$L_{1,2} \rightarrow A > L_{1,2} \rightarrow G$	T
$W \rightarrow A > H \rightarrow A$	T	$A \rightarrow G > L_{1,2} \rightarrow G$	T	$L_2 \rightarrow G > L_3 \rightarrow A$	?
$W \rightarrow E > H \rightarrow E$	T	$A \rightarrow E > E \rightarrow A$	T	$L_1 \rightarrow Pb > L_1 \rightarrow G$	?
$W \rightarrow L_{1,2} > H \rightarrow L_{1,2}$	T	$A \rightarrow L_2 > L_2 \rightarrow A$	?T	$G \rightarrow A > G \rightarrow Pb$	T
$W \rightarrow L > W \rightarrow A$	T	$E \rightarrow W > E \rightarrow H$	T	$G \rightarrow A > G \rightarrow W$	T
$W \rightarrow L > W \rightarrow G$	T	$E \rightarrow A > E \rightarrow G$	T	$G \rightarrow L_{2,3} > G \rightarrow W$	?T
$W \rightarrow A > W \rightarrow G$	T	$E \rightarrow A > E \rightarrow L$	T	$G \rightarrow A > G \rightarrow L_{1,2}$	?T
$W \rightarrow E > W \rightarrow Pb$	T	$E \rightarrow W > E \rightarrow L$	T	$G \rightarrow A > A \rightarrow G$	T
$W \rightarrow E > W \rightarrow G$	T			$G \rightarrow L > L \rightarrow G$	T
$W \rightarrow E > W \rightarrow A$	?			$G \rightarrow H > G \rightarrow W$	?F

Table 2. Observed homeotic transformation and the code changes required for the code scheme in Fig. 2. A set of homeotic mutants causing the same transformation is represented by one member: (1) Antennapedia, Antennapedix, aristapedia, aristatarsia; (2) Ophthalmoptera,, eyes-reduced; (3) tetraltera, Metaplasia, Haltere mimic; (4) extrasexcombs, Extrasexcomb, reduplicated sex comb, sparse arista. (From Kauffman et al. [1978]. Sources key: (a) Gehring and Nöthiger [1973]; (b) Shearn et al. [1971]; (c) Ouweneel [1976]; (d) Lewis [1978]; (e) Gehring [1966] (see Part I); (f) Stepshin and Ginter [1972]; and (g) Postlethwait et al. [1972].

Mutant	Symbol	Transformation			Coordination	Code Change	Switches Required	Ref
Antennapedia(1)	Antp	antenna	→	leg 2	-	1010→1110	1	a,c
Pointed wing	Pw	antenna	→	wing	-	1010→1111	2	c
Nasobemia	Ns	antenna	→	leg 2	parallel	1010→1110	1	e
		eye	→	wing		1011→1111	1	f
dachsous	ds	tarsus	→	arista	-	1110→1010	1	c
Opthalmoptera(2)	OptG	eye	→	wing	-	1011→1111	1	c
Hexaptera	Hx	prothorax	→	mesothorax	-	1110→1111	1	c
podoptera	pod	wing	→	leg		1111→1110	1	c
tetraltera(3)	tet	wing	→	haltere	-	1111→0111	1	c
Contrabithorax	Cbx	wing	→	haltere	parallel	1111→0111	1	c
		leg 2	→	leg 3		1110→0110	1	d
Ultrabithorax	Ubx	haltere	→	wing	parallel	0111→1111	1	d
		leg 3	→	leg 2		0110→1110	1	c
tumorous head	tuhl,3	eye	→	genital	parallel	1011→0011	1	g
		antenna	→	genital	divergent	1010→0010	1	c
		antenna	→	leg		1010→1110	1*	g
lethal(3)III-10	1(3)III-10	haltere	→	wing	parallel	0111→1111	1	b
lethal(3)XVI-18	1(3)XVI-18	genital	→	antenna	divergent	0010→1010	1	c
		genital	→	leg		0010→0110	1*	b
lethal(3)703	1(3)703	antenna	→	leg	parallel	1010→1110	1	b
lethal(3)1803k	1(3)1803k	genital	→	leg	divergent	0010→0110	1	c
		genital	→	antenna	parallel	0010→1010	1*	c
		haltere	→	wing		0111→1111	1	c
proboscipedia	pb	proboscis	→	antenna	divergent	1000→1010	1	a
		proboscis	→	leg		1000→1100	1*	c
extrasexcombs(4)	ecs	leg 2	→	leg 1	convergent	1110→1100	1	c
		leg 3	→	leg 1		0110→1100	2	c
Polycomb	Pc	antenna	→	leg 2		1010→1110	1	c
lethal(4)29	1(4)29	leg 2	→	leg 1	convergent	1110→1100	1*	a
		leg 3	→	leg 1		0110→1100	2*	c

gene system, then a binary code model precludes either convergence or divergence. It is in this sense that the existence of these classes of homeotic mutants suggest that different pictures of the underlying genetic logic must be assumed to provide minimal models of homeotic transformation. For example, a variety of mutants transform second and third legs into first legs. Both the second and third leg cannot differ from one another and from the first leg by the binary state of a single gene, or genetic system. At a minimum, a genetic system with three alternative states must be assumed, coupled with the assumption that the mutant converts two of these states to the third. Because convergent and divergent homeotic mutants are minority classes among homeotics, and require more complex logic, it may prove useful to assume that those exhibiting either single transformation of parallel transformations are the general class, and the convergent and divergent classes special abberations from the norm. Then the general class can in principle be accounted for by a binary combinatorial code. Three state genetic systems can be obtained in several ways; one simple way is an appropriate coupling of two genetic circuits with two states each, to yield a three state device.

Since its formulation, a number of interesting mutants bearing on both the eigenfunction model and combinatorial model have been discovered. The successive eigenfunction model proposes that ever shorter wavelength patterns successively divide up the egg into smaller subdomains. Mutants of such a system might be expected to produce embryos which failed to undergo the complete sequence. Recently, a larval lethal mutant has been discovered which yields an embryo having 6 wide longitudinal segments rather than 12 narrow segments [T. Kaufman, 1980]. It appears that final partition of these 6, each into 2 segments, has not occurred. Mirror symmetrical abberations have now been discovered in a remarkable spectrum of spatial lengths on the _Drosophila_ embryo. Well known in both _Drosophila_ and _Smittia_, are mutants and experimental procedures which produce double abdomen embryos with mirror symmetry about an anterior abdominal or metathoracic segment [Kalthoff, 1979; Nüsslein-Volhard, 1979]. In these cases, the mirror symmetry extends over half the embryo. It is attractive to account for such transformations by assuming that the first master gene in Fig. 2 which records the anterior-posterior decision, is left uniformly in the posterior-0-state, yielding a mirror symmetric double abdomen [see also, Kalthoff, 1979]. Were the second master gene system, distinguishing ends from middle, to mutate to a constitutive 0 state, the embryo would consist of mirror symmetric posterior abdominal segments in the abdominal half of the embryo and mirror symmetric head segments in the anterior half of the embryo. A recent early lethal mutant, still only partially characterized [Nüsslein-Volhard and Wieschaus, 1980], has mirror symmetric posterior abdominal segments in the posterior half of the embryo, is missing the thoracic segments, and has grossly abnormal head structures whose details are difficult to elucidate because head segments are inverted inside the hatching larva. In the posterior half embryo, where mirror symmetry is clear, it ranges over 1/4 the length of the body plan. Another lethal mutant yields an embryo with four large segments, each bearing mirror symmetrical arrangements of ventral denticles [Nüsslein-Volhard and Wieschaus, 1980]. Here, mirror symmetry ranges locally over 1/8 the embryo length. Failure of the third digit in Fig. 2 to function normally, such that it remained constitutively in the 9 state would yield four wide domains. Mirror symmetry within some or all of these domains would be obtained if an additional longitudinal chemical pattern normally subdivided the 8 segments of Fig. 2 into the full 12 segments of the embryo, analogous to the mirror symmetries obtained if the second digit is constitutively 0 in Fig. 2. Finally, a mutant has been uncovered which is internally mirror symmetric in the cuticular patterns of all larval thoracic and abdominal segments [Nüsslein-Volhard and Wieschaus, 1980].

This broad spectrum of embryonic lengths over which mirror symmetry occurs is clearly suggestive of a variety of chemical patterns of successively shorter wavelengths, with successively more peaks and troughs, fitting onto the egg. Although provocative at this stage, it is obvious that the class of models I have discussed remains a macroscopic analogy. It should be stressed that some of the mutants are still incompletely characterized, and that a variety of other mechanisms might account for the patterns observed.

In summary, reaction-diffusion systems provide a means to successively subdivide a domain at a sequence of critical parameter values due to size, shape, diffusion constants or other parameters. The chemical patterns which arise are the eigenfunctions of the Laplacian operator on that geometry. The succession of eigenfunctions on geometries close to the wing, leg, haltere and genital discs yield sequential nodal lines reasonably similar to the observed sequence and symmetries and geometries of the observed compartmental lines. The identity is not perfect. The hypothesis that the egg is subdivided by such sequential eigenfunctions into subdomains yields a combinatorial epigenetic code which interprets a wide variety of homeotic mutants and transdetermination relations as transitions between neighboring developmental code words. The accord is good, but not perfect. Recent experiments and mutants on the egg confirm that an initial anterior-posterior compartmental boundary exists, that longitudinal compartments arise successively, and prior to a major dorsal-ventral line in the thoracic regions. Mutants reveal the existence of embryos with half the number of normal segments, and mirror symmetric segmental structures whose range varies from 1/2 egg, 1/4 egg, 1/8 egg, to mirror symmetry within individual longitudinal segments. These patterns at least suggest chemical gradients with varying numbers of longitudinal peaks and troughs.

REFERENCES

Babloyantz, A., and Hiernaux, J., 1975, Models for cell differentiation and generation of polarity in diffusion-governed morphogenetic fields, Bull. Math. Biol., 37:637.

Bryant, P., 1975, Pattern formation in the imaginal wing disc of Drosophila melanogaster: fate map, regeneration and duplication, J. Exp. Zool., 193:49.

Dubendorfer, K., 1977, Unpublished Ph.D. Thesis, University of Zurich.

French, V., Bryant, P.J., and Bryant, S.V., 1976, Pattern regulation in epimorphic fields, Science, 193:969.

Garcia-Bellido, A., Ripoll, P., and Morata, G., 1973, Developmental compartmentalization of the wing disk of Drosophila, Nature New Biol., 245:251.

Garcia-Bellido, A., 1975, Genetic control of wing disc development in Drosophila, in: "Cell Patterning," Ciba Foundation Symp. 29, Elsevier, Amsterdam.

Garcia-Bellido, A., and Merriam, J.R., 1969, Cell lineage of the imaginal discs in Drosophila gynandromorphs, J. Exp. Zool., 170:61.

Gehring, W., and Nöthiger, R., 1973, The imaginal discs of Drosophila, in: "Developmental Systems: Insects, Vol. 1," S.J. Counce and C.H. Waddington, eds., Academic Press, New York.

Gmitro, J.I., and Scivin, L.E., 1966, in: "Intracellular Transport," K.B. Warren, ed., Academic Press, New York.

Hadorn, E., 1966, Dynamics of determination, Symp. Soc. Dev. Biol., 25:85.

Hannah-Alava, A., 1958, Developmental genetics of the posterior legs in Drosophila melanogaster, Genetics, 43:878.

Harrison, R.G., 1921, On the relations of symmetry in transplanted limbs, J. Exp. Zool., 32:1.

Haynie, J., 1980, 21st Annual Drosophila Conference.

Herschkowitz-Kaufman, M., 1975, Bifurcation analysis of nonlinear reaction-diffusion equations. II. Steady state solutions and comparison with numerical simulations, Bull. Math. Biol. 37:589.

Hotta, Y., and Benzer, S., 1972, Mapping behavior in Drosophila mosaics, Nature, 240:527.

Janning, W., 1978, Gynandromorph fate maps in Drosophila, in: "Genetic Mosaics and Cell Differentiation," W.J. Gehring, ed., Springer-Verlag, Berlin.

Kalthoff, K., 1979, Analysis of a morphogenetic determinant in an insect embryo (Smittia Spec., Chironomidae, Diptera), in: "Determinants of Spatial Organization," S. Subtelny and I. Konigsberg, eds., Academic Press, New York.

Kauffman, S.A., 1973, Control circuits for determination and transdetermination, Science, 181:310.

Kauffman, S.A., 1979, The compartmental and combinatorial code hypotheses in Drosophila development, Bioscience, 29:581.

Kauffman, S.A., 1980, A cartesian coordinate model of positional information in epimorphic fields, Biophysical Journal, to appear.
Kauffman, S.A., Shymko, R.M., and Trabert, K., 1978, Control of sequential compartment formation in Drosophila, Science, 199:259.
Kauffman, S.A., and von Allmen, C., 1980, manuscript in preparation.
Kaufman, T., 1980, 21st Annual Drosophila Conference.
King, M.J., and Wiltse, J.C., 1958, Derivative zeros and other data pertaining to Mathieu functions, Johns Hopkins Radiation Laboratory Technical Report No. AF57, Baltimore, Md.
Lawrence, P.A., 1973, The development of spatial patterns in the integument of insects, in: "Developmental Systems: Insects, Vol. 2," S.J. Counce and C.H. Waddington, eds., Academic Press, New York.
Lawrence, P., and Morata, G., 1977, The early development of mesothoracic compartments in Drosophila, Dev. Biol., 56:40.
Lewis, E.B., 1964, Genetic control and regulation of developmental pathways, in: "The Role of Chromosomes in Development," M. Locke, ed., 23rd Symposium of the Society for the Study of Development and Growth, Academic Press, New York.
Lewis, E.B., 1978, A gene complex controlling segmentation in Drosophila, Nature, 276:565.
McLachlan, N.W., 1947, "Theory and Applications of Mathieu Functions," Clarendon, Oxford.
Meinhardt, H., 1977, A model of pattern formation in insect embryogenesis, J. Cell Sci., 23:117.
Merriam, J.R., 1978, Estimating primordial cell numbers in Drosophila imaginal discs and histoblasts, in: "Genetic Mosaics and Cell Differentiation," W. Gehring, ed., Springer-Verlag, New York.
Nicolis, G., and Prigogine, I., 1977, "Self-organization in Nonequilibrium Systems," Interscience, New York.
Nöthiger, R., 1972, The larval development of imaginal discs, in: "Results and Problems in Cell Differentiation," V.C.H. Ursprung and R. Nöthiger, eds., Springer, Germany.
Nüsslein-Volhard, C., 1979, Maternal effect mutations that alter the spatial coordinates of the embryo of Drosophila melanogaster, in: "Determinants of Spatial Organization," S. Subtelny and I. Konigsberg, eds., Academic Press, New York.
Nüsslein-Volhard, C., and Wieschaus, E., 1980, Mutations affecting segment number and polarity in Drosophila, Nature, 287:795.
Ouweneel, W.I., 1976, Developmental genetics and homeosis, Advances in Genetics, 18:179.
Postlethwait, J.H., Bryant, P., and Schubiger, G., 1972, The homeotic effect of "tumorous head" in Drosophila melanogaster, Dev. Biol., 29:337.
Sander, K., 1975, Pattern specification in the insect embryo, in: "Cell Patterning," Ciba Foundation Symp. 29, Elsevier, Amsterdam.
Sander, K., 1977, Current understanding of cytoplasmic control centers, in: "Insect Embryology," S.W. Visscher, ed., Montanna State University Press, Montana.

Shearn, A., Rice, T., and Garen, A., 1971, Imaginal disc abnormalities in lethal mutants of Drosophila, Proc. Natl. Acad. Sci. USA, 68(10):2594.

Steiner, E., 1976, Establishment of compartments in the developing leg imaginal discs of Drosophila melanogaster, Wilhelm Roux's Archiv. Dev. Biol., 180:9.

Stephshin, V.P., and Ginter, E.K., 1972, A study of the homeotic genes Antennopedia and Nasobemia of Drosophila melanogaster, Genetika, 8:93.

Trainor, L., and Goodwin, B., 1980, preprint.

Turing, A.M., 1952, The chemical basis of morphogenesis, Philos. Trans. R. Soc. London, Ser. B., 237:37.

Turner, F.R., and Mahowald, A.P., 1976, Scanning electron microscopy of Drosophila embryogenesis. I. The structure of the egg envelope and the formation of the cellular blastoderm, Dev. Biol., 50:95.

Turner, F.R., and Mahowald, A.P., 1977, Scanning electron microscopy of Drosophila melanogaster embryogenesis. II. Gastrulation and segmentation, Dev. Biol., 57:403.

Wieschaus, E., and Gehring, W., 1975, Clonal analysis of primordial disc cells in the early embryo of Drosophila melanogaster, Dev. Biol., 50:249.

Wolpert, L., 1971, Positional information and pattern formation, in: "Current Topics in Developmental Biology, Vol. 6," A.A. Moscana and A. Monroy, eds.

SELECTION AND EVOLUTION IN MOLECULAR SYSTEMS

Peter Schuster

Institut fuer Theoretische Chemie und Strahlenchemie
Universität Wien
Austria

TABLE OF CONTENTS

SELECTION AND EVOLUTION IN MOLECULAR SYSTEMS

Peter Schuster

Institut fuer Theoretische Chemie und Strahlenchemie
Universität Wien, Austria

I. MOLECULAR SELF-REPLICATION

The capability of self-replication is one of the unique features of living beings. Outside biology we encounter this property only occasionally. Examples of such exceptions are autocatalytic reactions in chemistry which play an important role in the formation of chemical oscillations and waves as well as in vapor phase combustion. In this class of chemical reactions the numbers of small molecules, radicals or ions double or, in general, multiply. There are also examples of self-replication in the intellectual world like the highly sophisticated programs of learning computers or some complicated mathematical games. In this contribution we shall summarize a series of studies on molecular self-replication by means of chemical kinetics. This work has been initiated in 1971 by the comprehensive paper by Eigen [1971]. Later on, several publications continued work along these lines [Eigen and Winkler-Oswatitsch, 1975, 1980; Eigen, 1976; Eigen and Schuster, 1977, 1978 a,b; Schuster, 1980; Schuster and Sigmund, 1980]. Finally, we shall apply the concept of self-organization through self-replication to two vividly discussed problems in biological evolution, to the emergence of life from prebiotic mixtures of molecules and to the self-regulation of behaviour in animal societies.

At first, we shall focus on molecules and we have good reasons to do so. Molecular biology has revealed and continues to reveal the fascinating details of polynucleotide replication [Watson, 1977; Kornberg, 1974]. Polynucleotides, in particular, ribonucleic acid (RNA) and 2-deoxyribonucleic acid (DNA) are the only known biomolecules which fulfill the structural requirements for self-replication. These consist in a regular repetitive structure of the "backbone" and a unique 1:1 complementary assignment of the side chains (Fig. 1). The assignments are well known as "Watson-Crick" base pairs, G ≡ C and A = U, which are stabilized by patterns of hydrogen bonds. Polynucleotide replication can occur in two different ways: (1) indirect or complementary replication and (2) direct replication. In the former process a negative copy is synthesized first which on replication yields a new copy of the original sequence. Direct replication of a double helix involves the so-called "fork" mechanism. Two new strands are synthesized at a time and each of the two daughter molecules consists of one old and one new strand. Both mechanisms, as we shall see later, are highly complicated and involve many intermediate steps.

A. Autocatalysis and Steady States

The simplest example of autocatalysis is represented by the folllowing second order reaction:

$$A + B \xrightarrow{f} 2A \; ; \; \frac{da}{dt} = \dot{a} = fab \qquad (1.1)$$

By a and b we denote the concentrations of A and B respectively. In case there is no shortage of B or more precisely under the condition of buffered concentration of B, $b = b_o$, integration of (1.1) is trivial and yields a solution curve of exponential growth

$$a = a_o \exp(fb_o t) \quad . \qquad (1.1a)$$

By a_o we denote the concentration of A at $t = 0$. Exponential growth can occur for off equilibrium only. Close to equilibrium we shall always observe relaxation behaviour. In our concrete example we introduce the backward reaction to make the autocatalytic process reversible:

$$A + B \underset{f_2}{\overset{f_1}{\rightleftharpoons}} 2A \; ; \; \dot{a} = f_1 ab - f_2 a^2 \quad . \qquad (1.2)$$

Again we run the reaction at buffered concentration of B : $b = b_o$. The equilibrium concentration of B, $\bar{b}$, then has the same value, $\bar{b} = b_o$. Figure 2 shows a characteristic solution curve of (1.2). We recognize the switch from exponential growth to relaxation behaviour.

In order to see what is unique about autocatalytic reactions, we look at equation (1.2) by means of irreversible thermodynamics. For this purpose we recall the definitions of chemical forces and fluxes. In the case of a general one step reaction we have:

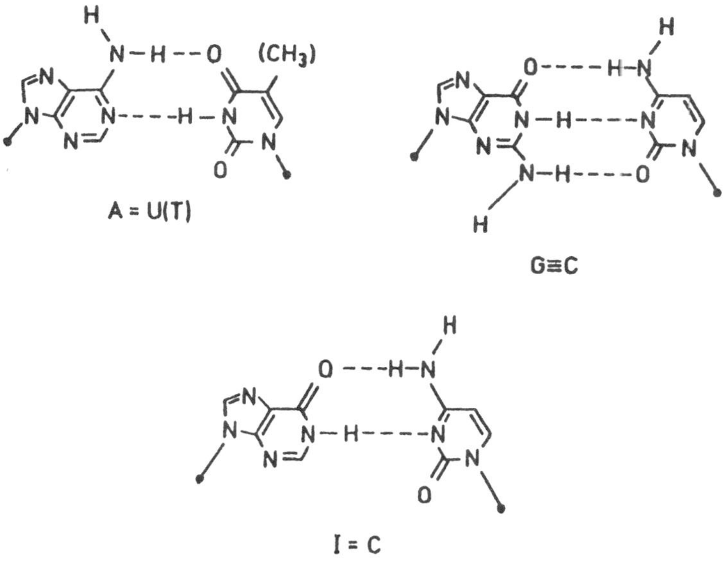

A = U(T)

G≡C

I = C

Fig. 1. Complementarity of purine and pyrimidine bases in polynucleotides and the principles of indirect and direct replication. The purine bases are: A = adenine, G = guanine, I = hypoxanthine. The pyrimidines: U = uracil, T = thymine and C = cytosine. The base pairs occurring in DNA are A = T and G ≡ C. U replaces T in RNA. There are occasionally also I = C base pairs in RNA, namely in transfer-ribonucleic acids (t-RNA's).

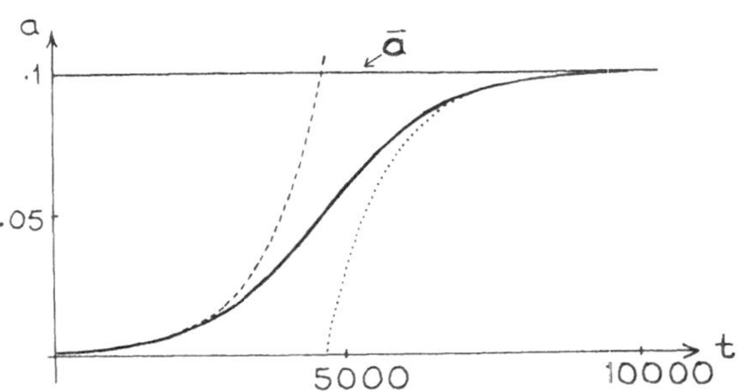

Fig. 2. A characteristic solution curve of Eq. (1.2) at buffered concentration of B, $b = b_o = \bar{b}$. Pure exponential growth and relaxation to equilibrium are shown as broken lines. The constants chosen are $f_1 = 1[t^{-1}C^{-1}]$, $f_2 = 0.01[t^{-1}C^{-1}]$, and $\bar{b} = 0.001[C]$ in arbitrary time and concentration units.

$$\nu_A A + \nu_B B + \ldots \underset{k_2}{\overset{k_1}{\rightleftharpoons}} \nu_C C + \nu_D D + \ldots \tag{1.3}$$

$$X_{ch} = R\left\{\frac{1}{\nu_A}\ln\frac{a}{\bar{a}} + \frac{1}{\nu_B}\ln\frac{b}{\bar{b}} + \ldots - \frac{1}{\nu_C}\ln\frac{c}{\bar{c}} - \frac{1}{\nu_D}\ln\frac{d}{\bar{d}} - \ldots\right\} \tag{1.4}$$

and

$$J_{ch} = -\frac{1}{\nu_A}\dot{a} = -\frac{1}{\nu_B}\dot{b} = \frac{1}{\nu_C}\dot{c} = \frac{1}{\nu_D}\dot{d} = k_1 a^{\nu_A} b^{\nu_B} \ldots - k_2 c^{\nu_C} d^{\nu_D} \tag{1.5}$$

Herein ν_I denotes the stoichiometric coefficient for substance I, R is the gas constant and X_{ch} and J_{ch} represent the chemical force and flux respectively. In autocatalytic reactions, where the same substance appears on both sides of the equation for the reaction, i.e. as both, reactant and product, the flux refers to the net production:

$$\nu_A A + \ldots \rightleftharpoons \nu'_A A + \ldots \; ; \; J_{ch} = \frac{1}{\nu'_A - \nu_A}\dot{a} \quad .$$

Now, we compare an ordinary chemical reaction, e.g. the dissociation process

$$B \underset{k_2}{\overset{k_1}{\rightleftharpoons}} 2A \tag{1.6}$$

with the autocatalytic reaction (1.2). For reaction (1.6) we find under the condition of buffered concentration of B, $b = b_o = \bar{b}$,

$$X_{ch} = -\frac{R}{2}\ln\frac{a}{\bar{a}}, \text{ and } J_{ch} = \frac{1}{2}(k_1 b_o - k_2 a^2) \tag{1.6a}$$

In case of the autocatalytic process we obtain by the same formalism:

$$X_{ch} = -R\ln\frac{a}{\bar{a}}, \text{ and } J_{ch} = a(k_1 b_o - k_2 a) \quad . \tag{1.2a}$$

There is a fundamental difference between (1.2a) and (1.6a) which we realize immediately by looking at Figure 3. Reaction (1.6) shows the characteristic behaviour of ordinary chemical reactions. In analogy to Hooke's law we have a linear range around equilibrium where the flux is proportional to the force as required by the theory of small displacement. Further off equilibrium the flux is smaller than required by the linear relation and reaches a maximum

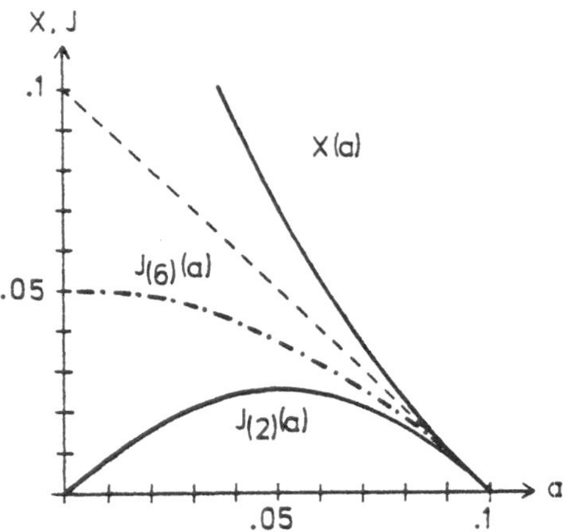

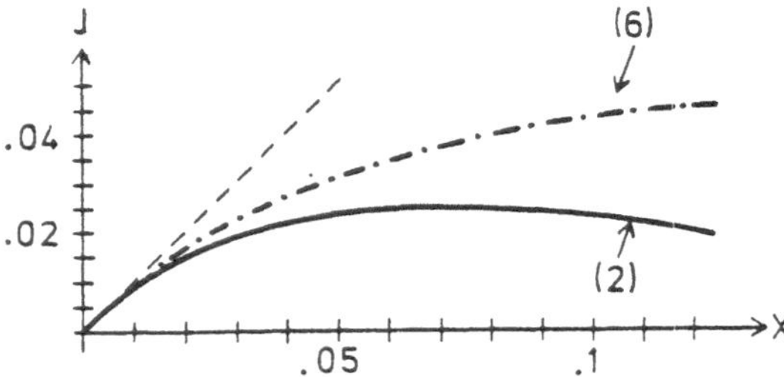

Fig. 3. The relation between chemical forces X(a) and fluxes, $J_{(1.2)}(a)$ and $J_{(1.6)}(a)$ in reactions (1.2) and (1.6). The units were chosen such that the tangent around the equilibrium point equals unity.

value at $a = 0$. The autocatalytic process (1.2) behaves differently in the range far off equilibrium. Below the critical concentration, $a_{cr} = \bar{a}/2$ in our example, force and flux have slopes of opposite sign. Hence, the flux increases as the force decreases. We encounter an analogy to the well known instability in electronics called a negative current-voltage-characteristic, which is found with electric arcs or electronic tubes. An increase in the concentration of the product causes an increase of the reaction rate which in turn leads to more product. The result is exponential growth until the system reaches certain limits, e.g. it exhausts the reservoir of B and/or reaches equilibrium. Considering elementary step reactions only it is easy to visualize that autocatalysis is a necessary and sufficient condition for the occurrence of a negative force-flux-characteristic. An autocatalytic process by definition is a reaction in which the same substance appears on both sides of the reaction equation with different stoichiometric coefficients. Hence we can always factorize the flux, viz.,

$$J_{ch} = \frac{1}{\nu'_A - \nu_A}\dot{a} = \frac{1}{\nu'_A - \nu_A} a^{\nu'_A - \nu_A} F(a,b,...) \tag{1.7}$$

Equation (1.7) has two zero values, $a = 0$ and $F(\bar{a},\bar{b},...) = 0$. The latter being the condition for equilibrium. J_{ch} is necessarily positive at values between $0 < a < \bar{a}$ and since the rate equation of an elementary step reaction is a simple two term polynomial (1.5) it has to have a maximum at some value of a. Other elementary step reactions except autocatalytic do not follow rate equations in which the flux is proportional to one of the reaction products and hence cannot meet the condition of a negative force-flux-characteristic.

How does one keep an autocatalytic reaction far off equilibrium? In order to achieve this goal we have to introduce a kind of flow which keeps the system in a steady state. We apply a many step mechanism in which synthesis and degradation of the autocatalyst occur via different reaction paths. Coupling to a "recycling" process (1.8c) keeps the system off equilibrium. For the sake of simplicity we assume all reactions to be irreversible:

$$A + B \xrightarrow{f} 2A \tag{1.8a}$$

$$A \xrightarrow{d} C \tag{1.8b}$$

$$C \xrightarrow{g(E)} B \tag{1.8c}$$

We introduce a rate constant g(E) of the recycling process which depends on some external input. In the simplest case E is provided as thermal energy, electromagnetic radiation etc. by an external energy source. Since there is no material flow into the system we have the conservation law

$$a + b + c = C_o \quad . \tag{1.8d}$$

The steady states of (1.8) can be studied by straightforward analysis. For the sake of convenience we introduce a concentration space, which in our case is $\mathbb{R}^3_+$, since concentrations have to have non-negative values (Figure 4). We obtain two solutions P_1 and P_2, the stability of which is analyzed by linearization.

$$P_1 = (0,C_o,0) \; ; \; \omega_1^{(1)} = -g \; , \; \omega_2^{(1)} = C_o - \frac{d}{f}$$

and

$$P_2 = \left(\frac{g}{g+d}\left(C_o - \frac{d}{f}\right), \frac{d}{f}, \frac{d}{g+d}\left(C_o - \frac{d}{f}\right)\right) ,$$

$$\omega_{1,2}^{(2)} = -\frac{1}{2}\left(g + \delta \pm \sqrt{(g+\delta)^2 - 4\delta(g+d)}\right)$$

where

$$\delta = f\,\overline{a} = \frac{fg}{g+d}\left(C_o - \frac{d}{f}\right) .$$

The fixed point P_1 is asymptotically stable iff $c_o < d/f$ and P_2 is asymptotically stable iff $\delta > 0$ which means $C_o > d/f$. The positions of P_1 and P_2 as functions of C_o are shown in Figure 5. Two aspects of these results are relevant for our further consideration: (1) the existence of a non-vanishing stationary concentration of A and C requires a total concentration C_o which exceed the threshold value $C_o = d/f$ and (2) the stationary concentrations of A and C are proportional to the rate constants g and d respectively. The larger the recycling rate and the slower the decomposition process, the more autocatalyst is available at the stable stationary state P_2. Interestingly the rate constant f of the autocatalytic step has no influence on the relative stationary concentration of A. This is a consequence of the negative force-flux-characteristic discussed above: the autocatalytic process exhausts the available reservoir of B completely, the concentration of which is determined by the threshold value.

Biology, in principle, follows reaction schemes related to equation (1.8) although they are enormously complicated as far as the details are concerned: the synthesis of the genetic material,

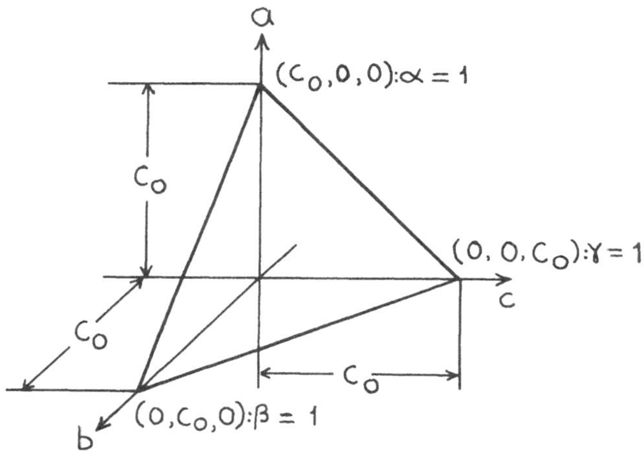

Fig. 4. The three-dimensional concentration space and the concentration simplex S_3 used to discuss the stationary solutions of Eq. (1.8). The simplex corresponds to normalized concentrations: $\alpha = a/C_o$, $\beta = b/C_o$, $\gamma = c/C_o$ and hence $\alpha + \beta + \gamma = 1$.

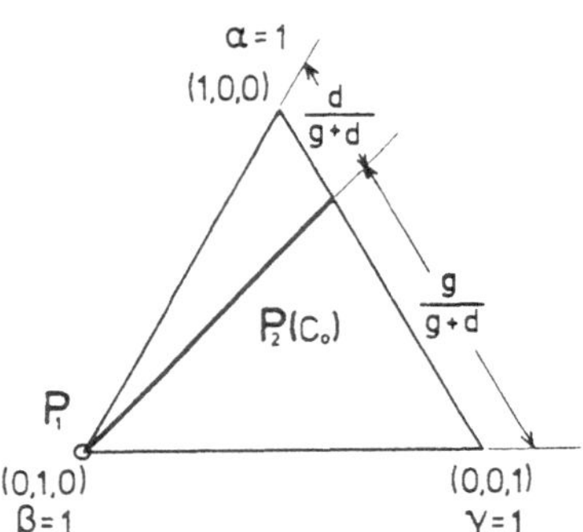

Fig. 5. The position of the two fixed points of Eq. (1.8) as a function of C_o. The relative values of the rate constants were chosen to be g:d = 3:1.

the polynucleotides, which act as autocatalysts (A), requires activated monomers the nucleotide triphosphates (B). Polynucleotides are degraded to yield "energy-poor" material, e.g. the nucleotide monophosphates (C). This energy-poor material is activated again by an energy consuming recycling reaction, the pyrophosphorylation of the monophosphates. In the biosphere any recyling process has to use ultimate energy sources from outside the cycle like the sunlight or internal heating from the Earth. "In vitro" experiments can replace the recycling process by continuous (flow reactor) or stepwise (serial transfer experiments) renewal of the material consumed (see section I.C.).

In reality, as we shall see in the next section, any autocatalytic process known so far is much more complex than equations (1.2) or (1.8) suggest. As far as polynucleotide replication is concerned, each of the three reactions corresponding to (1.8a), (1.8b) and (1.8c) consists of many individual steps. Nevertheless, the "overall" kinetics will come close to (1.8) under many realistic conditions.

B. Polynucleotide Replication

Before we can consider polynucleotide replication in more detail we have to study two simple models which are of very general character. The first involves the action of an enzyme called RNA polymerase. The enzyme binds to the RNA molecule and replicates this template digit per digit. Schematically we can describe this process by a two step mechanism

$$I + E \underset{k_2}{\overset{k_1}{\rightleftharpoons}} I \cdot E \qquad (1.9a)$$

$$I \cdot E + \sum_{\lambda=1}^{4} \nu_\lambda A_\lambda \xrightarrow{f} I \cdot E + I \qquad (1.9b)$$

Herein we denote the RNA molecule by I, the enzyme by E and the polynucleotide-enzyme complex by I· E. The four energy-rich monomers which are to be incorporated into the newly synthesized RNA-molecule are symbolically characterized by A_i, $i = 1,\ldots,4$, e.g. $A_1 \equiv$ ATP, $A_2 \equiv$ UTP, $A_3 \equiv$ GTP and $A_4 \equiv$ CTP. The following symbols are used for the concentrations:

$$[I] = x, \; [E] = e \text{ and } [I \cdot E] = y \; .$$

We shall assume buffered concentrations of the activated monomers: $[A_i] = A_i^\circ$; $i = 1, \ldots, 4$. Therefore they do not enter as variables into the kinetic equations. We may account for them by redefining the rate constant f,

$$f' = fF(\overset{\circ}{A}_1,\overset{\circ}{A}_2,\overset{\circ}{A}_3,\overset{\circ}{A}_4)$$

and thus obtain the following rate equations:

$$\dot{x} = f'y + k_2y - k_1xe \tag{1.9c}$$

$$\dot{y} = k_1xe - k_2y \tag{1.9d}$$

$$\dot{e} = k_2y - k_1xe \tag{1.9e}$$

There is an apparent linear dependence: $\dot{y} = -\dot{e}$ or $y + e = \text{const.}$, which corresponds to the catalytic action of the enzyme. The total concentration of E, $e_o = e + y$ remains constant. The total concentration of the template I, $x_o = x + y$, is steadily growing, since

$$\dot{x}_o = \dot{x} + \dot{y} = f'y > 0 \tag{1.9f}$$

In order to find the nature of the growth function we make a straightforward approximation. We assume quasiequilibrium of complex formation or, in other words, f' is considered to be so small that reaction (1.9a) is practically at equilibrium. Then we find

$$y = \frac{x_o + e_o + K^{-1}}{2}\left(1 - \sqrt{1 - \frac{4x_oe_o}{(x_o + e_o + K^{-1})^2}}\right) . \tag{1.10}$$

The expression for y can be analyzed easily if the equilibrium constant $K = k_1/k_2$ fulfills the relation $K^{-1} > e_o$. Then we have the two asymptotic cases:

(1) low concentrations of I: $x_o << e_o$,

$y \simeq K\, x_o\, e_o$ and hence $\dot{x}_o = f'Ke_o\, x_o$

We obtain exponential growth. The mechanism (1.9) thus contains a "hidden" autocatalytic reaction.

(2) high concentrations of I: $x_o > K^{-1} > e_o$,

$y \simeq e_o$ and hence $\dot{x}_o = f'e_o = \text{const.}$

At the condition of "enzyme saturation", as one usually says, we get linear growth. The enzyme produces at the maximum speed since all enzyme is converted into the complex I•E under these conditions. Frequently, an approximation of the Michaelis-Mentcn type to equation (1.10) is very useful:

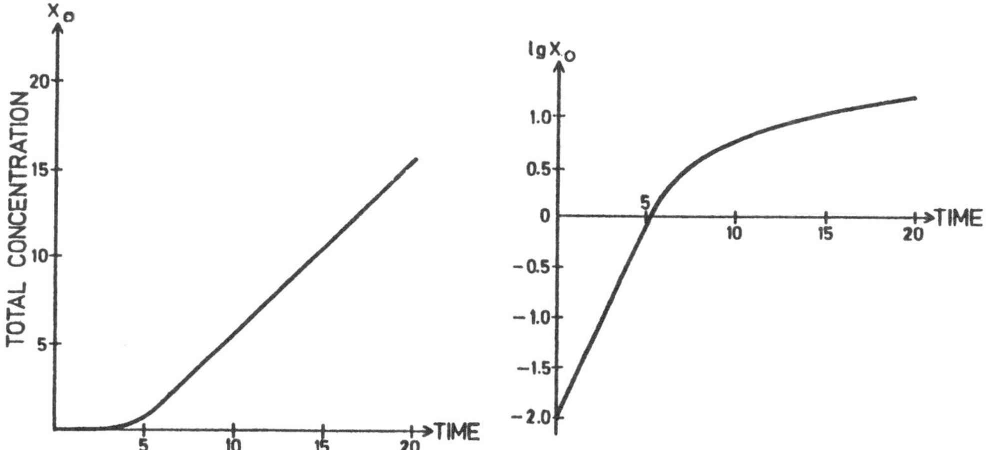

Fig. 6. An integration of Eq. (1.9). The rate constants and initial conditions applied are: $f' = 1$, $k_1 = 1000$, $k_2 = 10$, $x(0) = 0.01$, $y(0) = 0$, $e_o = 1$. Note that exponential growth changes in into linear growth at the point of saturation: $x_c \sim e_o$. All quantities are given in arbitrary time and concentration units.

$$y \simeq \frac{x_o e_o}{x_o + e_o + K^{-1}} \quad . \tag{1.11}$$

It leads directly to the exact expressions in both asymptotic limits, $\lim x_o \to 0$ and $\lim x_o \to \infty$. Figure 6 shows a computer plot of equation (1.9) without approximations. We can easily recognize exponential and linear growth behaviour in the asymptotic limits.

The second simple model refers to complementary replication. We assume the presence of two strands of RNA which are complementary to each other and denote the plus strand by I^+ and the minus strand by I^-. Both are replicated by the same enzyme E.

$$E + I^+ + \sum_{\lambda=1}^{4} \nu_\lambda^{(-)} A_\lambda \xrightarrow{\ell_1} E + I^+ + I^- \tag{1.12a}$$

$$E + I^- + \sum_{\lambda=1}^{4} \nu_\lambda^{(+)} A_\lambda \xrightarrow{\ell_2} E + I^- + I^+ \tag{1.12b}$$

Again we assume buffered concentration of the energy-rich monomers. Using the results at low concentrations obtained in the previous model we get a system of two coupled, linear differential equations [Eigen, 1971]:

$$\dot{x}_1 = \ell_1' x_2 \tag{1.12c}$$

$$\dot{x}_2 = \ell_2' x_1 \tag{1.12d}$$

Herein the variables are defined as the concentration of the plus and minus strand, $[I^+] = x_1$ and $[I^-] = x_2$, respectively. The new rate constants are

$$\ell_1' = \ell_1 F_1(A_1^o, A_2^o, A_3^o, A_4^o) e_o \quad \text{and} \quad \ell_2' = \ell_2 F_2(A_1^o, A_2^o, A_3^o, A_4^o) e_o \ ,$$

and e_o is the total concentration of enzyme. Equations (1.12 c,d) are integrated easily after an appropriate linear transformation,

$$X = \sqrt{\ell_2'}\, x_1 + \sqrt{\ell_1'}\, x_2 \quad \text{and} \quad \zeta = \sqrt{\ell_2'}\, x_1 - \sqrt{\ell_1'}\, x_2 \ :$$

$$\dot{X} = \sqrt{\ell_1' \ell_2'}\, X \longrightarrow X(t) = X(0) \exp(+\sqrt{\ell_1' \ell_2'}\, t) \tag{1.13a}$$

$$\dot{\zeta} = -\sqrt{\ell_1' \ell_2'}\, \zeta \longrightarrow \zeta(t) = \zeta(0) \exp(-\sqrt{\ell_1' \ell_2'}\, t) \quad . \tag{1.13b}$$

We see immediately that the relative concentrations of I^+ and I^- approach a constant value:

$$\lim_{t\to\infty} \zeta(t) = 0 \text{ and hence } \frac{x_1(t)}{x_2(t)} \longrightarrow \frac{\sqrt{\ell_1'}}{\sqrt{\ell_2'}}$$

The plus-minus ensemble grows exponentially with a rate constant of $\sqrt{\ell_1' \ell_2'}$.

We mention two experimental systems on RNA replication, which were run under conditions comparable to our model systems. Schneider et al [1979] reported detailed kinetic studies on poly(A)-poly(U) synthesis in a stirred flow reactor. The enzyme used is RNA polymerase from Escherichia coli. The experimental setup and the mechanism proposed is shown in Figure 7. The overall kinetics of this ten-dimensional system is characterized by exponential growth of the plus-minus ensemble as predicted by equation (1.13a). Moreover, they found multiple steady states and critical slowing down near the bifurcation point under certain experimental conditions. It is worth noticing that characteristic non-linear phenomena were observed experimentally in the dynamics of polynucleotide replication.

The kinetics of phage RNA replication has been studied in vitro by Biebricher et al [1980]. Q_β replicase was used as the enzyme. The authors investigated the kinetics of RNA synthesis at constant total concentration of enzyme and buffered nucleotide-triphosphate concentrations. The polynucleotide synthesis shows three distinct phases of growth: at low concentrations exponential increase in total RNA concentration is observed which changes into linear growth at moderate concentrations, and finally levels off to saturation or zero growth at large excess of RNA (Figure 8). The authors suggest a complicated multistep mechanism as shown in Figure 8. It is somewhat more detailed than, but nevertheless closely related to, that mentioned before. It consists of three sets of reaction steps dealing with initiation of polymerization, chain elongation and reactivation of the enzyme. The mechanism is able to explain all the kinetic details observed.

For our purpose here we remember that RNA synthesis following complicated many-step mechanisms fulfills the same exponential growth law at the low concentration limit as the simple autocatalytic reaction (1.1) does.

C. The Evolution Reactor

In order to make a problem as complex as molecular self-organization accessible to a formal mathematical treatment we have to minimize mechanistic diversity and environmental influence.

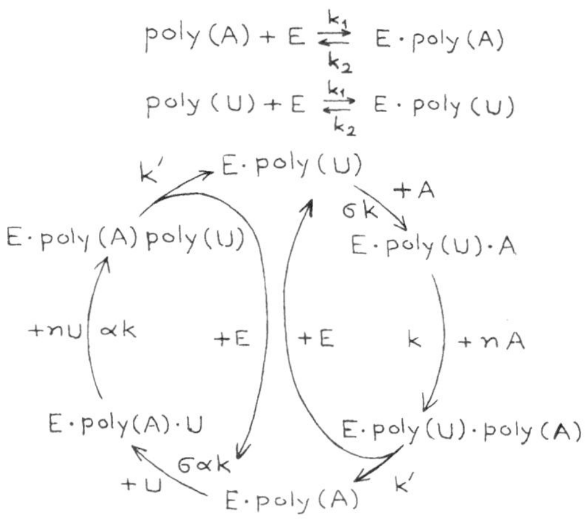

$$\text{poly(A)} + E \underset{k_2}{\overset{k_1}{\rightleftarrows}} E\cdot\text{poly(A)}$$

$$\text{poly(U)} + E \underset{k_2}{\overset{k_1}{\rightleftarrows}} E\cdot\text{poly(U)}$$

k' E·poly(U)
+A σk
E·poly(A) poly(U) E·poly(U)·A
+nU αk +E +E k +nA
E·poly(A)·U E·poly(U)·poly(A)
+U $\sigma\alpha k$ E·poly(A) k'

Fig. 7. RNA synthesis in the poly(A) - poly(U) system according to Schneider et al [1979].

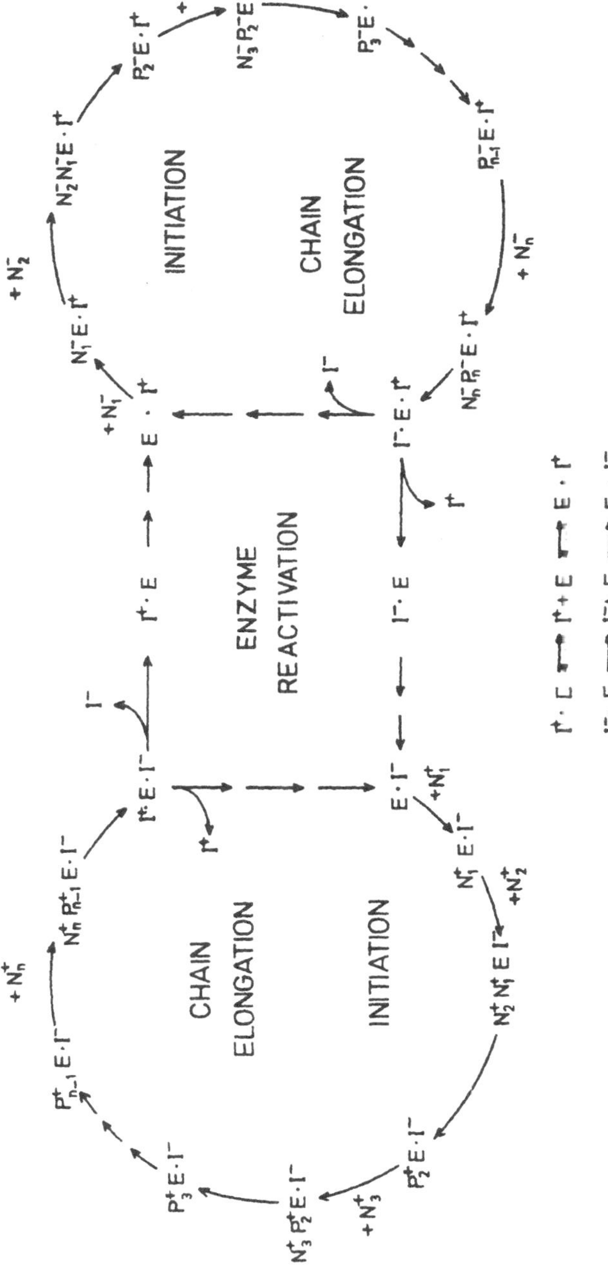

Fig. 8. RNA-synthesis in the in vitro Q_β replication system [Biebricher et al, 1980].

A kind of dialysis reactor shown schematically in Figure 9 and described in detail by Küppers [1979a] is well suited for this purpose. It allows one to control the concentrations of all low molecular weight material, in particular those of the four triphosphates ATP, UTP, GTP and CTP as well as those of the degradation products AMP, UMP, GMP and CMP. Keeping the constraint, the concentrations of macromolecules remain the exclusive variables. Further control may be applied to the total concentration of polynucleotides,

$$C = [I_1]+[I_2]+\ldots+[I_n] = \sum_{i=1}^{n} x_i ,$$

by means of a dilution flux, ϕ.

For further analysis we define growth functions Γ_i which describe the evolution of the polynucleotide in the absence of any constraint or dilution flux. The equations we have to study then are of the following form:

$$\dot{\underline{x}} = \underline{\Gamma}(\underline{x}) - \underline{x}\,\frac{\phi(t)}{C(t)} \qquad (1.14)$$

For short we use a state vector $\underline{x}$ which describes the concentrations of all polynucleotides considered:

$$\underline{x} = (x_1,x_2,\ldots x_n);\quad \underline{x}\in\mathbb{R}^n;\quad x_i \geq 0,\ i = 1,\ldots,n \qquad (1.15)$$

$\underline{\Gamma}(\underline{x})$ is the corresponding vector function:

$$\underline{\Gamma}(\underline{x}) = (\Gamma_1(\underline{x}),\ \Gamma_2(\underline{x}),\ \ldots,\ \Gamma_n(\underline{x})) . \qquad (1.16)$$

Dilution flux and total concentration are mutually dependent:

$$\phi(t) = \sum_{j=1}^{n} \Gamma_j(\underline{x}) - \dot{C} \qquad (1.17a)$$

$$C(t) = C(0) + \int_0^t \left\{ \sum_{j=1}^{n} \Gamma_j(\underline{x}) - \phi(\tau) \right\} d\tau \qquad (1.17b)$$

Stationarity with respect to total concentration is obtained in the reactor if

$$C = C_o,\ \dot{C} = 0 \longrightarrow \phi = \phi_o = \sum_{j=1}^{n} \Gamma_j(\underline{x})$$

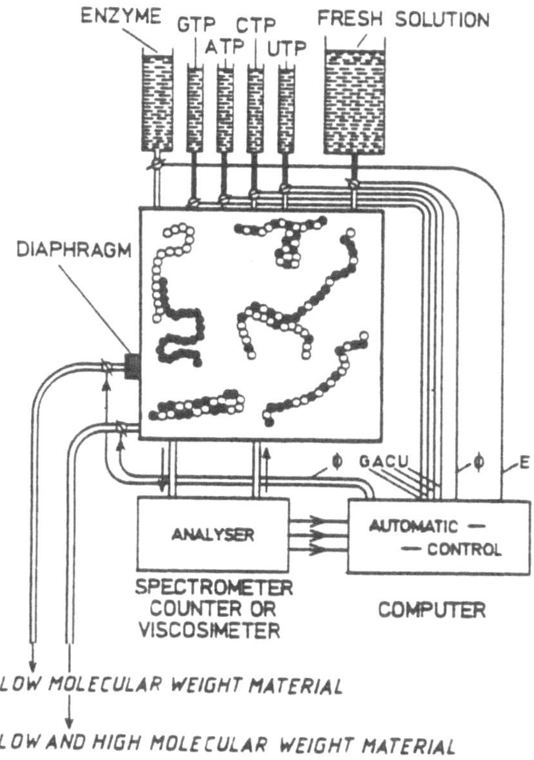

Fig. 9. The dialysis reactor for evolution experiments. This kind of flow reactor consists of a reaction vessel which allows for temperature and pressure control. Its walls are impermeable to polynucleotides. Energy-rich material (ATP, UTP, GTP and CTP) is poured from the environment into the reactor. The degradation products (AMP, UMP, GMP and CMP) are removed steadily. Material transport is adjusted in such a way that the concentration of monomers is constant in the reactor. A dilution flux ϕ is installed in order to remove the excess of polynucleotides produced by multiplication. Thus the sum of the concentrations:

$$[I_1] + [I_2] + \ldots + [I_n] = \sum_{i=1}^{n} x_i = C ,$$

may be controlled by the flux ϕ. Under "constant organization" ϕ is adjusted such that the total concentration $C = C_o$ is constant.

is fulfilled and hence we obtain the following equation for the evolution of the concentrations at constant total concentration or "constant organization" as the condition has been called previously [Eigen, 1971].

$$\dot{\underline{x}} = \underline{\Gamma}(\underline{x}) - \underline{x}\,\frac{1}{C_o}\sum_{j=1}^{n}\Gamma_j(\underline{x}) \quad . \tag{1.18}$$

The stability of this stationary state can be analyzed easily:

$$\dot{C} = \sum_{j=1}^{n}\Gamma_j(\underline{x})\left\{1 - \frac{C}{C_o}\right\}$$

The fixed point at $C = C_o$ has an eigenvalue

$$\omega_C = -\frac{1}{C_o}\left(\sum_{j=1}^{n}\Gamma_j(\underline{x})\right)_{C=C_o}$$

and hence is asymptotically stable iff

$$\left(\sum_{j=1}^{n}\Gamma_j(\underline{x})\right)_{C=C_o} > 0 \quad . \tag{1.19}$$

Thus, "constant organization" is a stable steady state in case the sum of all growth functions is positive or, in other words, if we have an overall net growth which can be compensated by the flux ϕ.

It is straightforward to derive an analogous equation which holds for growing systems. We introduce new variables, $\zeta_i = x_i/C$, which map the entire concentration space on the unit simplex:

$$\underline{\zeta} = (\zeta_1,\zeta_2,\ldots,\zeta_n) \in S_n\left\{\underline{\zeta} \in \mathbb{R}^n:\ \zeta_i \geq 0, \sum\zeta_i = 1\right\} . \tag{1.20}$$

By means of (1.14) and (1.17a) we find

$$\dot{\underline{\zeta}} = \frac{1}{C(t)}\left\{\underline{\Gamma}(\underline{x}) - \underline{\zeta}\sum_{j=1}^{n}\Gamma_j(\underline{x})\right\} . \tag{1.21}$$

In order to be able to make predictions we have to specify the growth functions Γ_i. Let us assume first that we are dealing with homogeneous functions of degree λ in $\underline{x}$. Then

$$\underline{\Gamma}(\underline{x}) = \underline{\Gamma}(C\underline{\zeta}) = C^{\lambda}\,\underline{\Gamma}(\underline{\zeta})$$

is valid and we obtain

$$\dot{\underline{\zeta}} = C^{\lambda-1}\left\{\underline{\Gamma}(\underline{\zeta}) - \underline{\zeta}\sum_{j=1}^{n}\Gamma_j(\underline{\zeta})\right\} . \tag{1.22}$$

This equation contains a very important and useful statement: (Provided $C > 0$ which is trivially fulfilled) the differential equation (1.18) at $C_o = 1$, i.e. the equation for "constant organization" on the simplex S_n, and equation (1.22) have identical trajectories. Hence, the growing system in internal coordinates approaches the same ω-limits as the corresponding stationary system does. Moreover, the trajectories of (1.18) do not depend on C_o in case they are projected onto the unit simplex.

As long as we are dealing with mass action kinetics, the growth functions Γ_i will have a rather simple form and can be represented by a polynomial expansion:

$$\Gamma_i = k^{(i)} + \sum_{j=1}^{n} k_j^{(i)} x_j + \sum_{j=1}^{n}\sum_{\ell=1}^{n} k_{j\ell}^{(i)}\, x_j x_\ell + \ldots \tag{1.23}$$

The power of an individual term in this series refers to the molecularity of the polymerization process as far as macromolecules are involved. Homogeneity of the growth functions means that all macromolecules under consideration replicate with the same growth characteristics.

Now, we are in a position to discuss several examples. At first we study the competition between self-replicating elements, the growth function of which corresponds to the simple autocatalytic reaction (1.1) which was found to be valid for enzyme catalyzed RNA replication at the low concentration limit. As in the previous cases we assume buffered nucleotide triphosphate concentrations. The excess production (E_i) of a given polynucleotide is obtained as the difference between the formation and the degradation rate which are denoted by f_i and d_i respectively:

$$\Gamma_i = (f_i - d_i)x_i = E_i x_i ; \; i = 1,\ldots,n \quad . \tag{1.24}$$

The mean excess production of the entire system denoted by

$$\overline{E} = \frac{1}{C}\sum_{i=1}^{n} E_i x_i \tag{1.25}$$

is compensated by the dilution flux ϕ_o at the constraint of constant organization,

$$\overline{E} = \frac{1}{C_o}\phi_o \ .$$

The analysis of the corresponding equation

$$\dot{x}_i = x_i(E_i - \overline{E}); \ i = 1, \ldots, n \tag{1.26}$$

is straightforward (see e.g. Eigen [1971]). Selection takes place in the reactor: the polynucleotide with the largest value of E_i, $E_m = \max(E_i;\ i = 1,\ldots,n)$, survives the competition. A characteristic example is shown in Figure 10. The function $\overline{E}(t)$ plays an important role. It increases steadily and approaches an optimum at the steady state:

$$\lim_{t \to \infty} \overline{E}(t) = E_m \ ; \ E_m = \max(E_i, i=1,\ldots,n) \tag{1.27}$$

Implicitly, we assume all E_i values to be different.

It is also worth-while to look at equation (1.26) by means of fixed-point analysis. For this purpose we consider the mapping onto the unit simplex S_n. The long term behaviour of (1.26) is determined by n fixed points at the corners of S_n. Due to the conservation law $\Sigma x_i = C_o$, we expect n-1 eigenvalues of the Jacobian matrix which determine the stability of the corresponding fixed point:

$$\begin{aligned}
P_1 &= (C_o,0,\ldots,0) \ ; & \omega_j^{(1)} &= E_j - E_1, \ j = 2,3,\ldots,n \\
P_2 &= (0,C_o,\ldots,0) \ ; & \omega_j^{(2)} &= E_j - E_2, \ j = 1,3,\ldots,n \\
& & & \\
P_n &= (0,0,\ldots,0,C_o); & \omega_j^{(n)} &= E_j - E_n, \ j = 1,2,\ldots,n-1
\end{aligned} \tag{1.28}$$

From (1.28) we derive immediately that there exists only one sink, $\omega_j < 0 \ \forall j$. It corresponds to the polynucleotide with the largest value of E_i, E_m. Since all eigenvalues are real the sink is of nodal character and we do not expect oscillations to occur in this system. An example for three variables is given in Figure 11.

Selection within a system of polynucleotides replicating with different rates has been demonstrated by the serial transfer experiments of Spiegelman [1971]. His results and some more recent works on the Q_β-system are discussed in great detail in the review article by Küppers [1979a].

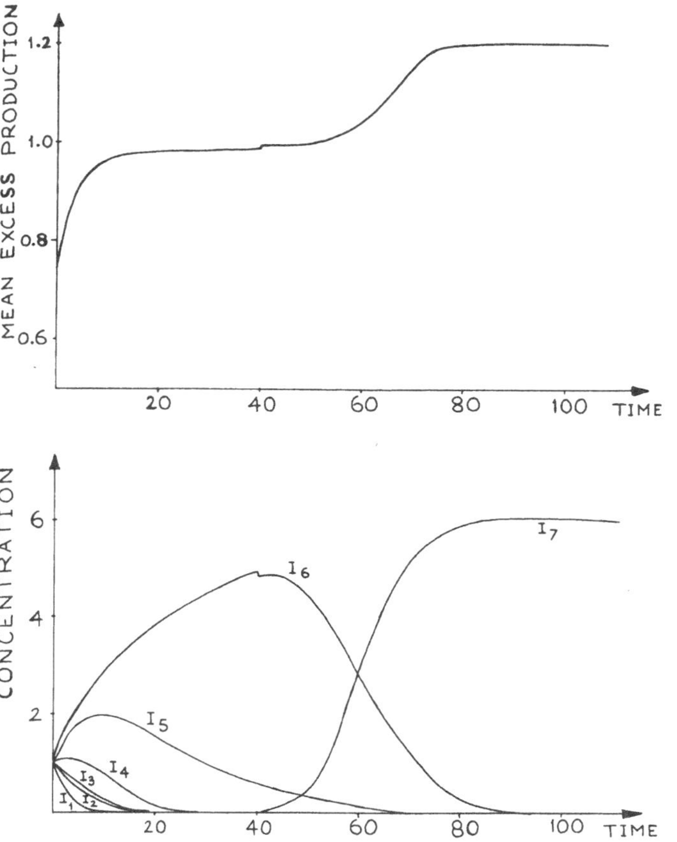

Fig. 10. Solution curves for the differential equation $\dot{x} = (E_i - \bar{E})x_i$, $i = 1, \ldots, 7$ with $E_1 = 0.4$, $E_2 = 0.6$, $E_3 = 0.65$, $E_4 = 0.8$, $E_5 = 0.95$, $E_6 = 1.0$ and $E_7 = 1.2$;

$$\bar{E} = \sum_i E_i x_i / \sum_i x_i \ .$$

Initial conditions: $x_1(0) = x_2(0) = \ldots = x_6(0) = 1$, $x_7(0) = 0$. The mean excess production $\bar{E}$ starts from an initial value of $\bar{E} = 0.733$ and increases steadily up to $\bar{E} \sim 1.0$ as the population becomes homogeneous: $x_1 = x_2 = \ldots = x_5 \sim 0$, $x_6 \sim 6$, $x_7 = 0$. One easily recognizes the cause for the increase in $\bar{E}$: the less efficiently growing polynucleotides are eliminated. They disappear in the same sequence as their excess production E_i increases. At $t = 40$ we introduced a fluctuation $x_7 = 0 \rightarrow x_7 = \delta$. The appearance of a more efficiently growing polynucleotide destabilizes the steady state. $\bar{E}$ increases further and approaches the value $\bar{E} = 1.2$ when the population becomes homogeneous again: $x_1 = x_2 = \ldots = x_6 \sim 0$, $x_7 \sim 6$.

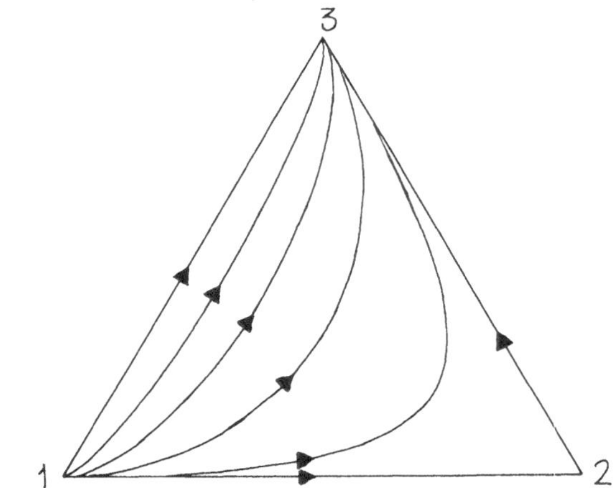

Fig. 11. Fixed-point map for independent competitors at constant organization according to equation (1.26). The numerical values chosen are $E_1 = 1$, $E_2 = 2$, $E_3 = 3$. A pure state of species 3 represents the only stable long-time-range solution of the system. With the exception of the two edges $\overline{12}$ and $\overline{23}$ all trajectories start in P_1 and have P_3 as ω-limit.

II. DARWINIAN SYSTEMS

No physical process can occur with absolute accuracy. In reality we have to deal always with certain frequencies of copying errors. Copying gives rise to new sequences of polynucleotides. We call these erroneous copies mutants of the original sequence. In this section we shall analyze the problem of self-replication with errors following along the lines of previous work on this topic [Eigen, 1971; Thompson and McBride, 1974; Jones et al, 1976; Eigen and Schuster, 1977; Epstein and Eigen, 1979; Küppers, 1979b, Ebeling and Feistel, 1978].

A. Replication with Errors

In order to illustrate the nature of the problem, we consider a simple stochastic model [Schuster and Sigmund, 1980]. As shown in Figure 12, every correct sequence is assumed to create a progeny of Σ copies per generation and have a lifetime of just one multiplication cycle. All error copies, i.e. mutants, are lethal. They have no offspring. The question we are going to ask is whether such a chain of consecutive replications has a non-zero probability to extend to $t \to \infty$ or whether the polynucleotide sequence under consideration will die out after a finite number of generations. Let us assume that every digit is replicated with a single digit accuracy q. This single digit accuracy is the probability of incorporation of the correct digit at a given position in the sequence. The errors are assumed to occur independently and with equal probability for every digit. Thus a sequence of ν digits will be replicated with an accuracy

$$Q = q^{\nu} . \tag{2.1}$$

We call Q the quality factor for the replication of a sequence of ν digits.

The problem of error propagation in this model can be formulated and analyzed as a multiplicative chain (see [Bartlett, 1978]). The number of offspring is a random variable X with binomial distribution. Thus, the probability to have k viable descendants from a single copy is of the form

$$\mathrm{Prob}(X = k) = \binom{\Sigma}{k} Q^{k} (1 - Q)^{\Sigma - k} . \tag{2.2}$$

This distribution has an expectation value of

$$\langle k \rangle = m = \Sigma Q = \Sigma q^{\nu} . \tag{2.3}$$

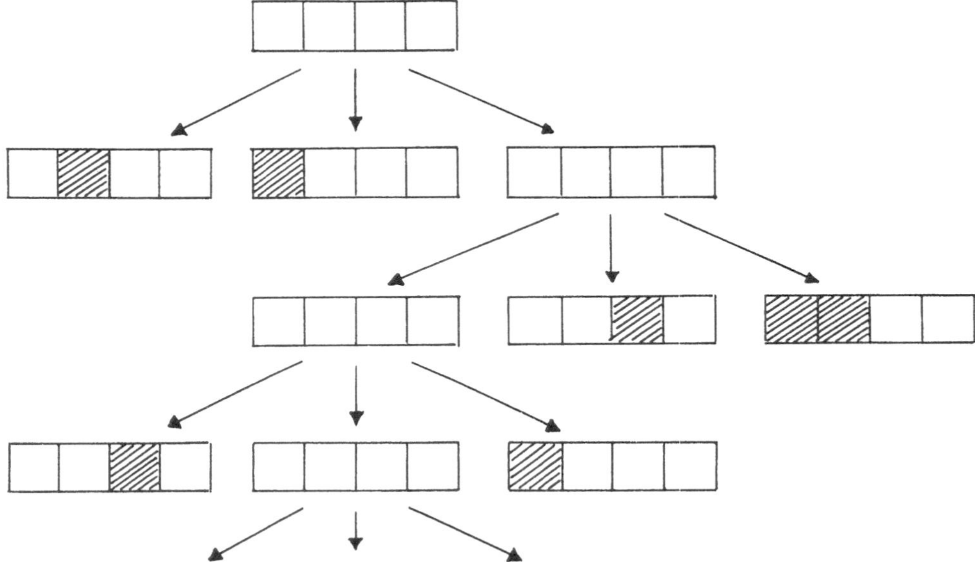

Fig. 12. A simple stochastic model for the self-replication with errors. In the concrete example we have chosen $\nu = 4$, $\Sigma = 3$ and $q = 5/6$.

From the theory of multiplicative chains it follows that $m \leq 1$ is a necessary and sufficient condition for certain extinction. Hence, the condition for a non-zero probability of survival has the simple form:

$$Q\Sigma = m > 1 \quad \text{or} \quad Q > \Sigma^{-1} \quad .$$

There is a sharply defined minimum accuracy of replication,

$$Q_{min} = \Sigma^{-1} ,$$

below which the system certainly goes extinct. For a given value of q this minimum corresponds to a maximum chain length of the polymer

$$Q_{min} = q^{\nu_{max}} \quad \text{or} \quad \nu_{max} = -\frac{\ln\Sigma}{\ln q} \sim \frac{\ln\Sigma}{1-q} \quad \text{if } q \sim 1 \ . \qquad (2.4)$$

A given polymerization mechanism is characterized by a certain value of q which in turn sets a critical value to the number of digits or the information content that can be transmitted from generation to generation by the copying mechanism. An increase in the number of digits exceeding ν_{max} thus requires an improvement of the replication machinery.

This simple stochastic model revealed one important feature of self-replication with errors: the content of information to be transferred is limited by the accuracy of replication. In order to gain some more quantitative insight, we have to drop the most restrictive and unrealistic assumption concerning lethality of mutants. Evolution can take place in an efficient way only if we allow for an ample mutant distribution with an almost continuous (or "quasi-continuous") gradation of properties. The resulting mathematical problem is very hard to solve at the level of stochastic processes but can be treated by means of differential equations.

B. The Concept of a "Quasispecies"

The deterministic approach to studying self-replication with errors starts from conditions as encountered in the evolution reactor which are mimicked by the differential equation (1.18) with a growth function

$$\Gamma_i = \sum_{j=1}^{n} \omega_{ij} x_j \ ; \ i = 1,2,\ldots n \quad . \qquad (2.5)$$

Equation (2.5) is a straightforward generalization of equation (1.24). In addition to correct self-replication, we allow also errors leading to mutants according to the reaction

$$I_j + \sum_{\lambda=1}^{4} \nu_\lambda^{(i)} A_\lambda \xrightarrow{\omega_{ij}} I_i + I_j ; \; i,j = 1,2,\ldots n ; \quad (i \neq j) \tag{2.6}$$

The constant ω_{ij} thus represents the rate at which the sequence I_i is obtained as an error copy of I_j. In assigning a rate constant to the process of mutation, we make two implicit assumptions: (1) that we are considering a sufficiently large ensemble of molecules and (2) that mutations occur frequently enough to reach stationary mutant distributions.

The "diagonal" rate constants require further specification

$$\omega_{ii} = f_i Q_i - d_i ; \; i = 1,2,\ldots,n \quad . \tag{2.7}$$

Q_i is the quality factor for the replication of the sequence I_i. As before, it describes the probability for correct replication of the whole polymer. The rate constant f_i gives the number of copies, correct and erroneous, which are synthesized per unit time on the template I_i, d_i is the rate constant for hydrolysis of these templates. The complete quantity ω_{ii} has been named the "value function" [Eigen, 1971] since it determines the selective value of the polynucleotide I_i. Equation (2.5) becomes identical with (1.24) or (1.26) if we put $Q_i = 1$ and $\omega_{ij} = 0 \; \forall \; j \neq i$ as required in the case of accurate replication.

Again we define an excess production $E_i = f_i - d_i$. From mass conservation we obtain the global relation

$$\sum_i f_i (1-Q_i) x_i = \sum_i \sum_{j \neq i} \omega_{ij} x_j \quad . \tag{2.8}$$

Accordingly, the mean excess production becomes identical to the dilution flux ϕ_o except for a factor C^{-1}, exactly as it was with error-free replication (1.25),

$$\overline{E} = \frac{1}{C} \sum_i E_i x_i = \frac{1}{C} \phi_o \quad .$$

The remaining differential equation

$$\dot{x}_i = [\omega_{ii} - \overline{E}(t)] x_i + \sum_{j \neq i} \omega_{ij} x_j ; \; i,j = 1,2,\ldots,n \tag{2.9}$$

can be analyzed by a change in variables and application of second-order perturbation theory [Thompson and McBride, 1974; Jones et al, 1976]. Again selection takes place in the evolution reactor provided the replication process is accurate enough. By successive elimination of the slowly growing molecules the mean excess production, $\bar{E}(t)$, increases steadily until it reaches an optimal value λ_{max}:

$$\lim_{t\to\infty} \bar{E}(t) = \lambda_{max} = \omega_{mm} + \sum_{j\neq m} \frac{\omega_{mj}\,\omega_{jm}}{\omega_{mm}-\omega_{jj}} \tag{2.10}$$

Actually, λ_{max} is the second-order approximation to the largest eigenvalue of the linear problem obtained from (2.9) by a non-linear transformation. By I_m we denote the master sequence, i.e. the polynucleotide which has the largest value function:

$$\omega_{mm} = \max(\omega_{ii}; i = 1,\ldots,n) \; .$$

The results of second order perturbation theory are good approximations to the exact solutions provided

$$\omega_{im} << (\omega_{mm} - \omega_{ii}) \; .$$

What has been selected actually by this optimization process? In order to answer this question we have to consider the eigenvector corresponding to λ_{max}. From perturbation theory we obtain the following expressions for the stationary concentrations $\bar{x}_i$:

$$\bar{x}_m = C\,\frac{\omega_{mm}-\bar{E}_{-m}}{E_m-\bar{E}_{-m}} = C\,\frac{Q_m-\sigma_m^{-1}}{1-\sigma_m^{-1}} \tag{2.11}$$

and

$$\bar{x}_i = \bar{x}_m\,\frac{\omega_{im}}{\omega_{mm}-\omega_{ii}} \; ; \; i=1,2,\ldots,n; \; i\neq m \; . \tag{2.12}$$

For short, we have defined the two quantities

$$\bar{E}_{-m} = \frac{\sum_{(i=1,i\neq m)}^{n} E_i x_i}{C - x_m} \quad \text{and} \quad \sigma_m = \frac{f_m}{d_m + \bar{E}_{-m}} \tag{2.13}$$

The latter quantity is called the superiority parameter σ_m. It is a measure of the kinetic superiority of the master sequence compared to the average of the other polynucleotides present in the evolution reactor. We distinguish two situations:

(1) $Q_m \leq \sigma_m^{-1}$: there is no stationary state of the evolution reactor. The accuracy of replication is too low and at every instant we find different sequences in the reactor. Since the number of possible polynucleotide sequences of only moderate length, $\nu > 50$ is much higher than the number of molecules in the evolution reactor, the system cannot reach a stationary distribution by random mutations.

(2) $Q_m > \sigma_m^{-1}$: the master copy is accompanied by a mutant distribution at the stationary state. The concentrations of the frequently occurring mutants are determined by equation (2.12). We call the whole ensemble of sequences a quasispecies because of the similarity to an actual biological species as far as the existence of mutants present in the stationary population is concerned.

As with the stochastic approach we obtain a lower limit for the accuracy of the copying process below which the transfer of information from generation to generation breaks down:

$$Q_m > Q_{min} = \sigma_m^{-1} \quad . \tag{2.14}$$

We introduce an average single digit accuracy $\bar{q}_m$ for the replication of a given polynucleotide with a chain length of ν_m:

$$Q_m = q_1 q_2 q_3 \cdots q_{\nu_m} = \bar{q}_m^{\,\nu_m} \quad . \tag{2.15}$$

Herein, q_1 represents the single digit accuracy for the incorporation of the complementary nucleotide at position 1, q_2 for that at position 2, etc. The quality factor q_i will depend on the nature of the nucleotide (A,U,G or C) and on its neighborhood in the sequence. For sequences of sufficient length, with similar nucleotide contents, in general the values for $\bar{q}$ can be expected to be similar.

The lower limit of the quality factor implies an upper limit for the chain length of the polynucleotide

$$\nu_m < \nu_{max} = -\frac{\ell n \sigma_m}{\ell n \bar{q}_m} \simeq \frac{\ell n \sigma_m}{1-\bar{q}_m} \quad . \tag{2.16}$$

The approximation $-\ell n \bar{q}_m \simeq 1-\bar{q}_m$ will be fulfilled in almost every case since $\bar{q}_m$ will be close to 1.

The dependence of $\bar{x}_m$ on the chain length of the polynucleotide is shown schematically in Figure 13. Close to the maximum value, ν_{max}, the stationary concentration of the master copy may be exceedingly small.

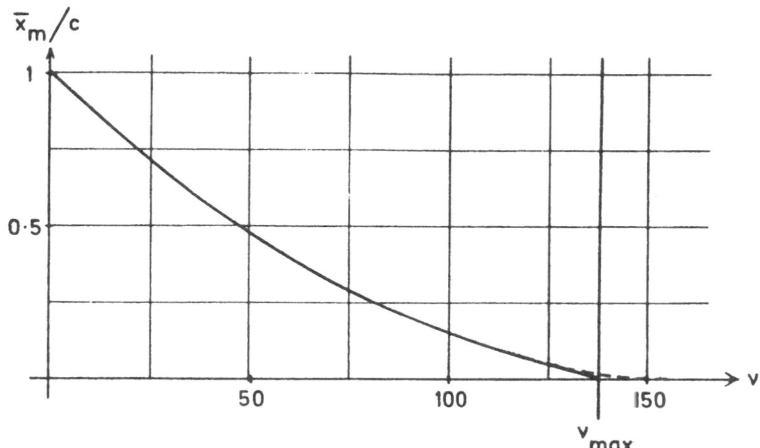

Fig. 13. The dependence of the stationary concentration of the master sequence in a quasispecies on the chain length ν of the polynucleotide. The single digit accuracy is assumed to be $\bar{q} = 0.99$ and a value of $\sigma_m = 4$ is chosen for the superiority parameter.

So far we have excluded systems in which two or more sequences have identical or almost identical value functions. In these cases the formalism of perturbation theory of degenerate states can be used to obtain approximate solutions.

C. Limitations to the Lengths of Genomes

Equation (2.16) may be used to discuss the limitations of replication processes in biology. The error threshold strongly depends on the average single digit accuracy of the replication process. Due to the logarithmic function, the influence of the superiority on the maximum chain length is not as pronounced. Some characteristic numerical examples are summarized in Table 1.

The top row of Table 1 gives some numbers typical for prebiotic, template induced, polynucleotide replication. The value taken for the single digit accuracy, $\bar{q} = 0.95$, corresponds well to the experimental results of Lohrmann et al [1980]. Theoretical estimates [Eigen, 1971; Eigen and Schuster, 1978b] derived from thermodynamic and kinetic data [Pörschke, 1977] led to very similar values between $\bar{q} = 0.90$ and $\bar{q} = 0.99$, the extremes corresponding to AU rich and GC rich sequences respectively. Under prebiotic conditions, the length of polynucleotides with defined sequences was limited to molecules up to the size of present day t-RNA's. Longer polynucleotides could form but their sequences were changing permanently and hence could not be used for the storage of information.

The accuracy of Q_β-RNA replication in bacterial cells and in "in vitro" systems has been determined by Weissmann and coworkers [Domingo et al, 1976; Batschelet et al, 1976]. Analyzing this experimental data by means of equation (2.16) yields a maximum digit content which comes very close to the actual genome length of this phage [Eigen and Schuster, 1978b]. A further consequence of the fact that the polynucleotides multiply at conditions close to the error threshold consists in a wide spread mutant distribution with a rather small percentage of master sequence, I_m. This prediction has been verified by Weissmann and co-workers by means of another set of experiments [Sabo et al, 1980].

Error frequencies in DNA replication have been studied extensively by Loeb and his group [Kunkel and Loeb, 1979; Kunkel et al, 1979]. An estimate of ν_{max} for bacteria based on his results again leads to values which are close to the known lengths of genomes in procaryots. The tremendous increase in accuracy compared to single stranded RNA replication is caused by the elaborate repair mechanism which eliminates errors that occurred in the primary replication process [Kornberg, 1974].

Table I: Single digit accuracy, superiority and error threshold in some chemical and biological systems.

Single digit accuracy $\overline{q}_m$	Error rate per digit $1 - \overline{q}_m$	Superiority σ_m	Maximum digit content ν_{max}	Biological examples
0.95	5×10^{-2}	2 20 200	14 60 106	Enzyme-free RNA replication, t-RNA precursor $\nu = 80$
0.9995	5×10^{-4}	2 20 200	1386 5991 10597	Single stranded RNA replication via specific replicases, Phage Q_β, $\nu = 4500$
0.999999	1×10^{-6}	2 20 200	0.7×10^6 3.0×10^6 5.3×10^6	DNA replication via polymerases and proof-reading, E. coli, $\nu = 4 \times 10^6$

In concluding this section we may suggest that nature approaches the limits of information content which is dictated by the mechanism of polynucleotide replication at least for procaryotic organisms. We have used the term Darwinian systems for this class of replication-mutation equations since they describe a phenomenon which comes very close to Darwin's original ideas: the polynucleotides considered optimise their efficiency in replication by the potent interplay of random variation and directed selection. In higher organisms at least, the mechanism which gives rise to biological variation is not simple mutation only, but also genetic recombination which opens up a new dimension of combinatorial richness.

III. COOPERATION OF SELF-REPLICATING ELEMENTS

In the previous sections we have shown that the capability of self-replication introduces competition and selection into molecular systems. Some of the results derived can be applied directly to biological systems which optimize most of their relevant properties by a closely related selection mechanism. By relevant, we mean important for survival. On the other hand, there are many well known examples of cooperation between self-replicating elements. In this section we try to search for simple mechanisms which are able to suppress competition. By "simple" we understand terms of low order in the polynomial expansion of the growth function $\underline{\Gamma}$ according to equation (1.23). The term linear in $\underline{x}$ always leads to competition, eventually to the formation of a quasispecies. Since we are searching for controlable interactions the fortuitous simultaneous presence of two or more polynucleotides in a mutant distribution is not sufficient. Thus we have to consider the second order terms. In order to facilitate the analysis we shall neglect the terms which are due to mutations:

$$\Gamma_i = \sum_{j=1}^{n} k_{ij}^{(i)} x_i x_j = x_i \sum_{j=1}^{n} f_{ij} x_j \quad . \tag{3.1}$$

Consequently, we can omit the dispensible superscript. All growth functions are homogeneous and we can restrict ourselves to the case $C_o = 1$.

A. A Theorem of Cooperation

The differential equation we are going to study now,

$$\dot{x}_i = x_i \left(\sum_{j=1}^{n} f_{ij} x_j - \phi_o \right); \quad \phi_o = \sum_{k=1}^{n} \sum_{\ell=1}^{n} f_{k\ell} x_k x_\ell \; ; \quad i=1,\ldots,n \tag{3.2}$$

deserves some interest since it is of importance in several fields: catalytic interaction of self-replicating macromolecules, population genetics, ecology of predator-prey systems and social behaviour of animals. Before we discuss the qualitative nature of solutions of equation (3.2) we shall devise an interesting equivalence.

Equation (3.2) is invariant to the addition of a constant to every column of the matrix of rate constants $F = \{f_{ij}\}$ [Hofbauer et al, 1980]: Taking

$$F = \{f_{ij} = k_j + d_{ij}\}$$

and substituting into (3.2) yields

$$\dot{x}_i = x_i\left(\sum_{j=1}^{n} d_{ij}x_j - \phi_o'\right); \quad \phi_o' = \sum_{k=1}^{n}\sum_{\ell=1}^{n} d_{k\ell}x_k x_\ell \; ; \quad i=1,\ldots,n \tag{3.2a}$$

We can remove this source of arbitrariness by setting all diagonal elements equal to zero:

$$F = \begin{bmatrix} f_{11} & f_{12} & \cdots & f_{1n} \\ f_{21} & f_{22} & \cdots & f_{2n} \\ \vdots & \vdots & & \vdots \\ f_{n1} & f_{n2} & \cdots & f_{nn} \end{bmatrix} \rightarrow B = \begin{bmatrix} 0 & f_{12}-f_{22} & \cdots & f_{1n}-f_{nn} \\ f_{21}-f_{11} & 0 & \cdots & f_{2n}-f_{nn} \\ \vdots & \vdots & & \vdots \\ f_{n1}-f_{11} & f_{n2}-f_{22} & \cdots & 0 \end{bmatrix} \tag{3.3}$$

In the following discussion it will turn out to be useful to distinguish a special case from the general situation. We call (3.2) "non-hyperbolic" iff $f_{ij} \geq f_{jj} \; \forall \; i,j = 1,\ldots,n$. Hence, we have

$$B = \left\{ b_{ij} = f_{ij} - f_{jj} \geq 0 \right\} \tag{3.4}$$

Recently, Hofbauer [1980] was able to prove equivalence between equation (3.2) and the general Lotka-Volterra equations. First, we make use of the invariance property of F:

$$F \rightarrow A = \left\{ a_{ij} = f_{ij} - f_{oj} \right\} = \begin{bmatrix} 0 & 0 & \cdots & 0 \\ f_{1o}-f_{oo} & f_{11}-f_{o1} & \cdots & f_{1n}-f_{on} \\ \vdots & \vdots & & \vdots \\ f_{no}-f_{oo} & f_{n1}-f_{o1} & \cdots & f_{nn}-f_{on} \end{bmatrix}$$

with

$$i,j = 0,1,\dots,n.$$

Equation (3.2) has the form (note that we started from n+1 variables!):

$$\dot{x}_o = -x_o\phi_o \text{ and } \dot{x}_i = x_i\left(\sum_{j=1}^{n} a_{ij}x_j - \phi_o\right);\ i = 1,2,\dots,n. \tag{3.2b}$$

Introduction of new variables $y_i = x_i/x_o$, $i = 1,2, \dots,n$ yields the new differential equation

$$\dot{y}_i = y_i\left(a_{io} + \sum_{j=1}^{n} a_{ij}y_j\right)\left(\frac{1}{1+\sum_{j=1}^{n} y_j}\right);\ i = 1,2,\dots,n. \tag{3.5}$$

Equation (3.5) is identical with the Lotka-Volterra equation except for the common time-dependent factor

$$\left(1 + \sum_{j=1}^{n} y_j\right)^{-1} .$$

Hence, both differential equations have identical trajectories. This equivalence relation turned out to be useful also for practical reasons since some properties like the existence of limit cycles are easier to prove for equation (3.2) than for (3.5) as shown by Hofbauer [1980].

We return to the non-hyperbolic case of equation (3.2) and the problem of cooperation. We assign a directed graph to the differential equation defined by the matrix B, (3.3): the vertices of the graph correspond to the species I_i, $i = 1,\dots,n$. If the element $b_{ij} > 0$ we draw an edge $j \to i$. Two definitions from graph theory are important. A graph is called irreducible if each vertex i can be reached from each vertex j through a directed arc. It is said to be Hamiltonian if it contains a directed circuit (an arc which returns to its starting point) which covers all vertices of the graph without self-interaction (Figure 14). Furthermore, we need two definitions concerning the long term behaviour of dynamical systems. The term "exclusion" says roughly that at least one of the species vanishes after long-enough time, and hence can be translated as meaning that the ω-limits of the orbits describing the dynamical system are not disjoint from the boundary of the concentration simplex S_n. It is necessary to include also some degenerate cases in which the boundary can be reached by fluctuations along a dense manifold of fixed points (fluctuational ω-limit,

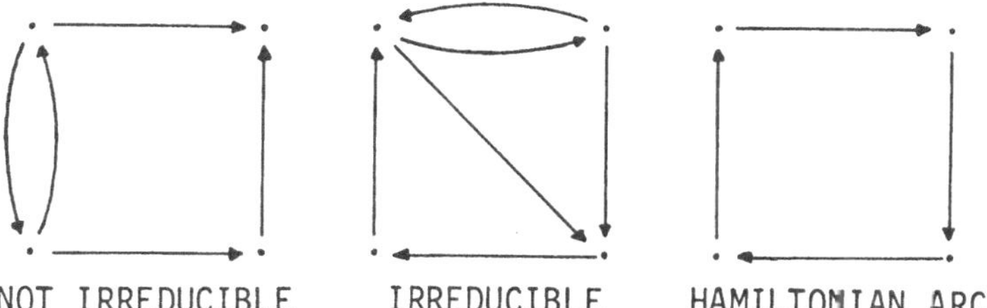

Fig. 14. Examples of not irreducible, irreducible and Hamiltonian graphs corresponding to the non-hyperbolic case of Eq. (3.2).

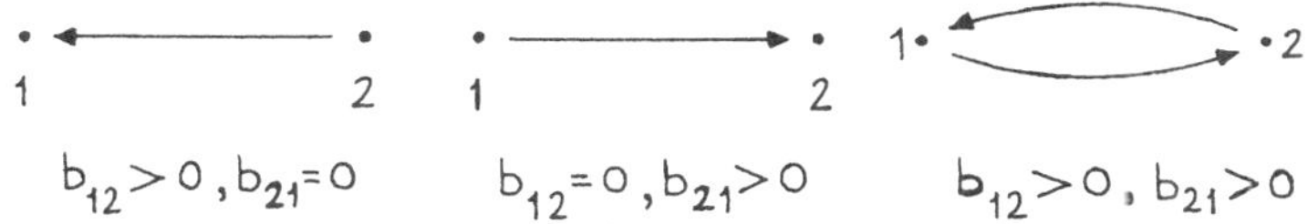

Fig. 15. The three cases of non-hyperbolic systems with n = 2.

see Hofbauer et al [1980]). "Cooperation", on the contrary, means that no species is going to vanish and hence, the ω-limits of the orbits cease to lie in the interior of S_n.

Recently, we were able to present a complete analysis of the low dimensional, non-hyperbolic systems with $n \leq 4$ [Hofbauer et al, 1980]. The case of $n = 2$ is trivial: we have cooperation if $b_{12} > 0$ and $b_{21} > 0$ and exclusion in the two other cases, $b_{12} > 0$, $b_{21} = 0$ and $b_{12} = 0$, $b_{21} > 0$ (Figure 15).

Up to permutations of the indices, there are 15 different graphs with $n = 3$ (Figure 16). The question of cooperation can be treated completely by two theorems:

Theorem 1: If the non-hyperbolic system (3.2) has a unique fixed point P in the interior of the simplex S_3 then it is cooperative and P is the ω-limit of every orbit in the interior of S_3.

Theorem 2: If the graph of the non-hyperbolic system (3.2) with $n = 3$ is not Hamiltonian, the system leads to exclusion. The proofs are given in Hofbauer et al [1980].

According to these two theorems there exist no stable limit cycles for the non-hyperbolic system with $n = 3$, which implies that there are no stable limit cycles in the corresponding Lotka-Volterra systems (3.5) with $n = 2$. Out of the 15 possible cases shown in Figure 16 we find cooperation only in the cases of systems 9, 12, 14 and 15.

The case $n = 4$ is somewhat more involved. We were able to derive a condition for exclusion.

Theorem 3: If the graph of the non-hyperbolic equation (3.2) with $n = 4$ is not irreducible, then the system leads to exclusion (Proof in [Hofbauer et al, 1980]).

Since all non-hyperbolic equations (3.2) of $n = 4$ with a Hamiltonian graph lead to cooperation [Schuster et al, 1979], we are left with 8 irreducible graphs without Hamiltonian arcs (Figure 17). Numerical solutions indicate that we always have exclusion and lend some weight to the conjecture that in order to be cooperative, the non-hyperbolic system must have a Hamiltonian graph.

In higher dimensions, $n \geq 5$, the analytic procedure proposed here becomes very clumsy, because the enormous number of individual cases leads to many different situations which have to be treated separately. Therefore we have to restrict our study to special cases. Two of them are of certain importance:

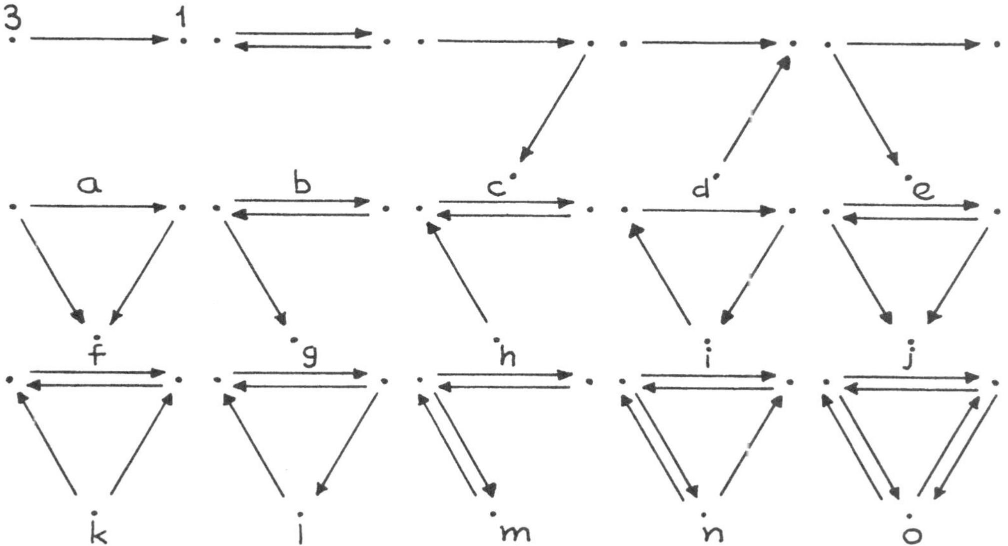

Fig. 16. Possible graphs of the non-hyperbolic Eq. (3.2) with n = 3.

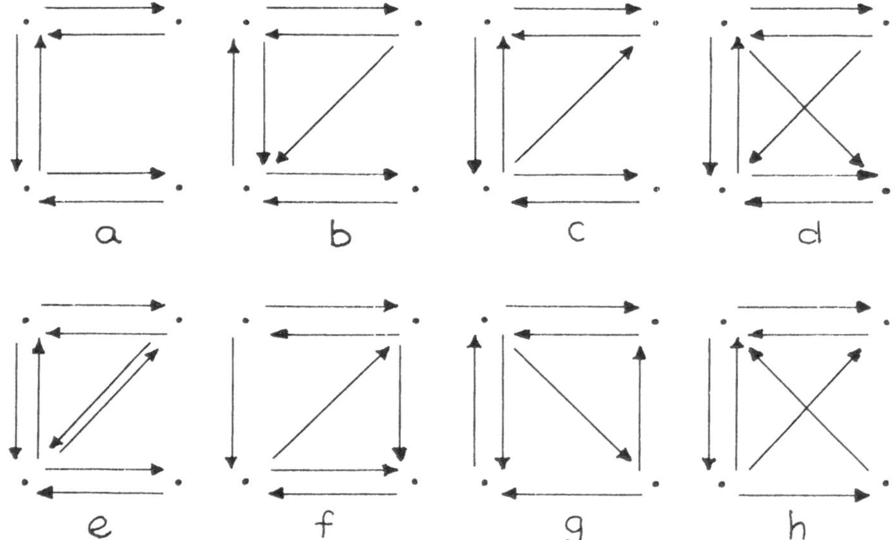

Fig. 17. Possible irreducible graphs without Hamiltonian arcs for the non-hyperbolic Eq. (3.2) with n = 4.

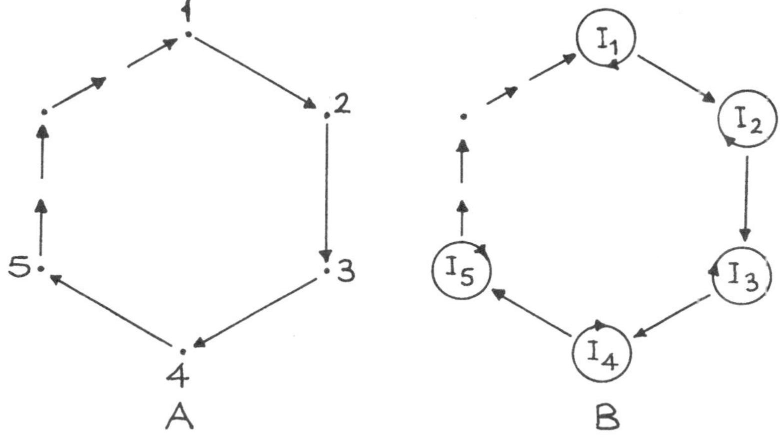

Fig. 18. Two graphical representations of the elementary hypercycle; A corresponds to the graphs in Figures 14-17, B indicates that every element of the cycle is capable of self-instructed replication.

(1). The case of cyclic symmetry: the matrix of coefficients of equation (3.2) is of the form

$$F = \begin{bmatrix} f_1 & f_2 & f_3 & \cdots & f_n \\ f_n & f_1 & f_2 & \cdots & f_{n-1} \\ f_{n-1} & f_n & f_1 & \cdots & f_{n-2} \\ \vdots & \vdots & \vdots & & \\ f_2 & f_3 & f_4 & & f_1 \end{bmatrix} \qquad (3.6)$$

In this case it is not necessary to restrict ourselves to the non-hyperbolic case. Some theorems on the occurence of Hopf-bifurcations in this system [Hofbauer et al, 1980] can be used together with the equivalence of (3.2) and (3.5) to prove the existence of stable limit cycles in Lotka-Volterra systems with $n \geq 3$ [Hofbauer, 1980].

(2). The elementary hypercycle: we consider a special case of the non-hyperbolic system (3.3) which contains only one non-zero coefficient in every row and the graph of which is Hamiltonian (Figure 18). The matrix of rate coefficients in this case has the simple form:

$$F = \begin{bmatrix} 0 & 0 & 0 & \cdots & 0 & f_1 \\ f_2 & 0 & 0 & \cdots & 0 & 0 \\ 0 & f_3 & 0 & \cdots & 0 & 0 \\ \vdots & \vdots & \vdots & & \vdots & \vdots \\ 0 & 0 & 0 & & f_n & 0 \end{bmatrix} \qquad (3.7)$$

The elementary hypercycle of dimension n, hence, is the simplest system belonging to the class defined by equation (3.2), which has a Hamiltonian graph with n vertices. It is cooperative according to a general theorem which holds for all dimensions n [Schuster et al, 1979].

B. The Concept of "Hypercycles"

Elementary hypercyles according to the last paragraph, are the simplest systems of a given dimension which ensure cooperation of their elements. These elements have the capability of self-instructed replication. The notion of a hypercycle goes back to Eigen [1971] who used it originally to define cooperative systems of polynucleotides which are coupled together by catalytic interaction. In a later paper [Eigen and Schuster, 1978 a], we presented a theoretical concept which treated this kind of catalytic interaction in a more formal way. In this section we shall discuss the basic properties of elementary hypercycles.

The differential equation corresponding to an elementary hypercycle

$$\dot{x}_i = x_i(f_i x_j - \phi_o); \; i = 1,\ldots,n; \; j = i-1+n\delta_{i1} \,, \tag{3.8}$$

is characterized by n + 1 fixed points. One of them, P_o, lies in the interior of the simplex S_n

$$P_o = \frac{1}{\left(\sum_{i=1}^{n} f_i^{-1}\right)} \left(f_2^{-1},\, f_3^{-1},\, \ldots,\, f_n^{-1},\, f_1^{-1}\right) \,, \tag{3.9}$$

the remaining n stationary points are the corners of S_n. Equation (3.8) can be subjected to a nonlinear transformation

$$y_i = \frac{f_j x_i}{\sum_{k=1}^{n} f_\ell x_k} \; ; \; i = 1,\ldots,n; \; j = i+1-n\delta_{in}; \; \ell = k+1-n\delta_{kn} \tag{3.10}$$

This transformation represents an isomorphic map of the simplex S_n onto itself.

$$S_n = \left\{(x_1,\ldots,x_n): x_i \geq 0,\; \sum_i x_i = 1\right\} \longleftrightarrow S_n \left\{(y_1,\ldots,y_n): y_i \geq 0,\; \sum_i y_i = 1\right\}$$

Thereby, P_o is transformed into the center of the simplex, C

$$P_o = \frac{1}{\left(\sum_{i=1}^{n} f_i^{-1}\right)} \left(f_2^{-1},\ldots,f_1^{-1}\right) \longleftrightarrow C = \left(\frac{1}{n},\ldots,\frac{1}{n}\right) , \tag{3.11}$$

and equation (3.8) takes the simple form

$$\dot{y}_i = y_i(y_j-\phi)\frac{1}{\left(\sum_{k=1}^{n} f_k y_\ell\right)}\,;\; i=1,\ldots,n;\; j=i-1+n\delta_{i1} \text{ and } \ell=k-1+n\delta_{k1}\,, \tag{3.12}$$

with

$$\phi = \sum_{k=1}^{n} y_k y_\ell .$$

The common factor in all n differential equations is strictly positive and hence corresponds to a change in the time axis only. Since we are interested in the trajectories and in long term behaviour, we can restrict our study to the equation

$$\dot{y}_i = y_i(y_j - \phi);\; i=1,\ldots,n;\; j=i-1+n\delta_{i1} \;. \tag{3.12a}$$

The barycentric transformation (3.10) thus simplifies the analysis substantially. Linearization and calculation of the eigenvalues of the Jacobian is straightforward. For the central fixed point we find

$$\begin{aligned} P_o &= \left(\frac{1}{n},\ldots,\frac{1}{n}\right);\; \omega_j^{(o)} = \exp\left(\frac{2\pi i}{n} j\right) = \lambda_j \\ \zeta_j^{(o)} &= \left(1,\lambda_j^{-1},\lambda_j^{-2},\ldots,\lambda_j^{-n+1}\right);\; j=1,\ldots,n-1 \end{aligned} \tag{3.13}$$

wherein $\zeta_j^{(o)}$ is the eigenvector corresponding to $\omega_j^{(o)}$.

According to (3.13) we find asymptotical stability of the central fixed point for n = 2 and 3. The case n = 4 is critical since we have Re $\omega_{1,3}^{(o)} = 0$ in the linearized system. For $n \geq 5$ we obtain Re $\omega_{1,n-1} > 0$ and hence the central fixed point is unstable. Qualitative analysis of (3.12a) by means of Liapounov functions and some geometrical operations [Schuster et al, 1978] clarifies the role of nonlinear contributions. Thereby, we obtained asymptotical stability of the central fixed point for n = 2,3 and 4. In the case of $n \geq 5$, numerical integration gives strong support to the existence of stable limit cycles. Moreover, the occurence of a Hopf-bifurcation in the analogous system with cyclic symmetry (3.6) reported by Hofbauer et al [1980] attributes further support to this conjecture.

Due to the equivalence of (3.12) and (3.8), this stability analysis holds for all elementary hypercycles. Schuster et al [1979] added a very general proof on the existence of at least one attractor in the interior of the concentration simplex, S_n, leading to cooperative behaviour. Depending on the number of elements, hypercycles may exhibit very different dynamic behaviour [Eigen and Schuster, 1978a]. Nevertheless, they share the basic feature of introducing cooperation into systems of otherwise competing self-replicating elements.

C. Physical Realization and Properties of Hypercycles

In this section we shall return to polynucleotides and ask how these molecules might interact catalytically. Formally, a second order term in kinetics can be traced back to some bimolecular interaction. Thus, we obtain an appropriate term in the case some polynucleotide acts as a catalyst in the replication of another polynucleotide:

$$I_i + I_j + \sum_{\lambda=1}^{4} \nu_\lambda^{(i)} A_\lambda \xrightarrow{f_{ij}} 2I_i + I_j \tag{3.14}$$

Polynucleotides, however, are poor catalysts for their own replication as far as replicase activity is concerned, although we have to admit that our present-day knowledge about the catalytic properties of polynucleotides is far from being complete. Thus, a reaction like (3.14) can be expected to be of rather low efficiency.

Is there another possibility of catalytic interaction between polynucleotides, perhaps one of a more indirect nature? The answer is yes. Indirect catalytic interaction between polynucleotides occurs in all organisms. The mechanism is translation of a polynucleotide sequence into a protein. A greater number of proteins is active in the regulation of gene activity, i.e. transcription, or gene replication. The evolution of any organism is characterized by a highly complicated dynamics of polynucleotide interactions via translation into proteins.

We shall discuss now a fairly simple model which introduces catalytic coupling between individual polynucleotides by means of translation into proteins. The proteins, denoted by E_i, obtained thereby are assumed to act as specific catalysts in polynucleotide replication. In the case where one introduces this kind of coupling into an elementary hypercycle, one obtains a complicated reaction scheme which has been called a hypercycle with translation and which has been studied in some detail [Eigen, 1971; Eigen and Schuster, 1978a; Eigen et al, 1980]. The reaction scheme is shown in Figure 19. The corresponding reaction equations are:

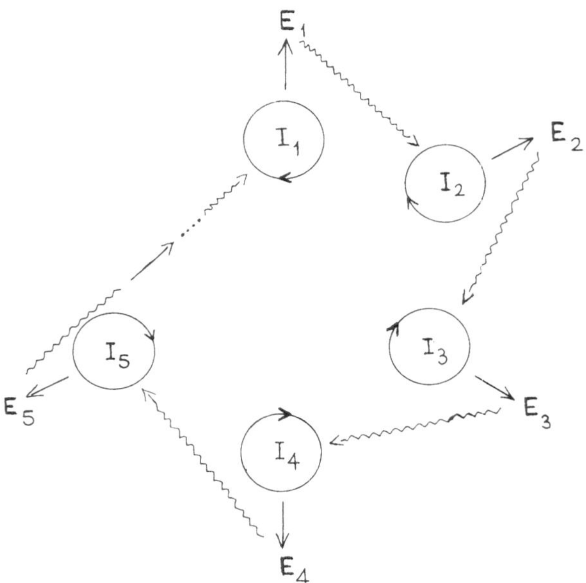

Fig. 19. Schematic diagram of a hypercycle with translation. Dimension: $2 \times n$ = n polynucleotides and n polypeptides.

$$I_i + E_j \underset{k_{-i}}{\overset{k_i}{\rightleftarrows}} I_i \cdot E_j \, ; \; i = 1,\ldots,n; \; j = i-1+n\delta_{i1} \qquad (3.15a)$$

$$I_i \cdot E_j + \sum_{\lambda=1}^{n} \nu_\lambda^{(i)} A_\lambda \xrightarrow{f_i} I_i + I_i \cdot E_j \, . \qquad (3.15b)$$

Polynucleotide synthesis according to reaction (3.15a,b) follows the same mechanism as discussed in equation (1.9). We assume buffered triphosphate concentrations and assign the following variables to the individual compounds present:

$$[I_i] = x_i; \quad [E_i] = e_i \text{ and } [I_i \cdot E_j] = y_i \quad .$$

Furthermore, we substitute

$$f_i' = f_i F_i(A_1^o, A_2^o, A_3^o, A_4^o)$$

and define the equilibrium constant

$$K_i = k_i / k_{-i} \; .$$

Mass conservation allows us to define total concentrations of polynucleotides and polypeptides:

$$x_i^o = x_i + y_i \text{ and } e_i^o = e_i + y_j; \; i=1,\ldots n; \; j=i+1-n\delta_{n1} \qquad (3.16)$$

For fast equilibration of the complex, the concentration y_i is related to the total concentrations x_i^o and e_j^o by

$$y_i = \frac{x_i^o + e_j^o + K_i^{-1}}{2} \left\{ 1 - \sqrt{1 - \frac{4 x_i^o e_j^o}{(x_i^o + e_j^o + K_i^{-1})^2}} \right\}; \; i=1,\ldots,n, \; j=i-1+n\delta_{i1} \, , \qquad (3.17)$$

in complete analogy to equation (1.10). Polypeptide synthesis is assumed to occur with the help of a common apparatus for the translation of the polynucleotide I_i:

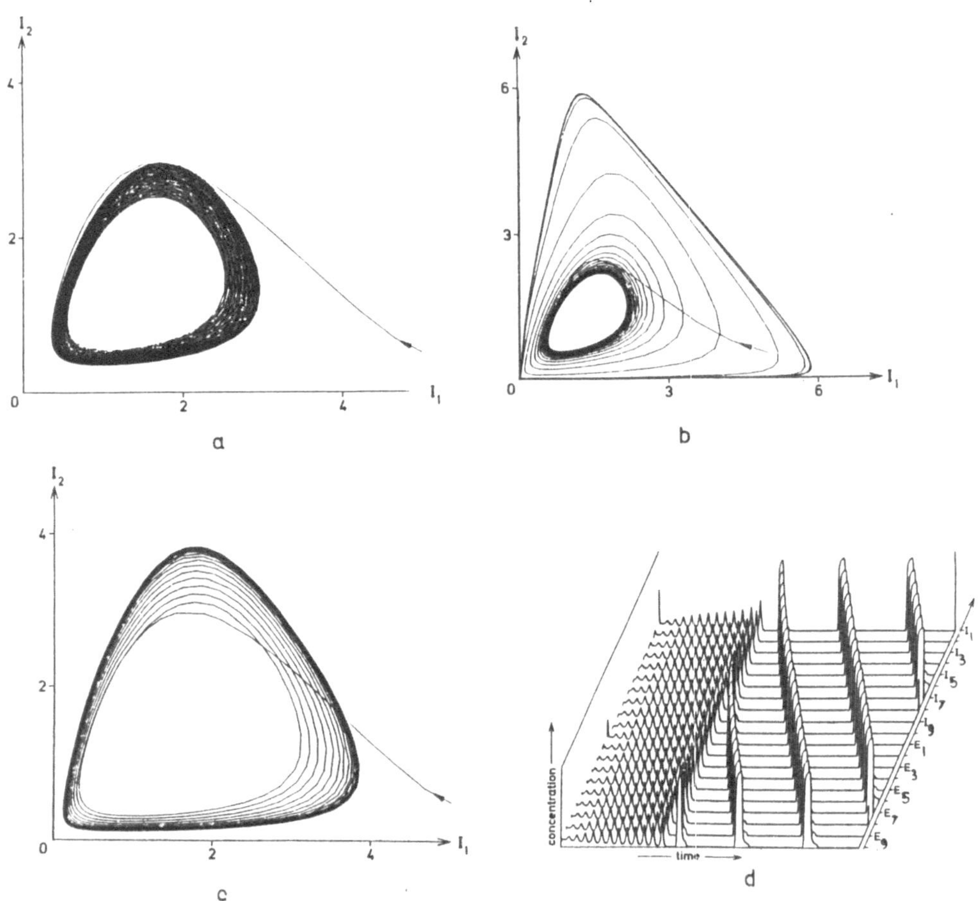

Fig. 20. Trajectories of the dynamical system for hypercycles with translation according to Eq. (3.21). a: Dimension 2×5, $K = 1.1$, $x_1^\circ = 5$, $x_2^\circ = \ldots = x_5^\circ = 0.5$; $e_1^\circ = \ldots = e_5^\circ = 1.0$; the value of K is slightly below the Hopf bifurcation. b and c: Dimension 2×6, $K = 0.2784$ and $K = 1.2$ respectively. $x_1^\circ = 5$, $x_2^\circ = \ldots = x_6^\circ = 0.5$; $e_1^\circ = \ldots = e_6^\circ = 1.0$. d: Dimension 2×10, $K = 0.026$, $x_1^\circ = 5$, $x_2^\circ = \ldots = x_{10}^\circ = 0.5$; $e_1^\circ = \ldots = e_{10}^\circ = 1.0$. Full concentration scale = 10 concentration units, full time scale = 1000 time units.

$$I_i + \sum_{\lambda=1}^{20} \mu_\lambda^{(i)} M_\lambda \xrightarrow{h_i} I_i + E_i \quad . \tag{3.18}$$

M_λ and $\mu_\lambda^{(i)}$ denote the activated amino acids and their stoichiometric coefficients respectively. Again we assume the concentrations of all compounds M_λ to be buffered:

$$h_i' = h_i H_i(M_1^o, \ldots, M_{20}^o) \quad . \tag{3.19}$$

We introduce two independent selection constraints in order to keep the total concentrations of the polynucleotides and polypeptides constant.

$$\sum_{i=1}^{n} x_i^o = c_o \quad \text{and} \quad \sum_{i=1}^{n} e_i^o = d_o \quad . \tag{3.20}$$

Under these conditions we obtain a system of 2n coupled differential equations:

$$\dot{x}_i^o = f_i' y_i - \frac{x_i^o}{c_o} \sum_{k=1}^{n} f_k' y_k ; \; i = 1, \ldots, n \quad , \tag{3.21a}$$

and

$$e_i^o = h_i' x_i - \frac{e_i^o}{d_o} \sum_{k=1}^{n} h_k' x_k ; \; i = 1, \ldots, n \quad . \tag{3.21b}$$

At the low concentration limit, lim $c_o \to 0$ and lim $d_o \to 0$, equation (3.21) shows the same qualitative features as the elementary hypercycle of dimension n [Eigen and Schuster, 1978a; Eigen et al, 1980]. Numerical integration revealed an interesting richness in the dynamics of equation (3.21) which depends on the dimension of the system. There is strong numerical evidence for the existence of a Hopf bifurcation in (3.21) with $n \geq 5$. The bifurcation seems to be supercritical for $n = 5$ and subcritical for $n > 5$. Some illustrative examples of numerical integration of (3.21) are shown in Figure 20.

In contrast to the systems considered in the previous sections, the physical realization of an appropriate experimental model system for hypercyclic coupling is exceedingly difficult and has not been achieved so far. This is not surprising since translation of polynucleotide sequences into protein requires the multicomponent translation machinery of the cell which is not easy to handle in an "in vitro" system.

Finally, we have to consider competition between hypercycles because a study of simultaneously present hypercycles will reveal an important property. In order to simplify the treatment we shall assume internal equilibration within each of the hypercyclically coupled subsystems. As we have seen before, elementary hypercycles have a stable steady state at P_o provided $n \leq 4$, see equation (3.9):

$$P_o = \frac{C}{\sum_{i=1}^{n} f_i^{-1}} (f_2^{-1}, f_3^{-1}, \ldots, f_1^{-1}) .$$

Thus, the hypercycle considered as an entity at internal equilibrium grows according to the following growth law:

$$\dot{C} = \sum_{i=1}^{n} \dot{x}_i = \sum_{i=1}^{n} f_i \bar{x}_i \bar{x}_j = \frac{1}{\sum_{i=1}^{n} f_i^{-1}} C^2 = \bar{f} C^2 . \qquad (3.22)$$

Competition in a system of m independently growing hypercycles can be described by the differential equation

$$\dot{C}_i = \bar{f}_i C_i^2 - \frac{C_i}{C_o} \phi_o ; \; i = 1, \ldots, m \text{ and } C_o = \sum_{i=1}^{m} C_i . \qquad (3.23)$$

Equation (3.23) is a simple (hyperbolic) special case of equation (3.2). All growth functions are homogeneous of the same degree ($\lambda = 2$) and we can set $C_o = 1$ without losing generality. The differential equation (3.23) has $(2^m - 1)$ fixed points on the simplex S_m. The m corners of S_m correspond to stable steady states each of which is surrounded by its own basin of attraction. An example with m = 3 is shown in Figure 21. Again we find that selection occurs, but the outcome of selection is not unique. It depends on the initial conditions, in particular on the point at which the trajectory started. After the system has approached one of the pure states it is very difficult to perturb it such that it can leave the basin of attraction and eventually reach another state. An enormous fluctuation is necessary to achieve this goal and the occurence of such a fluctuation is highly improbable. A system with hypercycles coupling thus has a strong tendency to conserve its characteristic properties, features and constituents. Evolution in systems with second or higher order growth functions does not fully optimize certain properties as it does in Darwinian systems. We may expect therefore to find "frozen accidents" in the case where evolution passed through a stage in which the replicating elements were coupled by higher order catalytic terms.

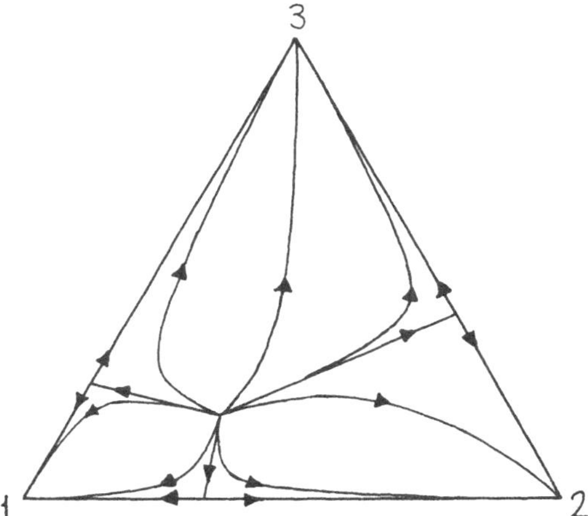

Fig. 21. Fixed points and trajectories of Eq. (3.23) with m = 3 and $\bar{f}_1 = 1$, $\bar{f}_2 = 2$, $\bar{f}_3 = 3$.

IV. PREBIOTIC EVOLUTION

Here we give a brief account on a model for prebiotic evolution which is based on the general results of molecular self-replication as well as the present-day knowledge of the conditions of the primordial earth (for further details see Eigen [1971], Eigen and Schuster [1978b] and Schuster [1981]). This model tries to bridge over the gap between prebiotic chemistry and the first primitive organisms. A sketch which makes it easier to follow the text is given in Figure 22.

Prebiotic chemistry has been studied extensively by means of simulation experiments [Miller and Orgel, 1974]. These investigations showed in a convincing manner that building blocks of biopolymers were abundant on the primordial earth. Polymer formation under prebiotic conditions is highly probable, but the most plausible sequences of reactions are presumably not known yet. The results of some experiments, in particular of those performed by Orgel and coworkers, are very promising and give strong support to the suggestion that polynucleotides of the size of present-day t-RNA molecules were formed in non-negligible amounts under prebiotic conditions. These molecules brought the capability of self-replication on the primordial scene. Template-induced polynucleotide synthesis is so much more efficient than spontaneous polymerization that the latter played no role after the first templates had been formed. At the same time a first process of optimization in the Darwinian sense starts. RNA molecules which replicate more efficiently are selected. Thereby, all the properties which are relevant for replication are steadily improved. These properties are determined by the sequences of bases, since the sequences in turn determine the stable secondary structures. One of the most important features is the melting temperature (T_M), this is the temperature at which double helices are paired to 50%. The value of T_M is essentially laid down by the relative GC content of the polynucleotide and the average lengths of double helical regions. Optimal fitness requires an intermediate melting temperature: If T_M is too high, a very small fraction of the polynucleotides is dissociated and thus ready for replication. The overall rate of synthesis, hence, will be fairly low. If T_M is too low, on the other hand, the majority of the molecules is present as single strands and therefore more easily accessible to hydrolysis. Then, the rate of degradation will be too high for an efficient net production.

After enzyme-free replication had been optimized the process of evolution temporarily comes to an end. That part of the prebiotic earth, which is accessible to polynucleotides soon becomes populated as densely as the production of suitable energy-rich starting material and the rate of hydrolysis permit. Differences

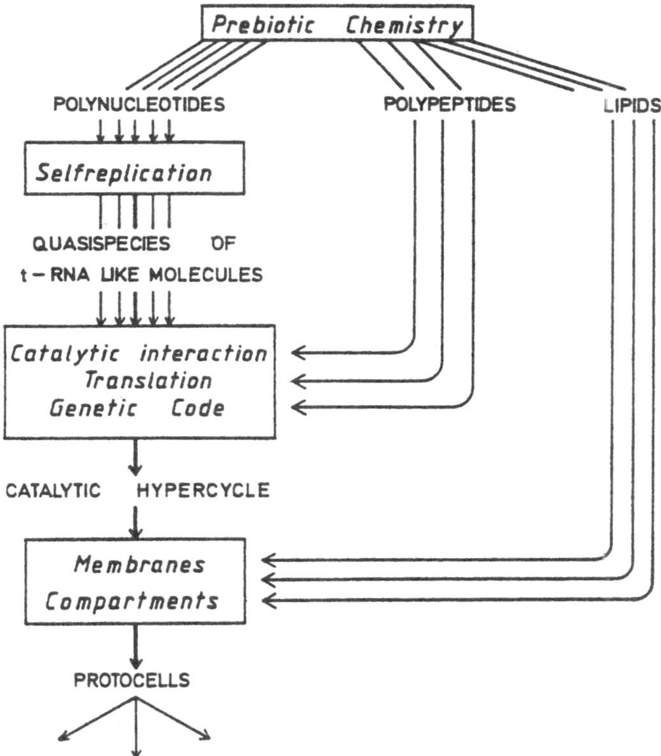

Fig. 22. A scheme for the logical sequence of steps in prebiotic evolution leading from unorganized macromolecules to protocells.

in the environment such as different temperature, different solid state catalysts etc., may introduce variations in the locally predominant polynucleotide sequences.

The error threshold discussed in the previous section sets a limit to the lengths of polynucleotides which is close to the size of the present-day t-RNA's. Here we can visualize the error threshold also from a different point of view. Within the quasi-species the master copy is selected against its own less efficient mutants. This selection breaks down when the maximum chain length is exceeded. Most biologists engaged in molecular evolution share the opinion that t-RNA molecules belong to the oldest class of biomolecules in living organisms. Attempts have been made to reconstruct phylogenetic trees from t-RNA's in different organisms and to learn about the degree of relationship between different t-RNA's in the same organism. We refer to a recent study [Eigen and Winkler-Oswatitsch, 1980], which aims at the reconstruction of a possible common ancestor of our t-RNA's. This hypothetical polynucleotide has a number of relevant properties: high GC content, a clear preference for a repetitive -RNY-, in particular -GNC-, pattern of period three. (By R we denote one of the purine bases A or G, by Y one of the pyrimidines, U(T) or C and finally, by N any of the four bases.) The extrapolated sequence allows a clover-leaf or a hairpin structure to be formed equally well. The existence of complementary symmetry at the 3'- and 5'- ends of the polynucleotide is of particular importance for replication. The minus strand then has an initial sequence identical with that of the plus strands which facilitates replication starting from one end.

What could the t-RNA like molecules do in order to overcome this dead end? Every possibility of further development points towards an increase in the content of information that could be stored in the sequence of the bases. One way to achieve this goal is to develop a more accurate mechanism of replication. Alternatively, cooperation of self-replicating molecules, each one present in a number of copies, would also allow for the increase in the total length of the transferable genetic information. For the second purpose, competition among the cooperating elements had to be avoided. At the same time it remains necessary to keep the property of selection against unfavourable mutants and other competitors. In Section III we saw a formal way to fulfill the two requirements by means of cyclic catalytic coupling. Any solution to the practical problem seems to combine both concepts for the increase of information: the t-RNA-like molecules can improve the accuracy of replication only in case they cooperate and start to design specific catalysts for their own needs. These specific catalysts are the proteins as we know from the actual outcome of prebiotic evolution. Thus, cooperation of small polynucleotides, the origin of translation and the structure of the genetic code are closely related problems. Various models for primitive translation have been conceived. We mention one which was published more recently and which continues earlier ideas [Crick et al, 1976]. We have modified this model slightly [Eigen and Schuster, 1978b]. Since an extensive discussion can be found there, we dispense here with all details. The essential features of our model are several predictions which are meaningful in the context of prebiotic chemistry and thus increase the plausibility of this approach:

(1) In order to guarantee sufficient stability of messenger-RNA - t-RNA complexes the primitive codons and anticodons would have to be rich in G and C.

(2) In order to avoid sliding of the t-RNA along the messenger during ribosome-free translation, one needs a repetitive pattern of period three [Crick et al., 1976]. Actually, a pattern of the type -RNY- has not yet been completely removed by mutations and does still exist as a kind of background noise in the sequences of present-day t-RNA's [Eigen and Winkler-Oswatitsch, 1980] and in the DNA of some viruses, bacteria and eucaryotes [Shepherd, 1980].

(3) Combining (1) and (2) in a straightforward way we obtain four codons which are presently used for the amino acids

glycine = G G C , alanine = G C C ,

aspartic acid = G A C and valine = G U C .

These four amino acids were the most abundant under prebiotic conditions. The corresponding adaptors, the t-RNA's with complementary anticodons are the four most closely related t-RNA molecules. They may well be descendants of the same quasispecies.

(4) Nucleotide binding probably was one of the most ancient properties of encoded polypeptides. Rossmann et al [1974] made an alignment study of various dehydrogenases and other nucleotide binding proteins in order to reconstruct a possible common ancestor protein. Walker [1977] analyzed the 21-site amino acid sequences of the nucleotide-binding surface and tried to identify the precursor amino acids coded into the ancestral surface which is assumed to be 3×10^9 years old. He concludes that valine and one or both of the aspartic acid - glutamic acid group would be likely precursors with valine predominating. Both constituents of the ancient recognition site are members of this group of the first four amino acids.

Our knowledge on the physics of polynucleotide-polypeptide interactions is fragmentary only. Therefore, all the details of the process and properties which might have played major roles in early translation are not known yet. The model considerations mentioned may have their most important impact on experimental research just in helping to ask relevant questions about the origin of translation. For our further considerations, let us take it for granted that the polynucleotides managed somehow to organize a primitive machinery for translation.

Using a primitive machinery for translation, polynucleotides were able to interfere in the replication of other polynucleotides by means of their translation products, the corresponding polypeptides. They might have exerted positive or negative, e.g. by coding a protein with specific replicase or nuclease activity. As we have seen in Section III, an indirect coupling via translation products is a realistic model for catalytic interaction in second order self-replication.

A translation machinery of present-day's complexity is an enormously involved chemical factory. In early evolution there was no need for such high perfection. In principle, some of the molecules could act two parts at the beginning: the early transmitters or precursors of our t-RNA's might have been early genes as well [Eigen and Winkler-Oswatitsch, 1980]. The first aminoacyl-synthetases, the enzymes which nowadays attach the correct amino acids to the corresponding t-RNAs and consequently carry a specific recognition site for a single polynucleotide, might have acted as specific polymerases as well [Biebricher, 1980]. These are just suggestions to indicate that an early translation using a primitive highly redundant code for a handful of amino acids could form from replicating polynucleotides. Admittedly, there is no sufficient experimental background available yet. We do not know precisely how probable such an event which brought the necessary parts together actually was.

Let us assume that a first, primitive translation machinery was formed by integration of information stored in a number of structurally independent genes which might well have originated from a single quasispecies. Their relative concentrations are controlled by hypercyclic organization. The first system that succeeded in completing its design had an incredible advantage. It increased the rate of polynucleotide synthesis many times over. At the same time, the accuracy of the replication process was increased as well. Longer genes became possible which in turn could code for better enzymes. The dynamical properties of such a replication - translation machinery in solution follow a higher order replication kinetics and do not allow mutants to grow even in case they are somewhat more efficient. Perfection can be

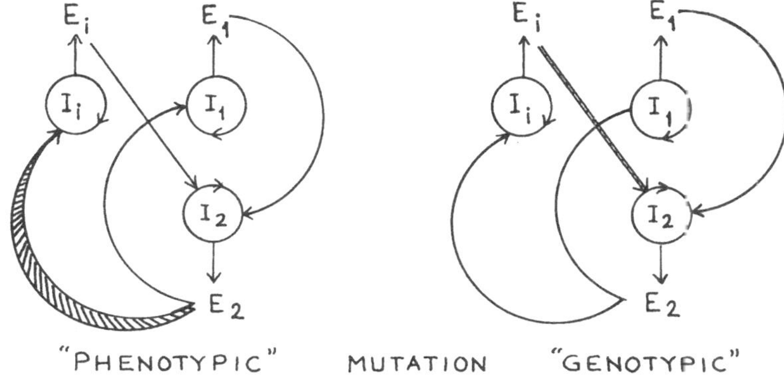

Fig. 23. Two idealized classes of mutations in primitive replication-translation systems. The "phenotypic" mutations lead to mutants (I_1') which are better targets for the specific replicase, whereas the properties of their translation products (E_1') are about the same as in the wild type (E_1). The "genotypic" mutant, on the contrary, is characterized by a better translation product but roughly unchanged recognition by the replicase.

achieved by enlargement of the feedback cycle, i.e. by admission of new members. More and more amino acids were incorporated into the encoded ensemble. Thereby, the catalysts were improved and ultimately replication became so accurate that gene multiplication was no longer necessary. Eventually, the independent genes were ligated to form a genome consisting of a single molecule.

Sooner or later, an inherent disadvantage started to overweigh the advantages of homogeneous solution. Any favourable genotypic mutation (Figure 23) is accessible to every beneficiary in the surrounding solution no matter whether it contributes an honest share to the common prosperity or not. Parasites may profit from the achievements of the integrated members. The best counteraction is spatial separation. In physical terms, formation of compartments and individualization appears as a logical consequence of the preceding development. Compartments, which are free of parasites are more prosperous and multiply faster than their companions which have to feed beneficiaries, will spread and finally win the competition. Compartmentalization at the same time brings the system back to first order replication kinetics since the compartment multiplies as an entity. Darwinian selection is restored. During the intermediate period of non-linear self-replication, the system was shifted from the level of individual molecules to the level of protocells which can be understood also as the level of first primitive but already integrated and individualized organisms. From now on, the diversity of environments may act on the individuals and thereby create the enormous variability in finish as we observe it in our studies on bacteria and blue-green algae.

Let us summarize our suggestions for prebiotic evolution. Selection sets in at the level of polynucleotides as soon as template-induced replication becomes important. Integration of self-replicating elements into a functional unit requires higher order kinetics in order to suppress competition between the individual members. This higher order kinetics is not compatible with Darwinian evolution. During such a phase of "non-linear" development, we do not expect optimization of the newly achieved properties. We expect, rather, to find "frozen accidents" or "once-forever" decisions. Some primitive polynucleotides joined their information content and formed together the coding capacity for a primitive translation machinery. Then, later in evolution, spatial isolation by compartment formation restored Darwinian behaviour. From now on variation may start again. But, the translation machinery as a whole is conserved as an entity and no longer subject to any serious change.

In principle, a formally similar mechanism may govern the other bursts in the evolution of the biosphere as well. The formation of eucaryotic organisms could be a result of a similar kind of symbiosis. Some procaryots had to share their metabolic properties and their coding capacities in order to create a machinery for organizing the complex genome and the apparatus for mitosis and meiosis observed with all higher organisms. Again it seems very likely that the common features observed with eucaryotes are a kind of "frozen accident", a remnant of a period lacking optimization.

We could try to push analogies further and enter the realm of pure speculations. This, however, is not the aim of this contribution which was essentially directed towards macromolecular self-organization. There is also another reason which forces us to stop the molecular description at a certain level, because it becomes more difficult to discover principles of development, the more highly organized the biological entities are. Jacob [1977] has formulated this problem so precisely that we need only to refer to his article: "evolution and tinkering". There is no reason for simplicity or intellectual elegance in nature. She does not design with the eyes of an engineer. The only thing that counts, and that is selected for is functional efficiency. Moreover, selection can act only on the things that are there at a certain moment. The basic principles therefore are very hard to detect and well hidden from an analyzing brain.

V. SOCIAL BEHAVIOUR OF ANIMALS

Some years ago Maynard-Smith [1974] developed a model for the self-regulation of social behaviour in animal societies. This model based on game theory soon became very popular because it could provide a successful explanation for one of the most conflicting concepts in animal behaviour: biologists invented concepts like "group selection" or the "benefit of the species" in order to explain the phenomenon of limited war or comment fighting between animals of the same species. These concepts are very hard to accept for the geneticists since natural selection essentially acts on individuals or, more precisely, on reproductive pairs of individuals. The game theoretical model, however, provides an explanation of apparently "altruistic" strategies like limited war by means of the individual's benefit. Maynard-Smith and Price [1973] introduced the notion of our "evolutionarily stable strategy" (ESS) on which the model centers.

An "evolutionarily stable strategy" (ESS) can be explained as the best reply when played against itself. There may exist another best reply against the ESS which then has to be the better response against this other strategy. The game theoretical model can be translated into differential equations [Taylor and Jonker, 1978; Hofbauer et al, 1979, Zeeman; 1979]. Most interestingly, and this relates the Maynard-Smith model to our approach to molecular self-organization, one obtains the differential equation (3.2). The variables in this equation are the probabilities to encounter a certain strategy in the society considered. An evolutionarily stable strategy corresponds to an asymptotically stable fixed point of equation (3.2). The converse, however, is not true: there are asymptotically stable states which cannot be obtained as ESS.

Many useful notions developed in game theory have little relevance in biological situations where the players to not obey any axiom of rationality. For example, the concept of Pareto equilibrium makes little sense here. The dynamical approach allows a better understanding of the effect of fluctuations. Another advantage of "game dynamics" consists in the accessibility of time-dependent phenomena, like the approach to the steady state, oscillations corresponding to stable limit cycles, the role of time averages, etc.

In a series of papers [Schuster et al, to be published] we tried to analyze these various dynamical phenomena which lead to self-regulation of behaviour in animal societies. One aspect of this study seems to be worth mentioning: despite a large number of impact parameters one frequently observes only a few dynamically different cases. This result may help to classify typical situations and to establish a semi-quantitative scale which eventually turns out to be useful in relation to experimental observations.

REFERENCES

Bartlett, M.S., 1978, "An Introduction to Stochastic Processes. 3rd ed.," Cambridge University Press, Cambridge, U.K.

Batschelet, E., Domingo, E., and Weissmann, C., 1976, The proportion of revertant and mutant phage in a growing population, as a function of mutation and growth rate, Gene, 1:27.

Biebricher, C., 1980, personal communication.

Biebricher, C., Eigen, M., and Luce, R., 1980, personal communication.

Crick, F.H.C., Brenner, S., Klug, A., and Pieczenik, G., 1976, A speculation on the origin of protein synthesis, Origins of Life, 7:389.

Domingo, E., Flavell, R.A., and Weissmann, C., 1976, In vitro site directed mutagenesis: generation and properties of an infectious extracistronic mutant of bacteriophage Q_β, Gene, 1:3.

Ebeling, W., and Feistel, R., 1978, On the Eigen-Schuster concept of quasispecies in the theory of natural self-organization, Studia Biophysica (Berlin), 71:139.

Eigen, M., 1971, Self-organization of matter and the evolution of biological macromolecules, Naturwissenschaften, 58:465.

Eigen, M., 1976, Wie entsteht Information? Prinzipien der Selbstorganisation in der Biologie, Ber. Bunsenges,f.Physikal.Chemie, 80:1059.

Eigen, M., and Schuster, P., 1977, The hypercycle. A principle of natural self-organization. A: Emergence of the hypercycle, Naturwissenschaften, 64:541.

Eigen, M., and Schuster, P., 1978a, The hypercycle. A principle of natural self-organization. B: The abstract hypercycle, Naturwissenschaften, 65:7.

Eigen, M., and Schuster, P., 1978b, The hypercycle. A principle of natural self-organization. C: The realistic hypercycle, Naturwissenschaften, 65:341.

Eigen, M., and Winkler-Oswatitsch, R., 1975, "Das Spiel," Piper-Verlag, München.

Eigen, M., and Winkler-Oswatitsch, R., 1980, Transfer-RNA an early adaptor. Was it the primordial gene too?, preprint.

Eigen, M., Schuster, P., Sigmund, K., and Wolff, R., 1980, Elementary step dynamics of catalytic hypercycles, Biosystems, 13:1.

Epstein, I.R., and Eigen, M., 1979, Selection and self-organization of self-reproducing macromolecules under constraint of constant flux, Biophys. Chem., 10:153.

Hofbauer, J., 1980, On the occurrence of limit cycles in the Volterra-Lotka differential equation, preprint.

Hofbauer, J., Schuster, P., and Sigmund, K., 1979, A note on evolutionarily stable strategies and game dynamics, J. Theor. Biol., 81:609.

Hofbauer, J., Schuster, P., Sigmund, K., and Wolff, R., 1980, Dynamical systems under constant organization. II: Homogeneous growth functions of degree p=2, SIAM J. Appl. Math., 38:282.

Jacob, B., 1977, Evolution and tinkering, Science, 196:1161.

Jones, B.L., Enns, R.H., and Rangnekar, S.S., 1976, On the theory of selection of coupled macromolecular systems, Bull. Math. Biol., 38:12.

Kornberg, A., 1974, "DNA Synthesis," Freeman and Co., San Francisco.

Kunkel, T.A., and Loeb, L., 1979, On the fidelity of DNA replication, J. Biol. Chem., 254:5718.

Kunkel, T.A., Meyer, R.R., and Loeb, L., 1979, Single-strand binding protein enhances fidelity of DNA synthesis in vitro, P.N.A.S. (USA), 76:6331.

Küppers, B.O., 1979a, Towards an experimental analysis of molecular self-organization and precellular Darwinian evolution, Naturwissenschaften, 66:228.

Küppers, B.O., 1979b, Some remarks on the dynamics of molecular self-organization, Bull. Math. Biol., 41:803.

Lohrmann, R., Bridson, P.K., and Orgel, L.E., 1980, An efficient metal-ion catalyzed template-directed oligonucleotide synthesis, Science, 208:1464.

Maynard-Smith, J., 1974, The theory of games and the evolution of animal conflicts, J. Theor. Biol., 47:209.

Maynard-Smith, J., and Price, G., 1973, The logic of animal conflicts, Nature, 246:15.

Miller, S.L., and Orgel, L.R., 1974, "The Origins of Life on the

Earth," Prentice Hall, Engelwood Cliffs, N.J.

Pörschke, D., 1977, Elementary steps of base recognition and helix-coil transitions in nucleic acids, in: "Chemical Relaxation in Molecular Biology," I. Pecht and R. Rigler, ed., Molecular Biology, Biochemistry and Biophysics, 24:191. Springer-Verlag, Heidelberg.

Rossmann, M.G., Moras, D., and Olsen, K.W., 1974, Chemical and biological evolution of a nucleotide-binding protein, Nature, 250: 194.

Sabo, D., and Weissmann, Ch., 1980, personal communication.

Schneider, F.E., Neuser, D., and Heinrichs, M., 1979, Hysteretic behaviour in poly(A) - poly(U) synthesis in a stirred flow reactor, in: "Molecular Mechanisms of Biological Recognition," M. Balaban, ed., Elsevier-North-Holland, Amsterdam.

Schuster, P., 1981, Prebiotic evolution, in: "Biochemical Evolution," H. Gutfreund, ed., Cambridge University Press, Cambridge, U.K.

Schuster, P., and Sigmund, K., 1980, Self-organization of biological macromolecules and evolutionary stable strategies, in: "Dynamics of Synergetic," H. Haken, ed., Springer-Verlag, Berlin, Heidelberg.

Schuster, P., Sigmund, K., and Wolff, R., 1978, Dynamical systems under constant organization. I: Topological analysis of a family of nonlinear differential equations - a model for catalytic hypercycles, Bull. Math. Biophysics, 40:743.

Schuster, P., Sigmund, K., and Wolff, R., 1979, Dynamical systems under constant organization. III: Cooperative and competitive behaviour of hypercycles, J. Diff. Equ., 32:357.

Shepherd, J.C.W., 1980, Periodic correlations in DNA sequences and evidence for their evolutionary origin in a comma-less genetic code, preprint.

Spiegelman, S., 1971, An approach to the experimental analysis of precellular evolution, Quarterly Review of Biophysics, 4:215.

Taylor, P., and Jonker, L., 1978, Evolutionarily stable strategies and game dynamics, Math. Biosciences, 40:145.

Thompson, C.J., and McBride, J.L., 1974, On Eigen's theory of the self-organization of matter and the evolution of biological macromolecules, Math. Bioscience, 21:127.

Walker, G.W.R., 1977, Nucleotide-binding site data and the origin of the genetic code, Biosystems, 9:139.

Watson, J.D., 1977, "Molecular Biology of the Gene, 3rd ed.," W.A. Benjamin Inc., Menlo Park, California.

Zeeman, E.C., 1979, Population dynamics from game theory, in: "Proceedings of an International Conference on Global Theory of Dynamic Systems," North-Western University, Evanston, Illinois.

ESCAPE FROM DOMAINS OF ATTRACTION FOR SYSTEMS PERTURBED BY NOISE*

Donald Ludwig

Department of Mathematics
University of British Columbia
Vancouver, B.C.
V6T 1W5

* I thank M. Levandowsky and B. White for numerous helpful discussions on these matters. This research was supported by NSERC of Canada under Grant No. A-9239.

TABLE OF CONTENTS

ESCAPE FROM DOMAINS OF ATTRACTION FOR SYSTEMS PERTURBED BY NOISE

Donald Ludwig

Department of Mathematics
University of British Columbia
Vancouver, B.C.
V6T 1W5

I. INTRODUCTION

Deterministic theories have been remarkably successful in interpreting and explaining the world, although more precise formulations will always involve random effects. An outstanding example is classical mechanics, whose utility is hardly impaired by the existence of quantum effects. Intuitively, we may think of the deterministic trajectories as being smeared out by the random effects. Even if the smearing is large, the qualitative behavior may still be captured by the deterministic result.

An exception to this pleasant situation occurs in the rare events when stochastic effects combine to overwhelm the deterministic effects, at least temporarily. An example is quantum mechanical tunnelling, where occasionally a potential barrier is crossed, as in radioactive decay. Similar phenomena occur in chemistry, where thermal fluctuations can cause dissociation of stable molecules [Kramers, 1940]. It is shown below that Wright's model of evolution of species may be viewed as an analogous escape problem. Although the waiting time for such an escape may be very long, the actual escape occurs very quickly. This mathematical point does not seem to be generally appreciated: it is the main objective of the following exposition.

Section II is devoted to the one dimensional theory. The Ornstein-Uhlenbeck process is treated at some length, because all of the calculations can be performed explicitly. It is shown that the waiting time for an escape grows exponentially as the size of the perturbing fluctuations decreases. However, the last excursion from the deterministic attractor requires a time which only grows logarithmically. This phenomenon is termed the "Great Leap Forward", as was proposed by Vent-tsel and Freidlin [1970].

Section III sketches the higher dimensional theory. The most likely path of escape may be found by the methods of Hamilton-Jacobi theory. The resulting asymptotic theory is based upon the WKB method. The connection of these ideas with Holling's "resilience notion" is briefly indicated. Finally, Wright's evolutionary model is presented. The phenomenon of the Great Leap Forward supports a saltational view of evolution, when viewed on a geological time scale.

II. ONE DIMENSIONAL THEORY

A. The Wiener Process

The Wiener process provides the simplest approach to stochastic phenomena with a continuous time variable. First consider a discrete model, where X_n represents the position of a particle after n steps. The size of each step is given by $\{Z_n\}$, $n = 1,\ldots,N$, where Z_n are independently and identically distributed random variables. It will do no harm to think of Z_n as being normally distributed, with mean 0 and variance σ^2. The X_n satisfy

$$X_1 = Z_1 \ , \quad X_n - X_{n-1} = Z_n \ , \qquad n = 2,3,\ldots \ . \tag{2.1}$$

Therefore

$$X_N = \sum_{n=1}^{N} Z_n \ . \tag{2.2}$$

Since the $\{Z_n\}$ are independent, we have

$$\mathcal{E}[X_N] = \sum_{n=1}^{N} \mathcal{E}[Z_n] = 0 \ , \tag{2.3}$$

$$\mathcal{E}[X_N^2] = \sum_{n=1}^{N} \mathcal{E}[Z_n^2] = N\,\sigma^2 \ ; \tag{2.4}$$

the cross terms which should appear in (2.4) disappear.

Now the right-hand side of Eq. (2.4) may be fixed at a value t, while $N \to \infty$. Then the central limit theorem implies that X_n will approach a normally distributed variable W(t), which is the

Wiener process. Further details are given in Karlin and Taylor [1975]. The density function for this process is given as follows:

$$\Pr[x \le W(t) < x + dx] = v(t,x)dx = \frac{1}{\sqrt{2\pi t}} e^{-x^2/2t} dx. \tag{2.5}$$

Moreover, v satisfies the forward or Fokker-Planck equation

$$\frac{\partial v}{\partial t} = \frac{1}{2}\frac{\partial^2 v}{\partial x^2} . \tag{2.6}$$

One can formulate an escape problem for the Wiener process: suppose that the process is to be stopped if $W(t) = x_1$ or $W(t) = x_2$. Let

$$u(t,x) = \Pr[W(\tau) \ne x_1, W(\tau) \ne x_2, \text{ for all } \tau \le t | W(0) = x]. \tag{2.7}$$

Then

$$u(0,x) = 1 \text{ if } x_1 < x < x_2 \tag{2.8}$$

$$u(t,x_1) = 0, \ u(t,x_2) = 0 . \tag{2.9}$$

It can also be shown that u satisfies the backward equation

$$\frac{\partial u}{\partial t} = \frac{1}{2}\frac{\partial^2 u}{\partial x^2} . \tag{2.10}$$

See Feller [1966] for details. The problem (2.8)-(2.10) may be solved by separating variables. It is an easy consequence that $u(t,x) \to 0$ as $t \to \infty$. That is, the probability of hitting the boundary approaches one as $t \to \infty$.

B. The Ornstein-Uhlenbeck Process

A simple equation with an asymptotically stable equilibrium is

$$\frac{dx}{dt} = -x . \tag{2.11}$$

In this case all trajectories are attracted to the origin. What happens if random effects are included? In analogy with (2.1), one may modify (2.11) to be

$$X_n - X_{n-1} = -\frac{1}{N} X_{n-1} + \sqrt{\varepsilon}\ Z_n , \tag{2.12}$$

where $\{Z_n\}$ are as described above.

If $N \to \infty$, (2.12) passes into the stochastic equation

$$dX = -Xdt + \sqrt{\varepsilon}\, dW , \tag{2.13}$$

which is to be interpreted in the Itô sense. The Itô interpretation is appropriate because of the forward difference which appears in (2.12). Further details and discussion are given in Turelli [1977]. The solution of (2.13) is the Ornstein-Uhlenbeck process.

Suppose that boundaries are erected at $x = \pm\ell$. In analogy with (2.7), let

$$u(t,x) = \Pr[X(\tau) \neq \pm\ell, \text{ for all } \tau \leq t | X(0) = x] . \tag{2.14}$$

Then

$$u(0,x) = 1 \text{ if } -\ell < x < \ell , \tag{2.15}$$

$$u(t,\pm\ell) = 0 , \tag{2.16}$$

$$\frac{\partial u}{\partial t} = \frac{\varepsilon}{2}\frac{\partial^2 u}{\partial x^2} - x\frac{\partial u}{\partial x} . \tag{2.17}$$

It follows as before that $u(t,x) \to 0$ as $t \to \infty$. Escape occurs with probability one, even though the deterministic flow is towards the origin. Moreover, this result holds regardless of the size of ε, as long as $\varepsilon > 0$.

At first sight, it might seem that the stability of the origin can be destroyed by arbitrarily small stochastic perturbations. This is not really so, because the time taken to escape may be very large. The expected time to escape is given by

$$T(x) = \int_0^\infty t\left(-\frac{\partial u}{\partial t}\right) dt = \int_0^\infty u(t,x)dt . \tag{2.18}$$

It follows that

$$\frac{\varepsilon}{2}\frac{d^2T}{dx^2} - x\frac{dT}{dx} = \int_0^\infty \left(\frac{\varepsilon}{2}\frac{\partial^2 u}{\partial x^2} - x\frac{\partial u}{\partial x}\right)dt = \int_0^\infty \frac{\partial u}{\partial t}\, dt = -1 . \tag{2.19}$$

Moreover, from (2.18) and (2.16),

$$T(\pm\ell) = 0 . \tag{2.20}$$

The parameter ε may be scaled out of the problem by setting

$$y = \frac{x}{\sqrt{\varepsilon}} , \quad L = \frac{\ell}{\sqrt{\varepsilon}} \quad . \tag{2.21}$$

Thus, considered as a function of y, T satisfies

$$\frac{1}{2} T'' - yT' = -1 \tag{2.22}$$

$$T(\pm L) = 0 . \tag{2.23}$$

This equation may be integrated in the form

$$T' = -e^{y^2} \int_0^y e^{-\eta^2} d\eta, \text{ since } T'(0) = 0 \tag{2.24}$$

from the symmetry. If y is large, T' is approximated by

$$T' \sim - \sqrt{\pi}\, e^{y^2} . \tag{2.25}$$

Thus the derivative of T grows very rapidly as y increases, i.e., as $\varepsilon \to 0$. The approximation (2.25) leads to an approximation of T in terms of the Dawson integral:

$$D(y) = e^{-y^2} \int_0^y e^{\eta^2} dy . \tag{2.26}$$

The function D(y) is bounded, and $yD(y) \to 1/2$ as $y \to \infty$. See Abramowitz and Stegun [1964] for the details. Thus we obtain the approximation

$$T(y) \sim \frac{\sqrt{\pi}}{2} \left(\frac{e^{L^2}}{L} - \frac{e^{y^2}}{y} \right) , \tag{2.27}$$

if L and y are both large, i.e., if $y \neq 0$ and $\varepsilon \to 0$. Thus, although escape is certain, the expected time to escape is of exponential order in $1/\varepsilon$. It is also apparent from (2.27) that T has a boundary layer at $x = \pm\ell$.

C. The Great Leap Forward

One can infer something about the qualitative behavior of the Ornstein-Uhlenbeck process from the preceding discussion. If ε is small, the deterministic flow will confine the process to the vicinity of the origin for long periods of time. Rare excursions will occur, but the average interval between such excursions increases sharply with the amplitude of the excursion. Finally, one of these excursions will reach the boundary, no matter how far away it might be. How long will that final excursion take?

In order to investigate this question, consider the constrained escape problem, where only those paths are considered which do not hit the origin. Let

$$u(t,x) = \Pr[x(\tau) = \ell \text{ for some } \tau \leq t, \text{ but } x(s) \neq 0 \text{ for all } s \leq t]. \tag{2.28}$$

Then u satisfies (2.17) as before but

$$u(0,x) = 0 \text{ if } 0 < x < \ell \ , \tag{2.29}$$

$$u(t,0) = 0 \ , \ u(t,\ell) = 1 \ . \tag{2.30}$$

As $t \to \infty$, $u(t,x) \to w(x)$, where w satisfies

$$\frac{\varepsilon}{2} \frac{d^2 w}{dx^2} - x \frac{dw}{dx} = 0 \ , \tag{2.31}$$

and the boundary conditions (2.30). The solution of (2.30), (2.31) is

$$w(y) = \frac{\int_0^y e^{\eta^2} d\eta}{\int_0^L e^{\eta^2} d\eta} \ . \tag{2.32}$$

In terms of the Dawson integral (2.26)

$$w(y) = e^{y^2 - L^2} \frac{D(y)}{D(L)} \sim 2 L D(y) e^{y^2 - L^2} \ . \tag{2.33}$$

It is apparent that w has a boundary layer at $x = \ell$; the probability of direct escape decreases sharply as x decreases from ℓ.

Now, in analogy with (2.18), let h(x) be given by

$$h(x) = \int_0^\infty t \frac{\partial u}{\partial t} dt \ . \tag{2.34}$$

The conditional expectation of the time to hit $x = \ell$ without hitting $x = 0$ is given by

$$T_c(x) = h(x)/w(x) \ . \tag{2.35}$$

A calculation similar to (2.19) shows that

$$\frac{\varepsilon}{2}\ \frac{d^2h}{dx^2} - x\,\frac{dh}{dx} = -w(x)\ . \tag{2.36}$$

It follows from (2.34) and (2.30) that

$$h(0) = 0,\ h(L) = 0\ . \tag{2.37}$$

In view of (2.33) and (2.36),

$$h'' - yh' \sim -2L\ e^{y^2-L^2}\ D(y)\ . \tag{2.38}$$

This equation integrates in the form

$$h' \sim -2L\ e^{y^2-L^2}(F(y)-C)\ , \tag{2.39}$$

where the constant C must be determined from (2.37). The function F(y) is given by

$$F(y) = \int_0^y D(\eta)\,d\eta\ ;\ F(y) \sim \frac{1}{2}\,\ln y \text{ if } y \text{ is large}\ . \tag{2.40}$$

A further investigation of (2.39) produces

$$h(y) \sim 2L\ e^{-L^2}\int_0^y e^{\eta^2}(C-F(\eta))\,d\eta\ . \tag{2.41}$$

Thus, in view of (2.37), C must satisfy

$$C = \frac{\displaystyle\int_0^L e^{\eta^2}F(\eta)\,d\eta}{\displaystyle\int_0^L e^{\eta^2}\,d\eta} \sim \ln L\ . \tag{2.42}$$

Since $F(\eta) \geq 0$, h may be estimated by

$$h(y) \leq 2LC\ e^{-L^2}\int_0^y e^{\eta^2}\,d\eta\ . \tag{2.43}$$

In view of (2.33) and (2.35),

$$T_c(x) \leq C \sim \ln L\ . \tag{2.44}$$

Thus the expected time for the last leap away from 0 is approximately the same time as is required to move towards 0 with the deterministic flow. This phenomenon of a Great Leap Forward was observed in a more general context by Vent-tsel' and Freidlin [1970]. It accords with the popular conception of gambling games. Losses may occur fairly steadily if the "house" has an edge. However, it is possible to "break the bank" by a relatively short stretch of extraordinary good luck.

III. HIGHER DIMENSIONAL THEORY

A. The Stochastic Equation

A system of ordinary differential equations may be written in the form

$$\frac{dx^i}{dt} = b^i(x), \quad i = 1,\dots n \ , \tag{3.1}$$

where $x = (x^1,\dots,x^n)$. A stochastic version of (3.1) is

$$dX^i = b^i(X)dt + \sqrt{\varepsilon} \sum_{k=1}^{K} \alpha_k^i(X)\,dW^k \ , \tag{3.2}$$

where $W^1,\dots,W^K$ are independent Wiener processes. This equation will be interpreted in the Itô sense, in analogy with (2.13). It follows from (3.2) that

$$\mathcal{E}(dX^i) = b^i(X)dt \ , \tag{3.3}$$

$$\mathcal{E}(dX^i dX^j) = \varepsilon\, a^{ij}(X)dt \ , \tag{3.4}$$

where

$$a^{ij}(X) = \sum_{k=1}^{K} \alpha_k^i(X)\ \alpha_k^j(X) \ . \tag{3.5}$$

The density for $X(t)$ is given by

$$\Pr[x^i \le X^i(t) < x^i + dx^i,\ i = 1,\dots,n] = v(t,x)dx^1\dots dx^n. \tag{3.6}$$

In this case, the forward (Fokker-Planck) equation is

$$\frac{\partial v}{\partial t} = \sum_{i,j=1}^{n} \frac{\varepsilon}{2} \frac{\partial^2}{\partial x^i \partial x^j} (a^{ij}(x)v) - \sum_{i=1}^{n} \frac{\partial}{\partial x^i} (b^i(x)v) \ . \tag{3.7}$$

The connection between (3.2) and (3.7) is developed in Gikhman and Skorohod [1973] and Strook and Varadhan [1979]. The backward equation which corresponds to (3.2) is

$$\frac{\partial u}{\partial t} = \sum_{i,j=1}^{n} \frac{\varepsilon}{2} a^{ij}(x) \frac{\partial^2 u}{\partial x^i \partial x^j} + \sum_{i=1}^{n} b^i(x) \frac{\partial u}{\partial x^i} . \tag{3.8}$$

The equations (3.7) and (3.8) are adjoints of one another; neither is self-adjoint except in trivial cases.

B. The Most Likely Path of Escape

Now suppose that (3.2) holds in a domain $\mathcal{D}$, which is a domain of attraction for the dynamics (3.1), i.e., the deterministic flow is inwards at every point of Γ, the boundary of $\mathcal{D}$. On the basis of the one-dimensional results, one would expect that $X(t)$ would remain in $\mathcal{D}$ for long periods, but would eventually reach Γ. Assuming that $X(0) = x$ and $X(t)$ lies on Γ, what is the most likely path that X will take?

According to (3.2), the density for a single increment dX is a multi-dimensional Gaussian:

$$\Pr[\zeta^i \leq dx^i < \zeta^i + d\zeta^i, i=1,\ldots,n] =$$

$$= C \exp\left[-\frac{1}{2\varepsilon dt} \sum_{i,j=1}^{n} a_{ij}(x)(\zeta^i - b^i dt)(\zeta^j - b^j dt)\right] \tag{3.9}$$

where (a_{ij}) is the inverse of the covariance matrix (a^{ij}). The factor C involves the determinant of (a_{ij}). The likelihood that $X(t)$ will follow the path $q(t)$ is the same as the likelihood that each increment dX will be given by $\dot{q}\,dt$. Since the increments are independent random variables, the likelihood of this event is

$$\mathcal{L}[X = q] = C_1 \exp\left[-\frac{1}{\varepsilon}\int_0^t L(q,\dot{q})dt\right] , \tag{3.10}$$

where

$$L(q,\dot{q}) = \frac{1}{2} \sum_{i,j=1}^{n} a_{ij}(q)(\dot{q}^i - b^i(q))(\dot{q}^j - b^j(q)) . \tag{3.11}$$

Therefore if $\varepsilon \to 0$, the most likely path will be the one which satisfies the following variational problem:

minimize $\int_0^t L(q,\dot{q})dt$ among all paths q which satisfy

$$q(0) = x, \; q(t)\in\Gamma. \tag{3.12}$$

The Hamilton-Jacobi theory may be applied to (3.12). The Hamiltonian H(p,q) is given by

$$H(p,q) = \frac{1}{2} \sum_{i,j=1}^{n} a^{ij}(q)p_i p_j - \sum_{i=1}^{n} b^i(q)p_i \, , \tag{3.13}$$

where

$$p_i = \sum_{j=1}^{n} a_{ij}(q)(\dot{q}^j - b^j(q)) \, , \tag{3.14}$$

i.e.,

$$\dot{q}^i = \sum_{j=1}^{n} a^{ij}(q)p_j + b^i(q) \, . \tag{3.15}$$

Let

$$\varphi(t,x) = \min_{\substack{q(0)=x \\ q(t)\in\Gamma}} \int_0^t L(q,\dot{q})dt \, . \tag{3.16}$$

Then ϕ satisfies the Hamilton-Jacobi equation

$$\frac{\partial\varphi}{\partial t} + \frac{1}{2} \sum_{i,j=1}^{n} a^{ij}(x) \frac{\partial\varphi}{\partial x^i} \frac{\partial\varphi}{\partial x^j} + b^i(x) \frac{\partial\varphi}{\partial x^i} = 0 \, . \tag{3.17}$$

The trajectory (3.15) should not be confused with the deterministic trajectory (3.1). However, in special cases, one may be the reverse of the other. Suppose that b^i is a gradient field, i.e.,

$$b^i(x) = - \sum_{j=1}^{n} a^{ij}(x) \frac{\partial V}{\partial x^i} \, , \tag{3.18}$$

where V is some function of x. Then V is a Liapunov function for the flow (3.1), since then

$$\frac{\partial V}{\partial t} = - \sum_{i,j=1}^{n} a^{ij}(x) \frac{\partial V}{\partial x^i} \frac{\partial V}{\partial x^j} \leq 0 \, . \tag{3.19}$$

In such a case, (3.17) is satisfied if $\phi(x) = -2V(x)$. Then (3.15) becomes

$$\dot{q}^i = -2 \sum_{j=1}^{n} a^{ij}(x) \frac{\partial V}{\partial x^j} + \sum_{j=1}^{n} a^{ij}(x) \frac{\partial V}{\partial x^j} = -b^i(x) . \quad (3.20)$$

Thus the most likely escape path for a gradient flow is precisely opposite to the flow!

The preceding is a very special result, and results for gradient flows are generally misleading. Ludwig [1974] gives an example where the deterministic flow spirals in to an equilibrium, and the most likely escape path spirals out. The two trajectories cross each other many times since they have the same angular components, but opposite radial components.

Vent-tsel and Freidlin [1970] prove that the most likely escape path is given by (3.14)-(3.17), using probabilistic methods.

C. Asymptotic Methods

Comparison of (3.7) and (3.17) reveals that $\phi(t,x)$ gives the leading term in an asymptotic solution of (3.7). Analogy with the WKB method suggests that

$$v(t,x) = e^{-\frac{1}{\varepsilon}\varphi(t,x)} \sum_{k=0}^{\infty} \varepsilon^k Z_k(t,x) . \quad (3.21)$$

If the form (3.21) is substituted into (3.7) and like powers of ε are combined, the result is (3.17). The coefficients Z_k satisfy transport equations along the trajectory (3.15). For example,

$$\frac{\partial Z_o}{\partial t} + \sum_{i=1}^{n} \left(\sum_{j=1}^{n} a^{ij} \frac{\partial \phi}{\partial x^j} + b^i \right) \frac{\partial Z_o}{\partial x^i} + C Z_o = 0 , \quad (3.22)$$

where the coefficient C involves derivatives of the coefficients a^{ij} and b^i, and second derivatives of ϕ. Details are given in Ludwig [1974].

In a similar fashion, the backward equation (3.8) has asymptotic solutions of the form

$$u(t,x) = e^{\frac{1}{\varepsilon}\varphi(t,x)} \sum_{k=0}^{\infty} \varepsilon^k w_k(t,x) . \quad (3.23)$$

In fact, all of the calculations of Section II have their formal analogues in the more general situation. Matkowsky and Schuss

[1977], Schuss and Matkowsky [1979] have emphasized boundary layer methods to calculate the expected exit time. The subject has been surveyed and the asymptotic formulas have been compared with exact results by Williams [1978]. Mangel and Ludwig [1977] and Mangel [1979] have considered more complicated asymptotics, which are associated with unstable trajectories or nearby critical points of the flow. The asymptotic expansions are analogous to some which arise in diffraction theory.

There are many pitfalls in the asymptotic theory, and only the one-dimensional theory approaches completeness. In one dimension, the justification of the asymptotics involves "resonance" of a singularly perturbed boundary value problem. Recent progress has been made by Kopell [1979], de Groen [1980], and Williams [preprints].

In several dimensions, the method of Matkowsky and Schuss has been justified by Kamin [1979], for the special case (3.18). A proof of the more general case has been rumored, but was not available at the time of writing. The Greap Leap Forward is apparently proved in Vent-tsel and Freidlin [1970].

D. Applications

Application of the escape problem to physics and chemistry are surveyed in Schuss [1980]. Some possible biological applications are described in Ludwig [1974]. More elaborate applications might involve deterministic dynamics with several domains of attraction $\mathcal{D}_\ell$, $\ell = 1,\ldots L$. On a short time scale, the motion is confined to one such domain. On a longer scale, the process makes transitions from one domain to another, in analogy with a Markov chain model. On a still longer scale, the motion is ergodic, i.e., all of the possible domains will be visited regularly.

Such ideas are implicit in the notion of "resilience" proposed by Holling [1973]. An ecosystem might have dynamics which are characterized by such domains of attraction $\mathcal{D}_\ell$. The short time scale is that of population dynamics. If the population is confined to such a domain for very long, stabilizing selection (described below) will tend to reduce genetic variability in the respective populations. However, on an ecological time scale, some systems may regularly move from one domain to another. These systems will tend to be more diverse genetically, and possibly more rich in species. Therefore they may be able to adjust to sudden environmental or other changes more easily than a system more or less in stasis. On a still longer time scale, such sudden changes are inevitable. Therefore one would expect that systems which are able to persist over long periods would regularly exhibit large qualitative differences in behavior.

E. A Model for Evolution

The following is a conceptual model for the evolution of species, which is based upon Wright [1932]. A population or subpopulation (deme) may be described by its distribution of genotypes. The number of possible genotypes is astronomical; so the number of dimensions in the state space is in no way inferior to the number of dimensions in problems of statistical mechanics. The difference is that the interactions are more complicated in the genetic case. Nevertheless, the genotypic characterization of the population will be represented along a line.

Different populations may be differently adapted to their environments; this adaptation is represented along the vertical axis, as in Figure 1. The term "fitness" is not used here, since various attempts have been made to give it a technical meaning. The "adaptation" is a function of the distribution of genotypes, rather than a single genotype. In particular, certain distributions may be "co-adapted". Although the populations may be genetically diverse, the various combinations which arise are usually well-adapted to the environment. Such a situation is represented by a local maximum in Figure 1.

It follows that selective forces will tend to oppose movement away from such an adaptive peak. That is, genotypes which are not well adapted will tend to contribute less to future generations. These forces will be called "stabilizing selection". They allow a population to maintain its integrity in the face of mutations, or gene flow from other demes. This integrity is also preserved by various isolating mechanisms, which tend to discourage interbreeding. Such isolating mechanisms may be thought of as part of a co-adapted gene complex, since they will be maintained by selection if hybrids are infertile or less well adapted.

Although the process of selection presumably involves some random phenomena, one may think of the preceding description as essentially deterministic: the gene complex tends to cluster near an adaptive peak. More important random effects appear in the process of reproduction. The genotypic frequencies of each generation are obtained by a more or less random sampling of the genotypes of the previous generation. For instance, let p represent the proportion of some allele at a given locus. The number of possible alleles at that locus in the next generation will be 2N in a diploid population of size N. The number of occurences of the allele under consideration will have a binomial distribution with mean 2Np, and variance $2Np(1-p)$, in the absence of selection.

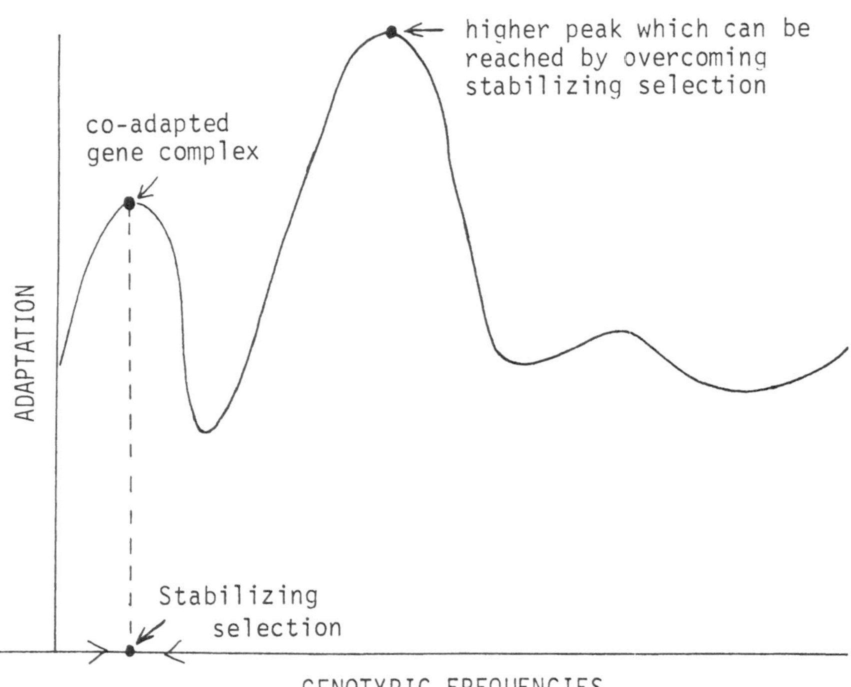

Fig. 1. Schematic representation of the "adaptation" function.

Hence the new allelic proportion will have mean p and variance $p(1-p)/(2N)$. Thus the amount of sampling variance will be inversely proportional to the population size. On the basis of this calculation, one would expect small populations to exhibit larger fluctuations in the distribution of genotypes than large ones.

The process of escape from a domain of attraction appears to be analogous to the process of moving from one adaptive peak to another. Such events should be rare, and they are most likely to occur for small demes, e.g., geographically isolated subpopulations. Although the waiting time for the escape to occur may be very long, the actual escape (the final deviation from the adaptive peak) will occur during a very short time, on the basis of the analogy with the Great Leap Forward. Once a new co-adapted gene complex has formed, selection should favor various isolating mechanisms, which eventually would lead to complete reproductive isolation, i.e., a new species.

Perhaps this analogy sheds some light on the old controversy concerning gradual vs. saltational evolution. According to the analogy, both sides are correct. Speciation is a gradual process, since it depends upon alteration of a gene complex in a deme. However, the actual splitting takes place relatively quickly. In fact, the process appears to be instantaneous on a geological or evolutionary time scale. This saltational aspect of evolution has important consequences for the interpretation of the fossil record. In particular, the notion of evolutionary tempo or rate must be reconsidered. These points are argued at some length in Gould and Eldredge [1977].

REFERENCES

Abramowitz, M., and Stegun, I.A., 1964, "Handbook of Mathematical Functions," Dover Pub., New York.

de Groen, P.P.N., 1980, The nature of resonance in a singular perturbation problem of turning point type, SIAM J. Math. Anal., 11:1.

Feller, W., 1966, "An Introduction to Probability Theory, Vol. II," John Wiley, New York.

Gikhman, I.I., and Skorohod, A.V., 1973, "Stochastic Differential Equations," Springer-Verlag, Berlin.

Gould, S.J., and Eldredge, N., 1977, Punctuated equilibria: the tempo and mode of evolution reconsidered, Paleobiology, 3:115.

Holling, C.S., 1973, Resilience and stability of ecological systems, Ann. Rev. Ecol. Syst., 4:1.

Kamin, S., 1979, On elliptic singular perturbation problems with turning points, SIAM J. Math. Anal., 10:447.

Karlin, S., and Taylor, H.M., 1975, "A First Course in Stochastic Processes," 2nd ed., Academic Press, New York.

Kopell, N., 1979, A geometric approach to boundary layer problems exhibiting resonance, SIAM J. Appl. Math., 37:436.

Kramers, H.A., 1940, Brownian motion in a field of force and the diffusion model of chemical reactions, Physica, 7:284.

Ludwig, D., 1974, Persistence of dynamical systems under random perturbations, SIAM Rev., 17:605.

Mangel, M., 1979, Small fluctuations in systems with multiple steady states, SIAM J. Appl. Math., 36:544.

Mangel, M., and Ludwig, D., 1977, Probability of extinction in a stochastic competition, SIAM J. Appl. Math., 33:256.

Matkowsky, B., and Schuss, Z., 1977, The exit problem, SIAM J. Appl. Math., 33:230.

Schuss, Z., 1980, Singular perturbation methods in stochastic differential equations of mathematical physics, SIAM Rev., 22:119.

Schuss, Z., and Matkowsky, B., 1979, The exit problem: a new approach to diffusion across potential barriers, SIAM J. Appl. Math., 35:604.

Strook, D.W., and Varadhan, S.R.S., 1979, "Multidimensional Diffusion Processes," Springer-Verlag, Berlin.

Turelli, M., 1977, Random environments and stochastic calculus, Th. Pop. Biol., 12:140.

Vent-tsel', A.D., and Freidlin, M.I., 1970, On small random perturbations of dynamic systems, Uspehi Mat. Nauk, 25:3. (Russian Math. Surveys, 25:1.)

Williams, M., preprints, Asymptotic one-dimensional exit time distributions, and Another look at Ackerberg-O'Malley resonance, Dept. of Math., Virginia Polytechs. Inst., Blacksburg, Virginia 24061.

Williams, R.G., 1978, The stochastic exit problem for dynamical systems, Ph.D. Thesis, Dept. of Appl. Math., Calif. Inst. of Tech.

Wright, S., 1932, The roles of mutation, inbreeding, cross-breeding and selection in evolution, Proc. VI Int. Cong. Genet., 1:356.

SEMINARS

ERROR PROPAGATION IN TRANSLATION AND ITS RELEVANCE TO THE NUCLEATION OF LIFE

Geoffrey W. Hoffmann

PSEUDOPOTENTIALS AND SYMMETRIES FOR GENERALIZED NONLINEAR SCHRODINGER EQUATIONS

J. Harnad and P. Winternitz

SUPERPOSITION PRINCIPLES FOR NONLINEAR DIFFERENTIAL EQUATIONS

R.L. Anderson, J. Harnad and P. Winternitz

ON SOME NONLINEAR SCHRODINGER EQUATIONS

H. Lange

ASYMPTOTIC EVALUATION METHODS OF NONLINEAR DIFFERENTIAL EQUATIONS NEAR THE INSTABILITY POINT

Masuo Suzuki

NONLINEAR SUPERPOSITION OF SIMPLE WAVES IN NONHOMOGENEOUS SYSTEMS

Alfred M. Grundland

A METHOD OF SOLVING NONLINEAR DIFFERENTIAL EQUATIONS

M.W. Kalinowski

ERROR PROPAGATION IN TRANSLATION AND ITS RELEVANCE TO THE NUCLEATION OF LIFE

Geoffrey W. Hoffmann

Department of Physics
University of British Columbia
Vancouver, B.C., V6T 1W5

Some of the enzymes produced by the translation of information stored in nucleic acid sequences participate as adaptors in the translation processes itself. The more errors there are in the adaptors doing the translation, the lower the quality (and therefore the lower the accuracy in action) of the adaptors that are produced. Conceivably the level of errors could escalate to give an "error catastrophe" [Orgel, 1963]. A mathematical model of this process has been formulated, based on a simple "all-or-none" classification of the amino acid residues of a typical adaptor residue [Hoffmann, 1974, 1975]. The model permits an estimate to be made of the threshold level of specificity, required by the components of a workable primitive translation apparatus. In the model, a certain number of the amino acid residues of an adaptor are presumed to be critical for the correct folding of the adaptor, (such that if an error is made there, all activity is lost), and a certain number are presumed to be critical for the specificity of adaptors (such that an error made in those sites results in an adaptor that makes correct each of the possible incorrect assignments with equal rates). The remaining amino acid residues are assumed to function as spacers, so that errors in those sites are without consequence for the functioning and specificity of the adaptors. The classification of residues permits the calculation of the accuracy in action q_i of a particular generation of adaptors, as a function of the accuracy in action q_{i-1} of the preceding generation. We then have

$$q_i = F(q_{i-1}, p) \qquad (1)$$

where F is an explicit nonlinear function, and p are parameters of the model. Steady states of the recurrence relation (1) are then specified by the condition

$$q_i = q_{i-1} \quad . \tag{2}$$

For small values of S, a parameter denoting the average specificity of the adaptors, (1) and (2) have only one solution. Translation is then essentially random, corresponding to the outcome of an error catastrophe. At higher values of S two additional solutions exist, including a stable steady state with a satisfactorily high accuracy q. The minimum value of S that yields the additional solutions can be calculated. For reasonable values of the parameters the result obtained corresponds to differences in free energy between binding correct and incorrect substrates by the adaptors of 2-3 Kcal/mole, suggesting that very sloppy components (by present day standards) would suffice to provide stable translation. A nucleation event involving the simultaneous chance occurence of several such sloppy adaptors consequently seems feasible.

Measurements of error propagation in present day bacterial systems have recently been made, and compared with this model and modifications of it [Gallant and Prothera, in press].

REFERENCES

Gallant, J. and Prothera, J., 1980 (in press), J. Theor. Biol.
Hoffmann, G.W., 1974, J. Mol. Biol., 68: 349.
Hoffmann, G.W., 1975, Ann. Rev. Phys. Chem., 26: 123.
Orgel, L.E., 1963, Proc. Nat. Acad. Sci. U.S.A., 49:517.

PSEUDOPOTENTIALS AND SYMMETRIES FOR GENERALIZED NONLINEAR SCHRÖDINGER EQUATIONS

J. Harnad and P. Winternitz

Centre de Recherche de Mathématiques Appliquées
Université de Montréal
Montréal, Québec, Canada

Applying the method of differential ideals and prolongations to equations of the type

$$iZ_t + Z_{xx} = f(Z,Z^*) ,$$

where $f(Z,Z^*)$ is any smooth function, we arrive at eight types of equations admitting "pseudopotentials" (in the sense of Wahlquist and Estabrook [1976]). For five of these, the integrability conditions are equivalent to conservation laws. For the remaining three, including the usual cubically non-linear Schrödinger equation, there are non-trivial pseudo-potentials, which may be interpreted as defining Bäcklund transformations or linear scattering equations. For the two cases other than the standard one, however, there is no parameter identifiable as an eigenvalue and no Lie symmetry (other than translation invariance) to generate such a parameter. For the standard case, we show that the real and imaginary parts of the parameter occuring in the Bäcklund transformation (or scattering equation) may be generated by composing a given transformation with the 2-parameter Lie symmetry group consisting of Galilean boosts and dilations.

Details may be found in Harnad and Winternitz (preprint CRMA-952).

REFERENCES

Estabrook, F.B. and Wahlquist, H.D., 1976, Prolongation structures of nonlinear evolution equations. II, J. Math. Phys., 17: 1293.

Harnad, J. and Winternitz, P., preprint CRMA-952, Pseudopotentials and Lie symmetries for the generalized nonlinear Schrödinger equations.

SUPERPOSITION PRINCIPLES FOR NONLINEAR DIFFERENTIAL EQUATIONS

R.L. Anderson

Department of Physics
University of Georgia
Athens, Georgia
U.S.A.

J. Harnad and P. Winternitz

Centre de Recherche de Mathématiques Appliquées
Université de Montréal
Montréal, Québec
Canada

Consider a system of n first order quasilinear ordinary differential equations

$$\frac{du^i}{dt} = \eta^i(u^1,\ldots,u^n,t) \quad 1 \leq i \leq n \ , \ u^i,t \in \mathbb{R} \ . \tag{1}$$

We shall say that this system of equations allows a superposition principle if its general solution can be expressed as a function of a finite number m of particular solutions $u_{(i)}$ and n constants c_k:

$$u^i(t) = F^i(u_{(1)},\ldots,u_{(m)} \ , \ c_1,\ldots,c_n) \tag{2}$$

S. Lie [1893] has shown that system (1) admits a superposition principle if and only if it can be rewritten as

$$\dot{u}^i \equiv \frac{du^i}{dt} = \sum_{k=1}^{r} Z_k(t) \ \xi^i_k(u^1,\ldots,u^n), \quad 1 \leq i \leq n \ , \tag{3}$$

where the vector fields

$$X_k = \sum_{i=1}^{n} \xi_k^i(u^1,\ldots,u^n) \frac{\partial}{\partial u^i} \ , \quad 1 \le k \le r \tag{4}$$

generate a finite dimensional Lie algebra.

Our contribution consists of the following:

A. We have developed three different methods of obtaining superposition principles explicity. These are:

1. Write the general solution of (3) in the form of an element of the Lie group G obtained by exponentiation of (4), acting on a constant vector, providing the initial conditions. Reconstruct the group element in terms of m particular solutions where $m.n \ge r$ and r is the dimension of the group.

2. Linearize and homogenize the system of equations by a transformation, making use of a sufficient number of particular solutions, and corresponding to the action of certain subgroups of the group G. Then write a linear superposition principle for the new set of equations and transform it back to the original variables.

3. Interpret the solutions $u_{(i)}$, $1 \le i \le m$ as points in a space on which the group acts. Take m solutions where m is such that we can form n group invariants involving $u_{(i)}$ and a general solution u which, for fixed values of $u_{(i)}$, are independent functions of u. The superposition principle is obtained by equating these invariants to n chosen constants $\{c_i\}$ and $\{u_{(i)}\}$.

B. We have applied these methods for specific Lie algebras realized by the vector fields (4), namely $s\ell(n+1,\mathbb{R})$ (leading to a system of "projective Riccati equations" [Anderson, 1980]), $O(p+1,q+1)$ with $p+q = n$ ("conformal Riccati equations") and the Euclidean Lie algebra $e(n)$ (we obtained a nonlinear superposition principle for a system of linear equations).

C. We have shown that the solutions $u_{(1)}(t),\ldots,u_{(m)}(t)$ in (2) are arbitrary up to certain independence conditions and hence do not constitute a preferred "fundamental set of solutions".

As examples, consider the case when the vector fields (4) generate the group $G = O(p+1,q+1)$ which we realize as the conformal group $C(p,q)$ of $n = p+q$ dimensional pseudo-Euclidean space $M_{p,q}$ with signature (p,q) and metric tensor g. The equations (3) are in this case the "conformal Riccati equations":

$$\dot{x}^\mu = a^\mu + \lambda x^\mu + \omega^\mu{}_\nu x^\nu + c^\mu(x^\nu x_\nu) - 2(c^\nu x_\nu)x^\mu \ , \quad 1 \le \mu \le p+q \ , \tag{5}$$

where a^μ, λ, ω^μ_ν and c^μ are known functions of t and the matrix ω satisfies $\omega g + g\omega^T = 0$. The nonlinear superposition principle is

$$x = x_a + \frac{\sum_{k=1}^{n} \alpha_k \left[\frac{x_k - x_a}{(x_k-x_a)^2} - \frac{x_b - x_a}{(x_b-x_a)^2}\right] + \frac{x_b - x_a}{(x_b-x_a)^2}}{\left[\sum_{k=1}^{n} \alpha_k \left[\frac{x_k - x_a}{(x_k-x_a)^2} - \frac{x_b - x_a}{(x_b-x_a)^2}\right] + \frac{x_b - x_a}{(x_b-x_a)^2}\right]^2} \tag{6}$$

where α_k are constants and x_a, x_b, x_k are n + 2 arbitrary particular solutions of (5) satisfying

$$(x_k-x_a)^2 \neq 0 \ , \ (x_b-x_a)^2 \neq 0$$

and

$$\det \left\{\frac{x_k - x_a}{(x_k-x_a)^2} - \frac{x_b - x_a}{(x_b-x_a)^2}\right\} \neq 0 \ . \tag{7}$$

If G = SL(n+1),$\mathbb{R}$) we realize it as the projection group P(n) of an n dimensional real vector space. The equations (3) in this case are the "projective Riccati equations" (we use a vector notation):

$$\dot{x} = (A - aI)x - (\gamma\cdot x)x + \beta$$

where A = A(t) is a traceless matrix, a = a(t) a scalar and β, γ, and x are vector functions of t. The nonlinear superposition principle providing a general solution x(t) can be written as

$$x = \frac{Bc + \rho}{\sigma^T c + b} \tag{8}$$

where c is a constant vector (n constants) and the coefficients b(t), $B^\mu_\nu(t)$, $\rho^\mu(t)$ and $\sigma^\mu(t)$ can be expressed in terms of n + 2 solutions, u, v, $x_1,\ldots,x_n$ as

$$b = (1 - \sum_{k=1}^{n} c_k)\det(x_1-v,\ldots,x_n-v)$$

$$\rho = bu \ , \ B^\mu_k = x^\mu_k \sigma_k \ \text{(no sum)} \tag{9}$$

$$\sigma^k = \det(x_1-v,\ldots,x_{k-1}-v,\ u-v,\ x_{k+1}-v,\ldots,x_n-v)$$

For all details and applications we refer to our article [Anderson, Harnad and Winternitz, 1980 to be published].

REFERENCES

Anderson, R.L., 1980, A nonlinear superposition principle admitted by coupled Riccati equations of the projective type. Lett. Math. Phys., 4:1.

Anderson, R.L., Harnad, J. and Winternitz, P., 1980 to be published (Preprint CRMA-980, Montreal), Nonlinear superposition principles for systems of ordinary differential equations.

Lie, S. and Scheffers, G., 1893,"Vorlesungen über Continuierliche Gruppen", B.G. Teubner, Leipzig.

ON SOME NONLINEAR SCHRÖDINGER EQUATIONS

H. Lange

Mathematisches Institut
Universität Köln
West Germany

The equation which is studied in this note is the following nonlinear Schrödinger equation

$$i\psi_t = \Delta\psi + \lambda v_q(\psi)^p \cdot \psi \tag{1}$$

where

$$\psi = \psi(x,t),\ x \in \mathbb{R}^3,\ t \in \mathbb{R},\ \lambda \in \mathbb{R},\ p,\ q \geq 1$$

and (with $r = |x|$)

$$v_q(\psi)(x,t) = (r^{-1} * |\psi|^{2q})(x,t) = \int_{\mathbb{R}^3} \frac{|\psi(y,t)|^{2q}}{|x-y|}\, dy\ .$$

For $\lambda < 0$, $p = q = 1$, Eq. (1) is the time-dependent version of the Hartree equation for the Helium atom where $v_1(\psi)$ is the Newtonian potential of the charge density $|\psi|^2$; in this case equations of type (1) may also be considered as equivalent models for the coupled Maxwell-Dirac equation with vanishing magnetic field. For $\lambda > 0$, $p = q = 1$, Eq. (1) was proposed for investigation by Ph. Choquard for the theory of one-component plasmas to describe an electron trapped in its own hole. In these cases equation (1) can be considered as the classical limit of the field equations which describe a quantum mechanical non-relativistic many-boson system interacting through a two-body potential ($\sim r^{-1}$); the equation is also used as the self-consistent field approximation for the same physical system. Schrödinger equations of type (1) with $p,\ q \geq 1$

arise in the situation where one takes into account higher-order many-body interactions in the Hamiltonian through a potential of the non-local form

$$\int_{\mathbb{R}^3}\cdots\int_{\mathbb{R}^3} dx_1\ldots dx_p V(x_1,\ldots,x_p;x)|\psi_1(x_1,t)|^{2q}\ldots|\psi_p(x_p,t)|^{2q}.$$

Specializing V and $\psi_1,\ldots,\psi_p$ one gets the model problem of type (1).

For equations of type (1) stationary solutions of the solitary wave type

$$\psi(x,t) = \varphi(x - v\cdot t)e^{i\mu t}$$

are of great interest and induce a corresponding nonlinear eigenvalue problem. For the case $p = q = 1$ and $v = 0$, these have been studied extensively for both signs of λ (the different signs of the nonlinear coupling are of some mathematical significance because of monotonicity properties). The global Cauchy problem for (1) was treated by Chadam and Glassey [1975] for $p = q = 1$, $\lambda < 0$ and an additive Coulombic potential; Glassey [1977] showed that the only solution to (1) for $p = q = 1$, $\lambda < 0$ with an asymptotic free state in the L^2-sense is identically zero, so that there is no non-trivial scattering theory in the usual sense for (1) in this case.

In this note we state results which have analogies in the theory of Schrödinger equations of the local form

$$i\psi_t = \Delta\psi + \lambda|\psi|^{p-1}\cdot\psi \tag{2}$$

(which were considered e.g. by Strauss [1974], Ginibre and Velo [1979], Pecher and v. Wahl [1979]. For big enough p, q (in fact, we must have $p \geq 3/2$ and $2p(q-1) \geq 1$) there exist global strong H^2-valued solutions to the initial-value problem for (1) and an appropriate nonlinear scattering theory for small initial data or small nonlinear coupling coefficient λ. These results may include also more general equations of type (1) with nonlinearities of the form

$$v_q(\psi)^p\cdot\psi$$

with

$$v_q(\psi) = f * |\psi|^{2q}$$

where f may be a general convolution kernel of the type $f(x) = g(r)\cdot r^{-\gamma}$ $(r = |x|)$, e.g. $f(x) = \exp(-\alpha r)\cdot r^{-1}$ which gives rise to an equation of interest in electron optics (see Karasev and Maslov [1979]; see also Ginibre and Velo [1979]).

The method in showing the above result uses a procedure given by Strauss [1974] in the treatment of equations of type (2) and nonlinear wave equations which consists of an iteration technique for a certain scattering norm involving the time decay for solutions of the free equation.

Open problems concerning equation (1) are the determination of the maximal parameters p^*, q^* such that we have well-posedness for (1) for $1 \leq p \leq p^*$, $1 \leq q \leq q^*$; or the determination of minimal p_o, q_o such that (1) is well-posed and has a nonlinear scattering theory for $p > p_o$, $q > q_o$ and small coupling coefficient (for the results above the minimal degree of homogeneity which is allowed is 5); furthermore, no "blow-up" theorems are known for (1) and no results on the corresponding nonlinear eigenvalue problems for $p, q > 1$.

REFERENCES

Chadam, J.M. and Glassey, R.T., 1975, Global existence of solutions to the Cauchy problem for time-dependent Hartree equations, J. Math. Phys., 16:1122.

Ginibre, J. and Velo, J., 1979, On a class of nonlinear Schrödinger equations. I. The Cauchy problem, general case; II. Scattering theory, general case, J. Functional Anal., 32:1.

Ginibre, J. and Velo, J., 1979, Preprint, Bologna, On a class of nonlinear Schrödinger equations with nonlocal interaction.

Glassey, R.T., 1977, Asymptotic behavior of solutions to certain nonlinear Schrödinger-Hartree equations, Comm. Math. Phys., 53: 9.

Karasev, M.V. and Maslov, V.P., 1979, Quasiclassical soliton solutions of the Hartree equations. Newtonian interaction with screening, Theoret. Math. Phys., 40:715.

Pecher, H. and Wahl, W.V., 1979, Time dependent nonlinear Schrödinger equation, Manuscripta Math., 27:125.

Strauss, W.A., 1974, Nonlinear scattering theory, in "Proc. NATO Advanced Study Institute, Denver, Colorado 1973", J.A. Lavita, J.P. Marchand, eds., Reidel, Dordrecht.

ASYMPTOTIC EVALUATION METHODS OF NONLINEAR DIFFERENTIAL EQUATIONS NEAR THE INSTABILITY POINT

Masuo Suzuki

Department of Physics
University of Tokyo
Hongo, Bunkyo-ku
Tokyo, Japan

The purpose of this paper is to present a general method [Suzuki, 1976a,b, 1977a,b,c, 1978, 1980, in press, to be published] to evaluate asymptotically solutions of nonlinear stochastic differential equations or Langevin's equations when the system is initially located at or near the unstable point. One of our keypoints is to notice the existence of the scaling regime in an intermediate time region, in which the temporal evolution of physical quantities is expressed by a certain scaling function of the so-called scaling variable $\tau = S(t, \varepsilon, \ldots)$ of time t and the relevant smallness parameter ε. Namely, we divide the whole time region into three regimes, i.e., initial, second (scaling), and final regimes. We consider the following nonlinear stochastic differential equation or Langevin's equation

$$\frac{dx}{dt} = \alpha(x) + \beta(x)\eta(t) , \tag{1}$$

where $\eta(t)$ is the Gaussian white noise satisfying

$$\langle\eta(t)\eta(t')\rangle = 2\varepsilon\delta(t-t') . \tag{2}$$

A typical example is the following laser model

$$\frac{dx}{dt} = \gamma x - gx^3 + \eta(t) . \tag{3}$$

Our general strategy is to introduce a nonlinear transformation and to simplify (or linearize) the transformed equation in a soluble form. The solution of it is transformed back in the original variable. It has a very interesting scaling property in the second regime and consequently it is called the scaling solution. Our general nonlinear transformation is

$$\xi = F^{-1}(e^{-\gamma t}F(x)) \; ; \; \gamma = \alpha'(x_o) > 0 \, , \tag{4}$$

where

$$F(x) = \exp \int_{a_o}^{x} \frac{\gamma}{\alpha(y)}\, dy \, , \tag{5}$$

and a_o is determined so that $F'(x_o) = 1$ at the unstable point x_o. This transformation is closely related to the characteristic curve (or solution) of the deterministic equation $dy/dt = \alpha(x)$ with the initial condition $x = \xi$. Now we apply this transformation to (1), and we have

$$\frac{d\xi}{dt} = G(\xi,t)\eta(t) \, , \tag{6}$$

where

$$G(\xi,t) = e^{-\gamma t}F'\left(F^{-1}(e^{\gamma t}F(\xi))\right)\beta\left(F^{-1}(e^{\gamma t}F(\xi))\right)/F'(\xi) \, . \tag{7}$$

Here we assume that

$$\beta(x) = x^p + (\text{higher order}) \, . \tag{8}$$

Then by extracting the most dominant part from $G(\xi,t)$, our desired asymptotic solution $\xi_{sc}(t)$ is found to be given by the equation

$$\frac{d}{dt}\xi_{sc}(t) = e^{(p-1)\gamma t}\,\xi_{sc}(t)^p\eta(t) \, . \tag{9}$$

Now the solution $\xi_{sc}(t)$ can be obtained explicitly. Thus, our scaling solution is also explicitly given by

$$x_{sc}(t) = F^{-1}\left(e^{\gamma t}F(\xi_{sc}(t))\right) \, . \tag{10}$$

For example, the scaling solution of (3) is easily given and the fluctuation $\langle x^2(t)\rangle$ is expressed by the integral

$$\langle x^2(t)\rangle_{sc} = \langle x^2\rangle_{st}\left[\frac{1}{1-e^{-2\gamma t}}\right]\frac{1}{\sqrt{2\pi}}\int_{-\infty}^{\infty} e^{-\xi^2/2}\left[\frac{\xi^2\tau}{1+\xi^2\tau}\right]d\xi \tag{11}$$

where

$$\tau = \frac{g\varepsilon e^{2\gamma t}}{\gamma^2}\left[\frac{\gamma\langle x^2(0)\rangle}{\varepsilon} + (1-e^{-2\gamma t})\right] \tag{12}$$

Here, τ is proportional to $g\varepsilon$ exp $(2\gamma t)$ for large t, and consequently it is called a scaling variable. The distribution function $P_{sc}(x,t)$ is also given by [Suzuki, 1980; Pasquale et al, 1979]

$$P_{sc}(x,t) = \frac{1}{\sqrt{2\pi\tau'}}\left[1 - \frac{g}{\gamma}x^2(1 - e^{-2\gamma t})\right]^{-\frac{3}{2}} \times$$

$$\times \exp\left[-\frac{x^2}{2\tau'\left[1-(g/\gamma)x^2(1-e^{-2\gamma t})\right]}\right], \tag{13}$$

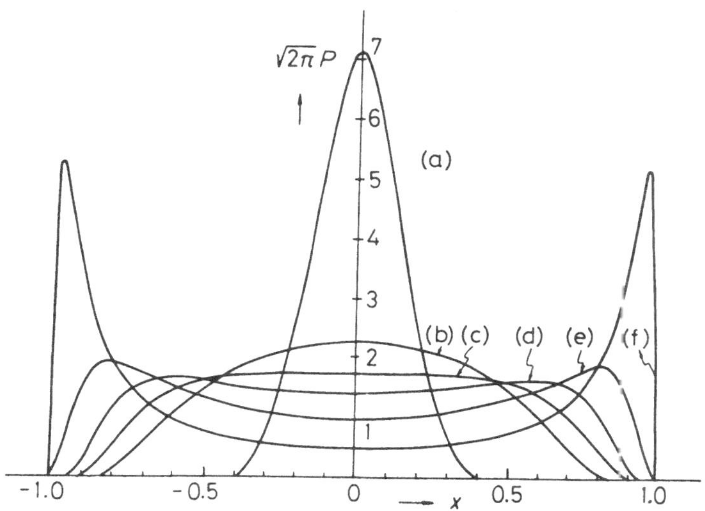

Fig. 1. Distribution function:
(a) τ = 0.02, (b) τ = 0.2, (c) $\tau = \tau_o$ = 1/3, (d) τ = 0.5, (e) τ = 1 and (f) τ = 4.

Suzuki, M., 1977c, Anomalous fluctuation and relaxation in unstable systems, J. Stat. Phys., 16:477.

Suzuki, M., 1978, Scaling theory of transient nonlinear fluctuations and formation of macroscopic order, Prog. Theor. Phys. Suppl., 64:402.

with $\tau' = \tau\gamma/g$. This is shown explicitly in Fig. 1. The onset time t_o is defined by the time at which P begins to change from a single peak to double peaks, and it is given in our case by

$$t_o = \frac{1}{2\gamma} \log \left[\frac{3\varepsilon g}{\gamma^2} \left(1 + \frac{\gamma}{\varepsilon} \langle x^2(0)\rangle\right) \right]^{-1} . \tag{14}$$

It is easily seen that t_o becomes larger and larger, as ε, g and $\langle x^2(0)\rangle$ becomes smaller and smaller. This shows the importance of the synergesim of nonlinearity g, random force ε, and initial fluctuation $\langle x^2(0)\rangle$ for the formation of macroscopic order.

This result has been confirmed recently by Caroli et al [1979, 1980] using the WKB method. The above scaling theory can be applied to many interesting systems [Suzuki, 1976a,b, 1977a,b,c, 1978, 1980, in press; Pasquale et al, 1979].

REFERENCES

Caroli, B., Caroli, C. and Roulet, B., 1979, Diffusion in a bistable potential: a systematic WKB treatment, J. Stat. Phys., 21: 415.

Caroli, B., Caroli, C. and Roulet, B., 1980, Growth of fluctuations from a marginal equilibrium, Physica, 101A:581.

de Pasquale, F., Tartaglia, P., and Tombesi, P., 1979, Transient laser radiation as a stochastic process near an instability point, Physica, 99A:581.

Suzuki, M., 1976a, Scaling theory of nonequilibrium systems near the instability point. I. General aspects of transient phenomena, Prog. Theor. Phys., 56:77.

Suzuki, M., 1976b, Scaling theory of nonequilibrium systems near the instability point. II. Anomalous fluctuation theorems in the extensive region, Prog. Theor. Phys., 56:477.

Suzuki, M., 1977a, Scaling theory of nonequilibrium systems near the instability point. III. Continuation to find region and systematic scaling expansion, Prog. Theor. Phys., 57: 380.

Suzuki, M., 1977b, Scaling theory of transient phenomena near the instability point, J. Stat. Phys., 16:11.

Suzuki, M., 1980, _Advances in Chem. Phys._, 46:195.
Suzuki, M., in press, _in_: "Proceedings of the XVII Solvay Conference on Physics", John Wiley and Sons, Inc., N.Y.
Suzuki, M., to be published, _in_: "Lecture Notes in Physics. Proceedings of the VIth Sites Conference on Statistical Mechanics", Springer-Verlag, Berlin.

NONLINEAR SUPERPOSITION OF SIMPLE WAVES IN NONHOMOGENEOUS SYSTEMS*

Alfred Michal Grundland†

Institute of Theoretical Physics
Warsaw University
00-681 Warszawa, Hoza 69
Poland

ABSTRACT

In this paper we develop a method of constructing the solutions of nonelliptic systems of nonhomogeneous quasilinear partial differential equations of the first order. This method comes from a concept of Riemann's invariants for homogeneous systems, but it can be generalized for a larger region of solutions than those described by Riemann's invariants. The base for the construction of solutions are called simple integral elements resulting from the algebraization of the initial system of equations. The notion of the simple state corresponding to the elementary solution of the nonhomogeneous system is introduced, and then some more complicated forms of solutions are proposed and the conditions for their existence are found. The last problem, similarly to the case of the homogeneous systems, is reduced to examining Pfaff's forms with the use of Cartan's theory of systems in involution. We propose the physical interpretation of the obtained classes of solutions as a description of the nonlinear superposition of simple waves propagating on a simple state. This method was illustrated by applying it to the Korteweg de Vries equation and to nonhomogeneous equations of magnetohydrodynamics.

---

* This work was partially supported by NSF Grant No INT 7.3.20002 A01, formerly GF-41958.

† Present address: Department of Physics, Institute of Geophysics, Warsaw University, 02-093 Warszawa, Pasteur 7, Poland.

A METHOD OF SOLVING NONLINEAR DIFFERENTIAL EQUATIONS*

M.W. Kalinowski

Institute of Philosophy and Sociology
Polish Academy of Sciences
00-330 Warsaw, Nowy Swiat 72
Poland

ABSTRACT

A new method of exactly solving nonlinear differential partial equations is developed. This method is applied to the equation of potential flow of a compressible gas. A new solution is obtained in the supersonic region which may be treated as an analogy of a non-linear stationary wave.

* The work was partially supported by N.S.F. contract INT 73-20002A01.

LECTURERS

Ablowitz, Professor M.	Department of Mathematics and Computer Science Clarkson College, U.S.A.
Fornberg, Professor B.	Department of Applied Mathematics California Institute of Technology, U.S.A.
Howard, Professor L.N.	Department of Mathematics Massachusetts Institute of Technology, U.S.A.
Kauffman, Professor S.A.	Department of Biochemistry and Biophysics The University of Pennsylvania U.S.A.
Kaup, Professor D.J.	Department of Physics Clarkson College, U.S.A.
Ludwig, Professor D.	Department of Mathematics University of British Columbia Canada
Miura, Professor R.M.	Department of Mathematics University of British Columbia Canada
Nicolis, Professor N.	Pool de Physique Université Libre de Bruxelles Belgium
Rinzel, Dr. J.	National Institutes of Health Bethesda, Maryland, U.S.A.
Schuster, Professor P.	Institut für Theoretische Chemie und Strahlenchemie Universität Wien, Austria
Scott, Professor A.	Department of Electrical and Computer Engineering University of Wisconsin-Madison U.S.A.
Zabusky, Professor N.J.	Department of Mathematics University of Pittsburgh, U.S.A.

Invited Lecturers and Directors
Banff NATO ASI, 17 - 29 August, 1980

From left to right, first row: Bengt Fornberg, Donald Ludwig, Gregoire Nicolis, Norm Zabusky, Al Scott, Robert Miura; second row: Mark Ablowitz, David Kaup, Stuart Kauffman, Lou Howard, Richard Enns, Peter Schuster, Billy Jones; missing: John Rinzel.

PARTICIPANTS

Barcus, Dr. L.C.	Dept. of Physics University of Lowell Lowell, Ma., U.S.A.
Bell, Dr. C.	Glasshouse Crops Research Inst. Littlehampton, West Sussex, U.K.
Bruce, Mr. D.	Dept. of Physics University of Edinburgh Edinburgh, Scotland
Buican, Mr. T.	Dept. of Physics University of British Columbia Vancouver, B.C., Canada
Clarke, Dr. B.	Dept. of Chemistry University of Alberta Edmonton, Alta., Canada
Crutchfield, Mr. J.	Dept. of Physics University of California, Santa Cruz Santa Cruz, Ca., U.S.A.
Doedel, Dr. E.	Dept. of Computer Science Concordia University Montreal, Que., Canada
Enns, Dr. R.H.	Dept. of Physics Simon Fraser University Burnaby, B.C., Canada
Elskens, Mr. Y.	Service de Chimie Physique II Université Libre de Bruxelles Brussels, Belgium
Evans, Dr. G.T.	Woods Hole Oceanographic Inst. Woods Hole, Mass., U.S.A.
Farmer, Mr. D.	Dept. of Physics University of California, Santa Cruz Santa Cruz, Ca., U.S.A.
Ferracin, Dr. A.	Istituto di Biologia Generale Universitá Degli Studi di Roma Rome, Italy

Froehling, Mr. H.	Dept. of Physics University of California, Santa Cruz Santa Cruz, Ca., U.S.A.
Galaretta, Mr. D.	University of Paris Sud Orsay, France
Gavish, Dr. B.	Dept. of Physics University of Illinois Urbana, Ill., U.S.A.
Gonzalez-Gascon, Dr. F.	Dept. de Fisica Teorica Universidad de Madrid Madrid, Spain
Gomes, Dr. J.	Dept. de Quimica Universidade do Porto Porto, Portugal
Granero, Dr. M.	Istituto de Fisica Universita di Parma Parma, Italy
Grasman, Dr. J.	Mathematisch Centrum Amsterdam, Holland
Grundland, Dr. A.M.	Institute of Theoretical Physics Warsaw University Warsaw, Poland
Haritos, Dr. A.	Zoological Laboratory and Museum University of Athens Athens, Greece
Harnad, Dr. J.	C.R.M.A. Université de Montréal Montréal, Que., Canada
Hemmer, Dr. P.	Institutt for Teoretisk Fysikk Universitetet I Trondheim Trondheim, Norway
Hilhorst, Dr. D.	Mathematisch Centrum Amsterdam, Holland
Hoffmann, Dr. G.	Dept. of Physics University of British Columbia Vancouver, B.C., Canada

Hubbard, Dr. J.C.	Dept. of Physics University of California, Irvine Irvine, Ca., U.S.A.
Jones, Dr. B.L.	Dept. of Physics Simon Fraser University Burnaby, B.C., Canada
Kalinowski, Dr. M.W.	Institute of Philosophy and Sociology Polish Academy of Science Warsaw, Poland
Karcz, Dr. I.	Geological Survey Jerusalem, Israel
Kathuria, Mr. Y.	Institute fur Angewandte Physik Darmstadt, West Germany
Kaufman, Dr. M.	Service de Chimie Physique II Université Libre de Bruxelles Brussels, Belgium
Kernevez, Dr. J.P.	U.T.C., B.P. 233 60206 Compiegne, France
Klinker, Mr. T.	Drittes Physikalisches Institut der Universität Göttingen Göttingen, West Germany
Knight, Mr. G.	141 Davisville Ave., Apt. 510 Toronto, Ont., Canada
Kono, Dr. M.	Dept. of Theoretical Physics Oxford University Oxford, England
Kotorynski, Dr. W.	Dept. of Mathematics University of Victoria Victoria, B.C., Canada
Lakshmanan, Dr. M.	Dept. of Mathematics The University of Manchester Institute of Science and Technology Manchester, England

Lange, Dr. H.	Mathematisches Institut Univ. Köln Köln, West Germany
Leung, Dr. H.K.	Physics Dept. National Central University Chung-li, Taiwan
Lie, Dr. S.	Dept. of Physics McMaster University Hamilton, Ont., Canada
Manitius, Dr. A.	C.R.M.A. Université de Montréal Montréal, Que., Canada
Mayer, Mr. T.	Dept. of Physics University of Texas Austin, Texas, U.S.A.
Mbaeyi, Mr. P.	Institut für Physikalische und Theoretische Chemie Universität Tübingen Tübingen, West Germany
Milinazzo, Dr. F.	Dept. of Mathematics Royal Roads Military College Victoria, B.C., Canada
Moleko, Mr. L.	Dept. of Physics University of Ottawa Ottawa, Ont., Canada
Müller, Dr. S.	Dept. of Chemistry Stanford University Stanford, Ca., U.S.A.
McNeil, Ms. L.B.	Dept. of Electrical and Computer Engineering University of Wisconsin-Madison Madison, Wisc., U.S.A.
Nerenberg, Dr. M.	Dept. of Applied Mathematics University of Western Ontario London, Ont., Canada
Noronha da Costa, Ms. Ana	Rua Carlos Calisto, no. 3-30 Esq, 1400 Lisboa Portugal

Olsson, Dr. M.	Dept. of Molecular Biology University of Aarhus Aarhus, Denmark
Osterberg, Dr. N.	Physisk-Kemisk Institut Lyngby, Denmark
Pathria, Dr. R.K.	Dept. of Physics University of Waterloo Waterloo, Ont., Canada
Pederson, Dr. H.	Central Institute for Industrial Research Oslo, Norway
Perrie, Dr. W.	Dept. of Mathematics University of British Columbia Vancouver, B.C., Canada
Pick, Dr. P.	240 W 98 St., 7E New York, N.Y. 10025 U.S.A.
Plischke, Dr. M.	Dept. of Physics Simon Fraser University Burnaby, B.C., Canada
Popielwicz, Dr. A.	Dept. of Theoretical Physics Oxford University Oxford, England
Ram, Dr. S.	Dept. of Physics University of Essex Colchester, England
Rangnekar, Dr. S.S.	Dept. of Physics Simon Fraser University Burnaby, B.C., Canada
Ray, Dr. J.	Dept. of Physics Clemson University Clemson, S.C., U.S.A.
Rehmus, Dr. P.	Dept. of Chemistry Stanford University Stanford, Ca., U.S.A.
Robinson, Dr. P.	School of Mathematics Bradford University Bradford, England

Rypdal, Dr. K.	Inst. of Mathematical and Physical Sciences University of Tromsö Tromsö, Norway
Savage, Dr. H.	Naval Surface Weapons Centre White Oak Lab Silver Spring, Md., U.S.A.
Sheats, Dr. J.	Dept. of Chemistry Stanford University Stanford, Ca., U.S.A.
Speranza, Dr. A.	Istituto di Fisica Via Irnerio, 46 40126 Bologna, Italy
Spiga, Dr. G.	Nuclear Eng. Laboratory University of Bologna Bologna, Italy
Stamnes, Dr. J.	Central Institute for Industrial Research Oslo, Norway
Suzuki, Dr. M.	Dept. of Physics University of Tokyo Tokyo, Japan
Tebaldi, Dr. C.	Istituto di Matematica Applicata Universitá de Bologna Bologna, Italy
Tevlin, Mr. P.	Dept. of Physics University of Toronto Toronto, Ont., Canada
Thompson, Mr. G.	2 Malmesbury Road South Woodford London E18 2NN, England
Totafurno, Mr. J.	Dept. of Physics University of Toronto Toronto, Ont., Canada
Trainor, Mr. C.	Dept. of Physics University of Toronto Toronto, Ont., Canada

Ucisik, Dr. A.H.	Metalurji Fakültesi Instanbul Teknik Üniversitat Tesvikiye, Instanbul, Turkey
Valluri, Dr. S.	School of Physics Georgia Inst. of Technology Atlanta, Geo., U.S.A.
Vasco, Mr. D.	Centre for Studies in Statistical Mechanics University of Texas Austin, Texas, U.S.A.
Veini, Dr. M.	Zoological Laboratory and Museum University of Athens Athens, Greece
Weiss, Dr. G.H.	Division of Computer Research and Technology National Institutes of Health Bethesda, Md., U.S.A.
Wesselius, Dr. W.	Dept. of Applied Mathematics Twente University of Technology Enschede, Holland
Winternitz, Dr. P.	C.R.M.A. Université de Montréal Montréal, Que., Canada
Yuksel, Dr. H.	Dept. of Physics Ankara University Ankara, Turkey

INDEX